Cosmology: The Physics of the Universe

Proceedings of the
Eighth Physics Summer School

Cosmology: The Physics of the Universe

The Australian National University
Canberra, Australia
16 January – 3 February 1995

Editors

B A Robson, N Visvanathan & W S Woolcock

The Australian National University
Canberra, Australia

World Scientific
Singapore • New Jersey • London • Hong Kong

Published by

World Scientific Publishing Co Pte Ltd
5 Toh Tuck Link, Singapore 596224
USA office: 27 Warren Street, Suite 401-402, Hackensack, NJ 07601
UK office: 57 Shelton Street, Covent Garden, London WC2H 9HE

British Library Cataloguing-in-Publication Data
A catalogue record for this book is available from the British Library.

COSMOLOGY: THE PHYSICS OF THE UNIVERSE

ISBN-13 978-981-02-2513-1
ISBN-10 981-02-2513-X

PREFACE

A summer school on **Cosmology: The Physics of the Universe** was held for three weeks from 16 January to 3 February 1995 in the Research School of Physical Sciences and Engineering, The Australian National University. This school was the eighth organised by the Department of Theoretical Physics. On this occasion the school was organised in collaboration with Mount Stromlo and Siding Spring Observatories and under the auspices of the National Centre for Theoretical Physics.

A total of 48 lectures were presented, covering various aspects of cosmology. Besides basic courses in classical cosmology and general relativity, topics were chosen from the mainstream of contemporary activity. The main aim of the school was to expose participants to the basic concepts and modern developments in cosmology in a manner not normally covered in text books.

The summer school was organised by a committee consisting of B.A. Robson (Convenor), M.R. Anderson, G.V. Bicknell, C.J. Burden, R.L. Dewar, K.C. Freeman, L.J. Tassie and W.S. Woolcock. Financial support was given by the Research School of Physical Sciences and Engineering and the Mount Stromlo and Siding Spring Observatories.

Although an attempt was made to make this proceedings volume complete, it was not possible to include the two lectures on "Numerical Cosmology" presented by P. Quinn.

We are grateful to Professor Erich Weigold, Director of the Research School of Physical Sciences and Engineering and Professor Jeremy Mould, Director of Mount Stromlo and Siding Spring Observatories, for their interest and support. We are indebted to all the lecturers, participants and members of the Department of Theoretical Physics for their helpful cooperation during the entire period of the school. We are particularly grateful to Mrs Heli Jackson and Ms Martina Landsmann for their considerable assistance in running the school and Ms Martina Landsmann and Ms Roberta Bernstein for their help in preparing the manuscript of the Proceedings.

Canberra
April 9 1996

B.A. Robson
N. Visvanathan
W.S. Woolcock

1 David Wiltshire
2 Michael O'Rourke
3 Richard Farmer
4 Richard Ellem
5 Mike Burgess
6 Toby Allen
7 Ron Hola
8 Martin Bineau
9 Lindsay Tassie
10 John Salmon
11 Steven Tingay
12 Daniel Mortlock
13 Andrew Haigh
14 Jeremy Mould
15 Malcolm Anderson
16 David Pfitzner
17 Neil Pinder
18 Ross Taylor
19 Gerard Börner
20 Bill Woolcock
21 Don Faukner
22 Graham Chapman
23 Ian Grivell
24 Sungeun Kim
25 Mark Allen
26 Jennifer Wigley
27 Angus Nicholson
28 Götz Hoeppe
29 Matthew Young
30 Martina Landsmann
31 Ariane Lançon
32 Filipe Lim
33 Alex Szalay
34 Chuen Toh
35 David Schramm
36 Conrad Burden
37 Frank Masci
38 Vladimir Lukash
39 Martin Porter
40 Alister Graham
41 Kala Perkins
42 Tim Wilkin
43 Haret Rosu
44 Joe Silk
45 Brian Robson
46 Rachel Webster
47 Vu Ho
48 Stephen Hardy

CONTENTS

INTRODUCTION TO GENERAL RELATIVITY

Malcolm Anderson
Department of Mathematics
Edith Cowan University

0. Prefatory Remarks

Together with quantum theory, general relativity is one of the twin pillars of modern physics. It arose from a need to combine two simple and successful theories, Newtonian gravity and special relativity, into a single system that was physically acceptable as well as mathematically consistent. The required synthesis was achieved by Einstein in 1916, and although it is considerably more complex than its precursor theories, general relativity has successfully weathered almost 80 years of experimental and observational tests.

Compared to quantum physics, the range of terrestrial applications of general relativity is very limited. As a result, the theory was for many years after its inception treated as an arcane if elegant branch of mathematical physics. Before the 1960s general relativity made no important contributions outside the areas of celestial mechanics, stellar structure and cosmology, and even in the first two cases it appeared only in the form of small corrections. Cosmology, a scientific discipline to which general relativity is perhaps most relevant, was regarded as largely speculative until the advent of systematic large-scale galaxy surveys in the 1950s.

Since the 1960s the scope of general relativity has been extended dramatically by the discovery of neutron stars, quasars, active galactic nuclei, the cosmic microwave background and gravitational lenses, and by continuing progress in the technology of deep-sky surveys and gravitational wave detection. In tandem with these developments, the theory itself has been broadened by the discovery and classification of large classes of exact solutions, the use of functional analytic methods to study the global structure of spacetime, the pioneering of numerical relativity, and a variety of attempts to unify gravitation with quantum mechanics.

General relativity is such a rich subject that it is impossible to explore more than a smattering of its content in the course of four lectures. I therefore apologise from the outset for the small and somewhat arbitrary selection of topics that I will be reviewing here, and for the cursory treatment most of them will receive. In particular, I will not make any mention of the applications of general relativity to cosmology, as this important topic is treated at some length by Gerhard Börner elsewhere in this volume. Anyone interested in studying the material in more detail should of course refer to the attached bibliography.

One feature of these lectures that deserves a note of explanation is the choice of mathematical formalism used. The fundamental equations of general relativity have been derived and rederived many times using a variety of techniques ranging from tensor calculus to the language of operators and forms. This has contributed to the undeserved but general perception that the theory is mathematically difficult.

In fact, the basics of the theory can be explained using nothing more than undergraduate calculus, with no prior knowledge of differential geometry or topology required. In writing out these lectures I have been mindful of the need to provide just such a treatment, even though in some cases it entails a sea of subscripts and superscripts of the type quite often condemned in more formal textbooks. There are certainly briefer and more elegant ways of presenting much of the material appearing here, but none (I believe) that is more widely accessible.

Finally, a word about the conventions used in these lectures. In order to make the connection with Newtonian gravity more transparent, I have chosen to work with an unusual set of "natural" units, in which $c = \hbar = 1$ but the gravitational constant G remains arbitrary and so appears explicitly in all the formulae. Also, the indices a, b, c,... run from 0 to 3, and the indices i, j, k,... from 1 to 3.

1. THE ESSENTIALS OF GENERAL RELATIVITY

1.1 A Review of Special Relativity

The single most important conceptual foundation of relativistic physics in the absence of gravitation is what is sometimes referred to as the *Principle of Special Relativity*: namely, that the laws of physics are the same for all inertial (i.e. unaccelerated) observers. Before writing down a more rigorous statement of this principle, a few preliminary definitions are in order.

The class of inertial observers is most conveniently defined in terms of the dynamics of freely-moving particles. In the absence of gravitation, a particle is said to be freely moving if it is not subject to any external forces (which would almost always be electromagnetic forces in the case of a macroscopic body). By definition, an *inertial* observer is one who measures the velocity of any freely-moving particle to be constant.

Referred to a particular inertial observer O, any *event* (i.e. point in space and instant in time) can be assigned a set of 4 co-ordinates:

$$(x^a) = (x^0, x^1, x^2, x^3) = (t, \mathbf{x})$$

where t is the elapsed time as measured by the observer, and the components of $\mathbf{x}$ are Cartesian spatial co-ordinates centred on O. If $x^a=x^a(s)$ is a smooth curve through spacetime then the *proper time* $\tau(s)$ along the curve from an arbitrary initial point $s=0$ is defined to be:

$$\tau(s) = \int_0^s \left[\left(\frac{dt}{ds}\right)^2 - \left(\frac{d\mathbf{x}}{ds}\right)^2 \right]^{1/2} ds = \int_0^s \left(\eta_{ab} \frac{dx^a}{ds} \frac{dx^b}{ds} \right)^{1/2} ds$$

where

$$(\eta_{ab}) = \mathrm{diag}\,(1, -1, -1, -1)$$

is a 4×4 diagonal matrix, the *Minkowski metric tensor*. [Here, the Einstein summation convention is in force, so that all repeated upper and lower indices are summed over their range.]

According to a subsidiary assumption, the *Chronometric Hypothesis*, the proper time $\tau(s)$ along the curve $x^a=x^a(s)$ is in fact the time that would be measured by an observer travelling along that curve. The infinitesimal element of proper time

$$d\tau^2 = \eta_{ab}\, dx^a\, dx^b$$

can be used to distinguish three special types of curves. A *timelike curve* has $d\tau^2>0$ everywhere, a *spacelike curve* has $d\tau^2<0$ everywhere, and a *null curve* has $d\tau^2=0$ everywhere. The trajectory of a material particle is always a timelike curve, while a photon or other massless particle in vacuum will trace out a null curve.

In particular, in the co-ordinate frame of an inertial observer the path $x^a=x^a(\tau)$ of any freely-moving massive particle satisfies the equation

$$\frac{d^2x^a}{d\tau^2} = 0$$

Conversely, any curve that satisfies this equation in an inertial frame is called a (non-null) *geodesic*. Null geodesics need to be characterised in a slightly different way, as $d\tau^2=0$ on a null curve and so τ is not a well-defined parameter. A null geodesic is any curve with a parametrisation $x^a=x^a(s)$ that satisfies the equations

$$\frac{d^2x^a}{ds^2} = 0 \qquad \text{and} \qquad \eta_{ab}\frac{dx^a}{ds}\frac{dx^b}{ds} = 0$$

in the rest frame of an inertial observer.[1]

We are now in a position to be able to formulate a mathematical statement of the Principle of Special Relativity. This requires that the laws of physics be invariant under co-ordinate transformations of the form:

$$x'^a = \Lambda^a_b\, x^b + \mu^a$$

where μ^a is a constant vector, representing co-ordinate translations, and $\Lambda^a{}_b$ is a 4×4 matrix, the *Lorentz matrix*, which incorporates boosts (i.e. velocity shifts) and rotations of the co-ordinates, and satisfies the constraints:

$$\eta_{ab}\,\Lambda^a_c\,\Lambda^b_d = \eta_{cd} \qquad (\ \Lambda \text{ is orthogonal}\ )$$

$$\Lambda^0_0 \geq 1 \qquad (\ \text{time reversal is forbidden}\ )$$

and

$$\det(\Lambda) = 1 \qquad (\ \text{spatial inversion is forbidden}\ )$$

A transformation of this type is called a *restricted* (inhomogeneous) *Lorentz transformation*, and is the most general transformation from one inertial frame to another.

The set of restricted Lorentz matrices forms a 6-parameter group: 3 parameters corresponding to the velocity **v** of observer O' relative to observer O, and 3 parameters corresponding to the Euler angles of a general spatial rotation. The full 6-parameter Lorentz matrix is too forbidding to write down here, but a pure boost has the form:

$$(\Lambda^a_b) = \begin{bmatrix} \gamma & -\gamma\mathbf{v} \\ -\gamma\mathbf{v}^T & \underline{\underline{I}} + v^{-2}\,\mathbf{v}^T\mathbf{v}\,(\gamma - 1) \end{bmatrix}$$

where $\gamma = (1-v^2)^{-1/2}$.

Two important features of the Lorentz transformation are that it maps geodesics to geodesics:

$$\frac{d^2x'^a}{d\tau'^2} = 0 \qquad \text{if and only if} \qquad \frac{d^2x^a}{d\tau^2} = 0$$

and that it preserves proper time:

$$d\tau'^2 = d\tau^2$$

both of which are of course in accordance with the Principle of Special Relativity.

---

[1]Note that the first of these equations is not in general satisfied under an arbitrary reparametrisation $s \to s'(s)$. Any parameter s that does preserve the equation is called an *affine parameter* for the geodesic.

1.2 Particle Dynamics with Gravity

The Newtonian theory of gravity presumes the existence of an absolute timescale that is shared by all inertial frames. It is therefore not invariant under Lorentz transformations and fails to satisfy the Principle of Special Relativity. However, it does provide clues about the low-velocity, weak-field behaviour of a more general, locally Lorentz-invariant theory of gravity.

Another clue is provided by the *Principle of Equivalence*. Experimentalists from Galileo in the 16th century through to Braginsky in the 1970s have been unable to uncover any evidence that the inertial mass of a body, which appears in Newton's First Law $\mathbf{F}=m_i\mathbf{a}$, is distinguishable from its gravitational mass, which determines the gravitational force $\mathbf{F}=m_g\mathbf{g}$ on the body when it is placed in a gravitational field of strength $\mathbf{g}$.[2]

There is thus firm experimental support for what is known as the Weak Principle of Equivalence: namely, that the laws of motion for a freely-falling body in a uniform gravitational field are the same as for an unaccelerated body far from gravitational sources. By assuming a stronger version of this Principle (which follows by replacing "the laws of motion" with "all physical laws" in the above statement), it is possible to construct a relativistic theory of gravitation that locally preserves the invariance properties of Special Relativity.

In the presence of a non-uniform gravitational field, there is typically no co-ordinate system in which the path of every freely-falling particle can be expressed as a linear function of the particle's time τ. The advantage of choosing a special class of co-ordinate systems, analogous to the inertial frames of Special Relativity, to describe space-time on a global level therefore disappears. In what follows, space-time will be described by a set of co-ordinates

$$(x^a) = (x^0, x^1, x^2, x^3)$$

which is assumed to be completely arbitrary other than providing a continuous one-to-one mapping of events into $\mathbb{R}^4$.

In view of the Principle of Equivalence, it is possible to choose a second set of co-ordinates $y^a(x^b)$ in the neighbourhood of any event p such that

$$\left.\frac{d^2y^a}{d\tau^2}\right|_p = 0$$

along the path $y^a(\tau)$ of any freely-falling particle through p, where

---

[2]The most famous experiment of this type is of course Eötvös's torsion-balance experiment of 1889.

$$d\tau^2 = \eta_{ab}\, dy^a\, dy^b$$

By invoking the Chain Rule, it is possible to rewrite the first of these, the *geodesic equation*, in terms of the original co-ordinates x^a in the form[3]

$$\frac{d^2x^a}{d\tau^2} + \Gamma^a_{bc} \frac{dx^b}{d\tau} \frac{dx^c}{d\tau} = 0$$

where the *affine connection* Γ^a_{bc} is defined to be

$$\Gamma^a_{bc} = \frac{\partial x^a}{\partial y^d} \frac{\partial^2 y^d}{\partial x^b\, \partial x^c}$$

Also, the infinitesimal element of proper time can be rewritten as

$$d\tau^2 = g_{ab}\, dx^a\, dx^b$$

where

$$g_{ab} = \frac{\partial y^c}{\partial x^a} \frac{\partial y^d}{\partial x^b} \eta_{cd}$$

is the *metric tensor* corresponding to the choice of co-ordinates x^a.[4] In terms of the metric tensor, the affine connection can be re-expressed as

$$\Gamma^a_{bc} = \frac{1}{2} g^{ad} \left(\frac{\partial g_{db}}{\partial x^c} + \frac{\partial g_{dc}}{\partial x^b} - \frac{\partial g_{bc}}{\partial x^d} \right)$$

where the *inverse metric tensor* g^{ad} is given by

$$g^{ad} = (\underline{\underline{g}})^{-1}_{ad} = \frac{\partial x^a}{\partial y^e} \frac{\partial x^d}{\partial y^f} \eta_{ef}$$

Strictly speaking, the geodesic equation stated above applies only to non-null geodesics. A null geodesic in general curvilinear co-ordinates is any curve $x^a(s)$ which satisfies the equations:

$$\frac{d^2x^a}{ds^2} + \Gamma^a_{bc} \frac{dx^b}{ds} \frac{dx^c}{ds} = 0 \qquad \text{and} \qquad g_{ab} \frac{dx^a}{ds} \frac{dx^b}{ds} = 0$$

On the face of it, it would appear that the problem of constructing geodesics in a general space-time has here simply been translated into the equally difficult problem of finding an inertial co-ordinate system (y^a) at each point in the space-time. In fact, since the affine connection and the proper time can be reconstructed from a knowledge of g_{ab} and its first derivatives alone, it is more usual to regard the metric tensor as the defining feature of the space-time. What results then is a *metric theory of gravity*, of which General Relativity is only one of many possible examples.

---

[3]Note that since this procedure can be repeated at every point on the geodesic, the subscript p has been dropped in what follows.

1.3 Gravitational Red Shift

An immediate consequence of the assumption that gravity can be described by a metric theory is that electromagnetic radiation should be red- or blue-shifted as it moves through a gravitational potential. In order to be able to test this prediction in the relatively weak gravitaional environment of the Earth, it is necessary to first of all examine the Newtonian limit of a general metric theory.

By a weak gravitational field is meant a patch of space-time in which it is possible to find co-ordinates $(x^a)=(t,\mathbf{x})$ that reduce the metric tensor g_{ab} to the Minkowski tensor η_{ab} plus a small perturbation:

$$g_{ab} = \eta_{ab} + h_{ab} \qquad \text{where} \qquad |h_{ab}| << 1$$

For a non-relativistic particle in a weak field,

$$\left|\frac{d\mathbf{x}}{d\tau}\right| << \frac{dt}{d\tau}$$

and so the non-null geodesic equation simplifies to

$$\frac{d^2x^a}{d\tau^2} + \Gamma^a_{00}\left(\frac{dt}{d\tau}\right)^2 \approx 0$$

where, if the gravitational field is assumed to be stationary (i.e. $\partial h_{ab}/\partial t=0$),

$$\Gamma^a_{00} \approx -\frac{1}{2}\eta^{ab}\frac{\partial h_{00}}{\partial x^b}$$

Hence, in the Newtonian limit, the non-null geodesic equation becomes:

$$\frac{d^2t}{d\tau^2} = 0 \qquad \text{and} \qquad \frac{d^2\mathbf{x}}{d\tau^2} = -\frac{1}{2}\nabla h_{00}$$

Comparison of this system of equations with the standard Newtonian formula

$$\frac{d^2\mathbf{x}}{dt^2} = -\nabla\phi$$

(where ϕ is the Newtonian gravitational potential) indicates that

$$g_{00} \approx 1 + 2\phi$$

in the stationary weak-field limit.

In view of this result, the element of proper time $d\tau$ in the rest frame of a non-relativistic particle in a weak field is

$$d\tau = (g_{ab}dx^a dx^b)^{1/2} \approx (g_{00})^{1/2}dt \approx (1+\phi)dt$$

If light is emitted with frequency ν_{em} at a point p_1 and received at a second point p_2, and $\Delta\tau$ is the local period between successive crests of the light, then the

---

[4]By virtue of the orthogonality of the Lorentz matrix, g_{ab} is invariant under any Lorentz transformation of the y^as, and so corresponds uniquely to the choice of the x^as.

shift $\delta\nu$ in the frequency of the light as measured by an observer at p_2 is given by[5]

$$1+\frac{\delta\nu}{\nu_{em}}=\frac{\Delta\tau_1}{\Delta\tau_2}=\left[\frac{g_{00}(p_1)}{g_{00}(p_2)}\right]^{1/2}\approx 1+\phi(p_1)-\phi(p_2)$$

For light emitted from atoms on the surface of the Sun, the fractional red shift as measured by Earth-bound telescopes is predicted to be:

$$\frac{\delta\nu}{\nu_{em}}\approx\phi_\odot=-\frac{GM_\odot}{R_\odot}=-2.12\times10^{-6}$$

It is in principle possible to measure a red shift of this size, but in practice the effect is all but washed out by convective motions in the solar plasma. A better test is provided by observations of the line spectra of white dwarfs, where the fractional red shift is predicted to lie between -10^{-5} and -10^{-4}, but in fact the best limits to date have come from rocket experiments performed by Vessot et al (1980), whose observed values of $\delta\nu$ agree with the predicted values to within 0.02%. This is an impressive result, but it should be stressed that gravitational red shift does not test General Relativity *per se*, but only the Principle of Equivalence.

1.4 Mass, Energy and Gravitation

In Newtonian gravity, the potential ϕ is generated by the combined effect of all the mass in the Universe through *Poisson's equation*:

$$\nabla^2\phi=4\pi G\rho$$

where ρ is the density of matter at each point in space-time, and G is the gravitational constant (currently measured at 6.670×10^{-8} dyn cm^2 g^{-2}). As we have seen, a metric theory of gravity has in effect not one but 10 potentials (the 10 independent components of the symmetric 4×4 matrix g_{ab}), and so it is to be expected that the relativistic generalisation of Poisson's equation will relate second derivatives of the metric tensor to some symmetric 4×4 matrix that describes the distribution of mass and possibly other forms of energy in the Universe.

In Special Relativity, the flux of energy and momentum through any point in space-time is conventionally described by just such a matrix, the *energy-momentum tensor*:

---

[5]Note here that according to the null geodesic equation $d^2t/ds^2\approx0$ the co-ordinate period Δt between successive light crests remains constant as the light propagates through the potential.

$$(T_{ab}) = \begin{bmatrix} T_{00} & T_{01} & T_{02} & T_{03} \\ T_{10} & T_{11} & T_{12} & T_{13} \\ T_{20} & T_{21} & T_{22} & T_{23} \\ T_{30} & T_{31} & T_{32} & T_{33} \end{bmatrix}$$

where the T_{00} component represents the energy density as measured by an observer O at rest relative to the co-ordinates x^a, the remainder of the first row represents the (negative) spatial components of the energy flux, the remainder of the first column represents the (negative) spatial components of the momentum density[6], and the 3×3 spatial sub-matrix represents the x^i-x^j components of the stress (i.e. the flux of x^i-momentum in the x^j-direction).

The energy-momentum tensor can take a wide variety of forms, depending on the number and type of particles and fields present, but in all cases it will be symmetric (i.e. $T_{ab}=T_{ba}$) and *Lorentz covariant*: that is, if T'_{ab} is the energy-momentum tensor measured by a second observer O' whose rest frame is related to the rest frame of O by the Lorentz matrix $\Lambda^a{}_b$ then

$$T_{ab} = \Lambda^c_a \Lambda^d_b T'_{cd}$$

Another feature of the energy-momentum tensor that will prove to be important when it comes to constructing the relativistic version of Poisson's equation is that it satisfies the *conservation equation*:

$$\eta^{ac} \frac{\partial T_{ab}}{\partial x^c} = 0$$

One of the simplest forms of energy-momentum tensor is that corresponding to a *perfect fluid*. In the rest frame of a perfect fluid, its energy-momentum components are:

$$T'_{00} = \rho \quad T'_{0i} = 0 \qquad \text{and} \qquad T'_{ij} = p\,\delta_{ij}$$

where ρ is now the rest energy density of the fluid, and p is its pressure.

The exact relationship between the energy density and pressure and other thermodynamic quantities such as temperature will depend on the microphysics of the fluid, but in the case of a gas of point particles each with a mass m_n, position $\mathbf{x}_n(t)$ and velocity $\mathbf{v}_n=d\mathbf{x}_n/dt$ the energy-momentum tensor is:

---

[6]In fact, the positive components of the energy fluxes and momentum densities appear directly in the contravariant form $T^{ab}=\eta^{ac}\eta^{bd}T_{cd}$ of the energy-momentum tensor, so that for example $T^{30}=-T_{30}$ is the z-component of the momentum density.

$$T_{00} = \sum_n m_n \gamma_n \delta^3(\mathbf{x} - \mathbf{x}_n) \qquad T_{0i} = -\sum_n m_n \gamma_n v_{n(i)} \delta^3(\mathbf{x} - \mathbf{x}_n)$$

and

$$T_{ij} = \sum_n m_n \gamma_n v_{n(i)} v_{n(j)} \delta^3(\mathbf{x} - \mathbf{x}_n)$$

where $\gamma_n=(1-v_n^2)^{-1/2}$. It is always possible to choose the co-ordinates so that $T_{0i}=0$ (this defines the local rest frame of the gas). If the velocity distribution is such that T_{ij} is isotropic, the gas will be a perfect fluid with

$$\rho = \sum_n m_n \gamma_n \delta^3(\mathbf{x} - \mathbf{x}_n)$$

and

$$p = \frac{1}{3}\sum_{i=1}^{3} T_{ii} = \frac{1}{3}\sum_n m_n \gamma_n v_n^2 \delta^3(\mathbf{x} - \mathbf{x}_n)$$

In the non-relativistic limit ($v_n^2 << 1$), p is negligible in comparison with ρ, which can now be interpreted as the rest mass density of the fluid; whereas in the relativistic limit ($v_n^2 \approx 1$) the canonical relativistic equation of state $p \approx {}^1\!/_3\rho$ is recovered.

In the case of a general perfect fluid, if **v** is the bulk velocity of the fluid as measured by an inertial observer O then the energy-momentum tensor T_{ab} in the rest frame of O can be calculated through the covariance relation

$$T_{ab} = \Lambda_a^c \Lambda_b^d T'_{cd}$$

where $\Lambda^a{}_b$ is the pure boost matrix described in §1.1. Hence,

$$T_{00} = \gamma^2(\rho + p v^2) \qquad T_{0i} = -\gamma^2(\rho + p) v_i$$

and

$$T_{ij} = p\,\delta_{ij} + \gamma^2(\rho + p) v_i v_j$$

These expressions can be written more compactly in the form:

$$T_{ab} = (\rho + p) u_a u_b - p\,\eta_{ab}$$

where $u^a=dx^a/d\tau=\gamma(1, \mathbf{v})$ is the local *4-velocity* of the fluid, and $u_a=\eta_{ab}u^b$.

In the presence of a gravitational field, the energy-momentum tensor will retain the same form in any locally-inertial co-ordinate system (y^a), with $u^a=dy^a/d\tau$. In transforming back to the general co-ordinate system (x^a), the 4-velocity $dy^a/d\tau$ is replaced by $dx^a/d\tau$, and the Minkowski metric η_{ab} by the general metric g_{ab}. Hence, in contravariant form, the energy-momentum tensor for a perfect fluid in any metric theory is:

$$T^{ab} = (\rho + p) u^a u^b - p\,g^{ab}$$

where $u^a=dx^a/d\tau$.

Similarly, the equation of energy-momentum conservation in a metric theory, which in locally-inertial co-ordinates about a point p reads:

$$\frac{\partial T^{ab}}{\partial y^a}\Big|_P = 0$$

becomes, in a general co-ordinate system x^a,

$$\nabla_a T^{ab} \equiv \frac{\partial T^{ab}}{\partial x^a} + \Gamma^a_{ac} T^{bc} + \Gamma^b_{ac} T^{ac} = 0$$

The *covariant derivative operator* ∇_a appearing on the left of this equation is the generalisation of the partial derivative operator $\partial/\partial y^a$. It has a number of useful properties, among them that

$$\nabla_a \phi = \frac{\partial \phi}{\partial x^a} \qquad \text{for any scalar function } \phi$$

and

$$\nabla_a g_{bc} = 0 \qquad \text{and} \qquad \nabla_a g^{bc} = 0$$

(which is clearly true in locally-inertial co-ordinates, and therefore true generally).

In the case of a perfect fluid, the energy-momentum conservation equation reads:

$$\nabla_a T^{ab} = u^b u^a \frac{\partial}{\partial x^a} (\rho + p) + (\rho + p) (u^b \nabla_a u^a + u^a \nabla_a u^b) - g^{ab} \frac{\partial p}{\partial x^a} = 0$$

where $\nabla_a u^b = \partial u^b / \partial x^a + \Gamma^b{}_{ac} u^c$. This, the equation of inviscid relativistic hydrodynamics, can be simplified by decomposing it into components parallel and normal to the covariant 4-velocity $u_b = g_{bc} u^c$. Since $u_b u^b = 1$, it follows that

$$u_b \nabla_a T^{ab} = u^a \frac{\partial \rho}{\partial x^a} + (\rho + p) \nabla_a u^a = 0$$

which is simply the relativistic version of the continuity equation $\nabla \cdot (\rho \mathbf{v}) = 0$. Substituting this identity back into the full conservation equation gives

$$(\rho + p) u^a \nabla_a u^b = (g^{ab} - u^a u^b) \frac{\partial p}{\partial x^a}$$

which is the relativistic version of the momentum equation $\rho(\mathbf{v} \cdot \nabla)\mathbf{v} = -\nabla p$.

1.5 Derivation of the Einstein Equation

In terms of the components of g_{ab} and T_{ab}, Poisson's equation reads:

$$\nabla^2 g_{00} = 8 \pi G T_{00}$$

As was mentioned earlier, it is to be expected that the relativistic generalisation of this equation will relate all 10 components of g_{ab} to the 10 components of T_{ab}. In fact, the assumption that leads directly to Einstein's theory is that Poisson's equation generalises to a relationship of the form

$$G_{ab} = 8\pi G T_{ab}$$

where G_{ab} is some linear combination of the second derivatives of the metric tensor g_{ab}.

The reason for restricting attention to the second derivatives of g_{ab} is of course that Poisson's equation contains only second derivatives of g_{00}, but it is in principle possible to include terms involving higher or lower derivatives of the metric tensor in various linear and non-linear combinations. Some theories of this type will be examined in §1.11. One of the difficulties in rejecting higher-derivative terms is that the scale-length of the gravitational potential in the neighbourhood of the Earth is so large (as the field is so weak) that the effect of any terms of this type would be virtually undetectable. It is therefore not very easy to place constraints on their contribution to G_{ab}. On the other hand, terms involving only lower derivatives of the metric tensor (as, for example, Einstein's infamous cosmological constant term, which is proportional to g_{ab}) would have important effects on a terrestrial scale unless their linear coefficients were exceptionally small. Needless to say, there is at present no real evidence for corrections of either type.

In locally-inertial co-ordinates y^a the most general symmetric, Lorentz-covariant linear combination of second derivatives of the metric tensor[7] is:

$$G_{ab} = \alpha \Delta g_{ab} + \beta \left(\frac{\partial^2 g^c_b}{\partial y^a \partial y^c} + \frac{\partial^2 g^c_a}{\partial y^b \partial y^c} \right) + \gamma \eta_{ab} \frac{\partial^2 g^{cd}}{\partial y^c \partial y^d}$$

$$+ \delta \frac{\partial^2 g^c_c}{\partial y^a \partial y^b} + \varepsilon \eta_{ab} \Delta g^c_c$$

where α, β, γ, δ and ε are undetermined constants,

$$\Delta = \eta^{ef} \frac{\partial}{\partial y^e} \frac{\partial}{\partial y^f}$$

and the indices on g_{ab} are everywhere raised with η^{cd}.

Since G_{ab} is proportional to T_{ab}, the equation of energy-momentum conservation requires that

$$0 = \eta^{ac} \frac{\partial G_{ab}}{\partial y^c} = (\alpha + \beta) \frac{\partial}{\partial y^a} \Delta g^a_b + (\beta + \gamma) \frac{\partial}{\partial y^b} \left(\frac{\partial^2 g^{cd}}{\partial y^c \partial y^d} \right)$$

$$+ (\delta + \varepsilon) \frac{\partial}{\partial y^b} \Delta g^c_c$$

and so $\alpha+\beta=\beta+\gamma=\delta+\varepsilon=0$. Hence,

---

[7]In locally-inertial co-ordinates about a point p, $g_{ab}=\eta_{ab}$ and $\partial g_{ab}/\partial y^c=0$ at p but the second derivatives of g_{ab} are not generally zero.

$$G_{ab} = \alpha\left(\Delta g_{ab} - \frac{\partial^2 g^c_b}{\partial y^a \partial y^c} - \frac{\partial^2 g^c_a}{\partial y^b \partial y^c} + \eta_{ab} \frac{\partial^2 g^{cd}}{\partial y^c \partial y^d} \right)$$
$$+ \delta\left(\frac{\partial^2 g^c_c}{\partial y^a \partial y^b} - \eta_{ab} \Delta g^c_c \right)$$

Furthermore, in the Newtonian limit derivatives with respect to $t=y^0$ are negligible, and so $\Delta \approx -\nabla^2$ and

$$8 \pi G T_{00} + 8 \pi G \sum_{i=1}^{3} T_{ii} = G_{00} + \sum_{i=1}^{3} G_{ii}$$
$$\approx - (\alpha + \delta) \nabla^2 g_{00} + (\delta - \alpha) \nabla^2 \sum_{i=1}^{3} g_{ii}$$

Since $\Sigma T_{ii} << T_{00}$ in the Newtonian limit, but $\nabla^2 \Sigma g_{ii}$ is in general not small compared to $\nabla^2 g_{00}$, Poisson's equation will be recovered only if $\alpha=\delta=-1/2$. Hence, the unique solution for G_{ab} is:

$$G_{ab} = -\frac{1}{2}\left(\Delta g_{ab} - \frac{\partial^2 g^c_b}{\partial y^a \partial y^c} - \frac{\partial^2 g^c_a}{\partial y^b \partial y^c} + \eta_{ab} \frac{\partial^2 g^{cd}}{\partial y^c \partial y^d} \right)$$
$$-\frac{1}{2}\left(\frac{\partial^2 g^c_c}{\partial y^a \partial y^b} - \eta_{ab} \Delta g^c_c \right)$$

On transforming back to the general co-ordinate system x^a, G_{ab} assumes the form of what is known as the *Einstein tensor*:

$$G_{ab} = R_{ab} - \frac{1}{2} g_{ab} R$$

where

$$R_{ab} = \frac{1}{2} g^{cd} \left(\frac{\partial^2 g_{bc}}{\partial x^a \partial x^d} + \frac{\partial^2 g_{ad}}{\partial x^b \partial x^c} - \frac{\partial^2 g_{ab}}{\partial x^c \partial x^d} - \frac{\partial^2 g_{cd}}{\partial x^a \partial x^b} \right)$$
$$+ g^{cd} g_{ef} \left(\Gamma^e_{ac} \Gamma^f_{bd} - \Gamma^e_{ab} \Gamma^f_{cd} \right)$$

is the *Ricci tensor*, and

$$R = g^{ab} R_{ab}$$

is the *Ricci scalar*. The relativistic version of Poisson's equation (the *Einstein equation*) therefore becomes:

$$R_{ab} - \frac{1}{2} g_{ab} R = 8 \pi G T_{ab}$$

An alternative form of this equation can be found by taking its trace[8]:

$$R = - 8 \pi G g^{ab} T_{ab} \equiv - 8 \pi G T$$

and then substituting back into the original Einstein equation to give:

---

[8]Note here that $g^{ab} g_{ab}=4$.

$$R_{ab} = 8\pi G\,(T_{ab} - \frac{1}{2} g_{ab} T)$$

In either form, the Einstein equation constitutes a set of 10 non-linear second-order partial differential equations for the 10 independent components of the metric tensor.[9]

1.6 The Schwarzschild Solution

Of the many exact solutions of the Einstein equation that have been discovered over the last 80 years, the Schwarzschild solution is undoubtedly the most famous. It was first derived in 1916, and describes the gravitational field outside a spherical body of mass M. It has numerous applications in celestial mechanics, and its success in being able to explain the anomalous precession of the perihelion of Mercury and to predict the deflection of starlight by the Sun ensured that general relativity was quickly accepted as the correct relativistic generalisation of Newton's theory.

A general static and spherically-symmetric gravitational field has a *line element* of the form:

$$d\tau^2 \equiv g_{ab}\, dx^a\, dx^b = A(r)\, dt^2 - B(r)\, dr^2 - r^2\,(d\theta^2 + \sin^2\theta\, d\phi^2)$$

where θ and ϕ are standard spherical polar co-ordinates on a spherical 2-surface of radius r, and A and B are undetermined functions of r. The components of the metric tensor g_{ab} and its inverse can be read off directly from the line element, to give:

$$g_{tt} = A(r) \qquad g_{rr} = -B(r) \qquad g_{\theta\theta} = -r^2 \qquad g_{\phi\phi} = -r^2 \sin^2\theta$$

and

$$g^{tt} = \frac{1}{A(r)} \qquad g^{rr} = -\frac{1}{B(r)} \qquad g^{\theta\theta} = -\frac{1}{r^2} \qquad g^{\phi\phi} = -\frac{1}{r^2 \sin^2\theta}$$

Hence, the non-zero components of the affine conection $\Gamma^a_{\ bc}$ are:

---

[9]It is not unusual for the Ricci tensor to be defined with the opposite sign from that given here. The Einstein equation then takes the form:

$$R_{ab} - \frac{1}{2} g_{ab} R = -8\pi G T_{ab}$$

or equivalently

$$R_{ab} = -8\pi G\,(T_{ab} - \frac{1}{2} g_{ab} T)$$

$$\Gamma^t_{tr} = \Gamma^t_{rt} = \frac{1}{2} A^{-1} A' \qquad \Gamma^r_{rr} = \frac{1}{2} B^{-1} B' \qquad \Gamma^r_{tt} = \frac{1}{2} B^{-1} A'$$

$$\Gamma^r_{\theta\theta} = - B^{-1} r \qquad \Gamma^r_{\phi\phi} = - B^{-1} r \sin^2\theta \qquad \Gamma^\theta_{r\theta} = \Gamma^\theta_{\theta r} = \Gamma^\phi_{r\phi} = \Gamma^\phi_{\phi r} = r^{-1}$$

and

$$\Gamma^\theta_{\phi\phi} = - \sin\theta \cos\theta \qquad \Gamma^\phi_{\phi\theta} = \Gamma^\phi_{\theta\phi} = \cot\theta$$

where a prime denotes d/dr.

After a relatively lengthy calculation, the non-zero components of the Ricci tensor R_{ab} are found to be:

$$R_{tt} = \frac{1}{2} B^{-1} A'' - \frac{1}{4} B^{-2} A' B' - \frac{1}{4} A^{-1} B^{-1} (A')^2 + r^{-1} B^{-1} A'$$

$$R_{rr} = - \frac{1}{2} A^{-1} A'' + \frac{1}{4} A^{-1} B^{-1} A' B' + \frac{1}{4} A^{-2} (A')^2 + r^{-1} B^{-1} B'$$

$$R_{\theta\theta} = 1 - B^{-1} + \frac{1}{2} B^{-1} r (B^{-1} B' - A^{-1} A')$$

and

$$R_{\phi\phi} = \sin^2\theta\, R_{\theta\theta}$$

In a vacuum (T_{ab}=0), the Einstein equation reads:

$$R_{ab} - \frac{1}{2} g_{ab} R = 0$$

or equivalently:

$$R_{ab} = 0$$

and so the Einstein equation in this case reduces to the set of three second-order equations $R_{tt}=R_{rr}=R_{\theta\theta}=0$.

Now, it is evident by inspection that

$$0 = A^{-1} R_{tt} + B^{-1} R_{rr} = r^{-1} B^{-1} (A^{-1} A' + B^{-1} B')$$

and so an immediate first integral of the Einstein equation is:

$$A (r) B (r) = \text{constant}$$

In fact, if the metric is to reduce to the metric of flat space (where A=B=1) at large distances, the constant of integration in this equation must be 1. Furthermore, the equation for $R_{\theta\theta}$ now reads:

$$R_{\theta\theta} = 1 - A - r A' = 0$$

and so

$$A (r) = 1 + \frac{\text{constant}}{r}$$

where the constant of integration is now determined by the requirement that the metric reduce to the Newtonian limit (in which $A=g_{tt}\approx 1+2\phi$) at large distances. Since $\phi=-GM/r$ outside a spherical body of mass M, it follows that

$$A(r) = 1 - \frac{2GM}{r}$$

and therefore

$$B(r) = \left(1 - \frac{2GM}{r}\right)^{-1}$$

Hence, the full Schwarzschild line element (which describes the unique static, spherically-symmetric vacuum solution of the Einstein equation) is:

$$d\tau^2 = \left(1 - \frac{2GM}{r}\right) dt^2 - \left(1 - \frac{2GM}{r}\right)^{-1} dr^2 - r^2 \left(d\theta^2 + \sin^2\theta\, d\phi^2\right)$$

1.7 Geodesics in the Schwarzschild Metric

In a general static and spherically-symmetric spacetime, the geodesic equation

$$\frac{d^2x^a}{ds^2} + \Gamma^a_{bc} \frac{dx^b}{ds} \frac{dx^c}{ds} = 0$$

(where $s=\tau$ in the case of a non-null geodesic but is an arbitrary affine parameter if the geodesic is null) becomes:

$$\ddot{t} + A^{-1} A' \dot{t} \dot{r} = 0$$

$$\ddot{r} + \frac{1}{2} B^{-1} B' \dot{r}^2 - r^{-1} B^{-1} \dot{\theta}^2 - r \sin^2\theta\, B^{-1} \dot{\phi}^2 + \frac{1}{2} B^{-1} A' \dot{t}^2 = 0$$

$$\ddot{\theta} + 2 r^{-1} \dot{\theta} \dot{r} - \sin\theta \cos\theta\, \dot{\phi}^2 = 0$$

and

$$\ddot{\phi} + 2 r^{-1} \dot{\phi} \dot{r} + 2 \cot\theta\, \dot{\phi} \dot{\theta} = 0$$

where a dot denotes d/ds.

As is the case with orbits in the Kepler potential, it is possible to choose the co-ordinates so that the orbit is confined to the equatorial plane ($\theta=\pi/2$). The equation for ϕ can then be once integrated to give:

$$r^2 \dot{\phi} = J \qquad \text{(a constant)}$$

where J coincides with the specific angular momentum of the particle in the Newtonian limit. Similarly, the equation for t has a first integral of the form:

$$A(r)\dot{t} = A_0 \qquad \text{(a constant)}$$

where, for marginally-bound orbits, $A_0=1$ (since then $\dot{t} \to 1$ as $r \to \infty$). For the sake of simplicity it will henceforth be assumed that the parameter s has been rescaled so that $A_0=1$, whether the orbit is marginally bound or not. This compels a certain degree of care in interpreting the various quantities that appear in the geodesic equations, but does not alter the basic form of the equations.

The equation for r can now be rewritten as:

$$B\ddot{r}+\frac{1}{2}B'\dot{r}^2-r^{-3}J^2+\frac{1}{2}A^{-2}A'=0$$

which when once integrated gives the *energy equation*:

$$B\dot{r}^2+r^{-2}J^2-A^{-1}=-E \qquad \text{(a constant)}$$

where E>0 on a timelike orbit, and E=0 on a photon orbit. For the purposes of calculating the physical shape of the orbit, it is more convenient to divided the energy equation through by $r^4\dot{\phi}^2=J^2$, thereby generating the *orbit equation*:

$$r^{-4}B\left(\frac{dr}{d\phi}\right)^2+\frac{1}{r^2}-J^{-2}A^{-1}=-J^{-2}E$$

So far the analysis has been applicable to any static, spherically-symmetric spacetime (and would, for example, describe photon propagation in a diffuse ball of fluid). When restricted to the Schwarzschild metric, the orbit equation reads:

$$r^{-4}\left(1-\frac{2GM}{r}\right)^{-1}\left(\frac{dr}{d\phi}\right)^2+\frac{1}{r^2}-J^{-2}\left(1-\frac{2GM}{r}\right)^{-1}=-J^{-2}E$$

This equation figured prominently in two early and famous tests of general relativity: the deflection of light by the Sun, and the precession of the perihelion of Mercury.

In the case of light deflection, consider a photon orbit (E=0) approaching a spherical body of mass M from an initial azimuthal direction $\phi=\phi_\infty$, as in Fig. 1.1. If the distance of closest approach of the photon is r_0 then, from the orbit equation,

$$J^2=r_0^2\left(1-\frac{2GM}{r_0}\right)^{-1}$$

and so the total angular deflection of the photon is:

$$\Delta\phi=2\int_{r_0}^{\infty}\left(1-\frac{2GM}{r}\right)^{-1/2}\left[\left(\frac{r}{r_0}\right)^2\left(\frac{r_0-2MG}{r-2MG}\right)-1\right]^{-1/2}\frac{dr}{r}-\pi$$

For a weak gravitational field ($GM/r_0<<1$), the angular deflection reduces to:

$$\Delta\phi\approx\frac{4GM}{r_0}$$

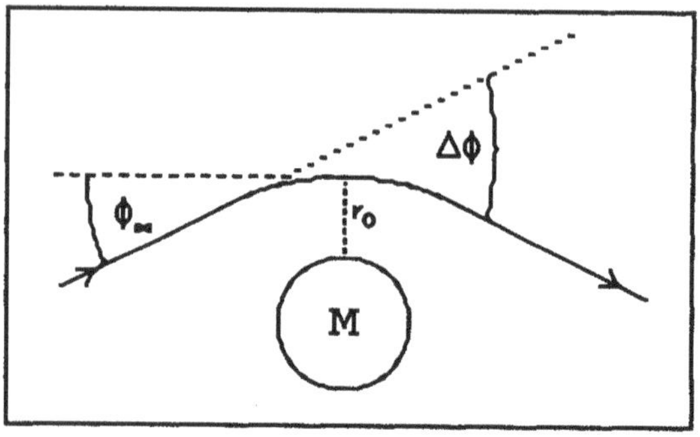

Fig. 1.1: Deflection of light in the Schwarzschild metric.

If the photon orbit just grazes the surface of the Sun then $r_0 = R_\odot$ and the predicted deflection angle is:

$$\Delta\phi \approx \frac{4GM_\odot}{R_\odot} \approx 1.75''$$

To date, the most accurate optical measurement of the deflection angle is 1.66"±0.19", which was made during the solar eclipse of 30 June 1973. However, much greater accuracy is possible at radio wavelengths, and the best interferometry measurement of the deflection angle to date agrees with the predicted value to within 0.1%.

The analysis involved in predicting the perihelion shift of Mercury is similar, but considerably more complicated. The important point is that, unlike bound orbits in the Kepler potential, bound orbits in the Schwarzschild metric do not trace out closed ellipses. Instead they trace out rosette-like patterns, where the point of closest approach to the central mass (the perihelion) varies from one passage to the next. In a weak gravitational field, the perihelion angle precesses by an amount:

$$\Delta\phi \approx \frac{6\pi GM}{(1-e^2)\,a} \qquad \text{radians/revolution}$$

where e is the eccentricity and a the semi-major axis of the orbit. Within the confines of the solar system, this general relativistic precession is detectable in the orbits of Mercury, Venus, the Earth and some asteroids, but it is the orbit of Mercury that provides the strongest test. The predicted value of Mercury's orbital precession is 42.98" per century[10], a figure which has been experimentally confirmed to an accuracy of 0.1%.

---

[10] The oft-quoted value of 43.03" per century is in fact wrong. See Clifford Will, *Theory and Experiment in Gravitational Physics*, Cambridge University Press, 1993.

1.8 The Riemann Tidal Tensor

The Principle of Equivalence asserts that it is possible to find a family of inertial frames at each point in a gravitational field. However, it is typically not possible to find a co-ordinate system that represents the measurements of inertial observers over a small but non-zero patch of spacetime. This breakdown of the correspondence between freely-falling and unaccelerated observers on a non-local scale is reflected in the existence of tidal forces. Tidal forces arise whenever neighbouring geodesics converge or diverge; any material body moving through a region of spacetime where the geodesics are not uniformly parallel will maintain its integrity only if its internal bonds (which are typically electromagnetic) are strong enough to counteract the convergence or divergence of its constituent particles. The resulting stress is the tidal force, and it is most conveniently described by the *Riemann tidal tensor*.

The Riemann tidal tensor is also useful in quantifying the effects of *parallel transport*. A vector property V^a of a particle (such as its spin S^a or its 4-momentum P^a) is said to be parallel transported if

$$\frac{dV^a}{d\tau}\Big|_p = 0$$

in locally-inertial co-ordinates y^a about each point p on the trajectory of the particle. In general co-ordinates x^a, the equation of parallel transport becomes:

$$\frac{DV^a}{D\tau} \equiv \frac{dV^a}{d\tau} + \Gamma^a_{bc}\frac{dx^b}{d\tau}V^c = 0$$

In particular, the equation for a (non-null) geodesic can be rewritten in the form:

$$\frac{D}{D\tau}\left(\frac{dx^a}{d\tau}\right) = 0$$

Consider now two nearby freely-falling particles following the geodesics $x^a(\tau)$ and $x^a(\tau)+\delta x^a(\tau)$ respectively, where $\delta x^a(\tau)$ is a small deviation. By virtue of the geodesic equation,

$$0 = \frac{D}{D\tau}\left[\frac{d}{d\tau}(x^a + \delta x^a)\right] - \frac{D}{D\tau}\left(\frac{dx^a}{d\tau}\right)$$

$$\approx \frac{d^2}{d\tau^2}(\delta x^a) + 2\Gamma^a_{bc}\frac{dx^b}{d\tau}\frac{d}{d\tau}(\delta x^c) + \frac{\partial\Gamma^a_{bc}}{\partial x^d}\frac{dx^b}{d\tau}\frac{dx^c}{d\tau}\delta x^d$$

to linear order in δx^a. In terms of the second covariant derivative of δx^a, this equation reads:

$$\frac{D^2}{D\tau^2}(\delta x^a) = -g^{ae}R_{ebcd}\frac{dx^b}{d\tau}\frac{dx^c}{d\tau}\delta x^d$$

where

$$R_{abcd} = \frac{1}{2}\left(\frac{\partial^2 g_{ad}}{\partial x^b \partial x^c} + \frac{\partial^2 g_{bc}}{\partial x^a \partial x^d} - \frac{\partial^2 g_{ac}}{\partial x^b \partial x^d} - \frac{\partial^2 g_{bd}}{\partial x^a \partial x^c} \right)$$
$$+ g_{ef} (\Gamma^e_{ad} \Gamma^f_{bc} - \Gamma^e_{ac} \Gamma^f_{bd})$$

is the Riemann tidal tensor. The quantity $D^2(\delta x^a)/D\tau^2$ is the relative acceleration of the two geodesics, and the equation describing it is known as the *equation of geodesic deviation*. If the two particles are carrying a parallel-transported vector quantity V^a then a similar calculation indicates that the deviation δV^a in V^a satisfies the equation:

$$\frac{D}{D\tau}(\delta V^a) = - g^{ae} R_{ebcd} \frac{dx^b}{d\tau} V^c \delta x^d$$

Comparison of the formula for the Riemann tensor with the expression for the Ricci tensor given in §1.5 shows that the two are closely related. In fact,

$$R_{ab} = g^{cd} R_{acbd}$$

However, the physical interpretations of the two tensors are somewhat different, and they should not be confused. The Ricci tensor is linearly related to the energy-momentum tensor by virtue of the Einstein equations, and so the constraint $R_{ab}=0$ is the defining property of a vacuum solution. By contrast, the equation $R_{abcd}=0$ is much stronger, and is satisfied only by a flat spacetime[11]. The fact that the tidal tensor can be non-zero even in a vacuum is crucial for the existence of gravitational waves (see §1.9).

The Riemann tensor has the following symmetry properties:

$$R_{cdab} = R_{abcd} \qquad R_{abcd} = - R_{bacd} = - R_{abdc} = R_{badc}$$

and

$$R_{abcd} + R_{adbc} + R_{acdb} = 0$$

which together reduce the number of its algebraically-independent components to 20. By way of comparison, the Ricci tensor has only 10 independent components; this again highlights the fact that the tidal tensor can be non-zero even when the Ricci tensor is zero[12].

Another important quantity related to the Riemann tensor is the *Weyl tensor* C_{abcd}, which is defined as the trace-free part of R_{abcd}:

[11] Minkowski spacetime is the only flat spacetime with the topology of R^4.

[12] In a 3-dimensional spacetime, by contrast, both the Riemann tensor and the Ricci would have only 6 independent components, and gravitational disturbances would be unable to propagate through a vacuum.

$$C_{abcd} = R_{abcd} - \frac{1}{2}(g_{ac} R_{bd} + g_{bd} R_{ac} - g_{ad} R_{bc} - g_{bc} R_{ad})$$

$$+ \frac{1}{6}(g_{ac} g_{bd} - g_{ad} g_{bc}) R$$

The Weyl tensor has 10 independent components, and provides a useful means for representing the degrees of freedom of the gravitational field that are not included in the Ricci tensor. It can be shown that there exist up to four distinct linearly-independent null vectors k^a satisfying the equation

$$k^b k^c k_{[e} C_{a]bc[d} k_{f]} = 0$$

where a square bracket denotes anti-symmetrisation (i.e. $A_{[ab]}=(A_{ab}-A_{ab})/2$). This forms the basis of what is known as the *Petrov classification* of solutions to the Einstein equation. The five Petrov classes are: I (4 distinct null vectors k^a), II (3 distinct null vectors, one of multiplicity 2), II-II or D (2 distinct null vectors, both of multiplicity 2), III (2 distinct null vectors, one of multiplicity 1, one of multiplicity 3), and IV or N (1 distinct null vector, of multiplicity 4)[13].

1.9 Gravitational Radiation

An important prediction of general relativity that has not yet been confirmed is the existence of *gravitational radiation*. In fact, the search for gravitational waves is currently the subject of an intensive worldwide research programme. At a fundamental level, the presence of gravitational waves reflects the fact that the gravitational field (i.e. the Riemann tidal tensor) in 4 dimensions has many more degrees of freedom than the distribution of energy-momentum that gives rise to it. It is therefore possible for there to be a net flux of energy and momentum through spacetime even in the absence of any matter fields, as is evidenced by the large number of known exact solutions to the vacuum Einstein equation describing (for example) plane-fronted or cylindrical ripples in the metric tensor.

Since the Einstein equation is non-linear, gravitational radiation typically gravitates as it propagates, and the behaviour of gravitational waves in the strong-field limit can therefore be very complicated. However, the sources of gravitational radiation that are most likely to be seen by terrestrial detectors (binary neutron stars and supernovae) are in general so weak that the radiation

---

[13] k^a is said to have multiplicity 2, 3 or 4 if

$$k^b k^c C_{abc[d} k_{e]} = 0, \qquad k^c C_{abc[d} k_{e]} = 0 \qquad \text{or} \qquad k^c C_{abcd} = 0$$

respectively.

can be adequately described by a linear approximation. The self-gravity of the waves is then neglected, and the resulting description is formally very similar to Maxwell's theory of electromagnetic radiation.

In a weak gravitational field, it is possible to find co-ordinates x^a in which

$$g_{ab} = \eta_{ab} + h_{ab} \qquad \text{where} \qquad |h_{ab}| << 1$$

The Ricci tensor then becomes, to linear order in h_{ab},

$$R_{ab} \approx -\frac{1}{2}\left(\Delta h_{ab} + h^{c}_{c,ab} - h^{c}_{a,cb} - h^{c}_{b,ca} \right)$$

where

$$\Delta = \eta^{cd} \frac{\partial}{\partial x^c} \frac{\partial}{\partial x^d} \qquad \text{and} \qquad h^{a}_{b,cd} = \frac{\partial^2 h^a_b}{\partial x^c \, \partial x^d}$$

The Einstein equation

$$R_{ab} = 8\pi G S_{ab} \qquad \left(\text{where} \quad S_{ab} \equiv T_{ab} - \frac{1}{2} T g_{ab}\right)$$

then reads:

$$\Delta h_{ab} + h^{c}_{c,ab} - h^{c}_{a,cb} - h^{c}_{b,ca} = -16\pi S_{ab}$$

Just as Maxwell's equations are invariant under gauge transformations of the vector potential, the linearised Einstein equation is invariant under a group of transformations of the metric function h_{ab}. If h_{ab} is any solution of the linearised Einstein equation then so too is:

$$\hat{h}_{ab} = h_{ab} - \varepsilon_{a,b} - \varepsilon_{b,a}$$

where the components of ε_a are four small but otherwise arbitrary functions. This gauge-invariance reflects the freedom available in choosing the quasi-Minkowskian co-ordinates x^a. In fact, the vector ε_a generates a co-ordinate transformation

$$x^a \rightarrow \hat{x}^a = x^a + \eta^{ab} \varepsilon_b$$

which maps one quasi-Minkowskian co-ordinate system to another.

If the vector ε_a is chosen so that

$$\Delta \varepsilon_a = h^{b}_{a,b} - \frac{1}{2} h^{b}_{b,a}$$

then h_{ab} is the metric corresponding to the so-called *harmonic gauge*, which is characterised by the constraint $g^{bc}\Gamma^a{}_{bc}=0$. In the harmonic gauge, the linearised Einstein equation has the form of an inhomogeneous wave equation:

$$\Delta \hat{h}_{ab} = -16\pi G S_{ab}$$

with retarded solution:

$$\hat{h}_{ab}(t, \mathbf{x}) = -4G \int d^3\mathbf{x}' \, \frac{S_{ab}(t - |\mathbf{x} - \mathbf{x}'|, \mathbf{x}')}{|\mathbf{x} - \mathbf{x}'|}$$

In vacuum (S_{ab}=0), the linearised Einstein equation has plane-wave solutions:

$$\hat{h}_{ab}(\hat{x}) = e_{ab} \exp(i k_c \hat{x}^c) + e^*_{ab} \exp(-i k_c \hat{x}^c)$$

where the *wave vector* k_c is null:

$$\eta^{cd} k_c k_d = 0$$

and the *polarisation tensor* e_{ab} is symmetric and satisfies the harmonic gauge condition:

$$k_b \eta^{bc} e_{ca} = \frac{1}{2} k_a \eta^{bc} e_{bc}$$

This last equation removes four of the 10 degrees of freedom in e_{ab}. Another four can be removed by noting that if x^a is any system of harmonic co-ordinates, then so is

$$\bar{x}^a = \hat{x}^a + i \beta^a \exp(i k_c \hat{x}^c) - i \beta^{a*} \exp(-i k_c \hat{x}^c)$$

where β^a is any small complex 4-vector. In the new harmonic co-ordinates,

$$\bar{e}_{ab} = e_{ab} + \eta_{ac} \beta^c k_b + \eta_{bc} \beta^c k_a$$

and so it is always possible to set four of the components of the polarisation tensor to zero by suitably choosing β^a. The polarisation tensor therefore has only two physical degrees of freedom (or polarisation states).

As an example, a plane wave travelling in the z-direction will have:

$$k^1 = k^2 = 0 \qquad \text{and} \qquad k^0 = k^3 = k > 0$$

and it is always possible to choose a gauge in which e_{11} and e_{12} are the only non-zero components of e_{ab}. The flux of energy and momentum carried by gravitational waves of this type can be calculated by substituting g_{ab} back into the full vacuum Einstein equation. All terms linear in h_{ab} will of course vanish from the equation, but there will remain a series of terms of quadratic and higher order in h_{ab} that need to be balanced by the inclusion of higher-order contributions to g_{ab}. The non-linear terms in h_{ab} therefore effectively behave as a source of energy and momentum for the higher-order contributions to g_{ab}. To quadratic order in h_{ab}, the effective energy-momentum tensor due to plane gravitational waves in our chosen gauge turns out to be:

$$< t_{ab} > = \frac{k_a k_b}{16 \pi G} \left(| e_{11} |^2 + | e_{12} |^2 \right)$$

As can be seen, this expression is formally very similar to the energy-momentum tensor associated with a plane wave in electromagnetic theory.

1.10 Alternative Theories of Gravity

As a brief postscript to this summary of the basic features of Einstein's theory, I will make mention of some of the alternative theories of gravity that have been proposed since the 1960s. It should be stressed that there is no reason to believe the classical description of gravity provided by general relativity is in any way inadequate, and that the alternative theories discussed below are at present nothing more than mathematical and historical curiosities.

One class of alternative theories that has been much studied are the so-called *scalar-tensor theories*. These are constructed by adding to general relativity a scalar function ϕ, a function $\omega(\phi)$ which couples ϕ to the metric tensor, and a cosmological function $\lambda(\phi)$. The Einstein equation is modified to read:

$$R_{ab} - \frac{1}{2} R\, g_{ab} - \lambda(\phi)\, g_{ab} = -8\pi \phi^{-1} T_{ab} + F(\phi, \omega, g_{ab})$$

(where F is a given function), and is supplemented by a second equation for ϕ. Standard general relativity is recovered in the limit $\omega \to \infty$. As can be seen, one feature of such a theory is that the (typically time-varying) function ϕ^{-1} supplants the gravitational constant G. The archetypal scalar-tensor theory is the Brans-Dicke theory (1961), which is characterised by $\lambda=0$ and a constant coupling function ω. More recent variants include Bekenstein's Variable Mass Theory (1977) and Barker's Constant G Theory (1978). Scalar-tensor theories cannot be experimentally distinguished from general relativity if the current value of the coupling function ω is greater than about 500.

A second family of alternative theories are the *vector-tensor theories*, which are generated by adding to Einstein's theory a vector field K_a and four parameters τ_1, τ_2, τ_3 and τ_4. The equation governing the metric tensor has the general form:

$$R_{ab} - \frac{1}{2} R\, g_{ab} = -8\pi G\, T_{ab} + \sum_{i=1}^{4} \tau_i F_i(K_a, g_{ab})$$

and is supplemented by an equation for K_a. Specific cases of vector-tensor theories have been considered by Will and Nordvedt (1972) and by Hellings and Nordvedt (1973), but overall they remain complex and little-studied.

The third class of alternative theories are the *bimetric theories*, which make use of a flat metric η_{ab} as well as the standard spacetime metric g_{ab}. The simplest bimetric theory is due to Rosen (1973), and is described by the field equation

$$\Delta g_{ab} - g^{cd} \eta^{ef} g_{ac'e} g_{bd'f} = 16\pi \left(\frac{g}{\eta}\right)^{1/2} \left(T_{ab} - \frac{1}{2} T g_{ab}\right)$$

where $g=\det(g_{ab})$ and $\eta=\det(\eta_{ab})$. More complicated versions include Rastall's theory (1976), which adds a vector field K_a; the BSLL theory (1973), which adds a tensor field B_{ab}; and the stratified theories of Lee, Lightman and Ni, which add a vector field, a tensor field and two scalar fields. Rosen's theory has the mathematical advantage of being singularity-free, but unfortunately contradicts the available experimental evidence in predicting that the pulse rate of a binary pulsar should accelerate rather than slow down.

2. SOME ADVANCED TOPICS IN GENERAL RELATIVITY

2.1 The Schwarzschild Black Hole

The Schwarzschild metric was originally conceived as the spacetime exterior to a spherical body, but with the realisation in the 1930s that the equation of state of degenerate matter could only support stars below a certain critical mass it gradually became clear that the metric also offered a description of the final state of a massive star.

The Schwarzschild line element was derived in §1.6, and has the form:

$$d\tau^2 = \left(1 - \frac{2GM}{r}\right) dt^2 - \left(1 - \frac{2GM}{r}\right)^{-1} dr^2 - r^2 \left(d\theta^2 + \sin^2\theta \, d\phi^2\right)$$

If the line element remains accurate down to $r=0$, then it is apparent that the metric has singularities both at $r=0$ and at the *Schwarzschild radius* $r=2GM\equiv r_S$. The second of these values was in fact identified by Laplace in 1799 as the radius of a body whose escape velocity, according to Newtonian gravity, is the speed of light.

The nature of the singularity at $r=r_S$ can be clarified by examining the fate of inward radial geodesics in the Schwarzschild metric. According to the geodesic equations derived in §1.7, if $J=0$ then

$$\frac{dt}{ds} = \left(1 - \frac{r_S}{r}\right)^{-1} \to \infty \qquad \text{as} \qquad r \to r_S$$

and

$$\left(\frac{dr}{ds}\right)^2 = 1 - E\left(1 - \frac{r_S}{r}\right) \to 1 \qquad \text{as} \qquad r \to r_S$$

while

$$d\tau^2 = \left[\left(1 - \frac{r_S}{r}\right)\left(\frac{dt}{ds}\right)^2 - \left(1 - \frac{r_S}{r}\right)^{-1}\left(\frac{dr}{ds}\right)^2\right] ds^2 = E\, ds^2$$

Thus $t \to \infty$ along any radial geodesic as it approaches $r = r_S$, but a particle tracing out such a geodesic will reach $r = r_S$ in a finite proper time τ. This indicates that t is a poor choice of time co-ordinate in the neighbourhood of the Schwarzschild radius.

A better-behaved system of co-ordinates can be found by first introducing advanced and retarded null co-ordinates u and v via the equations:

$$u = t + \bar{r} \qquad \text{and} \qquad v = t - \bar{r}$$

where

$$\bar{r} = r + r_S \ln (r - r_S)$$

The Schwarzschild line element then takes the considerably simpler form:

$$d\tau^2 = \left(1 - \frac{r_S}{r} \right) du\, dv - r^2 (d\theta^2 + \sin^2 \theta\, d\phi^2)$$

in which the metric tensor is now continuous but still singular at $r = r_S$.[14]

A further simplification occurs if u and v are replaced by so-called *Kruskal-Szekeres co-ordinates* t' and x', where:

$$t' = \frac{1}{2} [\exp (u / r_S) - \exp (- v / r_S)]$$

and

$$x' = \frac{1}{2} [\exp (u / r_S) + \exp (- v / r_S)]$$

The Schwarzschild line element then becomes:

$$d\tau^2 = \frac{4 r_S^2}{r} \exp (- r / r_S) (dt'^2 - dx'^2) - r^2 (d\theta^2 + \sin^2\theta\, d\phi^2)$$

where r is now just a function of t' and x', defined implicitly through the equation:

$$t'^2 - x'^2 = - (r - r_S) \exp (r / r_S)$$

The Kruskal-Szekeres line element is manifestly non-singular at $r = r_S$. The Schwarzschild radius thus marks the location of what is known as a *removable singularity*: a singularity in the metric tensor that disappears if the co-ordinates are chosen appropriately. By contrast, the singularity at $r = 0$ is a physical one, as it cannot be erased by a suitable transformation of co-ordinates. The easiest way of proving this is to note that certain scalar invariants of the Riemann tensor (for example, $R_{abcd}R^{abcd}$) diverge in the limit as $r \to 0$.

---

[14] u and v are referred to as null co-ordinates because lines of constant u, θ and ϕ or constant v, θ and ϕ are null.

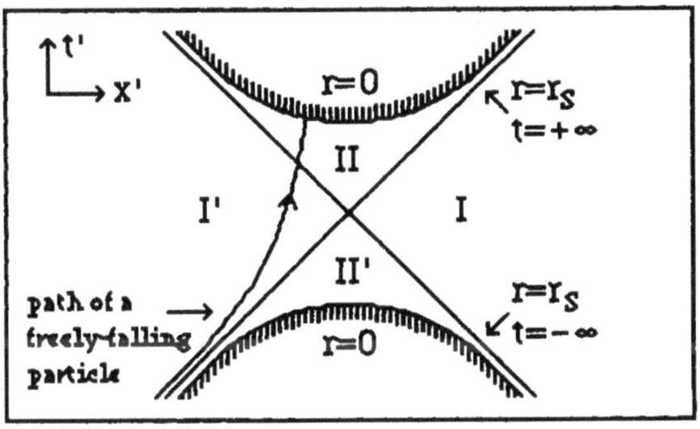

Fig. 2.1: The Schwarzschild metric in Kruskal-Szekeres co-ordinates.

If the co-ordinates θ and ϕ are suppressed, the structure of the Schwarzschild spacetime can be explored by plotting the salient features in the x'-t' plane, as shown in Fig. 2.1. Here, null lines run at 45° to the co-ordinate axes, and the Schwarzschild radius (where $t'^2-x'^2=0$) divides the spacetime into four distinct regions marked I, I', II and II'.The regions I and I' are isometric and asymptotically flat. In Kruskal-Szekeres co-ordinates the singularity at r=0 breaks up into the two branches of the hyperbola $t'^2-x'^2=r_S$, and forms the future boundary of region II (the "black hole") and the past boundary of region II' (the "white hole"). Any material particle or photon that enters region II can never subsequently leave it, and will inevitably strike the singularity in a finite proper time. Similarly, any particle that finds itself in the white hole region II' will leave it in a finite proper time and never subsequently re-enter it. The surface $r=r_S$, which is null, effectively forms a one-way membrane separating the asymptotically flat regions from the singular regions, and is referred to as the Schwarzschild *event horizon*.

An alternative way of representing the same information is by compactifying Fig. 2.1. This is achieved by introducing another pair of co-ordinates t" and x" defined by:

$$t'' = \frac{1}{2}\tan^{-1}\left(u'/r_S^{1/2}\right) + \frac{1}{2}\tan^{-1}\left(v'/r_S^{1/2}\right)$$

and

$$x'' = \frac{1}{2}\tan^{-1}\left(u'/r_S^{1/2}\right) - \frac{1}{2}\tan^{-1}\left(v'/r_S^{1/2}\right)$$

The resulting *Penrose diagram*, which is shown in Fig. 2.2, has the advantage of bringing points at infinity in to a finite co-ordinate distance. Once again null lines run at 45° to the co-ordinate axes. The points marked ι^- and ι^+ represent, respectively, past and future timelike infinity: the origin and destination of all timelike curves that remain outside the singular regions II and II'. The points marked i^0 designate spacelike infinity, which is inaccessible to all timelike and null curves, while the lines $\mathcal{J}^-$ and $\mathcal{J}^+$ represent past and future null infinity.

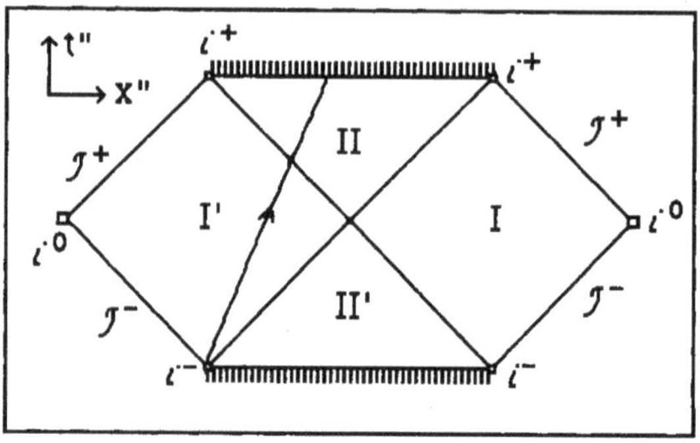

Fig. 2.2: Penrose diagram of the Schwarzschild metric.

It should be stressed that the Schwarzschild metric, which if a true representation of spacetime would predict the existence of a second asymptotically-flat universe (I') in addition to our own (I), does not portray the formation of a black hole as the result of stellar collapse. Instead it describes an eternal black (and white) hole that has no prior history. For a more realistic picture of the structure of a black hole, it is necessary to consider a spacetime which explicitly includes the collapsing material which presumably gives rise to the black hole.

2.2 Spherical Dust Models

A star in the final stages of collapse is thought to consist of a complicated mixture of degenerate plasma and high-density radiation which is typically rotating very rapidly and therefore highly non-spherical. To idealise such a star as a spherical ball of pressure-free fluid (dust) would seem to ignore much of the physics of the problem and shed very little light on the mechanics of a real collapse. However, it can be shown that the formation of a singularity is a generic feature of relativistic collapse, and does not depend on the existence of special symmetries or the specifics of the equation of state of the stellar material (see §2.6). The collapse of a sphere of dust is therefore important as the simplest case in which singularity formation can be examined in detail.

The energy-momentum tensor of a pressure-free fluid has the simple form:

$$T^{ab} = \rho\, u^a u^b$$

where $u^a=dx^a/d\tau$ is the 4-velocity of the fluid. From §1.4, the equations of energy-momentum conservation read:

$$u^a \nabla_a u^b = 0$$

and

$$\nabla_a (\rho\, u^a) = 0$$

The first of these equations indicates that dust particles follow geodesics, while the second expresses the conservation of mass.

If the co-ordinates $(x^a)=(t,r,\theta,\phi)$ inside the star are chosen so that r, θ and φ identify the elements of the fluid (i.e. lines of constant r,θ,φ are geodesics) and t is the proper time τ as measured by an observer comoving with the fluid, then $u^a=(1,0,0,0)$ and the line element for a spherical distribution of matter becomes:

$$d\tau^2 = dt^2 - e^{\mu(t,r)}\, dr^2 - R^2(t,r)\, (\, d\theta^2 + \sin^2\theta\, d\phi^2\,)$$

where μ and R are functions to be determined. The corresponding field equations are:

$$2\,\dot{R}' - R'\,\dot{\mu} = 0$$

$$e^{\mu}\,(\, 2\, R\ddot{R} + \dot{R}^2 + 1\,) - R'^2 = 0$$

and

$$e^{-\mu}\,(\, R^{-1} R'\mu' - R^{-2} R'^2 - 2\, R^{-1} \ddot{R}\,) + R^{-1}\dot{R}\,\dot{\mu} + R^{-2}\dot{R}^2 + R^{-2} = 8\,\pi\, G\,\rho$$

where a prime denotes ∂/∂r and an overdot ∂/∂t. In addition, the equation of mass conservation reads:

$$(\,\rho\, R\, e^{\mu}\,)^{\bullet} = 0$$

The actual form of the solution depends on the initial distribution of density with radius. In all cases, it is possible to find a first integral of the field equations, but one of the easiest cases to consider is a homogeneous ball of dust, which occurs when ρ is a function of t alone. If the solution is assumed to be separable, so that

$$e^{\mu} = T(t)\, f(r) \qquad \text{and} \qquad R = S(t)\, r$$

then

$$f = (\, 1 - k\, r^2\,)^{-1}, \qquad T = S^2 \qquad \text{and} \qquad \rho = \rho_0\, S^{-3}$$

where S satisfies the equation

$$\dot{S}^2 = k\,(\, S^{-1} - 1\,)$$

ρ_0 and k are constants, with $k=8\pi G\rho_0/3$, and the co-ordinates have been normalised so that S(0)=1 (which is equivalent to requiring that the fluid be initially at rest).

The resulting metric is formally identical to that of the closed dust-filled Friedmann-Robertson-Walker universe:

$$d\tau^2 = dt^2 - S^2(t)\left(\frac{dr^2}{1-kr^2} + r^2\,d\theta^2 + r^2\sin^2\theta\,d\phi^2\right)$$

except that, whereas in the Friedmann universe the radial co-ordinate r ranges from 0 to $1/\sqrt{k}$, in the case of stellar collapse r ranges from 0 to some boundary radius $a<1/\sqrt{k}$.[15] This last condition is necessary to ensure that the ball of dust does not initially lie inside its own Schwarzschild radius (see below).

In order to be able to describe the full spacetime, we need to match the interior Friedmann-like solution onto a Schwarzschild exterior solution. This can be done by rewriting the interior metric in the form:

$$d\tau^2 = A(R,\bar{t})\,d\bar{t}^2 - B(R,\bar{t})\,dR^2 - R^2(d\theta^2 + \sin^2\theta\,d\phi^2)$$

where R=S(t)r as before, and A and B are too complicated to reproduce here, but at the surface of the star (where r=a and so R=S(t)a) have the values:

$$A = 1 - \frac{k\,a^2}{S(t)} \qquad \text{and} \qquad B = \left(1 - \frac{k\,a^2}{S(t)}\right)^{-1}$$

Matching these expressions with the corresponding metric components at the boundary of the star in the Schwarzschild metric requires that:

$$M = \frac{k\,a^3}{2G} = \frac{4\pi}{3}\rho_0\,a^3$$

and so in particular $ka^2=r_S/a$, where r_S is the Schwarzschild radius of the star.

Recall that S=1 initially. According to the differential equation governing its behaviour, S decreases monotonically with the comoving proper time t and reaches 0 at a finite value of t. When $S=ka^2$, the surface of the star crosses the Schwarzschild radius and an event horizon forms. In terms of the co-ordinate t (which is the proper time of an observer at infinity), however, the formation of the event horizon occurs only in the limit as $t\to\infty$. The subsequent evolution of the star, and in particular the formation of the singularity in the last phase of the collapse, is therefore invisible to any external observer.

If $t=t_0$ is the proper time at which $S\to 0$, then $S\sim(t_0-t)^{2/3}$ and $\rho\sim(t_0-t)^{-2}$ as t approaches t_0. In other words, all the dust is simultaneously compressed to infinite density at $t=t_0$. This, it should be stressed, is not a generic feature even of spherical collapse. In other, inhomogeneous, dust models the inner layers

---

[15]The apparent singularity in the Robertson-Walker metric at $r=1/\sqrt{k}$ is removable. In fact, it can be shown that the expression multiplying $S^2(t)$ in the metric is the line element on the surface of a 3-sphere of unit radius in R^4.

typically strike the singularity before the outer layers do. Nonetheless, the Penrose diagram describing the spacetime is in all cases of the same basic form as Fig. 2.3. Here, only parts of regions I and II in the Penrose diagram of the pre-existing Schwarzschild hole (Fig. 2.2) are preserved. There is no white hole region, and no twin asymptotically-flat universe. However, there is still a black hole singularity, which any observer foolhardy enough to cross the event horizon would inevitably strike.

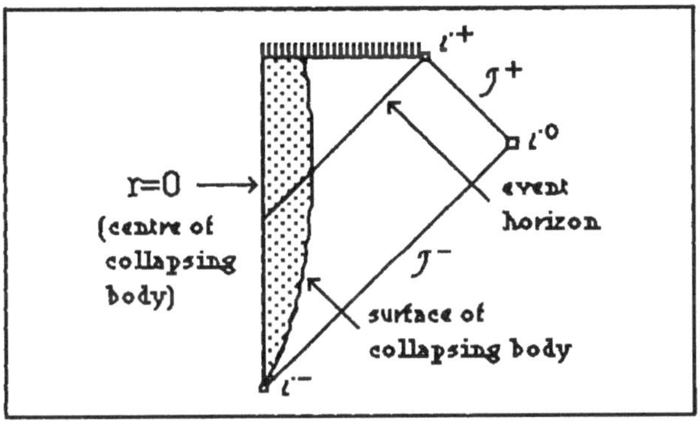

Fig. 2.3: Penrose diagram of a collapsing spherical body.

2.3 The Kerr Metric

The Kerr metric describes the spacetime exterior to a rotating axially-symmetric body. It was first discovered by Roy Kerr in 1963, and in asymptotically-flat co-ordinates has the line element:

$$d\tau^2 = \frac{r^2 + a^2\cos^2\theta - 2\,G\,M\,r}{r^2 + a^2\cos^2\theta}\,dt^2 + \frac{4\,G\,M\,r\,a\sin^2\theta}{r^2 + a^2\cos^2\theta}\,dt\,d\phi$$

$$-\left(r^2 + a^2 + \frac{2\,G\,M\,r\,a^2\sin^2\theta}{r^2 + a^2\cos^2\theta}\right)\sin^2\theta\,d\phi^2 - \frac{r^2 + a^2\cos^2\theta}{r^2 + a^2 - 2\,G\,M\,r}\,dr^2$$

$$-\left(r^2 + a^2\cos^2\theta\right)d\theta^2$$

Here, M is the mass of the central body, and J=Ma is its angular momentum. The line element clearly reduces to that of the Schwarzschild metric in the limit as the spin a goes to zero.

Apart from the usual co-ordinate singularity at the poles (θ=0,π), the Kerr metric has singularities at

$$r = r_{\pm} \equiv G\,M \pm (\,G^2\,M^2 - a^2\,)^{1/2}$$

(provided that $|a| \leq GM$), and at the point ($r=0$, $\theta=\pi/2$). The singularities at $r=r_-$ and $r=r_+$ are in fact removable, and mark the locations of the *inner* and *outer event horizons*, respectively. The physical significance of the two event horizons will become clearer once we examine the Penrose diagram for the Kerr metric, as we will shortly.

The singularity at ($r=0$, $\theta=\pi/2$) is not removable, but has a complicated geometric structure which is obscured by the asymptotically-flat co-ordinate system. It is more easily studied in Kerr-Schild co-ordinates $(\bar{t},x,y,z)$, which are defined by:

$$x + i\,y = (\,r + i\,a\,)\sin\theta\,\exp\left[\,i\left(\,\phi + \int \frac{dr}{r^2 - 2\,G\,M\,r + a^2}\right)\right]$$

$$z = r\cos\theta \qquad \text{and} \qquad \bar{t} = t - r + \int \frac{r^2 + a^2}{r^2 - 2\,G\,M\,r + a^2}\,dr$$

In terms of these co-ordinates, the singularity at ($r=0$, $\theta=\pi/2$) occupies the ring ($x^2+y^2=a^2$, $z=0$). Furthermore, it is possible to analytically continue the metric through the inside of the ring (i.e. through $z=0$ when $x^2+y^2<a^2$) to negative values of the radial co-ordinate r (see Fig. 2.4). For negative values of r, the metric describes a second asymptotically-flat universe with *negative* mass (as $g_{tt} \sim 1+2GM/|r|$ when $r \to -\infty$). Unlike the original, positive-mass universe, the second universe is not shielded from the singularity by a pair of event horizons.

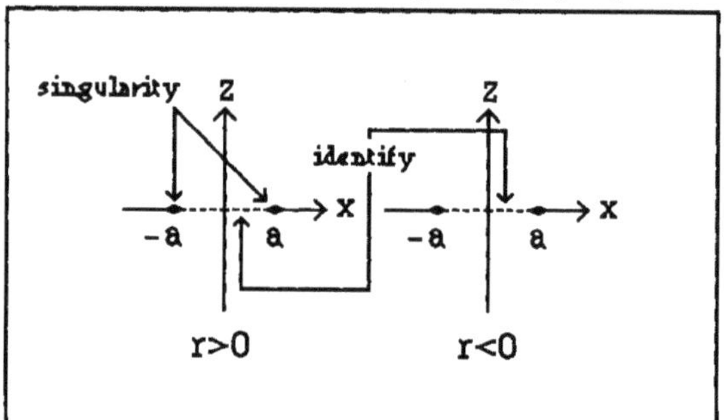

Fig. 2.4: The ring singularity at the centre of the Kerr metric.

As was the case with the Schwarzschild metric, it turns out that the maximal analytic extension of the Kerr metric encompasses more than a single black hole connected to a single asymptotically-flat exterior universe. The Penrose diagram for the maximal extension is shown in Fig. 2.5. Any massive or massless particle from the asymptotically-flat universe I that crosses the outer event horizon h_+ will inevitably reach one of the inner event horizons h_- or h_-' in a finite proper time. From there it will enter one of the two singular regions III or III'. Unlike the Schwarzschild singularity, however, the Kerr singularity is timelike and can be avoided. It is in principle possible for an observer in region III to examine the ring singularity closely without falling into it.

Once tired of the singularity, the observer would have the choice of passing through the interior of the ring into the asymptotically-flat negative-mass universe $\bar{I}$, or crossing the inner event horizon h_-'' and then proceeding inevitably into one of the two asymptotically-flat exterior universes I'' or I'''. By continuing this sequence indefinitely, it is apparent that the maximal extension of the Kerr metric consists of an endless chain of exterior universes isometric to I, interhorizon regions isometric to II, singular regions isometric to III, and negative-mass universes isometric to $\bar{I}$, as shown.

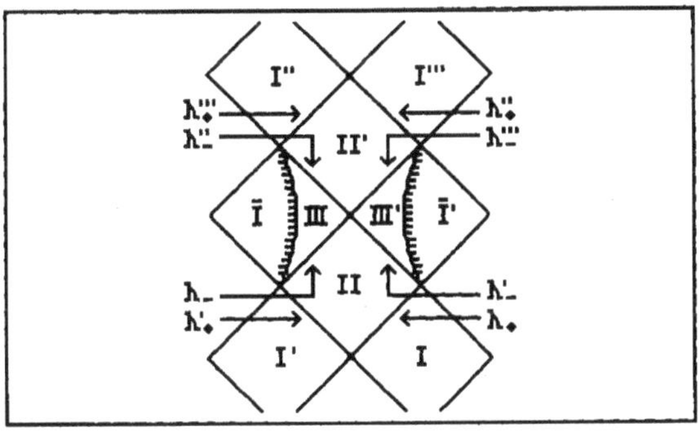

Fig. 2.5: Penrose diagram of the maximal extension of the Kerr metric.

Of course, the Kerr metric does not describe the type of rotating black hole that would be expected to form as a result of stellar collapse, but rather an eternal black hole with no prior history. Not surprisingly, solutions which model the collapse of a rotating fluid are found to form black holes with virtually none of the complicated causal structure of the maximal Kerr metric. Instead, their Penrose diagrams are very similar to that of the collapsing dust sphere illustrated in Fig. 2.3.

A further drawback of the maximal Kerr solution is that the inner event horizon at $r=r_-$ has been shown to be unstable to small perturbations. That is, if a particle of small but non-zero mass crosses the horizon, it (the horizon) typically collapses to form a singularity. This instability is related to the fact that the inner event horizon is a surface of infinite blue shift. Any observer who travelled from h_+ to h_- would see light from every epoch of the history of the asymptotically-flat exterior universe I' in a finite time, most of it compressed together into a "blue sheet" at h_-. The resulting concentration of energy would undoubtedly destroy the inner event horizon.

The maximal Kerr metric is therefore little more than a mathematical curiosity. However, down to the radius of the outer event horizon the Kerr solution remains a realistic and important tool for analysing spacetime outside a compact, rotating object. An interesting feature of this exterior spacetime is the *ergosphere*, an oblate spheroidal region defined by $r_+<r<r_e(\theta)$, where

$$r_e(\theta) = GM + (G^2M^2 - a^2\cos^2\theta)^{1/2}$$

The outer boundary of the ergosphere touches the outer event horizon at the poles, but is otherwise outside the horizon. The significance of the ergosphere is that the metric component g_{tt} is negative everywhere inside it, and therefore no material particle can remain at rest with respect to the spatial co-ordinates (r,θ,ϕ). In fact, $d\tau^2>0$ inside the ergosphere only if $a.dt.d\phi>0$. This is an example of an effect known as *frame dragging*: any material particle inside the ergosphere must rotate in the same direction as the spin of the central object.

Another, related, property of the ergosphere is that it is possible for a material particle to have negative net energy there. In general, the energy E of a particle with rest mass m travelling along the timelike curve $x^a(\tau)$ is:

$$E = m\, g_{ta}\frac{dx^a}{d\tau}$$

while its angular momentum L is:

$$L = -m\, g_{\phi a}\frac{dx^a}{d\tau}$$

Both E and L are constants of the geodesic equations and reduce to their non-relativistic analogues at spacelike infinity. If $\Omega=d\phi/dt$ is the angular velocity of the particle, then since g_{rr} and $g_{\theta\theta}$ are negative everywhere outside the event horizon, it follows from the form of the Kerr line element that:

$$m(g_{tt} + 2g_{t\phi}\Omega + g_{\phi\phi}\Omega^2)\dot{t}^2 \equiv (E - L\Omega)\dot{t} \geq 0$$

Furthermore, E is negative if

$$a\Omega < -\frac{a\, g_{tt}}{g_{t\phi}}$$

These two conditions are compatible only in the ergosphere, where $g_{tt}<0$. Also, the first condition implies that $E\geq L\Omega$ (and so $L\Omega<0$) and that, as the particle approaches the event horizon,

$$\Omega \to \Omega_H \equiv \frac{a}{r_+^2 + a^2}$$

The quantity Ω_H is conventionally referred to as the *angular velocity of the black hole*; the fact that all particles (whether massive or massless) rotate with angular velocity Ω_H at the horizon is simply an extreme manifestation of frame dragging.

The existence of negative-energy orbits[16] in the ergosphere can be used as the basis for a mechanism to extract energy from a Kerr black hole. If an object with net positive energy enters the ergosphere, it is in principle possible for it to break up into two pieces, one with negative energy and the other with the original energy of the object plus the energy deficit of the first piece. If the first piece falls into the event horizon and the second returns to the region outside the ergosphere, the overall effect is to transfer energy from the black hole to the outgoing piece.

To calculate the effect this mechanism has on the black hole itself, note that if the black hole has mass M and angular momentum J=Ma and swallows a particle with energy E and angular momentum L then the net changes in M and J are just $\delta M=E$ and $\delta J=L$. On the face of it, it would seem possible to continue extracting energy from the black hole until the mass M is reduced to zero. However, the process is limited by the requirement $E \geq L\Omega$, which effectively means that the black hole loses angular momentum more rapidly than it loses energy. Once the black hole has been reduced to a spinless state, the energy extraction mechanism ceases to be viable, as the Schwarzschild hole has no ergosphere.

To see this, define the *irreducible mass* M_{irr} of the hole through the equation:

$$2\,M_{irr}^2 = M^2 + (\,M^4 - G^{-2}\,J^2\,)^{1/2} \equiv \frac{M\,r_+}{G}$$

If M and J change by $\delta M=E$ and $\delta J=L$ then the change δM_{irr} in the irreducible mass is given by:

$$\begin{aligned} 4\,M_{irr}\,\delta M_{irr} &= (\,M^4 - G^{-2}\,J^2\,)^{-1/2}\,(\,2\,G^{-1}\,M^2\,r_+\,\delta M - G^{-2}\,J\,\delta J\,) \\ &= (\,M^4 - G^{-2}\,J^2\,)^{-1/2}\,\frac{M\,(\,r_+^2 + a^2\,)}{G^2}\,(\,E - L\,\Omega_H\,) \end{aligned}$$

and thus $\delta M_{irr} \geq 0$, as $E \geq L\Omega_H$ on the horizon. In other words, the irreducible mass of a black hole can never decrease as a result of energy extraction, and since $M_{irr} \leq M \leq \sqrt{2}M_{irr}$, the mass itself can never be reduced below a certain value. The maximum possible energy extraction occurs when an extreme Kerr hole ($|J|=GM^2$) is spun down to a Schwarzschild hole (J=0). If this is done adiabatically (i.e. with $\delta M_{irr}=0$), the initial and final masses M_i and M_f are related by $M_i=\sqrt{2}M_f$

[16]Negative-energy orbits should not be confused with *bound* orbits, which have E<m and are just as common in Newtonian gravity as in general relativity.

and so the total fraction of the initial mass that is available for energy extraction cannot exceed $1-1/\sqrt{2}\approx 29\%$.

The irreducible mass M_{irr} has a simple geometric interpretation which has had a profound influence on our understanding of the connection between general relativity and quantum mechanics. The *area* A of a Kerr black hole is defined to be the surface area of the outer event horizon, and is given by:

$$A = \int_{r=r_+} \sqrt{g_{\theta\theta}\, g_{\phi\phi}}\; d\theta\, d\phi = 16\,\pi\, M_{irr}^2$$

Thus the fact that the irreducible mass of a black hole can never decrease as a result of energy extraction is simply a restatement of what is known as the *second law of black hole thermodynamics*: namely, that the area of a black hole can never decrease as a result of classical interactions. The area A thus has a natural interpretation as an entropy-like variable, and it is possible to define a temperature-like variable, the *specific gravity* κ:

$$\kappa = \frac{r_+ - G\,M}{2\,G\,M\,r_+}$$

so that black holes obey analogues of all the classical laws of thermodynamics.

Explicitly, the four laws of black hole thermodynamics read:

Zeroth Law: κ is constant over the horizon of a stationary black hole

First Law:

$$dM = \frac{1}{8\,\pi}\,\kappa\, dA + \Omega_H\, dJ$$

Second Law: $\delta A \geq 0$ in any classical process

Third Law: It is impossible to achieve $\kappa=0$ (i.e. an extreme black hole, with $|J|=GM^2$) through any physical process.

Thus, in addition to the identifications $A=8\pi\alpha S$ and $\kappa=T/\alpha$ (where S is the entropy and T the temperature of the hole, and α is some constant), the mass M can be regarded as the internal energy of the hole, and $\Omega_H dJ$ as a measure of the work done on it by external forces. It was this intimate connection between black holes and thermodynamic systems that led Hawking to propose in 1974 that a black hole should radiate quantum mechanically, just like any other thermal body.

2.4 Hawking Radiation

The prediction that a black hole should radiate like a thermal body with temperature proportional to its surface gravity is based on an analysis of the behaviour of quantised fields in a curved background spacetime. It can be shown to apply to a wide range of massless and massive fields, as well as to linearised gravity itself, and to not depend on the existence or absence of special symmetries in the gravitational field (except that the field cannot be static). The basic features of the process can therefore be demonstrated by quantising a massless scalar field in the metric due to a collapsing spherical body.

Consider first of all a scalar field ϕ that obeys the massless Klein-Gordon equation

$$g^{ab}\nabla_a\nabla_b\phi \equiv g^{ab}\phi_{,ab} - g^{ab}\Gamma^c_{ab}\phi_{,c} = 0$$

In a spherically-symmetric gravitational potential ϕ can be expanded as a linear combination of spherical harmonics of the form $r^{-1}f(r,t)Y_{lm}(\theta,\phi)$; in the specific case of the Schwarzschild metric, for example, the function $f(r,t)$ satisfies the scattering equation

$$\frac{\partial^2 f}{\partial t^2} - \frac{\partial^2 f}{\partial \bar{r}^2} + \left(1 - \frac{r_S}{r}\right)\left[\frac{l(l+1)}{r^2} + \frac{r_S}{r^3}\right] f = 0$$

where, as before (see §2.1),

$$\bar{r} = r + r_S \ln(r - r_S)$$

and r_S is the radius of the black hole event horizon. In the case of a non-static spherical metric the scattering equation will in general be much more complicated.

At past null infinity $\mathscr{I}^-$ the scalar field ϕ will typically be a superposition of incoming monochromatic waves with time dependence $\exp[i\omega(t+r)]$, while at future null infinity $\mathscr{I}^+$ it will consist of a superposition of outgoing waves with time dependence $\exp[i\omega'(t-r)]$. Since the Klein-Gordon equation is linear, the outgoing waves will depend linearly on the incoming waves. The effect of the gravitational potential is therefore encapsulated in the transmission coefficients $\beta_{\omega\omega'}$ which relate the amplitudes of the incident and transmitted waves at the frequencies ω and ω'.

If the field ϕ is now quantised, and the background spacetime is assumed to have the global structure characteristic of a collapsing spherical body (Fig. 2.3) then the evolution of the field is fully determined by its initial data on $\mathscr{I}^-$. ϕ can then be decomposed in the form:

$$\phi = \sum_i (f_i \mathbf{a}_i + \bar{f}_i \mathbf{a}_i^\dagger)$$

where the functions $\{f_i\}$ and $\{\bar{f}_i\}$ together form a complete orthonormal set of incoming solutions to the massless Klein-Gordon equation, the set $\{f_i\}$ contains only positive-frequency solutions on $\mathscr{I}^-$, and $\mathbf{a}_i$ and $\mathbf{a}_i^\dagger$ are respectively creation and annihilation operators for particles in state i on $\mathscr{I}^-$.

Since the region external to the black hole in Fig. 2.3 is bounded in the future by the horizon and $\mathscr{I}^+$, it is expected that part of any incoming wave function would disappear across the horizon and part of it would recede to future null infinity. The field ϕ can therefore be decomposed in the alternative form:

$$\phi = \sum_i (\, p_i\, \mathbf{b}_i + \bar{p}_i\, \mathbf{b}_i^\dagger + q_i\, \mathbf{c}_i + \bar{q}_i\, \mathbf{c}_i^\dagger\,)$$

where the functions $\{p_i\}$ and $\{\bar{p}_i\}$ also form a complete orthonormal set of outgoing solutions to the Klein-Gordon equation (and $\{p_i\}$ contains only positive-frequency solutions on $\mathscr{I}^+$), $\mathbf{b}_i$ and $\mathbf{b}_i^\dagger$ are creation and annihilation operators for particles in state i on $\mathscr{I}^+$, and $\{q_i\}$, $\{\bar{q}_i\}$, $\mathbf{c}_i$ and $\mathbf{c}_i^\dagger$ have similar interpretations in terms of particle states on the horizon.[17] The data on $\mathscr{I}^+$ is completely determined by the initial data, and so

$$p_i = \sum_j (\, \alpha_{ij}\, f_j + \beta_{ij}\, \bar{f}_j\,)$$

for some set of coefficients α_{ij} and β_{ij}. If the incoming state is the $\mathscr{I}^-$ vacuum state $|0_-\rangle$ then the expectation value of the number operator for the ith outgoing mode on $\mathscr{I}^+$ is:

$$\langle 0_- |\, \mathbf{b}_i^\dagger\, \mathbf{b}_i \,| 0_- \rangle = \sum_j |\beta_{ij}|^2$$

Thus, provided that the effect of the gravitational potential is to scatter some of the negative-frequency solutions (anti-particle states) into positive-frequency solutions (particle states), the initial vacuum $|0_-\rangle$ will not appear to be a vacuum state to an observer at future null infinity. This is the origin of Hawking radiation.

It is easily seen that frequency scattering does not occur in the Schwarzschild metric or any other static metric, as the transmission coefficients $\beta_{\omega\omega'}$ are then simply proportional to $\delta_{\omega\omega'}$. Black hole evaporation therefore depends critically on the time-asymmetry in the metric due to the formation of the black hole in the first place. In order to provide a quantitative analysis of Hawking radiation from a non-rotating black hole it is thus necessary to solve the Klein-Gordon equation (or some other particle equation) in a collapsing spacetime of the type discussed in §2.2. Needless to say, this is very difficult to do explicitly, and the exact form of the outgoing radiation will typically depend on the fine details of the collapse.

---

[17]There is an ambiguity in distinguishing particle and anti-particle states (i.e. performing a unique operator decomposition) for particles whose wavelength $2\pi/\omega$ is greater than the characteristic length scale of the gravitational field, r_S, on the horizon. This ambiguity does not affect the calculation of the final state on $\mathscr{I}^+$, however, as the latter involves a sum over all the horizon states.

However, it is possible to calculate the asymptotic form of the spectrum as observed at late times on $\mathscr{I}^+$ (when the black hole has settled into its classical final state). The components of the field ϕ that reach $\mathscr{I}^+$ at late times either travel directly from $\mathscr{I}^-$ to $\mathscr{I}^+$ (in which case they undergo no frequency scattering), or pass through the centre of the collapsing body just before the formation of the event horizon (see Fig. 2.6). In the latter case, the amount of frequency scattering is effectively determined by conditions on the event horizon. Furthermore, since the incoming wave function is blue-shifted without limit as it approaches the horizon, the scattering equation can there be solved using the methods of geometric optics.

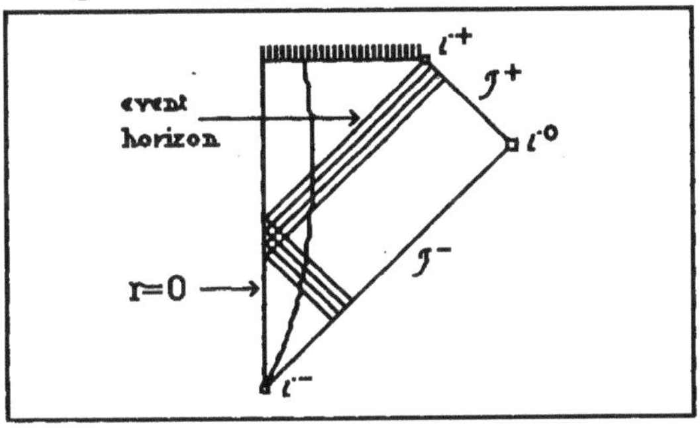

Fig. 2.6: The last of the scattered wave crests to reach future null infinity.

The derivation still remains relatively complicated and cannot be reproduced here [the interested reader should refer to S.W. Hawking, *Comm. Math. Phys.* **43**, 199 (1975) for full details], but the final result is that the expectation value of the energy density for outgoing modes with frequency ω on $\mathscr{I}^+$ is proportional to $[\exp(2\pi\omega/\kappa)-1]^{-1}$, where κ is the surface gravity of the black hole, as defined in §2.3. In other words, the spectrum of the radiated particles is thermal, with an effective temperature given by:

$$T = \frac{\kappa}{2\pi}$$

This is of course completely in accordance with the laws of black hole thermodynamics.

Just as was the case with energy extraction in the ergosphere of a rotating black hole, the source of the emitted energy is the mass of the black hole itself. At a heuristic level, Hawking radiation is due to vacuum fluctuations which create virtual particle-antiparticle pairs just outside the horizon. If one member of the pair crosses the horizon before the pair annihilates, it is possible for the other member to escape to infinity, with a consequent net decrease in the mass of the black hole.

Since the surface gravity (and so the effective temperature) of a black hole typically increases as its mass decreases, Hawking radiation can in principle lead to its complete evaporation. According to the Stefan-Boltzmann law the rate at which a radiating black hole loses mass is:

$$\frac{dM}{dt} \sim -A\,T^4$$

where A is the surface area of the black hole. For a Schwarzschild hole,

$$T = \frac{\kappa}{2\pi} = \frac{1}{8\pi G M} \qquad \text{and} \qquad A = 16\pi G^2 M^2$$

and so $dM/dt \sim -M^{-2}$. This equation is easily solved to give a total lifetime τ_0 to radiative evaporation of

$$\tau_0 \approx 10^{71}\,(M/M_\odot)\ \mathrm{s}$$

Thus, a solar-mass black hole would lose only a tiny fraction of its mass to Hawking radiation during a Hubble time ($\sim 10^{17}$s). Furthermore, since the effective temperature of a solar-mass black hole ($\sim 6 \times 10^{-8}$K) is so much smaller than the temperature of the microwave background (2.7K), any black hole of astrophysical provenance existing today would actually be expanding due to radiative absorption.

On the face of it, the existence of Hawking radiation seems to violate the second law of black hole thermodynamics, as the surface area of a Schwarzschild hole will shrink as its mass evaporates. And indeed, the Second Law as described in §2.3 applies only to classical processes. However, it is possible to reformulate the Second Law so as to take account of the entropy of the radiated particles: the sum of this entropy and the surface area of the black hole will never decrease under any circumstances.

If a black hole does evaporate by Hawking radiation, the quasi-classical approximations implicit in Hawking's analysis will break down once the mass of the black hole is comparable to the Planck mass ($\sim 10^{-5}$g). The subsequent evolution of the black hole would presumably be governed by quantum gravity, but one possibility is that the remaining mass-energy disintegrates in an explosion with an effective temperature of about 10^{32}K, leaving only empty space. The resulting Penrose diagram would then be similar to Fig. 2.7. The region in this diagram to the future of the singularity poses profound questions for classical physics, as it represents a breakdown of classical predictability. It can in principle be explored by an external observer, but evolves from a set of initial data which includes a point (the singularity) about which nothing is known in the absence of quantum gravity.

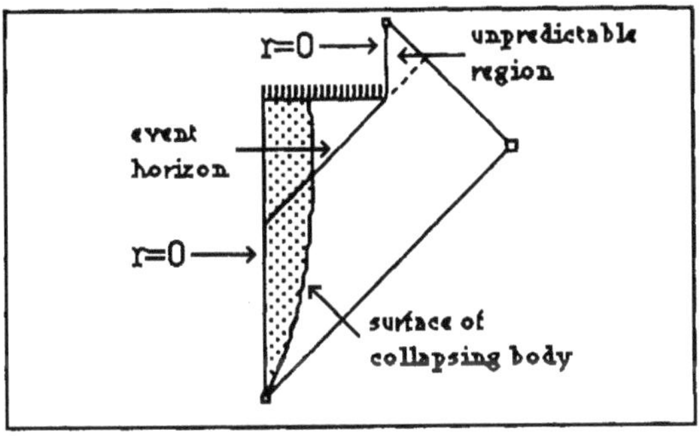

Fig. 2.7: Penrose diagram of an evaporating black hole.

2.5 Causal Structure and Singularity Theorems

In this, the final section, I will briefly explain the content of some of the singularity theorems that were derived in the 1960s, primarily by Penrose and Hawking, and which demonstrate rigorously that singularities are an inescapable feature of general relativity given a reasonable set of physical assumptions. Since a singularity represents a set of points in spacetime beyond which it is not possible to extend the laws of physics, the singularity theorems imply that general relativity is a fundamentally incomplete description of the universe and would need to be supplemented by a more general theory, presumably quantum gravity, in regions of high curvature.

First, some definitions are necessary. We have seen that the programme of general relativity is to describe spacetime in terms of a 4-dimensional manifold endowed with a twice-differentiable metric structure. Let S be any subset of spacetime. Then the *chronological future* $I^+(S)$ of S is defined to be the set of points that can be reached from S along a future-directed timelike curve. The *chronological past* $I^-(S)$ of S is defined similarly. S is said to be *achronal* if no timelike curve intersects S more than once; that is, if $I^+(S) \cap S = \emptyset$. The *domain of dependence* D(S) of a closed achronal set S is the set of points p for which either (i) every future-directed causal (i.e. timelike or null) curve from p intersects S, or (ii) every past-directed causal curve from p intersects S. The difference between $I^+(S)$, $I^-(S)$ and D(S) for an achronal set S is illustrated in Fig. 2.8.

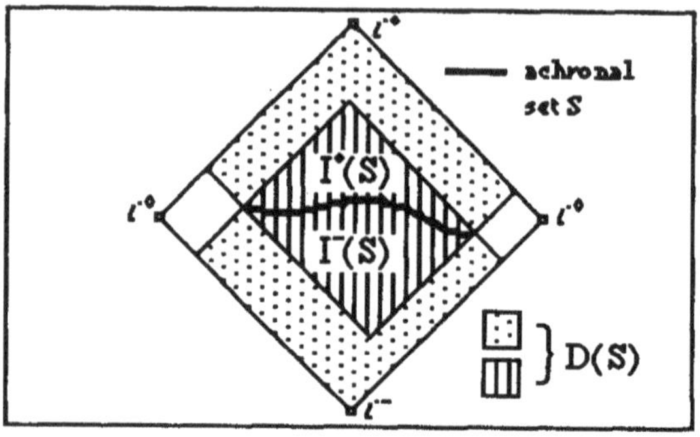

Fig. 2.8: Chronological past, chronological future and domain of dependence.

A *Cauchy surface* Σ is a closed achronal set with the property that $D(\Sigma)$ is the whole of spacetime (see Fig. 2.9). Thus, the entire past and future history of the spacetime can be reconstructed from a knowledge of the conditions on any Cauchy surface. It can be shown that a Cauchy surface must be a 3-dimensional spacelike[18] surface without a boundary. Not every spacetime contains a Cauchy surface: the evaporating black hole spacetime shown in Fig. 2.7 does not, for example. If the spacetime does contain a Cauchy surface, it is said to be *globally hyperbolic*.

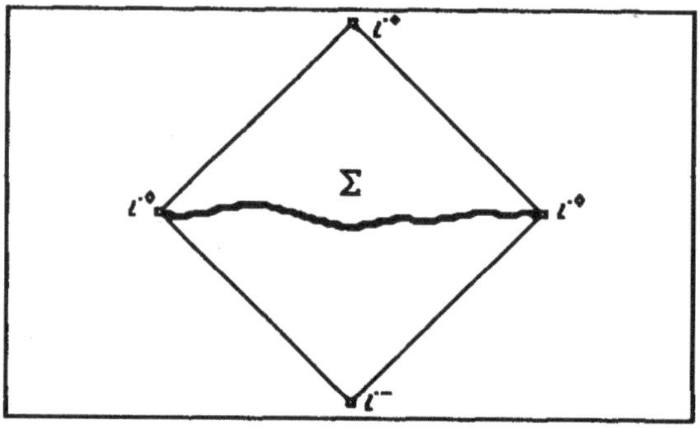

Fig. 2.9: A Cauchy surface.

A *congruence* is a family of curves passing through an open set S with the property that precisely one curve passes through each point in S. A *timelike congruence* is a congruence of timelike curves. If t^a is the unit timelike vector field (so that $g_{ab}t^a t^b = 1$) tangent to a particular timelike congruence, then the *expansion* θ of the congruence is defined to be:

---

[18]That is, any curve lying wholly within the surface is spacelike.

$$\theta = \nabla_a t^a \equiv t^a{}_{,a} + \Gamma^a_{ab} t^b$$

the *shear tensor* σ_{ab} of the congruence is:

$$\sigma_{ab} = \frac{1}{2}(\nabla_a t_b + \nabla_b t_a) - \frac{1}{3}\theta(g_{ab} - t_a t_b)$$

and the *twist tensor* ω_{ab} is:

$$\omega_{ab} = \frac{1}{2}(\nabla_a t_b - \nabla_b t_a)$$

Note that the expansion, shear and twist are respectively the trace, traceless symmetric and antisymmetric parts of $\nabla_a t_b$. If the curves in the congruence are all geodesics then the equation of geodesic deviation can be shown to lead to what is known as *Raychaudhuri's equation*:

$$\frac{d\theta}{d\tau} \equiv t^a \theta_{,a} = -\frac{1}{3}\theta^2 - \sigma_{ab}\sigma^{ab} + \omega_{ab}\omega^{ab} - R_{ab} t^a t^b$$

Now, if S is a 3-dimensional spacelike surface and the geodesics in the congruence all pass orthogonally through S then it can also be shown that $\omega_{ab}=0$ (this is a particular case of what is known as *Frobenius's Theorem*). Furthermore, σ_{ab} is by definition orthogonal to t^b [as $\nabla_a(t^b t_b)=0$] and so is tangent to the spacelike surface S. Hence, $\sigma_{ab}\sigma^{ab} \geq 0$.

So far, no restriction has been placed on the energy-momentum tensor T^{ab}. In practice, however, the range of possible forms of T^{ab} is limited by a variety of microphysical factors. A distribution of mass and energy is said to obey the *strong energy condition* if it satisfies

$$T_{ab} t^a t^b \geq \frac{1}{2} T^a_a$$

for any unit timelike vector t^b.[19] Needless to say, no field or material has ever been found that violates this condition (although a number of high-energy field theories have been postulated that do violate it). In the case of a perfect fluid, the strong energy condition reduces to the requirements $\rho+p\geq 0$ and $\rho+3p\geq 0$, which are certainly satisfied by ordinary astrophysical fluids.

If the congruence of timelike geodesics introduced earlier is assumed to pass through a region of spacetime where the strong energy condition is satisfied then

$$R_{ab} t^a t^b = 8\pi G(T_{ab} t^a t^b - \frac{1}{2} T^a_a) \geq 0$$

and Raychaudhuri's equation becomes:

---

[19]Other conditions that are sometimes imposed are the *weak energy condition:*

$$T_{ab} t^a t^b \geq 0$$

for any unit timelike vector t^a; and the *dominant energy condition*:

$$T^{00} \geq |T^{ij}| \qquad i,j = 1,2,3$$

in any orthonormal basis.

$$\frac{d\theta}{d\tau} \leq -\frac{1}{3}\theta^2$$

In other words, the congruence converges as it propagates, and it can be seen that the effect of assuming the strong energy condition is in some sense equivalent to assuming that matter is gravitationally attractive.

The last equation is easily solved, and gives:

$$\theta(\tau) \leq (\theta_S^{-1} + \frac{1}{3}\tau)^{-1}$$

where τ is the proper distance from S along the congruence, and θ_S is the value of θ on S. In particular, if the congruence is initially converging on S, so that $\theta_S<0$, then $\theta \to -\infty$ for some $\tau \leq 3/|\theta_S|$. A point p at which $\theta \to -\infty$ is said to be *conjugate* to the surface S. If p is conjugate to S then infinitesimally nearby geodesics which pass orthogonally through S will cross at p (see Fig. 2.10).

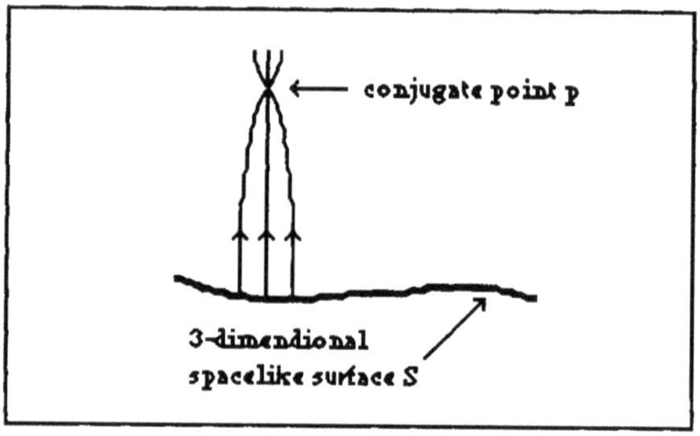

Fig. 2.10: A point conjugate to a spacelike surface S.

We are now in a position to be able to prove a singularity theorem first published by Hawking in 1967. A spacetime is defined to be *singular* if it contains at least one inextendible timelike or null curve of finite proper length[20]. According to Hawking's 1967 theorem, if

(i) the strong energy condition holds;

and (ii) spacetime contains a smooth Cauchy surface Σ, and the congruence of geodesics that passes orthogonally through Σ is expanding, so that $\theta \geq C > 0$ everywhere on Σ;

then no past-directed timelike curve from Σ can have length greater than $3/C$. In other words, if the Universe is globally hyperbolic and expanding then all timelike curves must emerge from an initial singularity (which in the

[20]A null curve is defined to have finite "proper length" if the elapsed value of any affine parameter s (defined in §1.1) is finite along it.

Friedmann-Robertson-Walker spacetimes at least would be identified with the Big Bang).

The proof of this proposition proceeds as follows. Note that the congruence of *past*-directed geodesics orthogonal to Σ has $\theta \leq -C < 0$ everywhere on Σ. Suppose by way of contradiction that there exists a past-directed timelike curve λ with initial point on Σ and length greater than $3/C$, and let p be any point on λ beyond $\tau = 3/C$. Then there is a curve γ from Σ to p whose length is the maximum of the lengths of all possible curves from Σ to p (see Fig. 2.11a).

The curve γ must be a geodesic (as geodesics extremise proper time), and furthermore can have no conjugate point between Σ and p, since otherwise it would be possible to round off a neighbouring geodesic γ' in such a way as to turn it into a curve from Σ to p with proper length greater than the length of γ (see Fig. 2.11b). A similar "rounding off" argument can be used to show that γ must pass orthogonally through Σ. γ is therefore a past-directed geodesic orthogonal to Σ with proper length greater than $3/C$, in contradiction to Raychaudhuri's equation, and the postulated timelike curve λ does not exist.

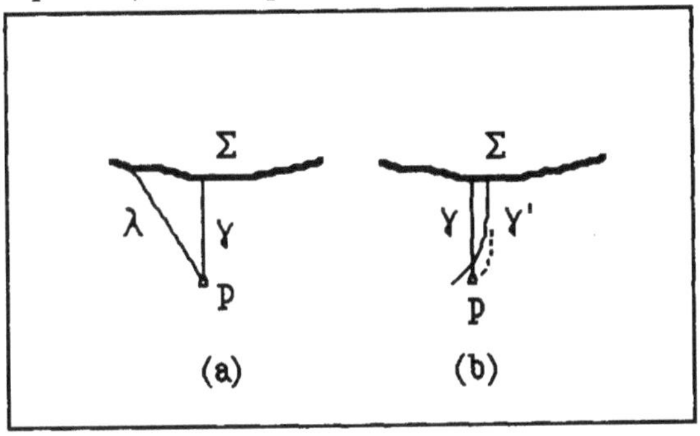

Fig. 2.11: (a) Maximum-length curve γ from Σ to p; (b) γ can have no conjugate point.

The requirement that spacetime be globally hyperbolic, which would be very difficult to confirm experimentally, can in fact be relaxed. If the Cauchy surface Σ is replaced by a compact, edgeless, achronal, smooth, 3-dimensional spacelike surface S with an expanding orthogonal geodesic congruence then it can be shown that there exists at least one incomplete past-directed timelike geodesic with initial point on S.

In addition to the theorem just proved, which concerns the existence of a past singularity in an expanding Universe, there is a family of singularity theorems which predict the formation of future singularities as a result of gravitational collapse. The earliest and simplest of these theorems was established by Penrose in 1965, and involves the notion of a *closed trapped surface*. A closed trapped surface T is a compact, smooth, spacelike 2-dimensional surface with $\theta<0$ for both the ingoing and outgoing congruences of null geodesics orthogonal to T. (For example, all spheres of constant t and $r<r_S$ in the Schwarzschild metric are closed trapped surfaces.)

Penrose's 1965 singularity theorem, which I will not prove here, states that if

(i) the strong energy condition holds;

(ii) spacetime contains a non-compact Cauchy surface;

and (iii) spacetime contains a closed trapped surface with $\theta \le C < 0$ everywhere;

then there exists at least one future-directed null geodesic with initial point on T and affine proper length no greater than 2/C. Thus at least one null geodesic from T ends at a singularity. Again, the assumption of global hyperbolicity can be eliminated by invoking other, less contentious assumptions.

Anyone who wishes to study these (and other) singularity theorems in more detail should refer to Hawking and Ellis's *The Large-Scale Structure of Spacetime*, which is fully cited in the Bibliography.

A SELECT BIBLIOGRAPHY

Historical Accounts

I. Nicolson, *Gravity, Black Holes and the Universe*, Wiley, 1981.
R. Torretti, *Relativity and Geometry*, Pergamon, 1983.

Phenomenological Accounts

M.G. Bowler, *Gravitation and Relativity*, Pergamon, 1986.
P. Rastall, *Postprincipia: Gravitation for Physicists and Astronomers*, World Scientific 1991.
W. Kopczynski and A. Trautmann, *Spacetime and Gravitation*, Wiley, 1992.
C. Will, *Theory and Experiment in Gravitational Physics*, Cambridge University Press, 1993.

Mathematical and General Accounts

S. Weinberg, *Gravitation and Cosmology*, Wiley, 1972.
C.W. Misner, K.S. Thorne and J.A. Wheeler, *Gravitation*, Freeman, 1973.

S.W. Hawking and G.F.R. Ellis, *The Large-Scale Structure of Spacetime*, Cambridge University Press, 1973.
P.A.M. Dirac, *General Theory of Relativity*, Wiley, 1975.
H.C. Ohanian, *Gravitation and Spacetime*, Norton, 1976.
R.K. Sachs and H. Wu, *General Relativity for Mathematicians*, Springer-Verlag, 1977.
W.L. Burke, *Spacetime, Geometry, Cosmology*, University Science Books, 1980.
H.A. Buchdahl, *Seventeen Simple Lectures on General Relativity Theory*, Wiley, 1981.
M. Carmeli, *Classical Fields: General Relativity and Gauge Theory*, Wiley, 1982.
R.L. Faber, *Differential Geometry and Relativity Theory*, Marcel Dekker, 1983.
R.M. Wald, *General Relativity*, University of Chicago Press, 1984.
J. Van Bladel, *Relativity and Engineering*, Springer-Verlag, 1984.
J.N. Islam, *Rotating Fields in General Relativity*, Cambridge University Press, 1985.
G.L. Naber, *Spacetime and Singularities*, Cambridge University Press, 1988.
J. Stewart, *Advanced General Relativity*, Cambridge University Press, 1990.
N. Straumann, *General Relativity and Relativistic Astrophysics*, Springer-Verlag, 1991.
S. Chandrasekhar, *The Mathematical Theory of Black Holes*, Oxford University Press, 1992.
R.A. Mould, *Basic Relativity*, Springer-Verlag, 1994.

<u>Referenced Papers</u>

Barker, B.M. (1978), *Astrophys. J.* **219**, 5.
Bekenstein, J.D. (1977), *Phys. Rev.* D **15**, 1458.
Brans, C. and Dicke, R.H. (1961), *Phys. Rev.* **124**, 925.
Hawking, S.W. (1967), *Proc. Roy. Soc. Lond.* A **300**, 187.
Hawking, S.W. (1975), *Comm. Math. Phys.* **43**, 199.
Hellings, R.W. and Nordvedt, K., Jr. (1972), *Phys. Rev.* D **7**, 3593.
Kerr, R.P. (1963), *Phys. Rev. Lett.* **11**, 237.
Lightman, A.P. and Lee, D.L. (1973), *Phys. Rev.* D **8**, 3293. (BSLL theory)
Penrose, R. (1965), *Phys. Rev. Lett.* **14**, 57.
Rastall, P. (1976), *Can. J. Phys.* **54**, 66.
Rosen, N. 91973), *J. Gen. Rel. and Grav.* **4**, 435.
Vessot, R.F.C. et al (1980), *Phys. Rev. Lett.* **45**, 2081.
Will, C.M. and Nordvedt, K., Jr. (1972), *Astrophys. J.* **177**, 757.

CLASSICAL COSMOLOGY

GERHARD BÖRNER
MPI für Astrophysik,
Karl-Schwarzschild-Str. 1
Postfach 15 23
85740 Garching

1. Friedmann-Lemaître Models ("FL-Models")

In the context of GR a spacetime is a 4–dimensional differentiable manifold with a metric $g_{\mu\nu}$ which we take to have the signature $(+ - --)$. The metric tensor $g_{\mu\nu}$ determines distances, time intervals, and the causal order of events. Einstein's field equations relate $g_{\mu\nu}$ to the energy — momentum tensor $T_{\mu\nu}$:

$$R_{\mu\nu} - 1/2R\, g_{\mu\nu} + \Lambda g_{\mu\nu} = 8\pi\, GT_{\mu\nu} \quad . \tag{1.1}$$

$R_{\mu\nu}$ is the contracted curvature tensor, $R \equiv R^{\mu}_{\mu}$: trace. These quantities depend on $g_{\mu\nu}$ and its partial derivatives up to second order. Λ is the famous or infamous cosmological constant.

In matter–free space $T_{\mu\nu} = 0$. For a perfect fluid or a radiation field

$$T_{\mu\nu} = (\varrho + p)V_{\mu}V_{\nu} - p\, g_{\mu\nu} \quad ;$$

for an electromagnetic field

$$T_{\mu\nu} = \frac{1}{4\pi}\left(F_{\mu\lambda}F_{\nu\kappa}g^{\lambda\kappa} - \frac{1}{4}\, g_{\mu\nu}F_{\delta\tau}F_{\kappa\lambda}g^{\delta\kappa}g^{\lambda\tau}\right) \quad .$$

If $f(x,p)$ is the distribution function of an ideal gas, we write

$$T_{\mu\nu}(x) = \int p_\mu p_\nu f(x,p) d\pi \quad .$$

Most known exact solutions of (1.1) describe spacetimes of high symmetry. The FL–spacetimes have exact spherical symmetry about every point. This implies (e.g. see ref. 14) that the spacetimes are spatially homogeneous and that they admit, at least, a six–parameter group of isometries.

Flat Minkowski space and de Sitter space are examples of FL–spaces which have an even higher symmetry.

Coordinates can be chosen such that the line element reads

$$ds^2 = dt^2 - R^2(t) d\sigma^2 \quad , \tag{1.2}$$

where $d\sigma^2$ is the line element of a Riemannian 3–space of constant curvature K/R^2.

$$d\sigma^2 = d\chi^2 + f^2(\chi)(d\theta^2 + \sin^2\theta d\varphi^2) \quad , \tag{1.3}$$

where $f(\chi)$ depends on the sign of the curvature. By a suitable scaling of the "expansion factor" $R(t)$ the curvature can be normalised such that $K = +1$, -1, 0. The function

$$f(\chi) = \begin{cases} \sin\chi & \text{for } K = +1 \\ \chi & \text{for } K = 0 \\ \sinh\chi & \text{for } K = -1 \end{cases} \tag{1.4}$$

characterizes the topology of the spaces at constant time. The coordinate χ runs from 0 to ∞, if $K = 0$ or $K = -1$. If $K = +1$ it runs from 0 to π. The metric (1.3) requires via (1.1) that $T_{\mu\nu}$ is of the perfect fluid type

$$T_{\mu\nu} = (\varrho + p) V_\mu V_\nu - p\, g_{\mu\nu} \quad . \tag{1.5}$$

The density ϱ and the pressure p are functions of t only, and the flow lines with tangent vector V_μ are the curves $(\chi, \theta, \varphi) =$ const. These coordinates are called "comoving". In comoving coordinates $V_\mu = \begin{pmatrix} 1 \\ 0 \\ 0 \\ 0 \end{pmatrix}$, and $T_{\mu\nu} = \mathrm{diag}(\varrho, p, p, p)$.

The expansion factor $R(t)$ changes with time like the spatial distance of any two particles comoving with the cosmic fluid, i.e. of two "galaxies" without peculiar motions. For two such particles with comoving coordinates (χ, θ, φ) and $(\chi_0, \theta, \varphi)$ the distance at a time t is $d = R(t)(\chi - \chi_0)$.

At some fixed time t_1 the area of a sphere around $\chi_0 = 0$ is $A_1 = \pi R^2(t_1)f^2(\chi)$. We can imagine testing the geometry of the space sections by comparing the radius $d/R(t_1)$ and the surface areas $A_1/R^2(t_1)$ of spheres centered on $\chi_0 = 0$. The expansion factor has been divided out to obtain relations valid at all times (or giving the 3–space geometry, when $R(t_1)$ is set equal to one.)

In the flat case, $K = 0$, the geometry is Euclidean, $A_1 \propto d^2$; for $K = -1$ we find $A_1 \propto \sin^2 h\chi$, i.e. the area increases with distance faster than in the Euclidean case. For $K = +1$ the increase is slower than in flat space, $A_1 \propto \sin^2 \chi$. A_1 reaches a maximum at $\chi = \pi/2$, and then decreases to zero for $\chi \to \pi$.

So far we have only considered local properties, but different global topologies are possible. $K = +1$, or locally spherical spacetimes are finite, and as we shall see later, these models will eventually recollapse. Models with $K = -1$, or $K = 0$ are infinite, if they are simply connected, and will expand forever.

In the case $K = 0$, the spatial sections are flat and (r, θ, φ) can be changed to Cartesian coordinates resulting in

$$ds^2 = dt^2 - R^2(t)\{dx^2 + dy^2 + dz^2\} \quad .$$

The simplest case is to assume the standard range for these coordinates $-\infty < (x, y, z) < +\infty$; the sections $t =$ const are then without boundary and of infinite volume. Another possibility, however, is the torus topology: there are numbers L_x, L_y, L_z such that a point P with coordinates (x, y, z) is identified with every point $Q = (x + nL_x, y + mL_y, z + pL_z)$, where n, m, p are arbitrary intergers. Each space section ($t =$ const.) is without boundary, but compact (finite and closed) (Fig.1).

The case $K = -1$ is similar. A natural topology is that of Euclidean 3–space for the ($t =$ const.) hypersurfaces. For $K = +1$ only compact topologies are possible.

An important point is that Einstein's theory of gravitation also describes the propagation of radiation. Light propagates along the null–geodesies of spacetimes. We can study this most easily by changing to a new time coordinate η

$$d\eta = dt/R(t) \quad . \tag{1.6}$$

Then

$$ds^2 = R^2(\eta)\{d\eta^2 - d\chi^2 - f^2(\chi)d\Omega^2\} \quad .$$

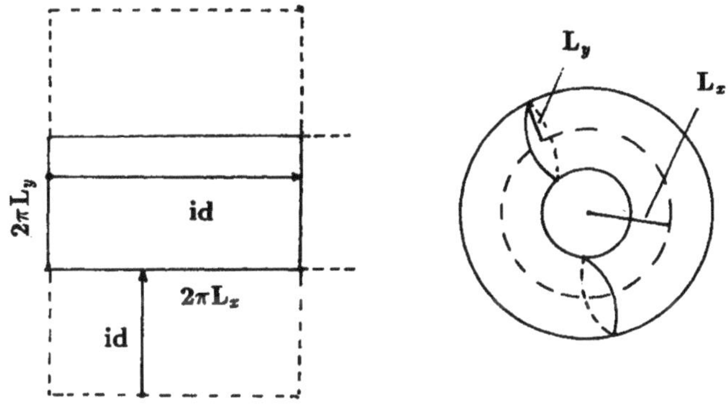

Fig.1 The torus topology arises from a suitable identification of points. Compact space sections can be constructed in this way from an infinite simply connected space-time.

These are called 'conformal' coordinates, $d\Omega^2$ is the 2–sphere line element $d\Omega^2 = d\theta^2 + \sin^2\theta d\varphi^2$. "Radial" propagation of signals (i.e. $\theta = \varphi =$ const.) gives $d\eta = \pm d\chi$ along the light cone; for rays towards an observer at $\chi = 0$ (Fig.2), $\chi = \eta_0 - \eta$, where η_0 denotes the present epoch. Two signals emitted at times η_s and $\eta_s + d\eta_s$ by a source following the mean motion at χ_s will be received at times η_0 and $\eta_0 + d\eta_0$ at $\chi = 0$. Evidently, $d\eta_0 = d\eta_s$; expressed by the cosmic time t this condition reads

$$\frac{dt_0}{R(t_0)} = \frac{dt_s}{R(t_s)} \quad ;$$

since $(dt)^{-1}$ corresponds to a frequency ω, we find that the emitted frequency ω_s differs from the observed one ω_0:

$$\frac{\omega_s}{\omega_0} = \frac{R_0}{R(t_s)} \quad .$$

This is the same as a shift of spectral lines

$$z = \frac{\lambda_0 - \lambda_s}{\lambda_s} = \frac{\omega_s}{\omega_0} - 1 = \frac{R_0}{R(t_s)} - 1 \quad .$$

We know from Slipher's and Hubble's observations that this shift is a redshift

$$1 + z = \frac{R_0}{R(t_s)} > 1 \quad . \tag{1.7}$$

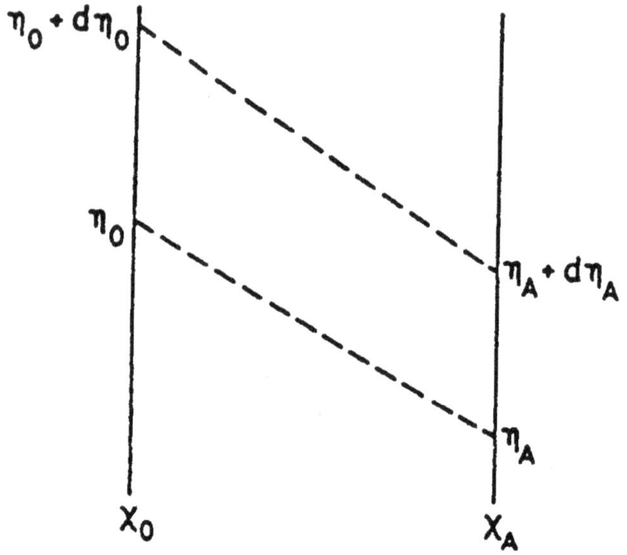

Fig.2 Radial lightrays in the (η, γ) frame from a source at χ_A to an observer at $\chi = 0$.

This redshift of the spectral lines arises from the change in the separation of source and observer, and may therefore be called a Doppler shift. In general space–times the frequency shift cannot be separated into a part due to gravitational effects and a part due to motion. For nearby galaxies we can easily see that (1.7) corresponds to the elementary, non–relativistic Doppler formula: Expanding $R(t) = R_0 + (t - t_0)\dot{R}_0$, we find

$$z = \frac{R_0}{(R_0 + (t - t_0)\dot{R}_0} - 1 \approx (t_0 - t)\frac{\dot{R}_0}{R_0} = d\frac{\dot{R}_0}{R_0} \equiv v_0 \quad .$$

Hence we have

$$v = z = dH_0 \quad , \tag{1.8}$$

where $H_0 \equiv \dot{R}_0/R_0$ is the "Hubble constant".

Equation (1.8) is "Hubble's Law", a linear relation between redshift and distance. First redshift measurements by Slipher and the distance determinations and redshift measurements in the 1920's by Hubble could be fitted to equation (1.8) which means that the system of galaxies is expanding.

There is a universal time coordinate in the FL–models. This cosmic time t corresponds to the proper time of observers following the mean motion V_μ. When we use these models to accommodate cosmological observations, we assume that they approximate in some sense to the real universe. We assume that there exists a mean motion at any time and at any point in the observable universe such that all cosmological properties are isotropic with respect to the reference frame carried along by that mean motion. The presently known measurements do not contradict this hypothesis; the isotropy of the 3K background gives, in fact, strong support to it.

We could now in principle go on and derive relations between the observable quantities of distant galaxies, assuming they populate a universe with an FL–geometry. The comparison between observations and theoretical predictions would then allow us to limit the range of possible FL–models without making use of the Einstein field equations of general relativity. We shall come back to this point in section 5. In view of the uncertainties in presently available data sets, it seems more practical to follow the strategy to obtain dynamical equations from the Theory of General Relativity (GR), and to parametrize the resulting cosmological solutions such that the parameters can be fitted to the observations.

1.1 The Field Equations

Inserting (1.2), (1.3), and (1.4) into the Einstein equations (1.1) leads after some rearranging of terms to 3 equations:

$$\ddot{R} = -\frac{4\pi}{3}(\varrho + 3p)\,GR + \frac{1}{3}\Lambda R \quad ; \tag{1.9}$$

$$\dot{R}^2 = \frac{8\pi}{3}\,G\varrho R^2 + 1/3\Lambda R^2 - K \quad ; \tag{1.10}$$

$$(\varrho R^3)^{\cdot} + p(R^3)^{\cdot} = 0 \quad ; \tag{1.11}$$

$\dot{\varrho}$, $\dot{R}$ stands for $\frac{d\varrho}{dt}$, $\frac{dR}{dt}$.

Whenever $\dot{R} \neq 0$ any two of (1.9) to (1.11) imply the third equation. These are the Friedmann equations[11)18)] and they determine the dynamics of the cosmological model, if an equation of state, e.g. $p = f(t, \varrho)$ or $p = f(\varrho)$ is given. The density ϱ and pressure p are just functions of cosmic time t, because of the assumed homogeneity and isotropy of the metric. Despite the coordinate freedom of GR, a universal time t can be defined by taking the surfaces $\varrho =$ const. as surfaces of constant time t.

At the present epoch $t = t_0$ the pressure is very small: $p/\varrho c^2 < 10^{-4}$. Let us put $p = 0$. Then from eq. (1.11)

$$\varrho R^3 = \text{const.} \quad . \tag{1.12}$$

This expresses the conservation of mass in a comoving volume $\propto R^3$. Define a mass M by $M \equiv \frac{4\pi}{3}\,\varrho R^3$. Then equation (1.10) can be written

$$\dot{R}^2 - \frac{2GM}{R} - \frac{1}{3}\Lambda R^2 = -K \quad . \tag{1.13}$$

This is the Newtonian equation of energy–conservation for a mass point m on a sphere containing the mass M, with a variable radius $R(t)$. The total energy $E \equiv -\frac{Km}{2}$ is the sum of the kinetic and potential energy, the term $\propto \Lambda R^2$ is a kind of oscillator energy (Fig.3).

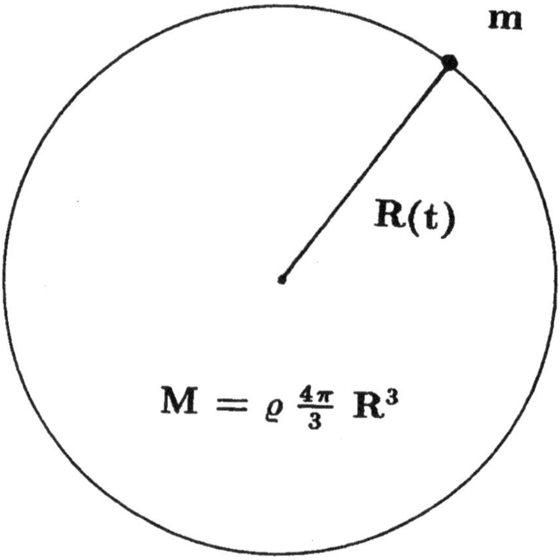

Fig.3 The Newtonian analogue of the Friedmann equation is the energy conservation equation of a mass point m on a sphere of radius $R(t)$ containing the mass M.

Another simple possibility is $p = (1/3)\varrho$, the equation of state for radiation, or for a gas of relativistic particles. Then

$$\varrho R^4 = \text{const.}$$

If ϱ is the energy density of a radiation field $\varrho = aT^4$, the temperature T satisfies $T \cdot R =$ const. With R decreasing towards the big bang, the temperature increases proportionally.

At the present epoch $t = t_0$, and for $p = 0$ eqs. (1.9) and (1.10) can be rewritten in terms of the "Hubble" constant $H_0 \equiv \dot{R}_0/R_0$, and the deceleration parameter $q_0 \equiv -(\ddot{R}\,R/\dot{R}^2)_0$ as

$$\begin{aligned} \frac{1}{3}\Lambda &= \frac{4\pi G}{3}\,\varrho_0 - q_0 H_0^2 \quad ; \\ \frac{K}{R_0^2} &= H_0^2(2q_0 - 1) + \Lambda \quad ; \end{aligned} \tag{1.14}$$

It is common to define a density parameter

$$\Omega_0 \equiv \frac{\varrho_0}{\varrho_c} \equiv \frac{8\pi G}{3H_0^2}\,\varrho_0 \quad . \tag{1.15}$$

This parametrisation brings out the fact that there is essentially only one scale in those models, namely the Hubble constant H_0. All other quantities can be given as dimensionless ratios. $\varrho_c = \frac{3H_0^3}{8\pi G}$ plays the role of a critical density. Eq. (1.14) shows that the sign of K is determined by the term $3\Omega_0 - 2q_0 - 2$. For $\Lambda = 0$ the relation is especially simple: $\Omega_0 = 2q_0$ then, and

$$\begin{aligned} K &= +1 \quad \text{for} \quad \Omega_0 > 1 \quad , \\ K &= 0 \quad \text{for} \quad \Omega_0 = 1 \quad , \\ K &= -1 \quad \text{for} \quad \Omega_0 < 1 \quad . \end{aligned} \tag{1.16}$$

Since the present mean density is, in principle, a measurable quantity, it is possible to determine whether our universe is a hyperbolic or spherical FL–model. It is perhaps interesting to note that a sphere has c/H_0 as its Schwarzschildradius: $M = \frac{4\pi}{3}\,\varrho_c\, r_s^3$, and $r_s = \frac{2GM}{c^2}$; hence $r_s = \frac{8\pi G}{3c^2}\, r_s^3\, \frac{3H_0^2}{8\pi G}$, or $r_s = c/H_0$.

Since the present mean density is close to ϱ_c, the observable universe is close to its Schwarzschild radius, i.e. gravity is really dominating at the present epoch.

If we set $\lambda_0 = \frac{\Lambda}{3H_0^2}$, we can rewrite eq. (1.14) as

$$\begin{aligned} \lambda_0 &= \Omega_0/2 - q_0 \quad ; \\ \frac{K}{R_0^2 H_0^2} &= \frac{1}{2}\,(3\Omega_0 - 2q_0 - 2) \quad ; \text{ or} \\ \frac{K}{R_0^2 H_0^2} &= (\Omega_0 + \lambda_0 - 1) \quad . \end{aligned} \tag{1.17}$$

We can, of course, define these quantities for all cosmic epochs. Then we describe the evolution of the universe by functions of time

$$H(t) = \frac{\dot{R}}{R} \quad ;$$
$$q(t) = -\frac{\ddot{R}R}{\dot{R}^2} \quad ;$$
$$\Omega(t) = \frac{8\pi G}{3H^2}\,\varrho \quad ;$$
$$\lambda(t) = \frac{\Lambda}{3H^2} \quad .$$

The Friedmann equations then take the form

$$q = \Omega/2 - \lambda \quad ;$$

and

$$\frac{K}{R^2} = H^2(\Omega + \lambda - 1) \quad .$$

Inserting eq. (1.17) and the relation $\varrho = \varrho_0(R/R_0)^{-3}$ — which holds for $p = 0$ — into eq. (1.10), we find

$$\left(\frac{\dot{R}}{R_0}\right)^2 = H_0^2\left\{1 - \Omega_0 - \lambda_0 + \Omega_0 R_0/R + \lambda_0\,\frac{R^2}{R_0^2}\right\} \quad .$$

Substituting $x \equiv R/R_0$

$$t - t_M = H_0^{-1}\int_{x_M}^{x}\left(1 - \Omega_0 - \lambda_0 + \frac{\Omega_0}{y} + \lambda_0 y^2\right)^{-1/2} dy \quad , \tag{1.18}$$

where t_M is a time in the past for which the approximation $p = 0$ still holds.

Since $t_M \ll t_0$, and $R_M \ll R_0$, we may write for the present epoch t_0

$$t_0\,H_0 = \int_0^1 (1 - \Omega_0 - \lambda_0 + \Omega_0/y + \lambda_0 y^2)^{-1/2} dy \quad . \tag{1.19}$$

For the simplest model, the "Einstein–de Sitter" model we have $\lambda_0 = 0$, and $\Omega_0 = 1$, then

$$t_0 H_0 = 2/3 \quad . \tag{1.20}$$

The time t_0 of the present epoch, i.e. the "age" of the universe is just $(2/3)H_0^{-1}$, 2/3 of the Hubble time. We see that for $\lambda_0 = 0$ the maximal value is reached by letting $\Omega_0 \to 0$. Then $H_0 t_0 = 1$.

Recent measurements of H_0 and t_0 (see below, and lectures by J. Mould, D. Schramm) indicate that $H_0 t_0$ must be greater than one. Looking hard at eq. (1.19), we can imagine that this would require a cosmological constant $\lambda_0 > 0$. In fact, by adjusting λ_0 we can reach any value for $H_0 t_0$ (see below).

1.2 The Singularity

Without solving the Friedmann equations it is seen from eq. (1.9) that the present expansion $\dot{R}_0 > 0$, together with the requirement $\varrho + 3p - \frac{\Lambda}{4\pi G} \geq 0$, i.e. $\ddot{R} \leq 0$, leads to a concave graph of $R(t)$ versus t. If the condition $\varrho + 3p - \frac{\Lambda}{4\pi G} \geq 0$ holds for all times, then $\ddot{R} \leq 0$ always, and $R(t)$ must necessarily be zero at some finite time in the past (Fig.4). Moreover, the time since then is bounded by H_0^{-1}, the "Hubble time" ($t_0 \leq H_0^{-1}$).

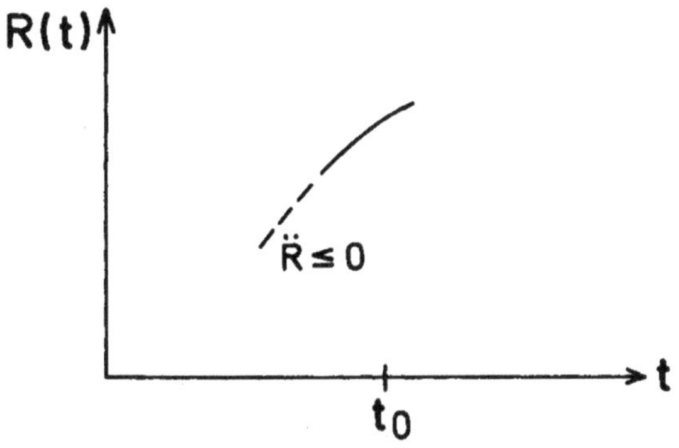

Fig.4 The qualitative graph of R vs. t, if $\ddot{R} \leq 0$ holds.

Let us call this instant $t = 0$. Thus $R(0) = 0$, and our present epoch t_0 is the cosmic time elapsed since the initial singularity. This "point", $t = 0$, is a true singularity of the spacetime: the energy density ϱ becomes infinite, when R goes to zero. One can construct an invariant from the components of the spacetime curvature tensor (even if $K = 0$) which will also diverge as $R \to 0$[14].

The "point" $R = 0$ does not belong to the spacetime, and it is not clear at present how close to $R = 0$ we should accept the validity of the classical theory of GR. Most physicists would agree that a classical description of the universe is reasonable up to temperatures and densities close to the limits of terrestrial laboratory experiments But it is a remarkable feature of some of the FL–cosmologies that they have such an initial singularity. It means that the universe we know came into being some finite time ago.

The condition $\varrho+3p-\frac{\Lambda}{4\pi G}\geq 0$ is sufficient, but not necessary, for the existence of a singularity. Observations do not unambiguously give $\varrho+3p-\frac{\Lambda}{4\pi G}\geq 0$ now (compare the discussion in section 2), but if this inequality was satisfied at some past epoch, this implies a singularity, provided $\varrho+3p$ was never smaller before that epoch.

The inequality $\varrho>0$ is certainly satisfied by the kinds of matter known to us. The inequality $\varrho+3p>0$ is certainly satisfied up to nuclear–matter density. Beyond that we must rely on theory and speculation. It could happen, e.g. that the effective potential energy $V(\varphi)$ of a scalar field was dominant in the very early universe. Such a "vacuum energy" can lead to an equation of state $\varrho=-p$, i.e. $\varrho+3p<0$. Then the conditions sufficient for a singularity are violated, and the singularity may be avoided altogether. $V(\varphi)$ will act like a cosmological "constant" which must be removed (transformed into heat) at some stage, because it would violate present astronomical limits (see section 2). We shall see below that there are FL–models without a singularity, if a special value of the cosmological constant Λ is chosen. But then $\ddot{R}$ is always positive, and it becomes virtually impossible to explain the background radiation or the cosmic abundance of helium.

The notion that the occurrence of the singularity might depend on the exact spherical symmetry and spatial homogeneity of the models is unfounded. Several interesting theorems have been proved[14)15)], which give a range of general conditions sufficient for the occurence of a singularity. It is clear that this does not depend on any symmetry of a spacetime. The inescapable conclusion of the existence of an initial singularity is that the description of the universe by a FL–model leads to the assumption of an early epoch of high density, when all the structures we see now — the stars and galaxies, etc. — did not yet exist.

What was there before the initial singularity? This question cannot be answered within the context of classical GR, because the concepts of this theory cease to be applicable at the singularity.

1.3 Exact Solutions

In this section I want to write down solutions for $R(t)$ in terms of the measurable quantities q_0, H_0, Ω_0. First of all consider the case $\Lambda=0$; then $\Omega_0=2q_0$, and we use q_0 to characterise the solutions. In general R and t will be given in terms of a parameter η. Only the zero–pressure approximation will be treated. This covers the largest part of the cosmic evolution. The early phase with $p = (1/3)\varrho$ has a slightly different behaviour, and this calculation is left as an exercise.

The integration of the Friedmann–equations proceeds in the following steps:

Take the case $p = 0$, $\Lambda = 0$.

The Friedmann equation

$$\dot{R}^2 + K = \frac{8\pi G}{3}\,\varrho\, R^2$$

can be rewritten in terms of a new time variable η with $d\eta = dt/R$ as

$$\begin{aligned} \left(\frac{\dot{R}}{R}\right)^2 &= \frac{(dR/d\eta)^2}{R^2} = \frac{8\pi G}{3}\,\varrho_0 \left(\frac{R_0}{R}\right)^3 - \frac{K}{R^2} \quad , \quad \text{or} \\ R^{-1/2}\,\frac{dR}{d\eta} &\equiv 2\,\frac{d}{d\eta}\,R^{1/2} = \left(\frac{8\pi G}{3}\,\varrho_0\, R_0^3 - KR\right)^{1/2} \quad . \end{aligned} \tag{1.21}$$

Integration of this relation leads to

$$\begin{aligned} \frac{1}{2}\,\eta &= \int_0^{\sqrt{R}} dR^{1/2}\left(\frac{8\pi G}{3}\,\varrho_0 R_0^3 - KR\right)^{-1/2} \\ &= \begin{cases} \sin^{-1}\left\{\frac{R^{1/2}}{(\frac{8\pi G}{3}\varrho_0 R_0^3)^{1/2}}\right\} & \text{for } K = +1 \\ R^{1/2}\left(\frac{8\pi G}{3}\,\varrho_0 R_0^3\right)^{-1/2} & \text{for } K = 0 \\ \sinh^{-1}\left\{R^{1/2}\left(\frac{8\pi G}{3}\varrho_0 R_0^3)\right)^{-1/2}\right\} & \text{for } K = -1 \quad . \end{cases} \end{aligned} \tag{1.22}$$

Since

$$\begin{aligned} \frac{8\pi G}{3}\,\varrho_0 R_0^3 &= \frac{2q_0}{H_0|2q_0 - 1|^{3/2}} \quad (K = \pm 1) \\ \frac{8\pi G}{3}\,\varrho_0 R_0^3 &= H_0^2 R_0^3 \quad (K = 0) \end{aligned}$$

we can replace this term in eq. (1.22), and invert the formulae

$$R = \begin{cases} q_0 H_0^{-1}(2q_0 - 1)^{-3/2}(1 - \cos\eta) & , \quad K = +1 \quad ; \\ \frac{1}{4}\,H_0^2 R_0^3 \eta^2 & , \quad K = 0 \quad ; \\ q_0 H_0^{-1}(1 - 2q_0)^{-3/2}(\cosh\eta - 1) & , \quad K = -1 \quad . \end{cases} \tag{1.23a}$$

Integrating $dt = R\, d\eta$ we obtain

$$t = \begin{cases} q_0 H_0^{-1}(2q_0 - 1)^{-3/2}(\eta - \sin\eta) & , \quad K = +1 \quad ; \\ \frac{1}{12}\,H_0^2 R_0^3 \eta^3 & , \quad K = 0 \quad ; \\ q_0 H_0^{-1}(1 - 2q_0)^{-3/2}(\sinh\eta - \eta) & , \quad K = -1 \quad . \end{cases} \tag{1.23b}$$

For $K = 0$ we have the especially simple relation $R(t)/R_0 = (\frac{3H_0t}{2})^{2/3}$.

Finally we have:

For the $\underline{K = +1 \text{ spherical model}}$

$$\begin{aligned} \frac{R}{R_0} &= q_0(2q_0 - 1)^{-1}(1 - \cos\eta) \\ H_0 t &= q_0(2q_0 - 1)^{-3/2}(\eta - \sin\eta) \quad ; \end{aligned} \tag{1.24}$$

where $0 \leq \eta \leq 2\pi$.

This describes an increase of $R(t)$ until a maximal expansion stage is reached, and from then on $R(t)$ decreases, the universe contracts. Since we measure expansion now, we are still in the ascending branch of the $R(t)$ graph in Fig.5.

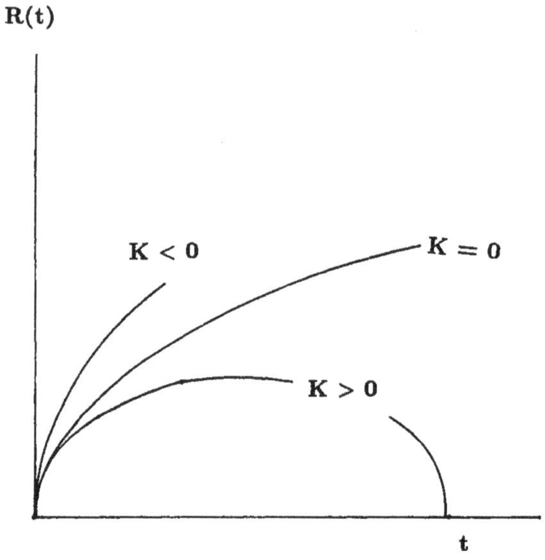

Fig.5 The graphs of R vs. t according to the Friedmann equations.

For the $\underline{K = -1 \text{ hyperbolic model}}$

we just have to replace $\cos\eta$ by $\cosh\eta$, $\sin\eta$ by $\sinh\eta$ and to change signs appropriately:

$$\begin{aligned} \frac{R}{R_0} &= q_0(1 - 2q_0)^{-1}(\cosh\eta - 1) \quad ; \\ H_0 t &= q_0(1 - 2q_0)^{-3/2}(\sinh\eta - \eta) \quad ; \\ & 0 < \eta < \infty \quad . \end{aligned} \tag{1.25}$$

This describes the ever increasing expansion factor $R(t)$ in Fig.5. The Einstein-de Sitter model $K = 0$, the flat model is the simplest. Here $2q_0 = 1$, and

$$\frac{R}{R_0} = \left(\frac{3}{2}H_0\right)^{2/3} t^{2/3} \quad . \tag{1.26}$$

The simple relation $t_0 H_0 = 2/3$ holds in this model, as we have seen already.

The Hubble parameter, i.e. the time–dependent quantity $\frac{\dot{R}}{R} = \frac{2}{3t}$ becomes infinite as $t \to 0$. This holds for the $K = -1, +1$ models as well. The universe starts with an infinite rate of expansion.

$\Lambda \neq 0$ solutions

In general the solutions with a nonzero cosmological constant can only be obtained by numerical integration. A few general features can be pointed out in a qualitative discussion.

It is convenient to introduce dimensionless quantities by setting

$$x \equiv R/R_0\ ; \quad \tau = t\,H_0 \quad .$$

Then the Friedmann–equation reads

$$\dot{x}^2 = \Omega_0/x + \lambda_0\, x^2 - \Omega_0 - \lambda_0 + 1 \quad . \tag{1.27}$$

We can discuss this equation like the one–dimensional Newtonian motion in a potential

$$V(x) = -\Omega_0/x - \lambda_0 x^2$$

with energy $1 - \Omega_0 - \lambda_0$. For $\lambda_0 < 0$ the potential $V(x)$ is monotonic, and all possible solutions evolve from an initial singularity into a final singularity for all values of $K(K/R_0^2H_0^2 = \Omega_0 + \lambda_0 - 1)$.

For $\lambda_0 > 0$ $V(x)$ has a maximum at $x_m = \left(\frac{\Omega_0}{2\lambda_0}\right)^{1/3}$. $V(x) < 0$ for all values of x, i.e. the $K = 0, -1$ solutions correspond to expansion from a singularity.

For $K = +1$ there is a critical value λ_c such that $\frac{dx}{d\tau} = 0$ at x_m.

λ_c can be found as the solution of the equation

$$(\lambda_c + \Omega_0 - 1)^3 = \frac{27}{4}\Omega_0^2\lambda_c \quad .$$

For $\lambda_0 < \lambda_c$ and $K = +1$ we still find expansion forever, starting from an initial singularity. When $\lambda_0 = \lambda_c$ and $K = +1$ there is the possibility of a static solution,

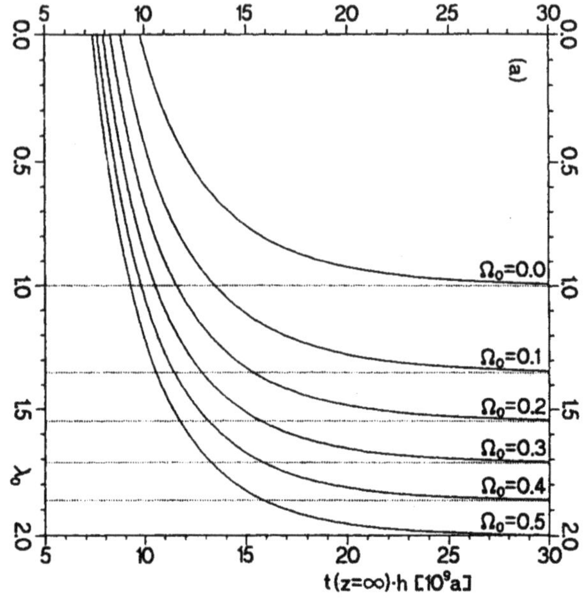

Fig.6 The normalized age of the universe (in units of) $(h^{-1}10^9 a)$ as it depends on λ_0 for different values of Ω_0 (taken from [10]).

the "Einstein Universe" with a constant radius $x_E = \left(\frac{\Omega_0}{2\lambda_c}\right)^{1/3}$. In Figs.6 and 7 we can see, how this radius defines a limiting case in the parameter space of these $\lambda_0 \neq 0$ models. A small change in λ_0 leads away from the static universe to an expanding solution. There is a solution starting from $x = 0$ and "creeping towards x_E, while another type of solution starts expanding from the finite radius x_E (the "Eddington–Lemaître–Universe"). If λ_0 is only slightly smaller than λ_c, an intermediate case between the Einstein and Eddington–Lemaître models arises: the cosmological model starts from $x = 0$, and then stays for a long time (depending on how close λ_0 is to λ_c) near x_E, with $\dot{x} \approx 0$, before it expands as in the E.–L. case (Fig.7). In Fig.6 the dependence of t_0 on λ_0 is shown for different values of Ω_0, while in Fig.7 $x(t)$ is displayed for various values of x_0, and for $\Omega_0 = 0.2$. It is clear from Figs.6 and 7 that we can reach any desired value for the product $H_0 t_0$, by appropriately choosing λ_0 close to λ_c. Since measurements of H_0 and t_0 may force us to introduce a cosmological constant to solve the "age problem" we should consider λ_0 as another fundamental cosmological quantity to be determined by observations.

For $\lambda_0 > \lambda_c$ there is the possibility of a model contracting from infinity to a minimal radius, and expanding again. This model, as well as the Eddington–Lemaître model, avoids the initial singularity. These models can be excluded by existing observational results.

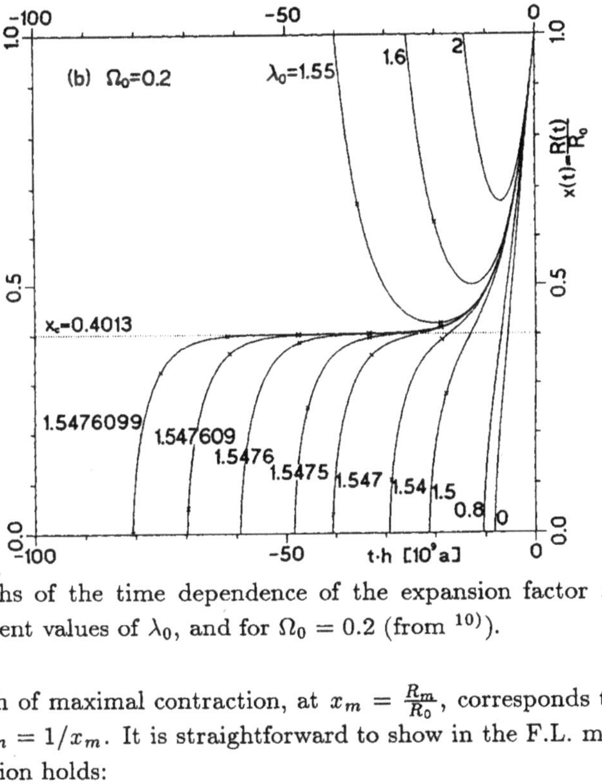

Fig.7 Graphs of the time dependence of the expansion factor $x(t)$ for different values of λ_0, and for $\Omega_0 = 0.2$ (from [10]).

The epoch of maximal contraction, at $x_m = \frac{R_m}{R_0}$, corresponds to a maximal redshift $1 + z_m = 1/x_m$. It is straightforward to show in the F.L. model that the following relation holds:

$$\Omega_0 = \frac{2x_m^3}{(1 - 3x_m^2 + 2x_m^3)} \quad . \tag{1.28}$$

Since quasars with redshifts up to almost 5 have been observed, we have $x_m < 0.2$, and therefore $\Omega_0 < 0.02$ in contradiction to observations.

For the oscillating model one can show[3] that the density parameter Ω_0 must satisfy

$$\Omega_0 \leq \frac{2}{z_m^2(z_m + 3)} \quad . \tag{1.29}$$

Since there are quasars with $z \approx 5$, we find $\Omega_0 \leq 0.01$ — again a contradiction to the measurements.

2. Fundamental Cosmological Quantities

2.1 The Cosmic Microwave Background (CMB) and the Spectrum of Primeval Density Perturbations

2.1.1 The CMB Spectrum

The really solid and convincing evidence to most (with a few tenacious exceptions) for a homogeneous and isotropic cosmological model comes from the measurements of a radiation field in the microwave range with a Planckian spectrum. The satellite COBE (COsmic Background Explorer) launched by NASA on Nov. 18, 1989 has measured this radiation field which was originally discovered in 1964 by Arno Penzias and Robert Wilson. The history of this discovery is told in Steven Weinberg's book "The First Three Minutes"[41]. In Fig.8 the spectrum measured by COBE is shown[24][25]. In fact Fig.8 shows the first such measurement, where the satellite pointed towards the North Galactic Pole. The observing time for this spectrum was 9 minutes. The error boxes lie very nicely on the solid line representing a Planck spectrum with a temperature of 2.73 K. The maximum of the intensity is at a wavelength of 2mm. Microwaves of this wavelength are absorbed by the Earth's atmosphere, and therefore the maximum and turnover of the spectrum can only be found in satellite experiments. The COBE measurements as well as previous groundbased or balloon based experiments have all established the high degree of isotropy, the black–body spectrum, and the absence of polarization. The correspondence is as close as one could hope to the simple hypothesis that the radiation is the relict of a homogeneous and structureless primeval explosion. Immediately after the discovery — Penzias and Wilson observed actually at a wavelength of $\lambda = 7.35$ cm — the observations have been interpreted as a cosmic background signal indicating the existence of a hot, dense early state of the universe. Matter and radiation were in thermal equilibrium in that epoch.

We must keep an open eye for other explanations of the microwave background which — like Hoyle's iron needle picture[1] — favour a more local origin. To date, however, no serious challenge to the cosmic interpretation as the relic radiation from the hot early phase of a FL model exists. The evidence has even become stronger in recent years, since the COBE team has analysed the data of their one year of spectral measurements. Many spectra in many different regions of the sky have been obtained, and the black–body shape of the CMB has been established in the most perfect way[25]. The temperature has been found to be

$$T = 2.726 \pm 0.005\ K \quad . \tag{2.1}$$

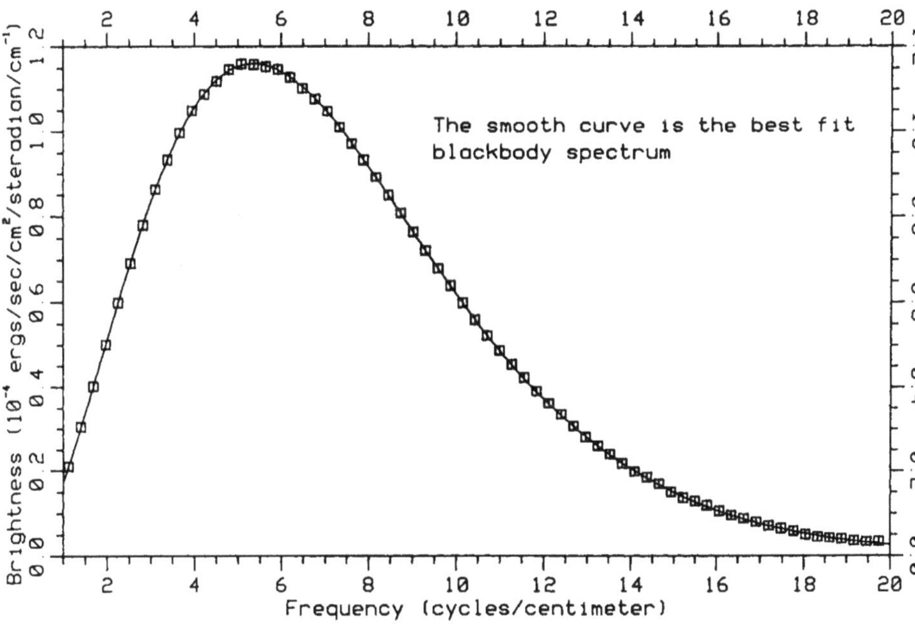

Fig.8 The spectrum of the microwave background radiation. The boxes are error bars of the measurements. The theoretical graph of a thermal Planck-spectrum is also shown. The best fit temperature is 2.726 ± 0.005K.

The black–body shape of the spectrum is apparently present with such high precision that limits can be set on other cosmic parameters. A fit to the Sunyaev–Zel'dovich form of the Comptonization effect[37] by a hot electron gas limits the parameter

$$y = \frac{\sigma_T k_B}{m_e c^2} \int n_e T_e \, dV \tag{2.2}$$

of the intergalactic medium (σ_T: Thomson scattering cross section; n_e, T_e, m_e: electron density, temperature, mass, resp.). The 3σ–limit is

$$|y| < 2.5 \times 10^{-5} \quad . \tag{2.3}$$

This rather strong limit can be used to set an upper limit on the fractional energy input $\frac{\Delta U}{U}$ to the CMB between $z = 10^5$ and the present epoch.

$$\frac{\Delta U}{U} < 10^{-4} \quad . \tag{2.4}$$

This bound is good enough to exclude or reduce to insignificance several of the less standard ideas in cosmology. To mention just a few examples:

1. The model of a smooth, hot intergalactic medium suggested to explain the X–ray background can be ruled out by this limit[43][44], since the intensity of the predicted X–ray background scales as y^2. Even for a very late heating time, the contribution of the hot, intergalactic medium would be limited to less than 1% of the observed value.

2. A model for galaxy formation by repeated supernova explosions can be ruled out.

3. The model advocated by Sir Fred Hoyle[1], in which an appropriate distribution of needle–like iron dust–grains produces the CMB by absorption and reemission of starlight will have difficulties too. First of all the absorption and emission processes must model a black–body spectrum almost perfectly in the range of 1/2mm to 1cm wavelength. Secondly there are distant objects (e.g. a galaxy with redshift $z = 2.2$) observed in the mm wavelength. This further limits the opacity that the iron–grains are permitted to have. There has not been a quantitative calculation which demonstrates that the iron–needle model can survive these restrictions.

Another interesting result[35] was obtained by the observation of radiation from CN molecules in a distant ($z = 1.78$) galaxy. The observations allow one to deduce that the higher levels, the excited states of the CN molecules are populated with an abundance which corresponds to a radiation field of a temperature of $(7.6 \pm 0.8)K$. This fits in very nicely with the standard cosmological interpretation of the CMB which has at $z = 1.78$ a temperature of $T = (1 + z)T_{\gamma 0} = 7.58K$.

One important fact about a Planck radiation field in a homogeneous FL universe is that its spectrum keeps the Planck-shape during the expansion. $T_\gamma R =$ const., but also $\nu \cdot R =$ const. for the frequency. Therefore the Planck–spectrum

$$\varrho_\gamma d\nu = 8\pi h \left(\frac{\nu}{c}\right)^3 d\nu \left[\exp \frac{h\nu}{k_B T} - 1\right]^{-1} \tag{2.5}$$

keeps its shape although the energy density $\varrho_\gamma = aT_\gamma^4$ decreases prop. to R^{-4} during the expansion ($a \equiv \frac{\pi^2}{15} \frac{k_B^4}{\hbar^3 c^3}$ is the Stefan–Boltzmann constant).

The number density of photons is

$$n_\gamma = \frac{1}{\pi^2} \frac{(k_B T_\gamma)^3}{(2\pi \hbar c)^3} 2\zeta(3) \quad , \tag{2.6}$$

where ζ is Riemann's ζ function. Thus

$$n_\gamma = \frac{30\zeta(3)}{\pi^4} \frac{a}{k_B} T_\gamma^3 \approx 0.37 \frac{a}{k_B} T_\gamma^3 \quad ; \tag{2.7}$$

The temperature of 2.7 K gives the number of photons now

$$n_{\gamma 0} \simeq 400/\mathrm{cm}^3 \quad . \tag{2.8}$$

The energy–density of the radiation is

$$\begin{aligned} \varrho_{\gamma 0} &= 0.25 eV/\mathrm{cm}^3 \quad , \quad \text{or} \\ \varrho_{\gamma 0}/c^2 &= 3 \times 10^{-34} \mathrm{g\ cm}^{-3} \quad . \end{aligned} \tag{2.9}$$

This is small compared to present estimates of the matter density (see section 2.3). If we take current estimates of the baryon–number density, we can give a number for the ratio of photons to baryons

$$\frac{n_{\gamma 0}}{n_{B0}} = 10^{9\pm 1} \quad . \tag{2.10}$$

The ± 1 in the exponent reflects the large uncertainty in the density determination. Except for epochs when photons were produced by heating processes, this ratio is a constant in FL cosmologies

$$\frac{n_\gamma}{n_B} = \frac{n_{\gamma 0}}{n_{B0}}$$

always, and thus is an initial condition for our universe.

We can replace this ratio by the entropy per baryon, since practically all the entropy is in the photon field now. The entropy density

$$\begin{aligned} s/k_B &= \frac{4}{3} a T_\gamma^3 \quad , \quad \text{and therefore} \\ s &= n_\gamma / 0.37 = 2.7 n_\gamma \\ \frac{s}{n_B} &= 10^{10} \quad \text{approximately} \quad . \end{aligned} \tag{2.11}$$

This large ratio, or equivalently the small number $\frac{n_B}{s} = 10^{-10}$ is a target of particle physicists, who want to derive this small number from physical processes starting with a symmetric initial state, in which $n_B = 0$.

Such a cosmic background radiation has important consequences within the cosmological model: since T_γ varies proportional to $R^{-1} \propto (1+z)$, the temperature in past epochs was much higher than now, and at redshifts around $z_R \simeq 1500$ it was high enough to ionize hydrogen ($T_\gamma \approx 4500$ K). Before that epoch of "recombination" as it is called (not very fittingly, because it was then that hydrogen atoms combined for the first time), there were many free electrons around, and the universe was essentially opaque. At earlier epochs radiation and matter were in thermal equilibrium, and they decoupled at the recombination time t_R. Since at t_R (or equivalently z_R) the mean free path of the photons increased drastically, the photons we observe come from a surface near $z_R \simeq 1500$ where they were last scattered. A thermal history of the universe in which a reionization at redshifts between 10 and 100 occurred, or an ionized intergalactic medium, would change this, and place the last scattering surface at a much smaller redshift.

Another interesting aspect is the different behaviour of matter and radiation with redshift: the energy–density of matter

$$\varrho_m = \varrho_{m0}(1+z)^3 \quad , \tag{2.12}$$

and of radiation

$$\varrho_\gamma = \varrho_{\gamma 0}(1+z)^4 \quad , \tag{2.13}$$

(since $\varrho_m\, R^3 =$ const., $\varrho_\gamma\, R^4 =$ const.).

We have that the ratio

$$\varrho_\gamma / \varrho_m = (1+z)\, \varrho_{\gamma 0} / \varrho_{m0}$$

changes with redshift. Matter dominates now, but at some time in the past, radiation was the dominating energy–density. Before a redshift z_{eq}, between 10^3 and 10^4, (the large uncertainty is due to the uncertainty in ϱ_{m0}) the universe was dominated by hot radiation. This radiation field eventually was hotter than the photospheric temperature of stars, and therefore structures known to us such as planets, stars, and galaxies would be dissolved, and disappear in a relatively smooth background of radiation, electrons, neutrinos, nucleons, and light nuclei. There must have been a plasma phase at early epochs in the cosmic evolution.

2.1.2 Anisotropies

The CMB is very isotropic, but the signal of a dipole–like anisotropy has been found already in early observations at 33 MHz using a U2–spy–plane[33], and has been confirmed by the COBE satellite. This dipole pattern is interpreted as the Doppler shift due to the peculiar motion of the sun, or equivalently the Local Group of Galaxies with respect to the rest frame of the background radiation. The CMB is viewed as an almost perfectly isotropic medium defining a preferred frame of reference. The expected dipole anisotropy due to the motion of the Local Group of galaxies has a $\cos\theta$ behaviour:

$$T(\theta) = T_0 \left(1 - \frac{v^2}{c^2}\right)^{1/2} \left(1 - \frac{v}{c}\cos\theta\right)^{-1} \simeq T_0 \left(1 + \frac{v}{c}\cos\theta\right) \quad . \tag{2.14}$$

The amplitude is (3.3 ± 0.3)mK in the direction $(\alpha, \delta) = (11.2^h \pm 0.2^h,\ -7^\circ \pm 2^\circ)$[43].

The velocity has a value of $v_{LG} = 627 \pm 22$km s^{-1}. The large magnitude and the direction of the velocity make it clear that the LG peculiar motion cannot be explained by the attraction by the Virgo cluster. The Virgo infall velocities are smaller — only ~ 200km s^{-1}— and in a different direction. Thus there must be a bigger global velocity field which explains the CMB dipole.

The search for smaller anisotropies produced by matter fluctuations is not easy. But in 1992 the COBE team announced the discovery of small fluctuations in the CMB, with $\Delta T = (30 \pm 5)\mu$K[34]. Theorists heaved a sigh of relief then, because for 28 years the hunt for the fluctuations had been in vain. But such fluctuations had been expected on the basis of the models for the formation of the large–scale structures traced out in the sky by the luminous galaxies. Such large–scale patterns (see the lectures by A. Szalay) are supposed to evolve during the expansion of the universe from initially small amplitude perturbations. Their density contrast grows due to the action of gravity, until they reach an overdensity which causes them to collapse by gravitational instability.

Since the discovery of the CMB it had been realized that these seed fluctuations should be visible on the last scattering surface of the CMB as small temperature anisotropies. The COBE result proves that this concept has a good chance of being right.

The analysis of the COBE sky–maps (Fig.9) of the first and second year showed good consistency, and provided statistically significant evidence for a rms–

sky variation on 10° scales of

$$\frac{\Delta T}{T} = 1.1 \times 10^{-5} \quad . \tag{2.15}$$

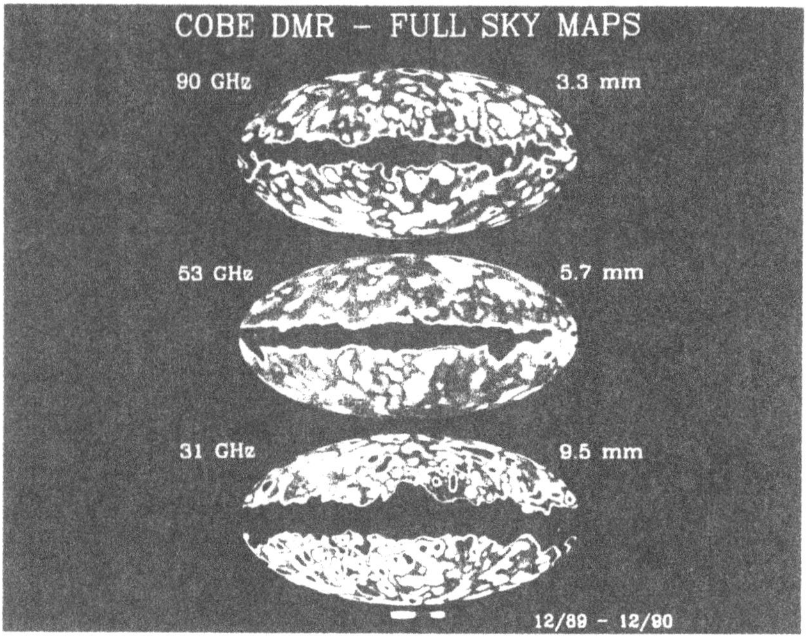

Fig.9 COBE all-sky maps of CMB temperature fluctuations at different frequencies. The maps are still noise dominated, but 4038 pixels (after a $|b| \leq 20^\circ$ cut, to remove most of the emission from the galaxy) of ΔT values provide a good basis for the statistical analysis.

A quadrupole component in the data is more difficult to estimate, and it seems to vary quite strongly between the first and second year. The quadrupole normalisation seems to depend strongly on the method of analysis of the temperature variations. The COBE instruments provide sky maps of $2.5^\circ \times 2.5^\circ$ pixels, i.e. 6144 pixels on the celestial sphere. The pixel size is determined from a smoothing of the measurement points with the 7° beam width of the antenna.

It is necessary to remove the galactic contribution, and this is done in a first step by cutting out an area of $|b| \leq 20^\circ$ in the sky. The remaining 4038 pixels are then analysed statistically. A large number of papers have been discussing this problem. I want to follow here the especially lucid representation of Tegmark and Bunn[38]. The analysis methods involve basic linear algebra, and things become transparent, if the various quantitites are written as vectors and matrices.

The CMB sky map is defined as the N–dimensional vector $\mathbf{x}$, defined as

$$x_i = \frac{\Delta T}{T}(\hat{n}_i) \quad , \tag{2.16}$$

where $\hat{n}_i$ is the unit vector in the direction of COBE pixel i. $N = 6144$ for an all–sky map, and $N = 4038$ after a 20° Galactic cut. TB write $\mathbf{x}$ as a sum of three terms

$$\mathbf{x} = Y\mathbf{a} + \boldsymbol{\varepsilon} + Z\mathbf{b} \quad , \tag{2.17}$$

which give the cosmological (Y) contribution, instrumental noise ($\boldsymbol{\varepsilon}$) and unknown multipole contributions (Z).

The spherical harmonics in matrix form are denoted by $Y_{i\lambda} \equiv Y_{\ell m}(\hat{n}_i)$, where TB combine ℓ and m into the single index $\lambda \equiv \ell^2 + \ell + m + 1 = 1, 2, 3 \ldots$ Assuming that the temperature differences are Gaussian distributed, $\mathbf{a}$ is a Gaussian random vector with zero mean, and with the diagonal covariance matrix

$$\langle a_\lambda a_{\lambda'} \rangle = \delta_{\lambda\lambda'} C_e \quad . \tag{2.18}$$

The angular power spectrum C_e can be related to the density fluctuations in linear approximation in some specific cosmological model. The addition theorem for spherical harmonics leads to the covariance matrix of the cosmological term

$$\langle Y\mathbf{a}(Y\mathbf{a})^T \rangle_{ij} = c(\hat{n}_i \cdot \hat{n}_j) \quad , \tag{2.19}$$

where the angular correlation function $c(\hat{n}_i \cdot \hat{n}_j) \equiv c(\cos\theta)$ is defined as

$$c(\cos\theta) = \frac{1}{4\pi} \sum_{\ell=0}^{\infty} (2\ell+1) C_e \, P_\ell(\cos\theta) \quad , \tag{2.20}$$

P_ℓ denoting the ℓth Legendre polynomial.

The N–dimensional noise vector $\boldsymbol{\varepsilon}$ is assumed to be Gaussian with $\langle \varepsilon \rangle = 0$. Its covariance matrix can be written as

$$\langle \hat{n}_i \hat{n}_j \rangle = \sigma_i \sigma_j C_n(\hat{n}_i \ldots \hat{n}_j) \quad , \tag{2.21}$$

where δ_i is the rms noise of pixel i. $C_n(\cos\theta)$ has been determined by Lineweaver *et al.* [21]. Since there are two instruments A and B aboard the satellite, the combined signal $\frac{A+B}{2}$ gives a noise "sky–map".

The third term in eq. (2.17) contains the unknown influence of the first ℓ_0 multipoles. Since the CMB temperature and the peculiar velocity of the Galaxy (the dipole) are not known a priori to the accuracies of $\Delta T \sim 30\mu K$, one must set $\ell_0 \geq 1$.

Systematic errors in the quadrupole may be suspected, and one might cautiously set $\ell_0 = 2$. The matrix Z consists of the first $(\ell_0 + 1)^2$ columns of Y, and $\mathbf{b}$ is a constant $(\ell_0 + 1)^2$–dimensional vector, whose value is a priori unknown. Since the noise and the cosmic signal are uncorrelated, TB[38)] give the pixel covariance matrix $M \equiv \langle \mathbf{x}\underline{x}^T \rangle - \langle \mathbf{x} \rangle \langle \underline{x}^T \rangle$ as

$$M_{ij} = c(\hat{n}_i \hat{n}_j) + \sigma_i \sigma_j C_n(\hat{n}_i \hat{n}_j) \quad . \tag{2.22}$$

The standard way to eliminate the influence of the $Z\mathbf{b}$ term is to "remove" the monopole, dipole, and perhaps the quadrupole from the data $\mathbf{x}$ before the analysis is carried out. This is in fact also a linear operation: define the matrix $\tilde{Z}$ as a matrix whose columns are orthonormal and span the same space as the columns of Z. This can be achieved either by an orthogonalization technique, or by defining

$$\tilde{Z} \equiv Z(Z^T Z)^{-1/2} \quad .$$

The $N \times N$ matrix $\tilde{Z}Z^T$ of rank $(\ell_0 + 1)^2$ acts as a projection operator onto the space spanned by the first ℓ_0 multipoles. The "removal" matrix D is given by

$$D = I - \tilde{Z}\tilde{Z}^T \quad .$$

The corrected data are defined as

$$\tilde{\mathbf{x}} \equiv D\mathbf{x} \quad .$$

Now $\langle \tilde{x} \rangle = DZ\mathbf{b} = 0$, so the effect of this term is taken care of. But after this procedure in general the orthogonality of the various multipoles is lost, i.e. $Y^T Y \neq I$. Thus the removal of the monopole and the dipole affects other multipoles, such as the octupole.

Assuming that the cosmic signal $\mathbf{a}$ is Gaussian, so is $\tilde{\mathbf{x}}$. Thus given an observed data set $\tilde{\mathbf{x}}$, the likelihood L is given by

$$-2 \ln L = \ln \det \tilde{M} + \tilde{x}^T \tilde{M}^{-1} \tilde{x} + \text{const.} \tag{2.23}$$

For a spectrum due to Sachs–Wolfe (see below) fluctuations from a power law density distribution $P(k) \propto k^n$, the coefficient C_ℓ is just a function of n, and $\langle Q \rangle$, the quadrupole normalization. So is $L(n, \langle Q \rangle)$, and the numerical calculation of TB is shown in Fig.10. Their results are

$$n = 1.11 \pm 0.29(0.92 \pm 0.32)\,, \quad \text{and} \quad \langle Q \rangle = (20.2 \pm 4.6)\mu K \tag{2.24}$$

including (excluding) the quadrupole term.

TB compare their direct method to other methods[44] which compress the data, and analyse only a few hundred statistical correlations. The agreement for n and $\langle Q \rangle$ is quite good.

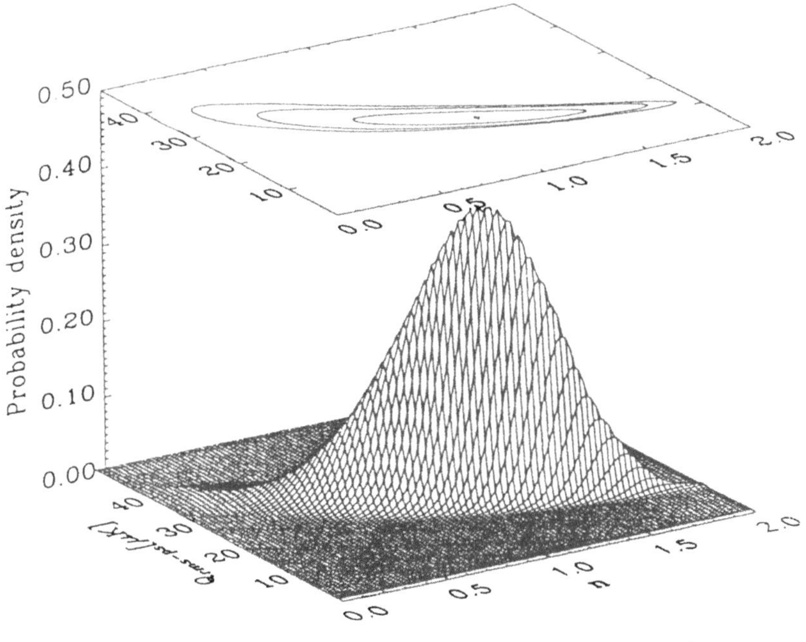

Fig.10 Fits to n and $\langle Q \rangle$ according to [38].

2.1.3 The Initial Power Spectrum

For a statistical description of the cosmic density the Fourier components of the density contrast

$$\delta = \frac{\varrho(\mathbf{x}, t) - \varrho_b(t)}{\varrho_b(t)}$$

are widely used. The basic assumption is that the distribution of mass is periodic, when sufficiently large cubic volumes V_u, fixed in comoving coordinates, are considered. The power spectrum of density fluctuations is defined as

$$P(k) \equiv \langle |\delta_k|^2 \rangle \quad , \tag{2.25}$$

where δ_k is the Fourier component of the density contrast with wave number $\mathbf{k}$:

$$\delta_k \equiv 1/V_u \int \delta(\mathbf{x})\, e^{i\mathbf{k}\cdot\mathbf{x}} d^3x \quad . \tag{2.26}$$

Here we wrote $\delta(\mathbf{x})$, since we consider the complete density contrast $\delta(\mathbf{x}, t)$ as consisting of 2 factors:

$$\delta(\mathbf{x}, t) = D_1(t)\, \delta(\mathbf{x}) \quad .$$

($D_1(t)$ is the solution of the time evolution equation of δ.)

The brackets $\langle \ldots \rangle$ always denote an ensemble average.

The power spectrum must be derived from physical processes in the early universe. Up to now only the model of the inflationary universe[20)] has produced at least an idea of how quantum fluctuations just after the big bang can lead to density perturbations later on. These schemes can certainly not be viewed as an exact derivation, but they make it plausible that a specific type of power spectrum arises: $P(k) \propto k$.

It has already been pointed out earlier[13)] that for a power law ansatz

$$P(k) = A_i\, k^n \tag{2.27}$$

("i" for initial) the index n is constrained to be close to 1.

The analysis of the CMB anisotropies (see preceding section, Fig.10) shows that for the very large scales at least, a power law index $n = 1$ gives a very good fit.

The value $n = 1$ has the consequence that in the matter dominated epoch fluctuations of different wavelength $\frac{2\pi}{k}$ have the same amplitude at the time when their scale $k^{-1}R(t)$ is equal to the horizon scale $\lambda_H = H^{-1}(t)$.

When the density contrast δ is small $|\delta| \ll 1$, the equation may be linearized, and when the pressure is neglected too, we have the following equation

$$\ddot{\delta} + 2\frac{\dot{R}}{R}\,\dot{\delta} - 4\pi G \varrho_b \delta = 0 \quad . \tag{2.28}$$

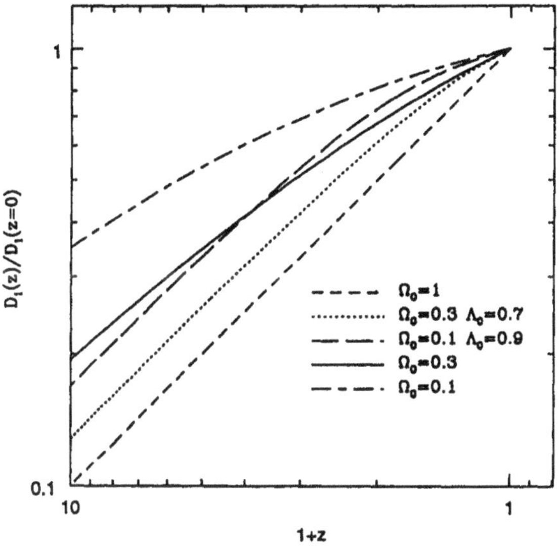

Fig.11 A plot of $D_1(t)/D_1(t_0)$ for various cosmological parameters.

This second-order differential equation has two independent solutions which are conventionally denoted by $D_1(t)$ and $D_2(t)$. $D_1(t)$ is a growing mode, and for $K = 0$ $D_1 \propto t^{2/3}$. In Fig.11 $D_1(t)$ is plotted for various sets of cosmological parameters. We can see from the curves that for $\Omega_0 < 1$ the linear growth is strongly reduced, as $R_0/R = 1 + z$ approaches 1.

Even for the strongest growth of $\delta(t)$, there is the following problem: perturbations in the baryonic matter density can only grow from the time of recombination t_R. The linear growth for $\Omega_0 = 1$ is thus $\frac{D_1(t_0)}{D_1(t_R)} = (1 + z_R) \simeq 1500$. At recombination $\delta(t_R)$ is limited by the anisotropies of the CMB. If we take the COBE result as typical, $\frac{\Delta T}{T} \simeq 10^{-5}$, then for adiabatic perturbations $\delta(t_R) = 3\ |\frac{\Delta T}{T}| = 3 \times 10^{-5}$, and $\delta(t_0) = (1+z_R)\,\delta(t_R) = 4.5 \times 10^{-2}$. The amplitude of the density perturbation stays small and does not grow to the nonlinear regime $\delta \simeq 1$. This means that the universe would stay almost homogeneous, and that the structures we observe could not have formed. One way out of this dilemma is to postulate dark, non-baryonic matter as an additional component of cosmic matter. Perturbations of nonbaryonic dark matter can grow from the epoch, when the energy densities of relativistic and non-relativistic matter are equal. The growth factor can then be increased by a factor of 10. In addition nonbaryonic dark matter couples to the photons only through the gravitational potential wells it forms, the "Sachs–Wolfe" effect. This looks like a promising approach to meet the limits imposed by the CMB.

A classical cosmological model consists of a smooth background configuration and a specific spectrum of perturbations.

2.2 The Age t_0 and the Hubble Constant H_0

These cosmic quantitites are discussed by J. Mould, so I just quote a few results. There has been talk of an "age" problem in cosmology for several years. Usually the argument runs as follows: let $\Omega_0 = 1$, and $\lambda_0 = 0$, then $H_0 t_0 = 2/3$, whereas astronomical observations require $H_0 t_0 > 1$. It is true that even within the 1σ error intervals it seems difficult to settle things at a value $H_0 t_0 = 2/3$. On the other hand the value of H_0

$$H_0 = (80 \pm 22)\mathrm{km\ s^{-1} Mpc^{-1}} \quad ,$$

as well as the best fits to the age of stars in globular clusters

$$t_0 = (15 \pm 3) \times 10^9 \mathrm{years}$$

allows one to accommodate values of $H_0 t_0 \simeq 1$.

If we take H_0 and t_0 near the upper end of their error interval, then the discrepancy becomes pronounced, and we must introduce the infamous cosmological constant $\lambda_0 \neq 0$. As we have seen in section 1 this solves the age problem. It seems sensible, from a pragmatic point of view, to take λ_0 in its own right and to limit it by observations. Although theorists do not like a $\lambda_0 \neq 0$, because it is an ugly quantity which cannot — at present — be derived from a fundamental theory, it is the easiest way out of difficulties with H_0 and t_0.

2.2.1 Some Remarks on H_0

It is perhaps useful to keep referring to a simple picture, such as the model with $K = 0$, $\Omega_0 = 1$, $p = 0$, and $\Lambda = 0$. In this Einstein–de Sitter model, the part of the universe that is observable has a coordinate radius $r_H = \int_0^{t_0} dt/R(t)$ (along the light cone $ds^2 = 0$; for radial directions then $dt = \pm R(t)dr$) and $r_H = \frac{3t_0}{R_0}$. Referred to the present epoch the metric distance is

$$d_H = R_0 r_H = 3t_0 = 2H_0^{-1} \tag{2.29}$$

(actually $d_H = 2c/H_0$, but often we set $c = 1$).

This is the radius of the observable universe at the present time. It is our "particle horizon" beyond which we cannot even in principle observe events.

Thus H_0^{-1} determines the size of the observable universe, regardless of the actual finite or infinite extent of the cosmological model. With $h = H_0/100$ in the usual units, the radius of the observable region, of the "Hubble sphere"

$$d_H = 6000h^{-1}\mathrm{Mpc} \quad . \tag{2.30}$$

It is quite instructive to picture for ourselves how far astronomical measurement have reached out into this sphere. An "angular size" distance in cosmological models can be defined by $r_A \equiv D/\delta$, where D is the linear extent and δ the angular diameter of an object (see section 5). In FL cosmologies

$$D = \int R(t)f(\chi)d\theta = R(t)f(\chi)\delta \quad ; \quad \text{hence}$$
$$r_A = R(t)f(\chi) \quad .$$

For a source at redshift z (and $\Omega_0 = 1$)

$$r_A \equiv d_e = \frac{d_H}{(1+z)}\left(1-(1+z)^{-1/2}\right) \quad , \tag{2.31}$$

where d_e is the metric distance at the time of emission. At the present epoch this distance corresponds to

$$d_0 = d_H(1-(1+z)^{-1/2}) \quad .$$

The time of emission $t_e = t_0(1+z)^{-3/2}$. The epoch of recombination at $z = 10^3$ covers 97% of d_H. A quasar with $z = 3$ has $d_0/d_H = 0.5$, and $t_e/t_0 = 0.125$. Deep surveys $z \simeq 0.3$ cover 12% of the Hubble radius. The CfA–survey with $z \leq 0.05$ explores just the "local neighbourhood": $d_0/d_H = 2.5 \times 10^{-2}$.

We can expect the unperturbed Hubble expansion only beyond the Local Supercluster at $z \geq 0.01$. The angular distance r_A can be expressed as a function of H_0, q_0, and z (for $\Lambda = 0$), and neglecting intergalactic scattering or absorption, the specific intensities at the emission I_{ν_e} and observation point I_{ν_0} are related by

$$I_{\nu_0} = I_{\nu_e}(1+z)^{-3} \quad .$$

Then the Friedmann equations lead to an $m - z$ relation:

$$m = M - 5\log H_0 + K(z) + 5\log\left\{\frac{1}{q_0^2}(1 - q_0 + q_0 z + (q_0 - 1)\sqrt{11 + 2q_0 z})\right\} + \text{const.} \tag{2.32}$$

$K(z)$ is a term containing bandwidth and color corrections which depend on the spectrum of the source and on the receiver sensitivity.

For small z

$$m \simeq M - 5\log H_0 + K(z) + 5\log z + 1.086(1 - q_0)z + \text{const.} \tag{2.33}$$

(even valid for $\Lambda \neq 0$). m and z can be measured directly, but M and $K(z)$ require certain more or less convincing assumptions on standard candles (M) and the spectrum ($K(z)$). Systematic errors are quite abundant in this procedure, and it will take some time, and many discussions, before the dust settles, and canonical values for these quantities are determined. As we have heard from J. Mould, the Hubble Space Telescope (HST) measurements of Cepheids in M100 lead to

$$H_0 = 80 \pm 22[\text{km s}^{-1}\text{Mpc}^{-1}] \quad ,$$

a value that comprises within its 1σ error bars many possible cosmological models — except perhaps the favourite one which has all the easy formulae, where

$$H_0 t_0 = 2/3 \quad .$$

No constraint on q_0 can be derived from these measurements. The differences in the $m-z$ relation due to q_0 are drowned in the uncertainties coming from the difficulties of the M-calibration of nearby galaxies, the selection of "standard candles", of accounting for evolution effects (time variability of M) and the influence of local peculiar velocity fields within the Local Supercluster.

2.3 The Density Parameter Ω_0

The mean density of the universe is a fundamental quantity in cosmology, but it is not known to high accuracy. Remember that in FL-models we have the relation

$$2q_0 = \Omega_0 - 2\lambda_0 \quad .$$

A precise measurement of q_0 (not possible at the moment, but perhaps in the future) would determine $\Omega_0 - 2\lambda_0$, and then limits on λ_0 would translate into limits on Ω_0. It may be, however, that λ_0 is even more difficult to measure than Ω_0 (see section 2.4 for a few remarks).

One general procedure to measure Ω_0 requires first a count of luminous objects to such an accuracy that a "mean luminosity density" $\langle \ell \rangle$ can be determined. A recent estimate of the mean luminosity density in the blue band of the spectrum $\langle \ell_B \rangle$[6] is given by

$$\langle \ell_B \rangle = (1.9 \pm 0.06) h \; L_\odot \; \mathrm{Mpc}^{-3} \quad . \tag{2.34}$$

Then

$$\text{“}\Omega_0\text{”} = \langle \ell_B \rangle \times M/L/\varrho_c; \; \varrho_c \equiv \frac{3H_0^2}{8\pi G} \quad . \tag{2.35}$$

The quantity M/L is the mass–to–light ratio of various kinds of objects, and the basic assumption is that the relation between light and mass in the universe is the same as in the objects measured. ϱ_c is the critical density defined earlier. If we express the M/L ratio in solar units $M/M_\odot/L/L_\odot$, then

$$\Omega_0 = (1450)^{-1} \; h^{-1} (M/L_B)_\odot \quad . \tag{2.36}$$

To reach $\Omega_0 = 1$, the M/L ratio of the objects considered should be

$$(M/L_B)_\odot = 1450h \quad . \tag{2.37}$$

The uncertainty in this formula is difficult to estimate, because systematic effects like galaxy evolution, obscuration and other photometric difficulties, might produce a much larger error in $\langle \ell_B \rangle$ than indicated in eq. (2.34). Also the population of objects used to measure M/L_B may not be representative for the whole universe. So the value in eq. (2.37) may easily be uncertain by a factor two. It is interesting to note that for dynamical measurements of M/L, the uncertainty in H_0 cancels, because $\langle \ell_B \rangle \sim H_0$, and $M/L \sim H_0$. Ω_0 is then independent of H_0.

This is not so for the optical starlight from galaxies which for an average $M/L \simeq 7$ gives

$$\Omega_* = 0.005h^{-1} \quad . \tag{2.38}$$

The rotation curves of spiral galaxies lead to an estimate of $M/L \simeq 20h$, and

$$\Omega_G = 0.014 \quad . \tag{2.39}$$

The mass directly associated with the optical regions of galaxies contributes less than 2% of the critical density $\Omega_0 = 1$.

Binary galaxies, galaxies with satellites have been used to determine the mass of a galaxy. The Kepler relation $\frac{GM}{r} = kv^2$ is used, and for a binary galaxy the characteristic radius is the projected separation, the velocity is the redshift difference. The dimensionless parameter k depends on model assumptions such as the shape of galaxy orbits, on the relative mass distribution and on selection and projection effects. The uncertainties are large, the values given $M/L \simeq (50-100)h$, i.e. $\Omega_0 = (0.03 - 0.06)$.

The situation is better for groups of galaxies, because a larger number of members reduces the scatter in k. While binaries are typically separated by $\leq$ 100kpc the typical size of groups is $\sim$ 1Mpc. Their dynamics allow one to infer the mass distribution on a much larger scale not tightly connected to an individual galaxy. There is, however, one great difficulty: groups are loose associations, and it is a problem to decide whether a galaxy belongs to a group or not. The collapse times of groups are very long, and they may be far from equilibrium[16)27)42)]. These uncertainties make the M/L determinations for groups quite unreliable. In the literature one can find a range of values $M/L = (50-200)h$, i.e. $\Omega_0 = (0.03-0.12)$.

Rich clusters of galaxies offer several possibilities to estimate their mass, and therefore M/L.

A) They are the largest virialized systems, and the virial theorem can be applied to the motion of the galaxies in the cluster. For a relaxed stationary object the virial theorem states that $2T + \Omega = 0$, where T is the kinetic, Ω the potential energy.

$$
\begin{aligned}
T &= \frac{1}{2}\Sigma M_i \dot{\mathbf{r}}_i^2 \equiv \frac{1}{2}\langle v^2\rangle M \quad ; \\
\Omega &= -\frac{GM^2}{R_v} \quad ; \quad \text{i.e.} \quad \langle v^2\rangle = \frac{GM}{R_v} \quad ;
\end{aligned}
\tag{2.40}
$$

M is the total mass of the cluster, R_v is the effective virial radius; $\langle v^2\rangle$ is actually a non–observable quantity, the sum of the mass–averaged velocity squares. Only the velocity in the direction of the line of sight is measurable. Some assumption is necessary to get an estimate of $\langle v^2\rangle$. Sometimes a luminosity weighted mean is taken. Often one uses $\langle v\rangle^2 = 3\sigma^2$, where σ^2 is the line of sight velocity dispersion, but this may be quite wrong — even by a factor of two.

The virial radius R_v is determined from the projected number of galaxies. For a spherical cluster the number of galaxies in the strip with minimal separation q and width dq from the center (Fig.12), $S(q)dq$, is related to the number density

$n(r)$ by

$$S(q) = 2\pi \int_0^{u(R)} n(r) u du \tag{2.41}$$

The cluster mass is

$$M = 2M_G \int_0^R S(q) dq \quad ,$$

where M_G is the mean galaxy mass. Now, the potential energy is $-G \int_0^M \frac{M(r)dM(r)}{r^2}$, and finally

$$\Omega = -2M_G^2 G \int_0^R S^2(q) dq \quad .$$

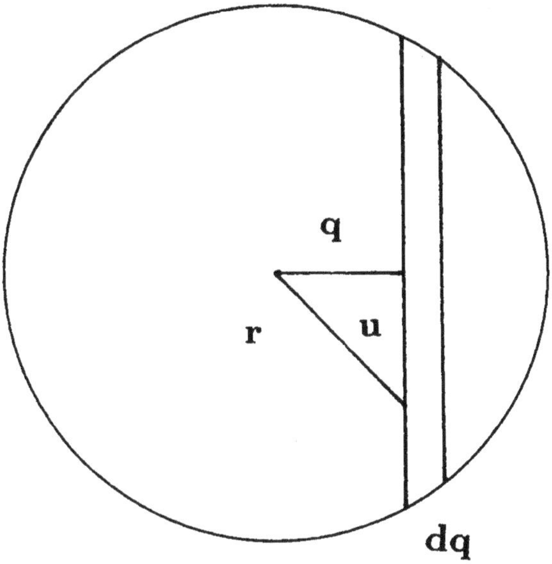

Fig.12 Calculation of the projected galaxy distribution.

Then we have

$$R_v = \frac{2(\int_0^R S dq)^2}{\int S^2 dq} \quad . \tag{2.42}$$

$\langle v^2 \rangle$ and R_v give the cluster mass via eq. (2.40). The method has been applied to various rich clusters. Especially the Coma cluster has a large number of measured redshifts, and various authors find $(M/L_B)_{\rm COMA} = 400h$, i.e. $\Omega_0 = 0.24$, if this is a typical value. The errors are just guesses of the uncertainties involved, and at present one usually takes

$$\Omega_0 = 0.1 \text{ to } 0.3 \tag{2.43}$$

as the result derived from the application of the virial theorem to clusters. The mass determinations can be roughly presented by the following formula

$$M_c \equiv 1.1 \times 10^{14} \left(\frac{\sigma_v}{10^3 \text{km s}^{-1}}\right)^2 \left(\frac{\theta}{30''}\right) \left(\frac{D}{1\text{Gpc}}\right) M_\odot \quad . \tag{2.44}$$

(θ is the angular extent analysed, D the distance).

B) Gravitational lensing has become a very useful tool for measuring cluster masses precisely[32]. The mass contained in a cluster can act as a gravitational lens on the light of distant galaxies. Sometimes the image appears as a giant circular arc. The radius of the arc determines the mass within. Very often the galaxies appear as tiny elliptical images, so–called "arclets". Typically there are about 70 arclets per $(\text{arcmin})^2$. The statistics of these arclets permits — in principle — a precise reconstruction of the cluster mass distribution, by an inversion of the lens equation. Typical values agree with the virial estimates

$$M_g(\theta_{\text{arc}}) = 1.1 \times 10^{14} \left(\frac{\theta_{\text{arc}}}{30''}\right)^2 \left(\frac{D}{1\text{Gpc}}\right) \quad . \tag{2.45}$$

There is also some evidence that clusters can have much more dark matter. The cluster MS1224 has 2 or 3 times more mass from the arclet reconstruction than obtained from the virial argument[9].

C) Cluster masses can also be determined from X–ray data with the additional bonus that the mass in baryons contained in the cluster can also be determined. The X–rays are emitted from a hot gas in the cluster at a temperature of $\sim 10^8$K. In Fig.13 the X–ray emission of the cluster A3528 as measured by the ROSAT X–ray satellite is shown. The hot gas must be kept in equilibrium

$$\varrho_{\text{gas}} \frac{dp}{dr} = -\frac{GM(r)}{r^2} \quad , \tag{2.46}$$

and the mass $M(r)$ within the radius r must be large enough to satisfy it. Both the density $\varrho(r)$, and the temperature $T(r)$ of the gas can be obtained directly from X–ray observations. The modelling of the X–ray emission is done under various simplifying assumptions on the cluster shape and its gravitational potential. The estimates of the gas density from the X–ray emission measure seem quite reliable, whereas temperatures can only be determined for spectra of extremely good resolution.

ROSAT observations[2)5)30)] lead to the conclusion that the hot gas producing the X-rays has about 5 times the mass of the galaxies in the cluster, while the fraction of the mass in baryons to the total mass is

$$\frac{M_b}{M_{\rm tot}} = 0.1 \text{ to } 0.4 \quad . \tag{2.47}$$

This can be translated into a limit on Ω_0, if we take into account the limit on Ω_b from nucleosynthesis, $\Omega_b < 0.015h^2$,

$$\Omega_{\rm tot} = (2.5 \text{ to } 10) \times \Omega_b < (0.04 \text{ to } 0.15) \quad . \tag{2.48}$$

This inequality seems to be a strong argument in favour of a low Ω_0 value, if clusters show the true cosmic ratio of $\Omega_b/\Omega_{\rm tot}$. Up to now this seems to be a generally accepted hypothesis. Even though we should be aware of the fact that besides regions of the universe which emit radiation, there are also dark regions which do not shine. So there still seems to be room for changes in the value of Ω_0.

An estimate of Ω_0 can also be obtained from the pairwise velocity dispersion of galaxies[26)]. This method estimates statistically the relative velocity of galaxy pairs. It has the advantage that group identification is not necessary, but it depends strongly on selection effects. The best estimate for Ω_0 from this approach[26)] is:

$$\Omega_0 = 0.28b^2 \left(\frac{\sigma}{400}\right)^2 \quad . \tag{2.49}$$

b here is a "biasing" parameter which describes a difference in the clustering of matter and of galaxies $\frac{\delta N_G}{N_G} = b\,\frac{\delta\varrho}{\varrho}$. In numerical simulations of Cold Dark Matter (CDM) $b \geq 2$ to achieve consistency with the galaxy correlation function.

Besides virialized parts of the cosmos we can also look at large–scale density fluctuations which are not in gravitational equilibrium. The difference in mass in large regions leads to large–scale peculiar velocity patterns which can give direct information on the underlying density field, and hence on Ω_0. There has been a lot of activity in this field, but the estimates of Ω_0 are still very imprecise. Values from $\Omega_0 = 0.1$ to $\Omega_0 = 1.3$ are all possible.

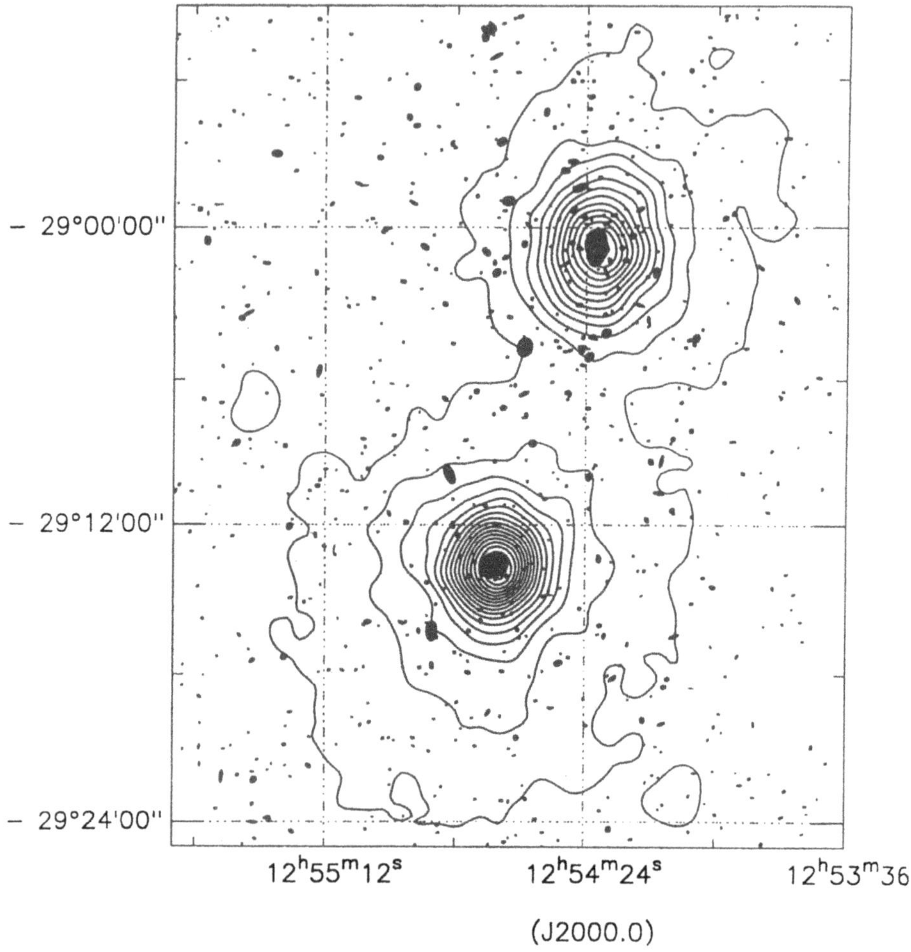

Fig.13 ROSAT PSPC image of the cluster A3528 which belongs to the Shapley Supercluster in the hard band (0.5-2.0 keV) (courtesy of S. Schindler[31)]). The value of the lowest contour and the contour spacing is 1.36×10^{-3} cts s^{-1} $arcmin^{-2}$. The X-ray image shows a double structure. Superposed on the X-ray contours are galaxy images from the COSMOS/UKSTU Southern Sky Object Catalogue[45)].

2.4 Estimates of λ_0

The big puzzle of the cosmological constant is the fact that $\frac{\Lambda}{8\pi G}$ is restricted by astronomical observations to be smaller or, at most, a little bit larger than ϱ_c, meaning that

$$\frac{\Lambda}{8\pi G} \leq (10^{-12}\mathrm{GeV})^4 \quad .$$

This is absolutely tiny, from the particle physics point of view: the GUT-vacuum energy density is $(10^{15}\mathrm{GeV})^4$, and one would naively estimate a cosmological constant Λ_{GUT} from that, $\frac{\Lambda_{\mathrm{GUT}}}{8\pi G} \simeq (10^{15}\ \mathrm{GeV})^4$, which is 10^{108} times larger than the value that can be tolerated at the present epoch in the universe (cf. section 4.5).

If we forget about this deep problem, and take the pragmatic view that Λ is just one more cosmological parameter to be determined by observations, we know that $\frac{\Lambda}{3H_0^2} \equiv \lambda_0$ is around 1, but apart from that it is difficult to restrict it further[4)23)40)]. We have seen in section 1 that models without a big bang can be excluded, and therefore $\lambda_0 < \lambda_c$ for positive λ_0.

A serious attempt at limiting the cosmological constant was made by using gravitational lensing theory[23)]. Observations of images of quasars with a narrow separation less than 2 arcseconds, and the statistics of such images produced by a conserved, nonevolving population of elliptical galaxies in a cosmological model with $\Omega_0 + \lambda_0 = 1$ lead to the following result: a best fit is obtained, if $\lambda_0 \leq 0.7$, and $\lambda_0 \leq 0.9$ with very high significance. Therefore a model where only baryonic matter is present $\Omega_0 = \Omega_B \leq 0.06$ is excluded. In that case $\lambda_0 \geq 0.94$ is required, and this cannot be reconciled with the data.

The limit holds only for $\Omega_0 + \lambda_0 = 1$. If we relax this condition, allowing $\Omega_0 + \lambda_0 > 1$, then no restriction on λ_0 apart from $\lambda_0 \leq \lambda_c$ can be derived.

3. The Hot Early Epoch

The hot–big–bang scenario is a natural consequence in a singular FL–model, where the expansion factor $R(t)$ approaches zero towards the limit $(t \to 0)$ of the cosmic time t. The pressure of a uniform radiation field implies a "hot" singularity, i.e. an ever increasing temperature as $t \to 0$.

Schemes that try to avoid the hot big bang have been suggested every now and then. They must produce the 3K background with its almost perfect blackbody shape from specific astronomical sources. Attempts at such unorthodox scenarios do not look very promising and have received little attention so far.

If we stick to the standard model, we may conclude that in the radiation–dominated epochs ($z > z_R$) the energy density evolved proportionally to the 4th power of the temperature (i.e. $\propto T^4$), or equivalently (since $RT =$ const.) proportional to R^{-4}. The curvature term $\frac{K}{R^2}$ and the cosmological constant Λ become negligible compared with the "potential energy" term ($\propto \varrho = aT^4$) in the Friedmann equation. To good approximation then

$$\dot{R}^2 = \frac{8\pi}{3} G\varrho R^2 \quad . \tag{3.1}$$

Since $\varrho R^4 =$ const., the expansion factor $R(t)$ satisfies $(\dot{R}R)^2 =$ const., hence $R \propto t^{1/2}$, and finally

$$\frac{32\pi}{3} G\varrho t^2 = 1 \quad . \tag{3.2}$$

3.1 Statistical Equilibrium Distributions

It is crucial in all attempts to describe the early universe to assume the existence of a phase of complete thermodynamic equilibrium at some very early epoch (e.g. around $T = 10^{11} K$, or earlier). This postulate implies an enormous simplification of the physics in the early universe; all the quantities of interest depend only on the temperature, and the chemical potentials. The previous history of the thermodynamic system need not be known.

In thermal equilibrium the number density n_i of particles of type i, as well as their energy density ϱ_i, at the temperature T can easily be derived[17][36][39]

$$n_i = \left(\frac{g_i}{2\pi^2}\right) \left(\frac{k_B T}{\hbar c}\right)^3 \int_0^\infty \left\{ \exp\left(\frac{E_i - \mu_i}{k_B T}\right) \pm 1 \right\}^{-1} z^2 dz \quad , \tag{3.3}$$

$$\varrho_i c^2 = \left(\frac{g_i}{2\pi^2}\right) \left(\frac{k_B T}{\hbar c}\right)^3 k_B T \int_0^\infty \left\{ \exp\left(\frac{E_i - \mu_i}{k_B T}\right) \pm 1 \right\}^{-1} z^2 \, \frac{E_i - \mu_i}{k_B T} \, dz \quad , \tag{3.4}$$

g_i is the number of spin degrees of freedom; i.e. the number of helicity states.

$$\begin{aligned} E_i &= \left\{ (pc)^2 + (m_i c^2)^2 \right\}^{1/2} \quad , \\ z &= \frac{|p| c}{k_B T} \quad , \end{aligned} \tag{3.5}$$

where the $+(-)$ sign in the integral applies to fermions (bosons). Antiparticles are treated as independent particle species.

The chemical potentials μ_i are unknown parameters in this state of thermal equilibrium. In the standard model their values have to be posed as an initial condition. In GUT theories there are no conserved additive quantum numbers, and thus all chemical potentials are put to zero, i.e. $\mu_i = 0$.

For bosons, the convergence of the integrals requires

$$|\mu_i| \leq m_i c^2 \quad .$$

Thus for mass zero bosons $\mu_i = 0$ anyway. If we set all chemical potentials equal to zero, $\mu_i = 0$, and we see from eq. (3.3) that the i–particles are present in thermal equilibrium for the range $m_i \ll T$. For $m_i \geq T$ their number is reduced drastically by the "Boltzmann" factor $\exp(-m/T)$.

For $T \geq m_i$ the i–particles are relativistic. Then, for bosons

$$n_B = (g_B/2)\, n_\gamma \quad ; \quad \varrho_B = (g_B/2)\, \varrho_\gamma \quad , \tag{3.6}$$

and for fermions

$$n_F = \frac{3}{8}\, g_F\, n_\gamma \quad ; \quad \varrho_F = \left(\frac{7}{16}\right) g_F\, \varrho_\gamma \quad . \tag{3.7}$$

n_γ, ϱ_γ are the number and energy density of the photons, respectively. g_B, g_F are the statistical degrees of freedom for bosons and fermions, respectively.

The total energy density

$$\varrho = \varrho_B + \varrho_F = \frac{1}{2}\, \varrho_\gamma (g_B + \frac{7}{8}\, g_F) \equiv \frac{1}{2}\, \varrho_\gamma\, g(T) \quad . \tag{3.8}$$

In $g(T)$ we count all particles in thermodynamics equilibrium, plus additional relativistic particles that have dropped out of equilibrium. Eq. (3.2) can be written as a relation between cosmic time and temperature

$$t_{\text{sec}} = 2.42 \times 10^{-6}\, g^{-1/2} T_{\text{GeV}}^{-2} \quad . \tag{3.9}$$

An instructive example is the equilibrium maintained by weak and electromagnetic interactions around $T = 1$ MeV. As far as we know the particles in equilibrium are photons (γ), electrons and positrons ($e^{\pm}$), and neutrinos (ν_e, ν_μ, ν_τ). We have

$$g_B = g_\gamma = 2 \quad ;$$

$$g_F = g_e + g_{\nu_e} + g_{\nu_\mu} + g_{\nu_\tau} = 10 \quad . \tag{3.10}$$

Hence

$$g(T \geq 1\text{MeV}) = \frac{43}{4} \quad , \tag{3.11}$$

$$t(1\text{MeV}) = 0.74s \quad . \tag{3.12}$$

The entropy in a comoving volume $V(\propto R^3)$ is

$$S = \frac{4}{3}\left(\frac{\varrho_I}{T}\right) V = \frac{2}{3} g_I(T)\ \varrho_\gamma / T\ V \quad ,$$

where the subscript I refers to particles in equilibrium. Below a temperature of 1MeV the neutrinos drop out of thermal equilibrium. Then the conservation of S has the consequence that the photon gas is heated up by $e^+ - e^-$ annihilation around $T = 0.5$MeV. The photon temperatures before ($T_{\gamma b}$) and after ($T_{\gamma a}$) the annihilation are related, because of the constancy of S:

$$g_I(T > 0.5\text{MeV}) = \frac{11}{4} \quad ,$$

$$g_I(T < 0.5\text{MeV}) = 1 \quad ,$$

and therefore

$$T_{\gamma b}^3 = \frac{4}{11} T_{\gamma a}^3 \quad . \tag{3.13}$$

The neutrinos are relativistic particles out of equilibrium around $T = 1/2$MeV, and thus remain unaffected by the annihilation. Their contribution to S stays the same $\frac{2}{3} g_\nu(T)\ \varrho_\gamma / T\ V$, but their temperature T_ν is lower than the photon temperature at $T < 0.5$MeV:

$$T_\nu = \left(\frac{4}{11}\right)^{1/3} T_\gamma \quad . \tag{3.14}$$

At the present time — since there are no further annihilation reactions — the neutrino temperature is

$$T_{\nu_0} = 1.9K \quad . \tag{3.15}$$

It may be of interest to consider quite generally particles X which fall out of equilibrium very early (with a g–factor g_E). These particles lose in every annihilation epoch compared to the system of particles in thermal equilibrium:

$$T_X = (g_A/g_E)^{1/3} T_\gamma \quad , \tag{3.16}$$

where g_A is the g–factor of the interacting system after the annihilation. If we assume that the X–particles stay relativistic until the final $e^+ - e^-$ annihilation (g_A), then

$$T_X = (g_{Af}/g_E)^{1/3} T_\gamma \quad ,$$

and the contribution of these X–particles to the energy density will be substantially reduced (by the factor $(g_{Af}/g_E)^{4/3}$) compared to a relativistic particle species whose temperature is equal to T_γ:

$$\varrho_X = (g_{Af}/g_E)^{4/3}\, g_X\, T_\gamma^4 a = \left(\frac{T_X}{T_\gamma}\right)^4 g_X\, \varrho_\gamma (\ll \varrho_\gamma) \quad . \tag{3.17}$$

For the system of photons and neutrinos below $T = 1/2\text{MeV}$ we have

$$\varrho_{\text{rel}} = \varrho_\gamma + \frac{7}{16} g_\nu \left(\frac{T_\nu}{T_\gamma}\right)^4 \varrho_\gamma = \varrho_\gamma \left(1 + \frac{7}{8}\left(\frac{4}{11}\right)^{4/3} N_\nu\right) \quad , \tag{3.18}$$

where N_ν is the number of neutrino types.

An additional component of nonrelativistic matter "n" leads to the following formula for the equality of relativistic and nonrelativistic matter: $\varrho_n = \varrho_{\text{rel}}$ at z_{eq} with

$$(1 + z_{\text{eq}}) = (\Omega_n h^2) \left(\frac{T_{\gamma 0}}{2.7}\right)^{-4} (1 + 0.227 N_\nu)^{-1} \times 4.2 \times 10^4 \quad . \tag{3.19}$$

For a dark matter component Ω_n, the density perturbations can grow from z_{eq}, and this can be (for $\Omega_n h^2 \approx 1$) more than 10 times the redshift at recombination $z_R \approx 1500$.

Dave Schramm explains in his lectures on nucleosynthesis, how the production of the light nuclei (especially helium) is affected by the number of relativistic particle species, and how eq. (3.18) can be used to obtain limits on N_ν from the theory of He^4 synthesis combined with abundance measurements of He^4.

The simple structure of the FL–models permits one to present the detailed thermal history of the universe as in Fig.14 starting from the present state and calculating backwards.

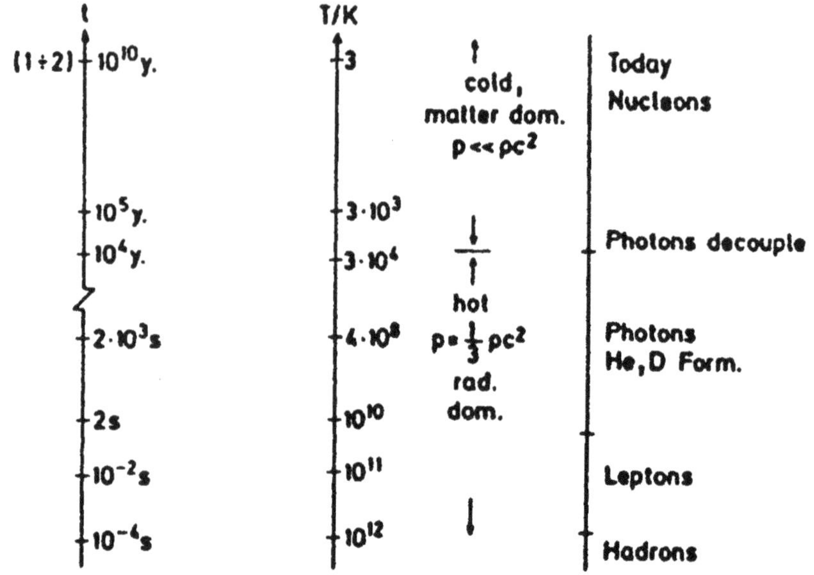

Fig.14 Graphs of the history of a typical big bang model.

4. Puzzles of the Standard Model and the Principle of Inflation

4.1 "Flatness"

The Friedmann equation at the present epoch

$$\frac{K}{R_0^2 H_0^2} = \Omega_0 - 1 + \lambda_0$$

consists of terms of the same order of magnitude, although at earlier or later times $\frac{K}{R^2(t)H^2(t)}$, $\Omega(t)$, and $\lambda(t)$ may greatly differ from each other. Ω_0 is larger than 0.1 and less than 2 (as far as we know), and a similar small interval of values is possible for $\frac{|K|}{R_0^2 H_0^2}$, and for λ_0. At very early times these terms are very different from each other. As $\varrho = \frac{\pi^2}{30} g(T)T^4$, and RT = const., the density changes as

$\varrho \propto R^{-4}$. For $R \to 0$ the $\Omega(t)$ term must have been completely dominant. For times in the future the cosmological constant will eventually dominate over the ϱ and $\frac{K}{R^2}$ terms.

One may wonder about this curious fact of a near balance of 3 such different terms. There is no doubt that to have the balance just at the present time is very beneficial for our existence, and for the existence of the world around us. Indeed, a balance between $G\varrho$ and $\frac{K}{R^2}$, for a $K = +1$ model, at a time of the order of the Planck–time for instance, would be disastrous. Similar to the "normal" $K = +1$ universe treated in section 1, we find in this case that the total time of existence before recollapse would be a few Planck–times. On the other hand, a balance for $K = -1$ at the Planck–time, with $\Omega_{pl} < 1$, would lead to such rapid expansion that at the present time Ω would be catastrophically small — matter condensations could not form in such a universe. It is definitely just as well for us that the curvature and the gravitational term are in near–balance at the present epoch.

This fact of a balance of three different terms can be reformulated as follows: it can be traced back to the large numerical value of the entropy in a comoving volume $\sim R^3(t)$. In a radiation–dominated early epoch the entropy density s is given as

$$\begin{aligned} s &= \frac{2\pi^2}{45} g(T) T^3 \quad , \quad \text{and} \\ S &= sR^3 \end{aligned} \tag{4.1}$$

is a constant throughout the whole evolution of the standard big–bang model.

At the present epoch the entropy density is dominated by the background of photons and neutrinos

$$s_0 = 7n_{\gamma_0} = 2.8 \times 10^3 \mathrm{cm}^{-3} \quad . \tag{4.2}$$

The entropy in a comoving volume $\propto R_0^3$ turns out to be

$$S = 2 \times 10^{87} \quad . \tag{4.3}$$

This number is an initial condition of the standard big–bang model. The fact that there are so many more photons than baryons in our universe is connected with that.

The large value of S has consequences for the early phases of a FL–model.

The Friedmann equation for $\Lambda = 0$,

$$H^2 R^2 |1 - \Omega| = |K|$$

can be written as

$$\frac{|1-\Omega|}{\Omega} \simeq 0.2|K|\, g^{-1/3}\, S^{-2/3} \left(\frac{M_{p\ell}}{T}\right)^2 \tag{4.4}$$

($M_{p\ell}$ is the "Planck-Mass". $M_{p\ell} = G^{-1/2} \simeq 10^{19}$GeV). Inserting numbers we find

$$\frac{|1-\Omega|}{\Omega} \simeq 10^{-59} \left(\frac{M_{p\ell}}{T}\right)^2 \quad . \tag{4.5}$$

Ω must have been close to 1 for early epochs, because S is so large.

For

$$\begin{aligned} T &= 1\text{MeV} \quad , \quad \frac{|\Omega-1|}{\Omega} \leq 10^{15} \quad ; \\ T &= 10^{14}\text{GeV} \quad , \quad \frac{|\Omega-1|}{\Omega} \leq 10^{-49} \quad . \end{aligned} \tag{4.6}$$

This feature of the standard model, that Ω is close to 1 at all times, has been called the "flatness–problem" — referring to the large radius of curvature of our universe — or the "entropy problem" — referring to the large value of S[12)19)28)]. It does not seem fair to speak of "problems" in the context of the standard big–bang model. In any solution of a differential equation there are certain specific properties of the initial data. If we compute backwards in time we just find the initial data, which are responsible for the state of affairs as we see it now.

But if we look at the present state of the universe as the consequence of certain initial data we might feel uneasy, if the initial data have to be extremely special — as e.g. for our universe requiring $S = 10^{87}$, or $\Omega = 1$ at early times. As physicists we would feel more at ease, if we could find an understanding of such specific conditions in terms of physical processes. Such a reduction of the initial data would be considered as improved insight into the early universe.

Is it possible to start with an S of order 1, and arrive via physical processes at the number $S = 10^{87}$ characteristic of the present epoch? The model of the inflationary universe indicates "yes" as the answer to this question.

4.2 The "Horizon Problem"

Another aspect of the standard model that is difficult to accept off–hand, is the horizon structure of the isotropic and homogeneous space–time. There are particle–horizons in these space–times, or in other words, there are at any epoch regions that have not had causal contact with each other. The size of the particle horizon at time t and $\chi = 0$ is

$$r_H(t) = R(t) \int_0^t \frac{ds}{R(s)} \quad . \tag{4.7}$$

This is the maximum size of a region in which causal relations can be established at the cosmic time t. In the standard model — during the radiation dominated phase — the expansion factor $R(t)$ is proportional to $t^{1/2}$, and $r_H = 2t$. For $t \to 0$ r_H shrinks much faster than $R(t)$. At very early epochs therefore, most of the regions within the typical dimension $R(t)$ are causally unrelated despite the homogeneity of the model.

This fact makes it impossible to give a physical explanation of the isotropy of the CMB. Since the decoupling time t_R the radiation background has evolved independently, and its present isotropy must therefore originate from a similar isotropy at t_R. But we see — within small fluctuations of $\sim 10^{-5}$ — the same radiation–temperature in regions of the sky that had not established causal contact with each other at the epoch t_R. The distance (measured at t_R) of two regions (A, B) on the "cosmic photosphere" which lie in opposite directions (Fig.15) is

$$2R(t_R) \int_{t_R}^{t_0} \frac{dt}{R(t)} \quad .$$

This distance contains

$$n_{AB} = \int_{t_R}^{t_0} \frac{dt}{R(t)} \left(\int_{t_{pl}}^{t_R} \frac{dt}{R(t)} \right)^{-1} \tag{4.8}$$

horizon–lengths at time t_R. For a $K = 0$ model we estimate $n_{AB} = 60$.

The factor $\int_{t_R}^{t_0} dt/R(t)$ is determined by $R(t)$ after the decoupling time, a well–known expression. The denominator $\int_{t_{pl}}^{t_R} dt/R$ depends on the behaviour of $R(t)$ in the early universe. Changing from $R(t) \sim t^{1/2}$ to another functional form of $R(t)$ may lead to drastic changes in the denominator. The scale–factor $R(t)$ is increasing exponentially $\sim \exp Ht$ in an inflationary phase of the universe, leading to a large value for the denominator, and to $n_{AB} \ll 1$.

How is it possible for two regions separated by a large number of causality–lengths to acquire an identical temperature at t_R? The horizon structure of the standard model gives rise to the so–called "horizon problem". It is, of course, not really a "problem" of the isotropic and homogeneous FL–models. For such models the horizon–structure is a trivial consequence of the prescribed isotropy and homogeneity. The problem appears, if an explanation is sought of the observed isotropy. Approaches to a solution of this problem might consist in a description starting from a very general inhomogeneous and anisotropic structure of the universe. The hope is that one can then show how reasonable physical processes lead to an isotropic and homogeneous FL universe.

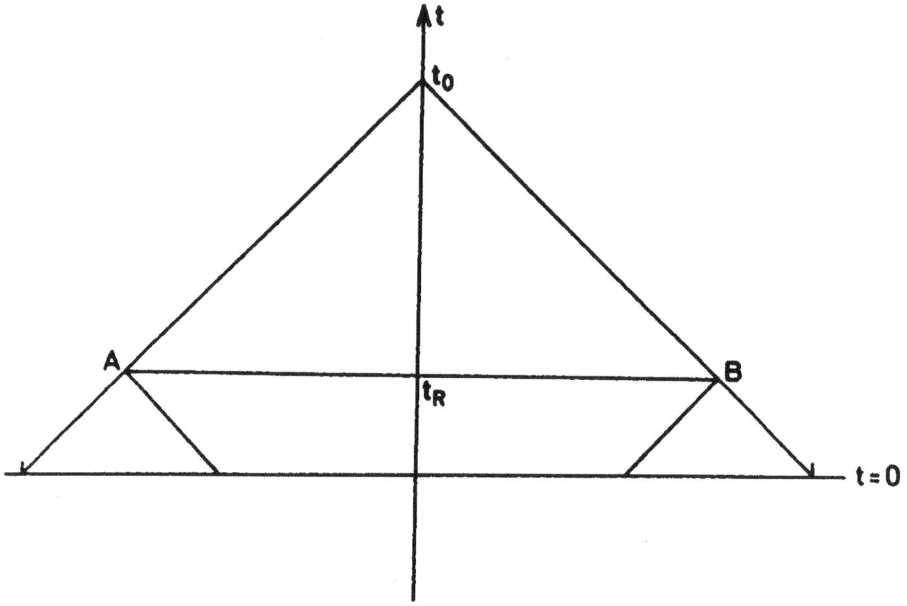

Fig.15 The cosmic photosphere at $t = t_R$ has points A and B, corresponding to directions in the sky 180 degrees apart, which did not have causal contact for times before t_R.

The horizon problem is again related to the value of the entropy S: the radius of the observable universe L_0 shrinks to a length $L(t)$ at an earlier epoch t, and $L_0 = (T/T_0)L(t)$. During the radiation–dominated era

$$r_H \simeq 2t \simeq (G\varrho)^{-1/2} \simeq M_{p\ell}T^{-2} \quad . \tag{4.9}$$

Therefore the ratio of the volume within L, and the horizon

$$\left(\frac{L(t_R)}{r_H(t_R)}\right)^3 \simeq (L_0T_0)^3\left(\frac{T_R}{M_{p\ell}}\right)^3 \simeq S\left(\frac{T_R}{M_{p\ell}}\right)^3 \quad . \tag{4.10}$$

This is a number $\simeq 10^5$, because S is so large. For $S \simeq 1$ it would be $\simeq 10^{-82}$.

So, the flatness and horizon problem both have their origin in the value of S the entropy in a comoving volume $\sim R^3$.

4.3 The Monopole Problem

A very interesting problem arises, when one tries to formulate some types of GUT–models of elementary particles in the early epochs of a FL–cosmology. Topological defects can occur when the GUT symmetry is spontaneously broken down to a smaller symmetry group. The correlation length ξ of the Higgs field is limited by the causality length in the early universe $\xi \leq r_H(t) = 2t$. A lower limit to the density of topological defects can thus be obtained, namely one per horizon volume.

For monopoles as topological defects with a mass $m_M \simeq 10^{16} m_p$ this limit leads to the "monopole catastrophe":

The mass–density of monopoles is

$$\varrho_M = n_M m_M \simeq 10^{14} \varrho_B \tag{4.11}$$

for a typical GUT–scale of $T_c \simeq 10^{14}$ GeV. This terrible dominance of monopoles would violate all observational limits on the mean density and the age of the universe. This is the monopole problem, the original motivation for the concept of the inflationary universe.

4.4 The Inflationary Universe — A Qualitative Account

The basic idea rests on the formulation of spontaneous symmetry breakdown in gauge theories. This is described by the introduction of so–called "Higgs fields": scalar fields φ with an interaction incorporated in an effective potential $V(\varphi)$. This concept is simply transferred from the thermodynamic description of a phase–transition, as in the case of ferromagnetism, for example — where the order parameter (corresponding here to the Higgs field φ) is the magnetization, and the effective potential is the Legendre transform of the free energy.

During the expansion of the universe the Higgs field φ develops from a high–temperature state to a stable ground state of lower energy. This happens in different ways in different models. The original (and simple) idea was that for high temperatures above a critical value $T > T_c$ the potential has a stable minimum at $\varphi = 0$ (the symmetric phase); for $T < T_c$ the state $\varphi = 0$ is metastable, and a state $\varphi = \sigma \neq 0$ becomes the new stable minimum (Fig.16).

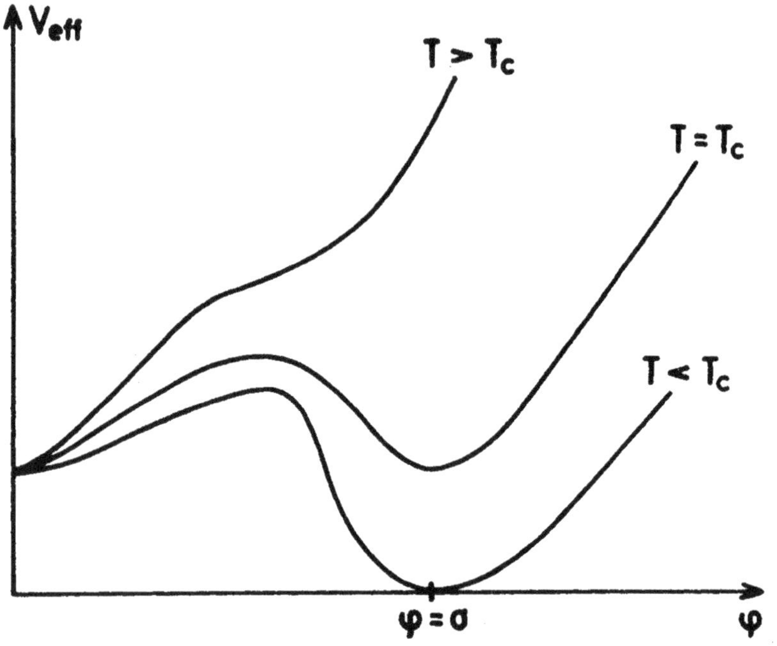

Fig.16 Qualitative graph of the potential $V(\varphi)$ of the classical Higgs-field φ. For $T > T_c$ the state of $\varphi = 0$ is stable. For $T < T_c$ a new stable minimum $\varphi \neq 0$ exists.

The energy difference $\Delta V = V(0) - V(\sigma)$ may become the dominant term in the Friedmann equation. Typical values are derived from the GUT energy scale

$$V(0) - V(\sigma) = (10^{56} \text{ to } 10^{60})(\text{GeV})^4 \simeq T_c^4 \quad .$$

Usually one sets $V(\sigma) = 0$, since $\varphi = \sigma$ should correspond to our present state with its very small cosmological constant. For $K = 0$ we have

$$H^2 = \frac{8\pi G}{3} \left(g(T) a T^4 + a V(0) \right) \quad . \tag{4.12}$$

Neglecting the T^4 term, we obtain a constant H:

$$H_v^2 \equiv \frac{8\pi G}{3} \, aV(0) \quad .$$

Then for some time (as long as the field φ stays at $\varphi = 0$) the expansion factor $R(t)$ increases exponentially $R(t) = R(t_i) \exp H_v(t - t_i)$. After a time interval $\Delta t = (t_f - t_i)$ the Higgs field ends up in $\varphi = \sigma$, and the energy density $V(0)$ is transformed into radiation. If this scheme works as claimed then the reheating up to T_c can produce entropy. In fact, the "entropy problem" is solved when the exponential, inflationary phase leads to an "inflation" of the scale–factor of at least

$$Z = \frac{R(t_f)}{R(t_i)} = 10^{29} \quad . \tag{4.13}$$

The reheating then leads from an initial entropy S_0 to an entropy

$$S = Z^3 \, S_0 = 10^{87} S_0 \quad .$$

The initial entropy can be $S_0 \simeq 1$, and the inflationary phase of the universe will produce the presently observed S. The time–interval Δt has to be long enough that $H_v \Delta t = 29 \ln 10$, hence

$$\Delta t = 7 \times 10^{-10} (\mathrm{GeV})^{-1} = 7 \times 10^9 t_{p\ell} \tag{4.14}$$

(for $T_c \simeq 10^{15}\mathrm{GeV}$).

Another model — the "chaotic" inflationary model[19)] relies on a chaotic distribution of scalar fields close to the Planck time. A mass term $m^2\varphi^2$ is already enough to initiate the inflationary expansion of a small domain of size H^{-1}, if the initial value $\varphi_0 = \frac{M_{p\ell}^2}{m}$ is larger than $M_{p\ell}$.

The classical field φ decreases with time, and for $\varphi < M_{p\ell}/5$ the exponential expansion stops. For small values of φ the original energy density $\propto M_{p\ell}^4$ is transformed into heat (random thermal modes). Again by appropriately chosen m and φ_0 very large growth factors for $R(t)$ can be obtained.

After the inflationary phase the curvature term $\frac{K}{R^2}$ is reduced by a factor of at least 10^{58}. This has the consequence that for reasonable initial conditions the mean density after inflation must be extremely close to the critical one; $\Omega = 1 \pm 10^{-58}$ at the end of inflation, and $\Omega = 1 \pm 10^{-6}$ at the present epoch. As we have seen, the astronomical observations are of sufficient uncertainty to leave $\Omega = 1$ as a viable possibility.

The horizon problem can also be resolved: a region of the size of the horizon at $t \simeq 10^7 t_{pl}$ (about $r_H \approx 10^{-25}$cm) grows to a size $r_H Z = 10^4$cm through the inflation time Δt. But the observable universe has only a radius of 10cm at that time, and is thus inside the dimensions of one lightcone. All regions that we can observe had — in principle at least — the chance of causal contact. The monopole problem is just inflated away, if monopoles were produced before the inflationary phase.

Any other objects are also thinned out exponentially — even the baryons. So the baryon number must be produced again, during or after the reheating.

The exponential expansion does not lead to a decrease in the vacuum energy–density $\varrho_v \simeq V(0)$. Therefore the energy of the expanding system increases tremendously. This, at first sight curious fact, has its roots in the unusual equation of state for the vacuum stuff

$$p_v = -\varrho_v \quad , \tag{4.15}$$

which acts like a substance with a strong tension. Positive energy means negative pressure. This equation of state satisfies the continuity equation

$$(\varrho R^3)^{\cdot} = -p(R^3)^{\cdot}$$

identically.

The expansion energy is gained from the negative pressure. Remember that the exponential expansion phase will only occur, if the energy–momentum density is dominated by the background vacuum energy–density.

To conclude, I want to say that the idea of the inflationary universe in general is very attractive, explains various cosmological puzzles in a natural way, but the detailed models do not inspire much confidence.

4.5 The Puzzle of the Cosmological Constant

Just how far away we are still from a fundamental cosmological theory is illustrated by the deep problem of the cosmological constant. The vacuum does not now support exponential expansion, so the cosmological constant at present must be close to zero. The observational limit is (e.g.[40])

$$\frac{\Lambda}{8\pi G} \leq (10^{-12}\mathrm{GeV})^4 \quad . \tag{4.16}$$

This is more than one hundred orders of magnitude below the GUT vacuum energy–density. Present–day elementary particle theories have transformed the vacuum into a complicated structure of quantum fluctuations and condensations — because of the many spontaneously broken symmetries. Why is this vacuum transparent to gravity? The changes of the energy–densities in the breaking of the GUT, the electroweak gauge symmetry, as well as of the chiral symmetry of QCD are of the order of $(10^{15}\mathrm{GeV})^4$, $(10^3\mathrm{GeV})^4$, and $(10^{-1}\mathrm{GeV})^4$ respectively, Even stronger, and unnatural from the physicist's point of view, than a complete transparency (i.e. $\Lambda = 0$), would be the introduction of a tiny nonzero value of the present–day cosmological constant. But there are reasons to do just that to solve the age problem in the standard model (see section 1). It is no wonder that physicists find this an ugly solution.

A story told occasionally by Wolfgang Pauli (Ch. Enz, A. Thellung) fits in this context: very early in his professional career he had asked himself the question whether the zero–point energy of the electromagnetic field could be gravitationally effective. Therefore he calculated the curvature radius of an Einstein universe with an energy–density given by the zero–point energy of the electromagnetic field with a cut–off at the classical electron radius. He was quite amused to find that this universe "would not even reach out to the Moon".

I am not sure that we have made a lot of progress since then.

5. Observational Relations and Tests of Models

The normal procedure in classical cosmology which we have followed so far too, is to parametrize the cosmological solutions of Einstein's equations in terms of a few fundamental quantities. Then observations are used to obtain fits to these few parameters — for FL–models with matter and radiation there are the quantities $(t_0, H_0, \Omega_0, T_{\gamma_0}, \Lambda, K, R_0)$ with relations between them from the dynamical equations. This seems the best way to proceed, especially because there are so few cosmological observations. It is, however, of interest in principle to investigate how far the cosmological model is determined by observational relations. In order to get anywhere with that approach the models considered have to be restricted considerably. One approach[7)8)] which has been worked out to some degree starts out from the isotropic and homogeneous metric of the FL–models, and examines in detail which observable relations could determine such a kinematic model. The dynamical equations of Einstein's field theory are not used, and thus their validity may even be tested in this approach. The main objective is to bring out clearly the information contained in the various observations, and the logical connections between them. I shall follow here the paper by Ehlers & Rindler[7)].

5.1 Observational Relations

The Robertson–Walker metric

$$\begin{aligned} ds^2 &= dt^2 - R^2(t)\,(d\chi^2 + f^2 d\Omega^2) \\ \text{with } f &= (\sin\chi, \chi, \sinh\chi) \text{ for } K = (+1, 0, -1) \quad , \\ \text{and } d\Omega^2 &= d\theta^2 + \sin^2\theta d\varphi^2 \quad , \end{aligned}$$

is used as in section 1.

This metric is isotropic, and therefore the information reaching us along the past lightcone is essentially determined by the information on one incoming light ray (Fig.17). Let us define $\chi = 0$ as the world line of our galaxy, t_0 as the present time. The area of a sphere $f =$ const. at time t is $4\pi R^2(t)\, f^2(\chi)$. This leads to the definition of an area distance

$$r_A = R(t)\, f(\chi) \quad . \tag{5.1}$$

The calculation of $r_A(z)$ in terms of q_0, H_0, and z can be done analytically for $\Lambda = 0$. For a galaxy at χ_1 emitting light at $R(t_1)$ corresponding to the redshift z, we have $r_A = \chi_1 R_1 = R_0\, \chi_1\, (1+z)^{-1}$, with

$$\begin{aligned} \chi_1 &= \int_{t_1}^{t_0} ds/R(s) = \int_{R_1}^{R_0} \frac{dR}{R\dot{R}} \\ &= \frac{1}{R_0 H_0} \int_{(1+z)^{-1}}^{1} \left[1 - 2q_0 + \frac{2q_0}{x}\right]^{-1/2} \frac{dx}{x} \quad . \end{aligned}$$

Thus

$$\chi_1 = \{z q_0 + (q_0 - 1)\left[-1 + \sqrt{2q_0 z + 1}\right]\} H_0^{-1} R_0^{-1} q_0^{-2} (1+z)^{-1}$$

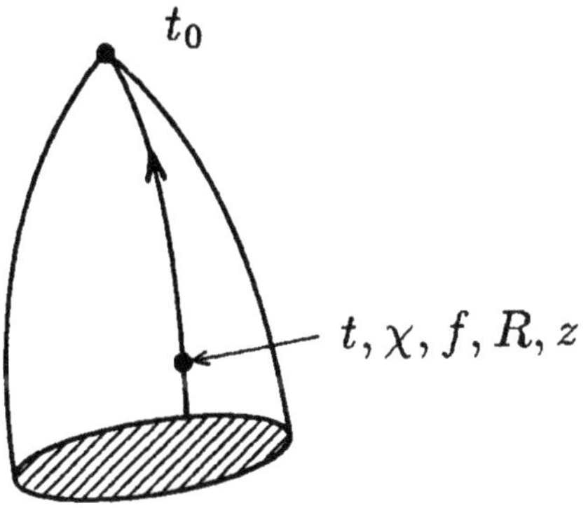

Fig.17 In FL-models information registered at t_0 rides essentially on one incoming light-ray, since the space-time is completely isotropic.

and

$$r_A = \chi_1 R_1 = H_0^{-1} q_0^{-2} (1+z)^{-2} \{ z q_0 + (q_0 - 1) \left[-1 + \sqrt{2 q_0 z + 1} \right] \} \quad .$$

On the incoming null ray

$$\frac{d\chi}{dt} = -1/R \quad . \tag{5.2}$$

Assuming that the universe was expanding, $\dot{R}(t) > 0$, for some time $t \le t_0$, at least, then t, χ, R, z are strictly monotonic functions of each other in that range. $f(\chi)$ can be expressed as a function of any one of these variables.

Of these variables only $z = \frac{R_0}{R} - 1$ is given directly by observations of distant galaxies. Other directly observable quantities are the bolometric flux S, the solid angle ω subtended by the object, and $N(z)$, $N(S)$, the number of galaxies with redshift $\ge z$ (or flux $\ge S$, respectively).

Each of the observables is related to a quantity describing an intrinsic property of the objects observed. S is related to the (bolometric) luminosity L of a standard galaxy, ω to the cross-sectional area A of a galaxy orthogonal to the line of sight, and N to the number n of galaxies per unit volume of space. L, A, and nR^3 change because of the evolution of galaxies. We consider $L(t)$, $A(t)$, and $n(t)$ as unknown functions of cosmic time t.

Within the FL-models the following relations hold

$$S = L(1+z)^{-2} R_0^{-2} f^{-2} / 4\pi \quad , \tag{5.3}$$

$$\omega = A(1+z)^2 R_0^{-2} f^{-2} \quad , \tag{5.4}$$

$$dN = n R_0^3 f^2 (1+z)^{-3} d\chi \quad . \tag{5.5}$$

These relations are the basis for all the different empirical tests which we will mention briefly in section 5.2.

From a given model, we can determine $t(z)$, $f(z)$. Then, if $L(t)$ is known, $L(t(z))$ can be calculated, and eq. (5.3) gives an $S - z$ relation predicted by the model.

Similarly $N(z)$ and $\omega(z)$ relations may be derived. Let us reverse the question, and ask, whether observed relations $S(z)$, $\omega(z)$, and $N(z)$ determine the cosmological model. The answer is no. Any 3 functions $S(z)$, $\omega(z)$, and $N(z)$ are compatible with a given function $R(t)$, and a given curvature index K: the model $[R(t); K]$ determines $z(t)$, $f(t)$, and finally $S(z)$ gives $L(t)$, ω gives $A(t)$, and $N(z)$ leads to $n(t)$. The given model is completed by the observable functions, but not restricted.

It is interesting to look at the possibility of knowing one pair of functions, e.g. $[S(z), L(t)] - S(z)$ from observations, $L(t)$ from theory. At the present time this is still a speculation, but future more detailed observations and better understanding of the physical processes in galaxies may lead to such knowledge. This case leads much further. Given $L(t)$, we obtain from eq. (5.3) $(1+z)^2 S(z)$ as a function of t. Hence we have $z = z(t, \chi; R_0; K)$ and from it $\chi = \chi(t; R_0, K)$ by solving $\frac{d\chi}{dt} = -\frac{1}{R_0}(1+z)$.

Finally we obtain R as a function of t from

$$R(t : R_0, K) = -\left(\frac{d\chi}{dt}\right)^{-1} \quad . \tag{5.6}$$

For any choice of R_0 and K the function $R(t)$ is thus determined. To find R_0 and K we need the pair $[N, n]$ and one of the other data pairs. To see this, the following result from differential geometry is used[7]: the Gaussian curvature κ of a 3–space of constant curvature is given by

$$\kappa = \frac{20\pi}{3} \lim_{A \to 0} \frac{36\pi V^2 - A^3}{A^4} \quad . \tag{5.7}$$

At the present epoch $\kappa_0 \equiv \frac{K}{R_0^2}$ for a FL–model. V is the volume, A the surface area of a geodesic sphere $f = $ const. at t_0.

$$A = 4\pi R_0^2 f^2 \quad ;$$

$$V = \int n^{-1}(t)(1+z)^3 N'(z)\, dz \quad .$$

This could be integrated, if we knew $t(z)$, so as to have $n(t)$ as a function of $z : n(t(z))$.

To get $t(z)$, we take a look again at the relations (5.3) to (5.5). By eliminating $R_0^2 f^2$ pairwise, we arrive at model–independent identities:

$$4\pi(1+z)^4\, S/\omega = L/A \quad , \tag{5.8}$$

$$\omega(z)\, N'(z)\, dz = -n(t)\, A(t)\, dt \quad , \tag{5.9}$$

$$4\pi(1+z)^4\, S(z)\, N'(z)\, dz = -n(t)\, L(t)\, dt \quad . \tag{5.10}$$

Now, suppose we knew $[S(z), L(t)]$ and $[N(z), n(t)]$. By integrating eq. (5.10) we get $t(z)$ and then determine the model. The function $A(t)$ would then also be determined, and can serve as a test on the basic assumptions, e.g. homogeneity.

If we can find κ_0 independently, e.g. by measuring the parallax of one galaxy, then $[S, L]$ together with the observables $\omega(z)$, $N(z)$ is sufficient to determine the model *and* the intrinsic properties $A(t)$, $n(t)$.

This line of arguments may be an interesting way also to test models of galaxy formation, once they become sufficiently detailed.

5.2 Cosmological Tests

The hope is to determine one favoured FL model from these tests. High redshift objects are most suitable for this purpose.

The assumptions usually made are

a) a conservation law for the galaxies

$$n_G R^3 = \text{const.} \quad ,$$

and b) the existence of a luminosity function[29)]

$$\Phi(L)\, dL = \Phi^*(L/L_*)^\alpha \exp(-L/L_*)\, d(L/L_*) \tag{5.11}$$

giving the number of galaxies with luminosity between L and $L+dL$. It is assumed that α is a constant, but a z–dependence of L_* is taken into consideration $L_* = L_*(z)$.

But the effects of galaxy evolution, the uncertainties in the assumptions made about star formation, as well as merging effects lead to $n_G R^3$ not being constant, to luminosity evolution, to $\alpha = \alpha(z)$. The tests are screwed up by these effects.

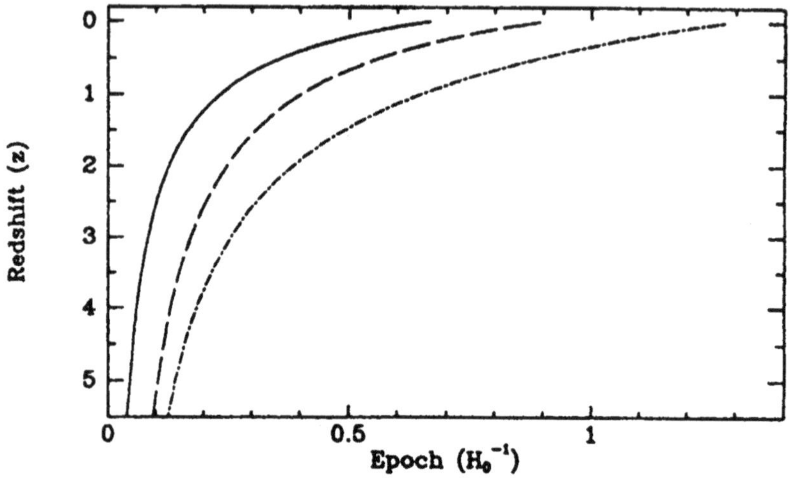

Fig.18 $t(z)$ for various FL-models: solid line ($\Omega_0 = 1$, $t_0 = 0$); dashed line ($\Omega_0 = 0.1$, $\lambda_0 = 0$); dot-dashed line ($\Omega_0 = 0.1$, $\lambda_0 = 0.9$).

A) At first we show the relation of time and redshift $t(z)$ for various FL models $(\Omega_0, \lambda_0) = ([1,0]$ — solid; $[0.1,0]$ — dashed; $[0.1,0.9]$ — dot — dashed) (Fig.18).
B) The angle $\theta(z)$ under which a source of linear extent D at redshift z appears, is given by $\theta(z)\, r_A = D$ (compare eq. (5.4)); i.e.

$$\theta(z) = D(1+z)\, R_0^{-1} f^{-1} \quad . \tag{5.12}$$

This is really the test of eq. (5.4) with $D = \sqrt{A}$, etc. For sources of standard size D, the angle $\theta(z)$ is plotted in Fig.19 for 3 different FL–models. To obtain $\theta(z)$, $R_0\, f(\chi)$ has to be computed in terms of z, H_0, q_0, Ω_0 — see section 1.

There has been a claim that certain short–lived jet structures in VLBI sources may define such a standard size[46)]. The claim was that the fit would lead to $\Omega_0 = 1$, $\lambda_0 = 0$. Subsequent observations have failed to substantiate this.

A similar test was tried for E–galaxies. A photometrically defined radius r_E satisfies

$$\log r_E = -0.58 M_B + \text{const.}$$

To get r_E within a factor 2, one needs to determined M_B to an accuracy of $\Delta M_B = 0.5$. So far there is no reasonable result from this test. For radiosources one has found the unexpected behaviour of $\theta(z) \propto 1/z$ even for very high redshifts.

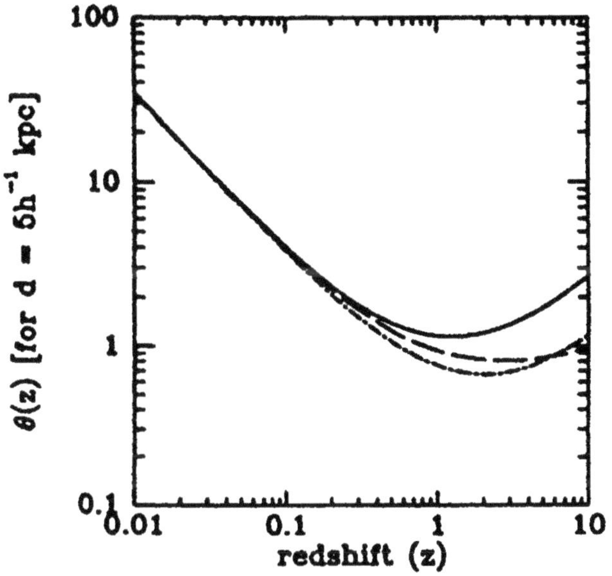

Fig.19 $\theta(z)$ for different models (parameters as in Fig.18).

This is the flat space Newtonian result. It is strange that evolution and selection effects should conspire in a way which leads to $\theta \propto 1/z$.

C) There is a strong dependence of $\frac{dN}{dz}$ or $\frac{dV}{dz}$ on the cosmological constant λ_0, as can be seen from Fig.20. There are claims that the counts of faint galaxies can be fitted much better by a $\lambda_0 \neq 0$ model. Again it is not yet clear whether this result stands up to further investigation. Normalization problems and evolution, especially the assumption of constant $n_G R^3$ have to be clarified.

D) Loh and Spillar [22]have devised an interesting test to estimate Ω_0 from a large and deep sample of galaxies. Their estimate involves assumptions about the luminosity function which are not easy to justify. Especially the condition $\alpha = const.$ is quite stringent. Relaxing it to a z-dependence of α can change their estimate of $\Omega_0 \simeq 1$ drastically to smaller values of Ω_0.

6. Final Remarks

The global structure and evolution of the cosmos is difficult to decipher, because we, the observers, belong to it, and because we can observe only a tiny —

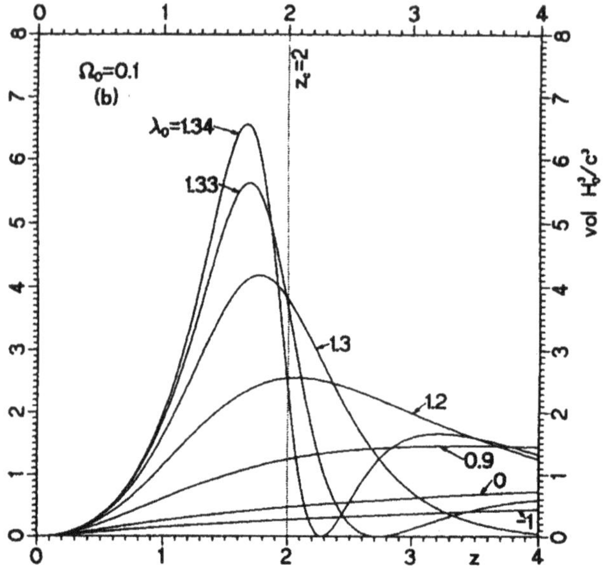

Fig.20 The volume element $\frac{dV}{dz}$ (normalized by H_0^{-3}) as a function of redshift z for $\Omega_0 = 0.1$ for different values of λ_0 [10].

cosmically viewed completely local — section in space and time. Therefore important quantities, as e.g. the Hubble constant, are not known very precisely, and many processes such as the evolution of galaxies are understood only in a preliminary way. We also do not know, whether the section we can observe is typical for the whole universe.

This problem can be discussed only in terms of a model. How could we talk of the "whole cosmos" without a theoretical picture of it? Another difficulty lies in the fact that experiments as in laboratory physics are not possible. You cannot experiment with stars and galaxies. There is only the gentle way of gaining knowledge by patient registration of the radiation and particle flow in telescopes and counters on earth or in space. These measurements are, of course, more and more uncertain as the "emitters" are more and more distant. Therefore the cosmologist designs a model of the world from theory and measurements, and only within this model can he interpret the observations. New observations can only be proposed within the framework of this model. The whole resembles a puzzle with quite a few parts missing.

Therefore one should not make off-hand changes in the basic assumptions, just to complete a part of the puzzle. The physical explanation of the universe rests on the simple principle, that the laws and physics found in lab experiments, are also valid in distant cosmic environments. This involves an extrapolation to areas which are not directly accessible to experiments. The pragmatic approach to cosmology sticks to known physical laws and tries to fit a "best model" by an optimal adaption to astronomical observations.

There are, of course, many interesting questions left out in such an approach, as for instance the question why the universe is the way it is, i.e. why different possibilities — according to the laws of nature — are not realized. Attempts to single out one special model by quantum–cosmological constraints, or the appeal to an anthropic principle which states that life as we know it can only exist in a very specially tuned universe like ours, try to go further in their explanation of the cosmos. These attempts have so far not produced any new, testable insights that go beyond the pragmatic approach.

The essential theoretical background of the classical models lies in Einstein's Theory of General Relativity, a theory of gravitation which has proved its validity in theory and in experiments. There are three main advantages of General Relativity as compared to Newton's theory:

1. The gravitational action of infinite mass distributions can be described without any problem. (In Newtonian theory the gravitational potential for such a case would be infinite at each point.)
2. The theory has solutions which correspond to simple models of the universe, e.g. closed finite space without boundary.
3. The propagation of light is incorporated in agreement with experiments.

Simple, highly–symmetric solutions of the theory are used to describe the uniformly expanding universe. Each of these Friedmann–Lemaître models is determined by the values of a few, in principle measurable quantities (such as the present expansion velocity of the system of galaxies, the mean density, the temperature of the background radiation field). As we have seen the cosmologically relevant observations can be accommodated reasonably well by these models, but the measurement uncertainties still allow a range of models which are even qualitatively different. Nevertheless we talk of a "Standard Big–Bang model", and we include in that term a class of models with several common typical properties: the universe has passed from an initial, rapid, explosion–like expansion through a hot and dense early phase to the present state. It is globally homogeneous and isotropic, and inhomogeneous structures, such as galaxies and clusters of galaxies, have formed from initially small density perturbations by gravitational instability. It is remarkable, how far we have come with these simple concepts. There does not exist up to now an alternative theory which can integrate all the observations about the global cosmic structure and evolution better or as well as the standard model. The existing alternate theories do not explain things better, they just shift the assumptions. On closer examination they have all the problems of the standard model, and several additional ones.

The belief in the standard model is also strengthened by a negative fact: there are observations which would disprove the model immediately, and which could be done with present technology, but which have not been made. These include, e.g.,

a) a stable, weakly interacting massive particle of mass ~ 17 keV,

b) a star with very little helium, or

c) an object which is $\sim 30 \times 10^9$ years old.

The problems and open questions of the standard model are two-fold: firstly we do not really know the properties of the dark matter, its composition and density, although the formation of structure in all current models is dominated by this component of cosmic matter. Thus these theories are built on shaky foundations, despite the good agreement with observations. Secondly, we do not understand how the initial perturbations of the density are created. This question can probably be answered only, when quantum processes in the very early universe, such as phase transitions and vacuum fluctuations are analysed exactly. This is connected with the fundamental unsolved problem, how to join the Theory of General Relativity and Quantum Theory to a whole, consistent theoretical structure. The earliest epochs of the universe and its singular beginning can only be analysed in a reasonable way, if we succeed in formulating this theory.

Modern cosmology displays the fascinating picture of a universe in dynamical evolution where the physics of elementary particles has decisive consequences for the global structure of space–time. This rapidly developing and changing field of research is good for more surprises and insights.

References

1. Arp H.C., *et al.* , 1990, *Nature* **346**, 807
2. Böhringer H., 1994, in "X-Ray Clusters of Galaxies" NASI **441**, p.123 (Kluwer, Seitter W.C. (ed.))
3. Börner G., Ehlers J., 1988, *Astron. Ap.* **204**, 1
4. Caroll S.M., Press W.H., Turner E.L., 1992, *Ann. Rev. Astron. Ap.* **30**, 499
5. David L.P., *et al.* , 1994, *Ap. J.* **428**, 544
6. Efstathiou G., Ellis R.S., Peterson B.A., 1988, *M.N.R.A.S.* **232**, 431
7. Ehlers J., Rindler W., 1987, *Astron. Ap.* **174**, 1
8. Ellis G.F.R., *et al.* , 1985, *Phys. Rev.* **124**, 315
9. Fahlman G.G., *et al.* , 1994, *Ap. J.* **437**, 56
10. Feige B., 1991, Diplomarbeit, Univ. Münster
11. Friedmann A., 1992, *Z. f. Physik* **10**, 377
12. Guth A.H., 1981, *Phys. Rev.* **D23**, 347
13. Harrison E.R., 1970, *Phys. Rev.* **D1**, 2726
14. Hawking S.W., Ellis G.F.R., 1973, "The Large Scale Structure of Spacetime" (Cambridge Univ. Press)
15. Hawking S.W., Penrose R., 1970, *Proc. Roy. Soc. London* **A314**, 529
16. Henriksen M.J., Mamon G.A., 1994, *Ap. J.* **421**, L63
17. Jüttner F., 1928, *Z. f. Physik* **47**, 542
18. Lemaître G., 1927, *Ann. Soc. Sci. Bruxelles* **47A**, 49
19. Linde A.D., 1985, *Comments Ap. Space Phys.* **16**, 229
20. Linde A.D., 1990, "Particle Physics and Inflationary Cosmology" (Harvard Acad. Publ.)
21. Lineweaver C.H., *et al.* , 1994, preprint astro-ph/9403021
22. Loh E.D., Spillar E.J., 1986, *Ap. J.* **307**, L1
23. Maoz D., Rix H.-W., 1993, *Ap. J.* **416**, 425
24. Mather J.C., *et al.* , 1990, *Ap. J.* **354**, L37
25. Mather J.C., *et al.* , 1994, *Ap. J.* **420**, 439
26. Mo H.J., Jing Y.P., Börner G., 1993, *M.N.R.A.S.* **264**, 825

27. Ramella M., Geller M.J., Huchra J.P., 1989, *Ap. J.* **344**, 57

28. Sato K., 1981, *M.N.R.A.S.* **99b**, 66 ; *M.N.R.A.S.* **195**, 198

29. Schechter P.L., 1976, *Ap. J.* **203**, 297

30. Schindler S., 1995a, *M.N.R.A.S.* (in press)

31. Schindler S., 1995b, *Astron. Ap.* ,

32. Schneider P., Seitz S., 1995, *Astron. Ap.* ,

33. Smoot G.F., *et al.* , 1977, *Phys. Rev. Lett.* **39**, 898

34. Smoot G.F., *et al.* , 1992, *Ap. J.* **396**, L1

35. Songaila A., *et al.* , 1994, *Nature* **371**, 43

36. Steigman G., 1979, *Ann. Rev. Nucl. Part. Sci.* **29**, 313

37. Sunyaev R.A., Zel'dovich Ya.B., 1980, *Ann. Rev. Astron. Ap.* **18**, 537

38. Tegmark M., Bunn E.F., 1995, *Ap.J.* (in press)

39. Weinberg S., 1972, "Gravitation and Cosmology" (Wiley, N.Y.)

40. Weinberg S., 1989, *Rev. Mod. Phys.* **61**, 1

41. Weinberg S., "The First 3 Minutes" (Deutsch, London)

42. White S.D.M., *et al.* , 1993, *Nature* **366**, 429

43. Wright E.L., 1991, *Proc. Texas Symp.*, Brighton UK (1990)

44. Wright E.L., *et al.* , 1994, *Ap. J.* **420**, 1

45. Yentis D.J., *et al.* , 1992, in *"Digitised Optical Sky Surveys"* (Kluwer, eds.: MacGillivray H.T. and Thomson E.B.), p.67

46. Kellerman K.I,, 1993, *Nature* **361**, 134

Textbooks which contain a presentation of Classical Cosmology:

Number 14, 20, 39.

Börner G., 1993, *"The Early Universe"* (Springer, 3rd edition)

Peebles P.J.E., 1993, *"Physical Principles of Cosmology"* (Princeton Univ. Press)

Peebles P.J.E., 1980, *"The Large Scale Structure of the Universe"* (Princeton Univ. Press)

PARTICLE PHYSICS APPLIED TO COSMOLOGY

John Ellis

Theoretical Physics Division, CERN
CH-1211 Geneva 23

I The Early Universe and the Standard Model

I.1 The first second - a general overview

To understand the very early history of our Universe, we need to understand elementary particles and their interactions. This is because, according to standard Big Bang Cosmology, during the first second the typical thermal energies of the constituents of the Universe were above 1 MeV, i.e., above the rest mass of the electron and at temperatures where nuclei dissolve into protons and neutrons. Back at around 10^{-6} to 10^{-5} s after the start of the Big Bang , protons and neutrons would themselves have been dissolved into their constituent quarks. The previous event of note was around 10^{-10} s, when the weak and electromagnetic interactions would have had comparable strengths, and hence been effectively unified as in the Standard Model of particle physics. The physics involved so far is well established and tested at accelerators: going back earlier in time requires progressively more speculation. Sometime between 10^{-10} and 10^{-12} s, the particles of the Standard Model would have lost their masses as the electroweak vacuum was heated up and its symmetry restored. The matter of the Universe may have been created at any time between this electroweak phase transition and the phase transition around 10^{-35} s or so when the expected grand unified symmetry would have been restored. This may also be when cosmological inflation took place, according to which the present density of the Universe is essentially critical. This requires most of the stuff in the Universe to be non-baryonic Dark Matter, for which the most plausible candidates may be neutrinos, present in the Standard Model but endowed with masses by grand unification, and/or the lightest supersymmetric particles.

To understand the framework for this sequence of events, it is useful to review first some basic facts about mass, energy, temperature and time in the early Universe. Like any gas, the constituents of the Universe have generally cooled as it expanded, with temperature $T \propto a^{-1}$, where a is the scale factor, during the adiabatic expansion that has dominated most of the Universe's history. While the cosmic gas is composed of relativistic particles, which is the case for most of the first second, the age

$$t \propto a^2 \propto T^{-2} \tag{1.1}$$

Numerically, the temperature of the Universe would have been about 10^{10} K when its age was 1s. This corresponds to thermal energies of about 1 MeV, comparable to the electron mass 0.511 MeV, whereas the proton mass of 938 MeV (about 1 GeV) corresponds to $T \sim 10^{13}$ K. The basic time-temperature relation (1.1) can be written as

$$t(s) = O(1)/T(MeV)^2 \tag{1.2}$$

with the constant of proportionality depending on the number of particle species.

The mathematical framework for this physical picture is provided by the basic equations of Friedman-Robertson-Walker cosmology [1], in particular the equation

$$\left(\frac{\dot{a}}{a}\right)^2 + \frac{k}{a^2} = \frac{8\pi\ G_N}{3}\ \rho \tag{1.3}$$

which determines the Hubble expansion rate $H \equiv \dot{a}/a$. The second term on the left-hand side is the curvature term: the present Hubble expansion rate is close to that for a flat $k = 0$ Universe, for which the explanation may be inflation. On the right-hand side, we find Newton's constant $G_N \equiv 1/m_P^2$, where $m_P \simeq 1.22 \times 10^{19}$ GeV is the Planck mass, and ρ denotes the energy density. If $k = 0$, the Universe has the critical density $\rho = \rho_c$ and

$$H = \sqrt{\frac{8\pi\ G_N}{3}\ \rho_c} \tag{1.4}$$

where $\rho_c \simeq 2 \times 10^{-29} h^2 gcm^{-3}$, with $h = H/(100kms^{-1}Mpc^{-1})$.

For relativistic particles in thermal equilibrium [2],

$$\rho \simeq \frac{\pi^2}{30}\ T^4 g\ :\ g = \#_B + \frac{7}{8}\#_F \tag{1.5}$$

where in the Standard Model the number of boson particle states (see Lectures II and III)

$$\#_B \;=\; \underset{\text{(Higgs)}}{4} \;+\; \underset{\text{(Gluons)}}{(8 \times 2)} \;+\; \underset{\text{(electroweak gauge)}}{(4 \times 2)} \;=\; 28 \tag{1.6}$$

and the number of fermion particle states

$$\#_F = 30\ N_G \tag{1.7}$$

where N_G is the number of matter generations. We also recall that for relativistic particles the pressure

$$p = \frac{1}{3}\rho \tag{1.8}$$

and that the number densities of bosons and fermions are

$$n_{B,F}(\underline{k}, T) = \frac{\#_{B,F}}{(2\pi)^3}\ \frac{1}{\exp(|\underline{k}|/T) \mp 1} \tag{1.9}$$

We have been assuming that thermal equilibrium was established in the early Universe: is this plausible? To answer this question, we should compare the Hubble expansion rate $H \sim T^2/m_P$ with typical interaction rates

$$\Gamma \;\sim\; \underset{(1 \leftrightarrow 2 \text{ particles})}{\alpha T} \;,\; \underset{(2 \leftrightarrow 2 \text{ particles})}{\alpha^2 T} \tag{1.10}$$

where (for example) the latter is obtained by estimating a typical thermal cross-section $\sigma(T) \sim \alpha^2/T^2$, where α is the fine structure constant, and using the fact that any relativistic particle number density $n(T) \sim T^3$. Equilibrium is maintained if $H \lesssim \Gamma$, i.e.,

$$T \lesssim \alpha m_P \tag{1.11}$$

However, interactions do drop out of equilibrium if they are mediated by massive particles. For example, for weak interactions mediated by massive W bosons in the Standard Model , $\sigma_W(T) \sim \alpha^2 T^2/m_W^4$ when $T \ll m_W$. Hence

$$\Gamma_W(T) \equiv \sigma_W(T) n(T) \sim \frac{\alpha^2 T^5}{m_W^4} < H \sim \frac{T^2}{m_P} \tag{1.12}$$

when

$$T \sim (m_W^4/m_P \alpha^2)^{1/3} \sim 1 \text{ MeV} \tag{1.13}$$

Therefore the weak interactions are believed to have maintained equilibrium during all the first second, except perhaps at the very beginning of the Big Bang [3].

It is useful to recall the effective horizon size $a_H = 2c.t$ of the Universe at different epochs during its expansion. At the quark-hadron transition when $T_c^{QCD} \simeq 150$ MeV, $t_c^{QCD} \simeq 2 \times 10^{-6}$ s and $a_H \simeq 9$ km, which has now expanded to about 1/2 ly. At the electroweak transition when $T_c^{EW} \simeq 150$ GeV, $t_c^{EW} \simeq 2 \times 10^{-12}$ s and $a_H \simeq 6.3$ mm, which expanded to 6m at t_c^{QCD}, and has expanded to about 20 astronomical units now.

I.2 Electroweak unification

In the laboratory, we see as force cariers the massless photon γ and the massive $W^{\pm}(\sim 80$ GeV) and Z^0 (91.187(7) GeV). Correspondingly, electromagnetic forces have infinite range, whilst the weak forces have a short range $\sim 10^{-16}$ cm $\sim m_{W,Z}^{-1}$. Why are the γ and W/Z masses so different, given the similarities in their spins (1) and couplings ($\alpha_W \sim 1/30$ cf. $\alpha_{em} \sim 1/137$)? The elementary particles - quarks and leptons - on which the electroweak forces act are displayed below together with their masses:

$$\begin{array}{lcccccc} \text{Quarks} & \begin{pmatrix} u \\ d \end{pmatrix} & \begin{array}{c} \sim 5 \text{ MeV} \\ \sim 8 \text{ Mev} \end{array} & \begin{pmatrix} c \\ s \end{pmatrix} & \begin{array}{c} \sim 1.5 \text{ GeV} \\ \sim 0.15 \text{ GeV} \end{array} & \begin{pmatrix} t \\ b \end{pmatrix} & \begin{array}{c} \sim 140 \text{ GeV ?} \\ \sim 5 \text{ GeV} \end{array} \\ \text{Leptons} & \begin{pmatrix} \nu_e \\ e \end{pmatrix} & \begin{array}{c} \lesssim 7 \text{ eV} \\ \sim \frac{1}{2} \text{ MeV} \end{array} & \begin{pmatrix} \nu_\mu \\ \mu \end{pmatrix} & \begin{array}{c} \lesssim \frac{1}{4} \text{ MeV} \\ \sim 0.1 \text{ GeV} \end{array} & \begin{pmatrix} \nu_\tau \\ \tau \end{pmatrix} & \begin{array}{c} \lesssim 30 \text{ MeV} \\ \sim 1.78 \text{ GeV} \end{array} \end{array} \tag{1.14}$$

where the matter particles are paired in doublets of electroweak isospin. The upper limits on the neutrino masses are discussed in Lecture III, together with grand unified ideas about their possible magnitudes that suggest that ν_τ is a good candidate for the Hot Dark Matter discussed in Lecture V. LEP experiments tell us that there are no more than three light neutrino species, as assumed in conventional Big Bang nucleosynthesis calculations (see Lecture II) and hence presumably three charged leptons. LEP data test and verify the electroweak sector of the Standard Model with a precision of a few per mille, as discussed in Section I.3.

To accommodate the above-mentioned gauge bosons using SU(2)× U(1) gauge symmetry [4], we need two gauge boson kinetic terms

$$\mathcal{L}_G = -\frac{1}{4}\left(G_{\mu\nu a}G^{\mu\nu a}\right) - \frac{1}{4}\left(F_{\mu\nu}F^{\mu\nu}\right) \tag{1.15}$$

where

$$F_{\mu\nu} \equiv \partial_\mu B_\nu - \partial_\nu B_\mu \tag{1.16}$$

for the U(1) of the electroweak hypercharge, and where

$$G_{\mu\nu a} \equiv \partial_\mu W_{\nu a} - \partial_\nu W_{\mu a} + ig\epsilon_{abc}W_{\mu b}W_{\nu c} \tag{1.17}$$

for the SU(2) of electroweak isospin. Expanding the latter, we obtain three-boson and four-boson interactions, as well as the bilinear terms.

The fermion kinetic terms in the Standard Model take the form

$$\mathcal{L}_F = -\sum_{f,L,R} i\left[\bar{f}_L\,\gamma^\mu D_\mu f_L + \bar{f}_R\,\gamma^\mu D_\mu f_R\right] \tag{1.18}$$

where the covariant derivative

$$D_\mu = \partial_\mu - ig\,\frac{\tau_a}{2}\,W_{\mu a} - ig'Y\,B_\mu \tag{1.19}$$

Note that the gauge interactions conserve helicity, that the term proportional to τ_a is absent for the f_R, reflecting the violation of parity, and that the values of the hypercharges Y will be fixed later. The g and g' are the gauge couplings for the SU(2) and U(1) gauge group factors, respectively. Expanding the fermion kinetic terms, we find in addition to bilinear terms also trilinear terms, in which a fermion emits a gauge boson while conserving its helicity.

We also need a scalar field, called the Higgs field ϕ, which is a doublet of electroweak SU(2), and has the kinetic term

$$\mathcal{L}_\phi = -|D_\mu\phi|^2 \tag{1.20}$$

Expanding this, we obtain, in addition to the usual bilinear terms, also trilinear terms with one gauge boson coupled to two fields ϕ, and quadrilinear terms with two boson fields coupled to two fields ϕ. The scalar field ϕ is also coupled to the fundamental quarks and leptons via what are called Yukawa terms

$$\mathcal{L}_Y = -\sum_{f,f'} \left[h_{ff'} \bar{f}_L \phi f'_R + h^*_{ff'} \bar{f}'_R \phi^\dagger f_L \right] \tag{1.21}$$

Here $h_{ff'}$ is an arbitrary matrix in flavour space. Note that these ϕ interactions flip helicity from left to right. When the Higgs field ϕ acquires a vacuum expectation value (v.e.v.), which it will do shortly, these Yukawa terms may be expanded to give masses $m_f \propto h_{ff'}$, as well as Higgs-fermion-antifermion couplings.

The final part of the Standard Model Lagrangian is the Higgs potential

$$\mathcal{L}_V \equiv -V(\phi) \ : \ V(\phi) = -\mu^2 \phi^\dagger \phi + \frac{\lambda}{2} (\phi^\dagger \phi)^2 \tag{1.22}$$

This can be expanded to give a mass term and quadrilinear interactions. You can regard the field ϕ as resembling a glorified harmonic oscillator, with the second term in the potential corresponding to an anharmonic term. The energy of the theory is minimized by going to the lowest point in the effective potential, which determines the vacuum of the theory. Since the quadratic (harmonic) term has a negative coefficient, the origin $\phi = 0$ is unstable, and the minimum of the potential is located at

$$\langle 0|\phi|0\rangle = \langle 0|\phi^\dagger|0\rangle = v \begin{pmatrix} 0 \\ 1/\sqrt{2} \end{pmatrix} \ : \ v^2 = \frac{\mu^2}{2\lambda} \tag{1.23}$$

The Higgs potential is sketched in Fig. 1, where we see that there is a non-trivial set of minima which are all at the same height. All of these vacua are equivalent, and it is a choice of one of them which breaks the symmetry. The appearance of this v.e.v. means that the electroweak $SU(2) \times U(1)$ symmetry is spontaneously broken, since the Higgs field is not invariant under the electroweak symmetry transformations. It should be noted here that it is convention which puts the v.e.v. v of the Higgs field in the lower part of the doublet.

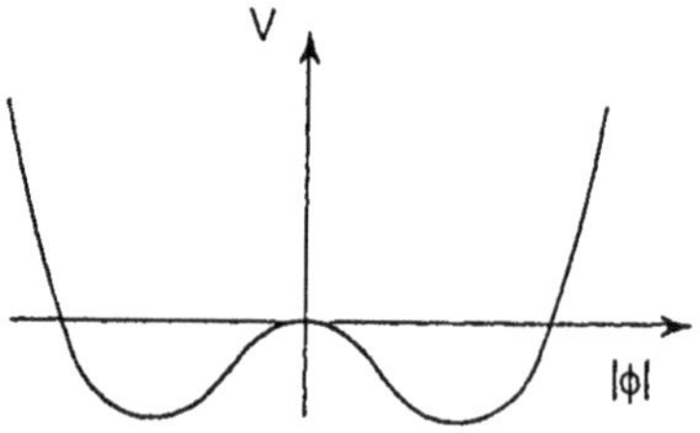

Figure 1: The form of the Higgs potential in the electroweak theory.

It is this symmetry breaking which provides the mechanism for mass generation in the Standard Model . If we substitute

$$\phi = \langle 0|\phi|0\rangle + \hat{\phi} \tag{1.24}$$

into the Higgs kinetic term $\mathcal{L}_\phi$, we find that the quadrilinear $\phi^2 v^2$ terms now give us quadratic terms

$$\mathcal{L}_\phi \ni -\frac{g^2 v^2}{2}\, W^+_\mu W^\mu - g'^2\, \frac{v^2}{2}\, B_\mu B^\mu + g g' v^2 B_\mu W^3_\mu - g^2\, \frac{v^2}{2}\, W^3_\mu W^3_\mu \tag{1.25}$$

The first of these gives a mass to the charged $W^\pm$ bosons

$$m_W = g\, \frac{v}{2} \tag{1.26}$$

The other three terms give masses to the neutral gauge bosons B_μ and W^3_μ. Recall that a gauge boson mass term $A_\mu A^\mu$ is not in gauge invariant by itself: however, since we have obtained these terms from what was originally a gauge-invariant kinetic term for the Higgs fields, this mechanism of mass generation is in fact gauge-invariant.

The neutral gauge boson masses require a bit more discussion than those of the charged bosons. We see in Eq. (1.25) above a matrix of terms coupling the fields B_μ and W^3_μ

$$\begin{pmatrix} \frac{g'^2 v^2}{2} & -\frac{g g' v^2}{2} \\ -\frac{g g' v^2}{2} & \frac{g^2 v^2}{2} \end{pmatrix} \tag{1.27}$$

It is easy to check that, when diagonalized, this matrix gives one massive combination

$$Z^0_\mu \equiv \frac{g W^3_\mu - g' B_\mu}{\sqrt{g^2 + g'^2}} \quad : \quad m_Z = \frac{1}{2}\, \sqrt{g^2 + g'^2}\;\; v \tag{1.28}$$

which we will soon interpret as the Z^0 boson explored at LEP and elsewhere, and one massless combination,

$$A_\mu \equiv \frac{g W^3_\mu + g' B_\mu}{\sqrt{g^2 + g'^2}} \tag{1.29}$$

which is the familiar photon of electrodynamics. We see that the photon is a combination of W^3_μ and B_μ, and therefore couples to the combination

$$Q_{em} = I_3 + Y \tag{1.30}$$

of weak isospin and hypercharge. This tells us how to choose the values of the hypercharges for the different fields: they must be chosen such that, when combined with the values of I_3, they together give the known electromagnetic charges of the elementary particles. The photon has a coupling of strength

$$e = \frac{g g'}{\sqrt{g^2 + g'^2}} \tag{1.31}$$

which is a combination of the SU(2) coupling g and the U(1) coupling g'. It is convenient to parametrize these couplings in terms of a mixing angle θ_W, which is related to the couplings g' and g by

$$\tan\theta_W = \frac{g'}{g}\ , \quad \cos\theta_W = \frac{g}{\sqrt{g^2+g'^2}}\ , \quad \sin\theta_W = \frac{g'}{\sqrt{g^2+g'^2}} \tag{1.32}$$

Referring back to the previous expressions for $m_{W^\pm}$ and m_{Z^0}, we find the following relationship between their masses

$$\sin^2\theta_W = 1 - \frac{m_W^2}{m_Z^2} \tag{1.33}$$

We shall shall call θ_W the electroweak mixing angle.

It is a useful exercise to check that the familiar low-energy four-fermion weak interactions can be obtained by the exchanges of massive $W^\pm$ and Z^0 bosons. In the case of the $W^\pm$ exchanges, we recover

$$\frac{1}{4}\ \mathcal{L}_{eff}^{CC} \equiv \frac{G_F}{\sqrt{2}}\left(J_\mu^+ J^{-\mu}\right) \tag{1.34}$$

where

$$J_\mu^+ = \bar{\nu}_L\gamma_\mu\ell_L + \dots \tag{1.35}$$

with the normalization

$$\frac{G_F}{\sqrt{2}} = \frac{g^2}{8m_W^2}\ :\ m_W = \frac{gv}{2} \tag{1.36}$$

This expression fixes the overall normalization of the v.e.v. of the Higgs field

$$v = \left(\sqrt{2}G_F\right)^{-1/2} \tag{1.37}$$

Similarly, the exchange of the massive Z^0 boson gives us an interaction of the form

$$\frac{1}{4}\ \mathcal{L}_{eff}^{NC} = \frac{G_F}{\sqrt{2}}\left(\rho\ J_\mu^{NC} J^{NC\mu}\right)\ :\ \rho = 1 \tag{1.38}$$

where the neutral current is given by

$$J_\mu^{NC} = J_\mu^3 - \sin^2\theta_W\ J_\mu^{em} \tag{1.39}$$

as can be seen from Eq. (1.28) above. The different overall coupling strength

$$\sqrt{g^2+g'^2} \tag{1.40}$$

and mass of Z^0 boson, Eq. (1.28), combine to give the same strength to the neutral current interactions as for the charged-current interactions, namely $\rho = 1$ in the language of Section I.2. This is a characteristic prediction of our assumption that the Higgs field ϕ has electroweak isospin $I = 1/2$. Other choices of isospin would have given us a different ratio for the $W^\pm$ and Z^0 boson masses, and hence for the relative strength of the charged-current and neutral-current interactions [5].

I.3 Testing electroweak unification

So far, our discussion of the Z^0 peak has been at the classical or "tree" level, which can be regarded as the first step in the systematic approximation scheme called perturbation theory. The expansion parameter in this perturbation theory is the electroweak analogue of the fine structure constant

$$\alpha_2 \equiv \frac{g^2}{4\pi} \simeq \frac{1}{30} \tag{1.41}$$

Up to now, we have just been calculating the tree-level amplitudes shown in Fig. 2, which are

$$A \sim \alpha_2 \tag{1.42}$$

and thus make contributions to the cross-sections that are of order α_2^2. The next step in the approximation scheme is to calculate one-loop quantum corrections like those shown in Fig. 3. The diagram in Fig. 3a is known as a vacuum polarization diagram, that in Fig. 3b is a typical vertex correction, that in Fig. 3c is called a box diagram. Adding these to the previous tree-level amplitudes A, we find

$$\sigma = |A + \delta A|^2 \simeq |A|^2 + 2\text{Re}(A\ \delta A^*) \tag{1.43}$$

providing a correction to the cross-section that is of order α_2^3.

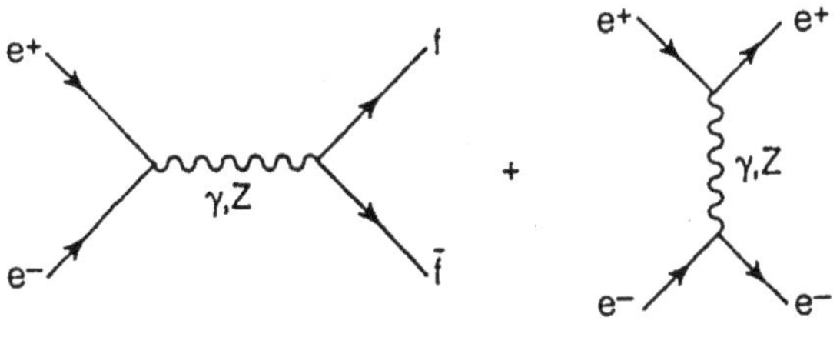

Figure 2: Tree diagrams contributing to $e^+e^- \to \bar{f}f$ at LEP.

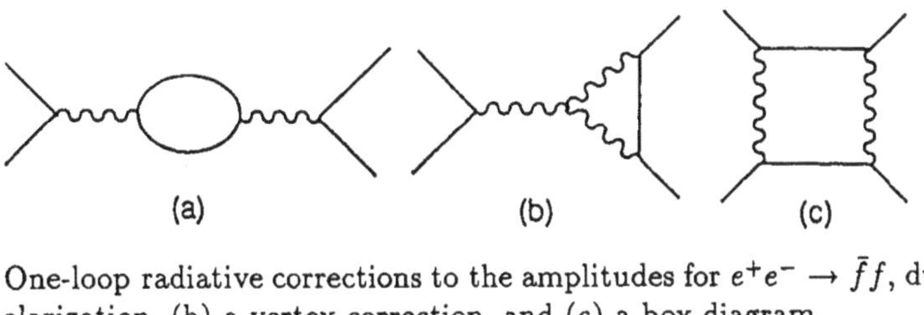

Figure 3: One-loop radiative corrections to the amplitudes for $e^+e^- \to \bar{f}f$, due to (a) vacuum polarization, (b) a vertex correction, and (c) a box diagram.

Do not worry, you are not expected to be able to calculate these diagrams. If you did, you would find that generic one-loop diagrams are infinite. In a renormalizable (i.e., calculable) theory such as the Standard Model [6], all these infinities can be absorbed into the physical values of a finite number of measurable input parameters. In the case of the electroweak sector of the Standard Model , these are $\alpha_{\rm em}$ (the fine structure constant), m_{Z^0} (measured very precisely at LEP), and G_μ (the Fermi μ decay constant). It is also often convenient to keep as an auxiliary parameter $\sin^2\theta_W \equiv 1 - m_W^2/m_Z^2$, which is closely related to the ratio of SU(2) and U(1) couplings, as discussed in Section I.2.

It was pointed out many years ago by Veltman [7] and others [8] that the one-loop diagrams were sensitive to the masses of unseen particles, such as the top quark and the Higgs boson. For example, the masses of the $W^\pm$ and Z^0 bosons are given by

$$m_W^2 \sin^2\theta_W = m_Z^2 \cos^2\theta_W \sin^2\theta_W = \frac{\pi\alpha}{\sqrt{2}G_\mu}\,(1+\Delta r) \tag{1.44}$$

where Δr is a quantum correction whose structure we now discuss. As has already been mentioned, the top quark is essential in the Standard Model . Indeed, the gauge symmetry structure of the electroweak theory would be lost, and renormalizability, i.e., calculability, with it, if the top quark were not present to complete the isospin doublet inhabited by the bottom quark. The difference between m_t and m_b is a measure of electroweak isospin symmetry breaking, which shows up at the one-loop level in vacuum polarization diagrams. They make a contribution to the above-mentioned quantum correction Δr of the form

$$\Delta r \ni \frac{3G_\mu}{8\pi^2\sqrt{2}}\; m_t^2 \tag{1.45}$$

in the limit $m_t \gg m_b$. This quadratic dependence on m_t gives one an experimental handle on the mass of the unseen top quark.

Similarly, since the Higgs boson is essential to spontaneous symmetry breaking, and the theory would not be renormalizable, i.e., calculable, without it, the one-loop diagrams are also sensitive to its presence, and hence its mass

$$\Delta r \ni \frac{\sqrt{2}G_\mu}{16\pi^2}\; m_W^2 \left\{\frac{11}{3}\ln m_H^2/m_W^2 - \ldots\right\} \quad \text{for} \quad m_H \gg m_W \tag{1.46}$$

Note, however, that the sensitivity to the Higgs mass is only logarithmic, which makes it more difficult to extract some information on it from the available electroweak data. Moreover, as seen in Fig. 4, the top and Higgs contributions to Δr can compensate each other. If, say, the quantum correction Δr were measured to have a value of 0.05, one could say that m_t was in the range 100-150 GeV, for m_H between 10 and 1000 GeV. However, one could not be more precise without additional information, either in the form of a determination of m_t or by having available other precision measurements sensitive to different combinations of m_t and m_H. As shown in Table 1, LEP does indeed provide us with a large number of precision electroweak measurements [9], including measurements of the total Z^0 decay width, cross-section measurements on the Z^0 peak, various forward-backward asymmetries, and determinations of individual partial Z^0 decay widths, which are sensitive to m_t and m_H in different ways.

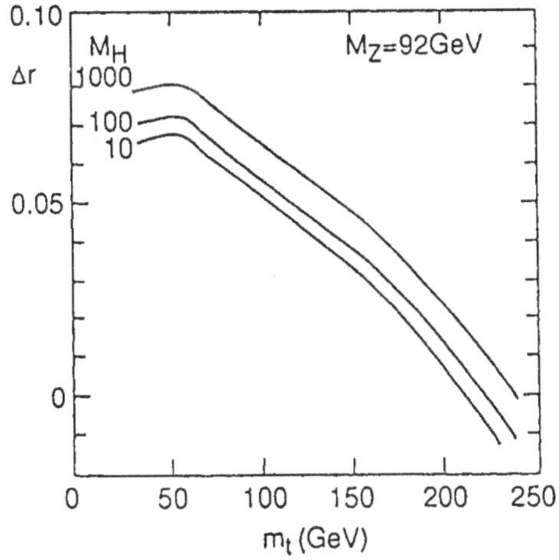

Figure 4: One-loop contributions to the radiative correction Δr (1.44) due to the t-quark and the Higgs boson, evaluated for $M_Z = 92$ GeV.

Table 1: Z Data

m_Z	=	91.187	$\pm$	0.007	GeV
Γ_Z	=	2.489	$\pm$	0.007	GeV
σ_h/σ_L	=	20.763	$\pm$	0.049	
σ_h	=	41.56	$\pm$	0.14	nb
A^{ℓ}_{FB}	=	0.0158	$\pm$	0.0018	
A^{b}_{FB}	=	0.094	$\pm$	0.006	
A^{τ}_{POL}	=	0.130	$\pm$	0.025	
$\Gamma_{b\bar{b}}/\Gamma_h$	=	0.2200	$\pm$	0.0027	
$\Gamma_{c\bar{c}}/\Gamma_h$	=	0.171	$\pm$	0.014	
α_s	=	0.123	$\pm$	0.007	
A_{LR}	=	0.100	$\pm$	0.044	

The current state of play in the sensitivity of precision electroweak measurements to m_t and m_H can be summarized as follows [10]. If one assumes (for the moment without justification, but see Lecture V) that $m_H = m_Z$, then the preferred range of m_t is

$$m_t = 160 \pm 20 \text{ GeV} \tag{1.47}$$

The preferred range of m_t is correlated positively with the value of m_H. Moreover, the data tend to prefer a small value of m_H, close to the present direct lower limit of about 65 GeV. Based on these indications from the precision electroweak data including quantum corrections, I personally would give 2-1 odds that $m_H < 300$ GeV.

I.4 The strong interaction sector of the Standard Model

According to the naïve quark model, strongly-interacting baryonic particles such as the nucleons (proton and neutron) are made out of three quark constituents each. For example, the proton contains two constituent u quarks and one constituent d quark, whilst the neutron contains two constituent d quarks and one constituent u quark. It is easy to see that according to this model, the quarks must have fractional charges

$$\begin{array}{l} 1 = Q_p = 2Q_u + Q_d \\ Q_u - Q_d = 1 \end{array} \quad \Rightarrow \quad \begin{array}{l} Q_u = \frac{2}{3} \\ Q_d = -\frac{1}{3} \end{array} \tag{1.48}$$

This naïve quark model leads to certain puzzles, which are displayed most clearly by the Δ baryon, which is an excited state of the nucleons. One of the Δ baryons is doubly-charged and, according to the naïve quark model, is made out of three constituent u quarks, all with their spins pointing in the same direction. However, we are accustomed from atomic and nuclear physics to expect that the lowest-lying states lie in a relative S-wave. In the case of the Δ^{++}, this would lead to an apparent paradox, since the three u quarks would have to be in identical states, which is forbidden by the Pauli principle. The resolution of this puzzle which is taken up by the Standard Model is that each flavour of quark comes in three identical varieties, whimsically called colours [11].

As we shall see later, the naïve quark model may not be the best guide to the composition of strongly-interacting particles. However, experiments give us many other reasons for believing in the colour degree of freedom. For example, at high centre-of-mass energies above about 2 GeV, the total cross-section for $e^+e^- \to$ hadrons is much larger than in the naïvequark model without colour

$$\frac{\sigma(e^+e^- \to \text{hadrons})}{\sigma(e^+e^- \to \mu^+\mu^-)} \simeq \sum_{u,d,s} Q_q^2 = \frac{2}{3} \tag{1.49}$$

Experiments indicate that between about 2 GeV and about 4 GeV, this ratio is close to 2, the ratio expected if each u, d and s quark flavour comes in three distinct colours.

The same pattern persists above 4 GeV, where the total $e^+e^- \to$ hadrons cross-section is about three times larger than that expected in the naïve quark model with charm c but without colour. A similar feature is seen in $\tau \to \nu_\tau +$ hadrons decays, which are about three times more frequent than $\tau \to \nu_\tau + e + \bar{\nu}_e$ decays, as would be expected if the τ decays into $\bar{u}d$ pairs with three different colours. An additional reason for believing in colour comes from the measured rate of $\pi^0 \to \gamma\gamma$ decay, which has an amplitude proportional to the number of colours. The experimental data indicate

$$N_c = 2.987 \pm 0.016 \qquad (1.50)$$

We therefore assume $N_c = 3$.

In the Standard Model , the strong interactions are described by a gauge theory analogous to that used for the electroweak interactions. It acts on the three colour degrees of freedom via a non-Abelian group called SU(3) in a theory known as Quantum Chromodynamics, or QCD. According to this theory, the quark kinetic terms are given by

$$\mathcal{L}_q = -\sum_q i(\bar{q}_L\gamma^\mu D_\mu q_L + \bar{q}_R\gamma^\mu D_\mu q_R) \qquad (1.51)$$

where

$$D_\mu = \partial_\mu - ig_3\frac{\lambda_a}{2}\, G^a_\mu \qquad (1.52)$$

is a gauge-covariant derivative, with g_3 the strong coupling, λ_a an SU(3) representation matrix, and the G^a_μ representing eight boson fields called gluons. Expanding these quark kinetic terms, we find, in addition to bilinear terms describing quark propagation, also vertices in which quarks emit or absorb a gluon while conserving helicity. Note that the q_L and q_R are placed in identical **3** representations of the strong colour SU(3) group. This means that they conserve parity P, charge conjugation C, and their combination known as CP.

In addition to the quark kinetic terms, QCD also contains gluon kinetic terms which take the form

$$\mathcal{L}_G = -\frac{1}{4}\,(G_{\mu\nu a}G^{\mu\nu a}) \qquad (1.53)$$

where

$$G_{\mu\nu a} = \partial_\mu G_{\nu a} - \partial_\nu G_{\mu a} + ig_3 f_{abc} G_{\mu b} G_{\nu c} \qquad (1.54)$$

Here g_3 is again the strong coupling, and the f_{abc} are the non-Abelian structure constants of the group SU(3). Expanding these gluon kinetic terms, we encounter bilinear terms and also trilinear and quadrilinear gluon self-interactions, analogous to those of the electroweak gauge bosons. The gluon kinetic terms written above also conserve P, C, and CP. However, there is an additional possible term in the QCD Lagrangian [12] which could violate P and CP.

One of the most remarkable features of QCD is its property of **asymptotic freedom**. As we have already seen, quantum effects are very important in field theory.

Specifically, in the case of QCD they cause the strong coupling g_3 to decrease as a function of the energy scale at which it is measured. At the leading one-loop order, the variation of the strong equivalent of the fine structure constant is described approximately by [13]

$$\alpha_3(Q) \equiv \frac{g_3^2(Q)}{4\pi} = \frac{12\pi}{(33 - 2N_q)\ln Q^2/\Lambda^2} + \cdots \qquad (1.55)$$

Here, N_q is the number of quark flavours with masses $m_q \ll Q$, the energy scale, and Λ is an intrinsic mass-energy scale of the strong interactions. The dots in Eq. (1.55) represent higher-order terms which have also been calculated, and do not alter the fact that the strong coupling becomes weaker at higher energies $Q \gg \Lambda$. By the uncertainty principle, such high energies correspond to short distances, and Eq. (1.55) tells us that quarks behave almost as free particles at sufficiently short distances. Figure 5 shows a compilation of data on $\alpha_3(Q)$ as a function of the energy scale Q, showing clear evidence that it decreases as the effective energy scale Q increases. These data indicate the following value for the intrinsic mass-energy scale of the strong interactions

$$\Lambda \simeq 200 \text{ MeV} \times 2^{0\pm1} \qquad (1.56)$$

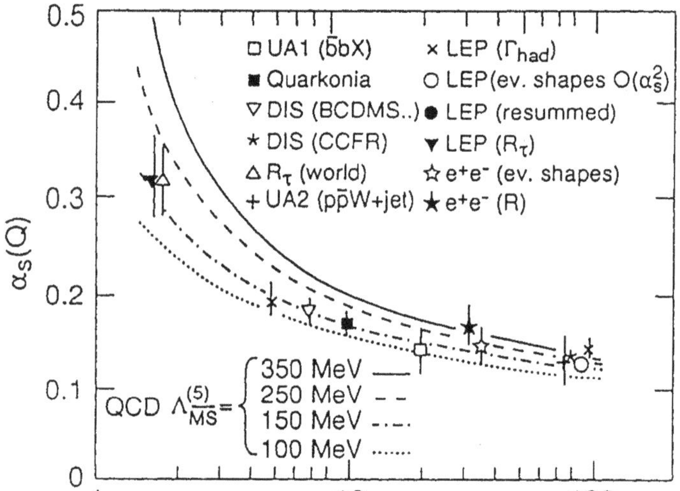

Figure 5: The decrease with energy Q of the strong coupling as measured in many different experiments.

Interpreted naïvely, Eq. (1.55) suggests that the strong coupling becomes really strong at energies $Q \lesssim 1$ GeV, with an apparent divergence at $Q = \Lambda$. The leading-order formula (1.55) cannot be trusted when the strong coupling becomes really strong, but this feature is highly suggestive of confinement.

Quarks and gluons have never been seen isolated. They are only known to exist confined within strongly interacting particles (hadrons) such as baryons and mesons. According to the naïve quark model, mesons such as the π can be regarded as $q\bar{q}$ bound states. The π^+, π^0, and π^- are the lightest hadronic particles, with masses below 140 MeV. Their lightness is related by chiral symmetry to that of the u and d quarks. According to the theory of QCD [14],

$$m_\pi^2 \simeq \frac{\langle 0|\bar{q}q|0\rangle}{f_\pi^2} \quad (m_u + m_d) \tag{1.57}$$

where $\langle 0|\bar{q}q|0\rangle$ is the value of the $q\bar{q}$ condensate in vacuo, and f_π is the pion decay constant which parametrizes the coupling of the charged pions to the $W^\pm$. Equation (1.57) tells us that

$$m_\pi^2 \sim \Lambda_{QCD} \times m_q \tag{1.58}$$

Putting in all the appropriate numerical factors, this relation is used to estimate [14]

$$m_u \sim 5 \text{ MeV} , \quad m_d \sim 8 \text{ MeV} \tag{1.59}$$

These quark masses are much smaller than those previously favoured in the naïve quark model, and are not so much larger than m_e. An analogous calculation based on the $K^+ = (u\bar{s})$ and $K^0 = (d\bar{s})$ masses yields the estimate $m_s \simeq 150$ MeV. As has already been mentioned, according to the naïve quark model, the baryons are interpreted as qqq bound states, and the observed masses $m_{p,n} \simeq 940$ MeV are interpreted as indicating that the constituent quark masses are of order 300 MeV. However, according to chiral symmetry in QCD [15], the baryon masses are not directly related to the light quark masses, Eq. (1.59), but instead

$$\begin{aligned} m_{p,n} &\sim \langle 0|\bar{q}q|0\rangle^{1/3} \sim \langle 0|G^a_{\mu\nu}G^{a\mu\nu}|0\rangle^{1/4} \\ &\sim \Lambda_{QCD} \end{aligned} \tag{1.60}$$

where $\langle 0|G^a_{\mu\nu}G^{a\mu\nu}|0\rangle$ is the gluon-gluon condensate in vacuo.

As already mentioned, both the $q\bar{q}$ condensate $\langle 0|\bar{q}q|0\rangle$ and the gluon-gluon condensate $\langle 0|G^a_{\mu\nu}G^{a\mu\nu}|0\rangle$ are non-zero at low temperatures and pressures. However, they are believed to decrease and go to zero at one or more phase transition temperatures of order 150 MeV, as mentioned in Section I.1. Just below this temperature, hadrons are described as a gas whose energy density is dominated by the π^+, π^0, π^- with relatively few p, n, since the latter have abundances suppressed by Boltzmann factors.

Figure 6 sketches the qualitative form of the boundary between this low-temperature hadron phase of QCD, and the quark/gluon phase which appears at high temperatures T and/or high values of the baryon chemical potential μ_B. The stable phase of strongly-interacting matter below and to the left of the indicated boundary is expected to consist of the familiar hadrons, whilst the stable phase above and to the right of the indicated boundary consists of deconfined quarks and gluons. It should be noted that superheating of the hadronic phase and/or supercooling of the quark/gluon phase would be possible if the transition between the two phases is first-order. This is expected to be the case for three light quark flavours, and may also be the case for the physical situation in which there are two very light quarks u and d and the strange quark s is estimated to have a mass around 150 MeV. Astrophysicists probe the quark-hadron phase boundary in two ways, as indicated in Fig. 6. There may be a quark-gluon phase in the core of certain neutron stars, at very high values of the baryon chemical potential μ_B and relatively low temperatures T. Of more direct interest at this School is the fact that the early Universe is believed to have cooled from a very high-temperature, low-μ_B phase down to low temperatures. In Lecture II, we will discuss in more detail the possible order of the quark-hadron phase transition, and its possible implications for Big Bang nucleosynthesis.

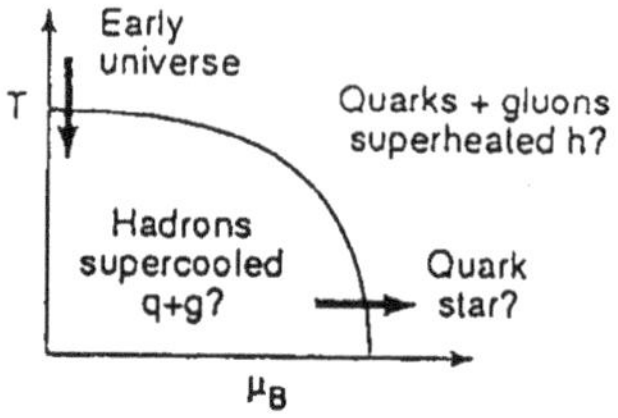

Figure 6: Qualitative picture of the phase boundary between quark and hadron matter.

II The Strong and Electroweak Transitions

II.1 General description of phase transitions

The most basic thermodynamic quantity [2] is the free energy F, which is defined in terms of the Hamiltonian H, the temperature $T \equiv 1/\beta$, and volume V by

$$e^{-F(T,V)/T} = Tr\ e^{-\beta H} \tag{2.1}$$

The free energy can be written in terms of the pressure p and the entropy S as follows:

$$F(T,V) = -Vp(T) = E(T,V) - T \cdot S(T,V) \tag{2.2}$$

At a phase transition, the free energy F (pressure p) is continuous, but changes occur in its derivatives and in one or more order parameters χ. We distinguish first-order and second-order phase transitions in which the change $\Delta\chi$ in the order parameter is either non-zero or zero, respectively. The generic natures of first- and second-order phase transitions are shown in Fig. 10. In the case of a first-order transition, Fig. 7a, the transition occurs as a jump in χ which results from quantum tunnelling through the indicated barrier. On the other hand, in a second-order phase transition, Fig. 7b, the order parameter χ varies continuously, and can be regarded as rolling classically down a hill from the unstable vacuum at $\chi = 0$ towards the stable vacuum at $\chi \neq 0$. The new vacuum with $\chi \neq 0$ forms via different processes in these two cases. In the case of a first-order transition, the dominant process is bubble nucleation, as indicated in Fig. 8a. A bubble region of the new vacuum is surrounded by the old vacuum with a wall in between. In the case of a second-order phase transition, the dominant mechanism is spinodal decomposition as illustrated in Fig. 8b. As the temperature T decreases, different parts of the medium may roll to different non-zero values of the order parameter χ, which are separated by regions of values of χ where the potential is unstable. A key feature of first-order phase transitions is that the correlation length ξ remains finite, whereas it goes to infinity

$$\xi \to (T - T_c)^{-p} \ : \quad p > 0 \tag{2.3}$$

in the case of second-order phase transition.

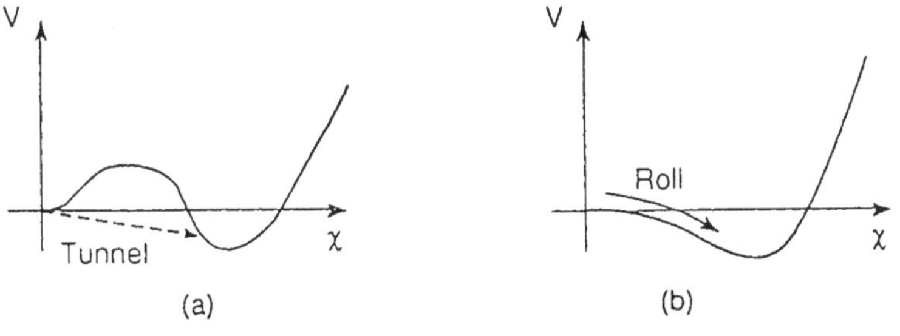

Figure 7: Sketch of the phase transition proceeding (a) by tunnelling in the first-order case, (b) by rollover in the second-order case.

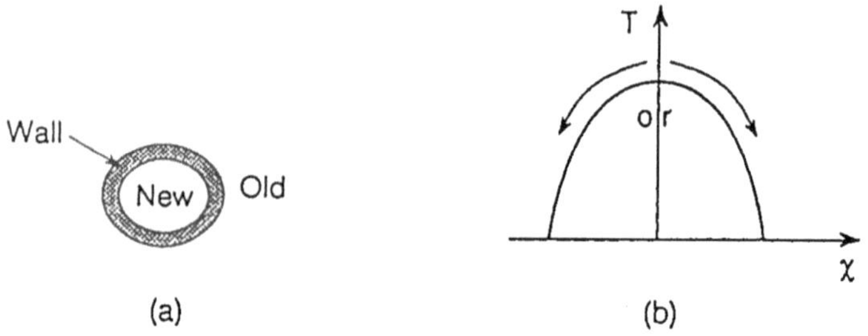

Figure 8: (a) In a first-order transition, bubbles of the new vacuum form, whilst (b) in a second-order transition there is a spinodal decomposition into regions with different values of the order parameter.

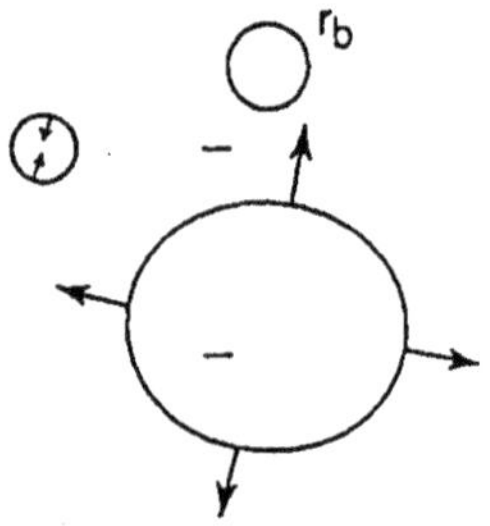

Figure 9: In a first-order transition, bubbles of the new vacuum are surrounded by a "cheese" made of the false vacuum. Bubbles smaller than the critical size r_b contract, whilst larger bubbles expand.

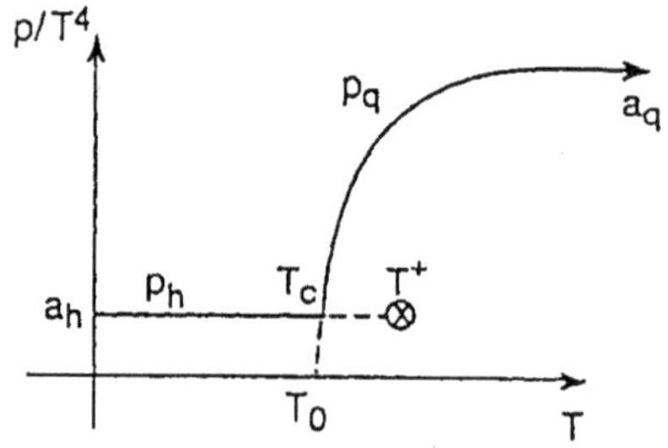

Figure 10: Qualitative picture of the temperature-pressure curves in the hadron and quark phases of QCD.

Since, as already mentioned, the quark-hadron phase transition may well have been first-order, we need to study the mechanism of bubble nucleation in more detail. The rate of bubble nucleation is given generically by an expression of the form

$$\Gamma \sim (\ldots) \exp\left(-\beta E_B\right) \tag{2.4}$$

where the exponent contains a factor of the classical bubble energy E_B, and we have indicated in parentheses a quantum prefactor that need not concern us here. Classical energetic considerations tell us that

$$E_B = \frac{-4\pi}{3} r^3 \cdot \Delta\rho + 4\pi\, r^2 \cdot \sigma \tag{2.5}$$

where the first is a volume term, r is the bubble radius, $\Delta\rho$ is the difference in energy density between the media inside and outside the bubble, and the second is a surface term with σ denoting the surface tension. The surface tension is just the contribution to the total free energy per unit of surface area which appears in the region of transition between the new and old vacua, represented by the wall in Fig. 8a. The classical bubble size r_B is obtained by extremizing the energy in Eq. (2.5)

$$\frac{\partial E_B}{\partial r} = 0 \ : \quad r_B = \frac{2\sigma}{\Delta\rho} \tag{2.6}$$

In this case, the classical bubble energy E_B is just given by

$$E_B = \frac{4}{3}\pi\sigma r_B^3 \tag{2.7}$$

Bubbles produced with $r < r_B$ will shrink and disappear after formation, whilst larger bubbles with $r > r_B$ will expand indefinitely. This yields the "Swiss cheese" picture of a phase transition shown in Fig. 9. The phase transition is completed when the expanding bubbles fill the whole of space.

II.2 Modelling the strong phase transition

As has already been mentioned, the pressure in the low-temperature hadron phase is determined essentially by the lightest degrees of hadronic freedom, namely the π^+, π^0, π^-

$$p_h = a_h T^4 \ : \quad a_h = \frac{\pi^2}{90} \times 3 \tag{2.8}$$

In addition, the e, μ, ν and γ, which are inert under the strong interactions, also contribute to the total pressure felt by the Universe. On the other hand, the pressure in the high-temperature quark phase is given by

$$p_q = a_q\, T^4 \ : \quad a_q = \frac{\pi^2}{90}\left[\frac{21}{2} \cdot N_q + 16\right] \tag{2.9}$$

where the first term comes from the $2(u,d)$ or $3(u,d,s)$ light quarks, and the second term is due to the eight gluons, each with two helicity states. In addition, there is the same contribution from the inert leptonic degrees of freedom. Comparing Eqs.

(2.8) and (2.9), we see that there is apparently no temperature at which the pressures in the hadron and quark phases can be equal, which we recall from the previous section was a necessary condition at the critical temperature T_c of a phase transition. However, we have not yet incorporated all the strong interaction dynamics which confines quarks and gluons. One simple way of incorporating its essence has been proposed [16], namely to add to the pressure a so-called bag constant

$$\Delta p_q = -B \tag{2.10}$$

The bag constant can be regarded as an effective cosmological constant present outside hadrons. Equating the standard expression for the energy-momentum tensor of a perfect fluid with that given by a bag constant,

$$\frac{p_h}{T^4} = a_h = a_q - \frac{B}{T^4} = \frac{p_q}{T^4} \tag{2.11}$$

we see that

$$T_c = \left(\frac{B}{(a_q - a_h)}\right)^{1/4} \tag{2.12}$$

in the presence of a bag constant. This is supposed to mimic the effect of the confinement potential. It is supposed to be present outside hadrons, but absent inside hadrons, so that quarks and antiquarks are bound inside mesons and qqq triplets are bound inside baryons. Adding this contribution to the pressure Eq. (2.9) in the quark phase, we see that it can become equal to that in the hadron phase given by Eq. (2.8)

$$T^{\mu\nu} = (\rho + p)u^\mu u^\nu - pg^{\mu\nu} = Bg^{\mu\nu} \tag{2.13}$$

where $u^\mu = (1,0,0,0)$. The resulting vacuum energy density and pressure are

$$\rho_{vac} = -p_{vac} = B \tag{2.14}$$

in this simple model.

The pressure curves in the hadronic and quark phases are shown as solid lines in Fig. 10. They cross at the critical temperature T_c where the hadron and quark pressures are equal. Also shown as dashed lines are possible superheating of the hadron phase up to a maximum superheating temperature T^+, and a supercooling curve for the quark phase down to a minimum supercooling temperature T^0. In the model of the previous paragraph, the quark/hadron phase transition is first-order, with the order parameter provided by the bag constant B. It follows from the discussion at the end of the previous section that a first-order phase transition is completed when the bubble nucleation rate becomes equal to the Hubble expansion rate. This condition can be written as

$$(H_c \simeq \frac{T_c^2}{m_P} = \frac{1}{t_c}) \simeq a_c^3 \; T_c^4 \exp(-\frac{E_B}{T}) \tag{2.15}$$

where the left-hand side is the Hubble expansion rate H_c at the critical temperature, the first factor on the right-hand side is the volume of the Universe at the critical temperature T_c (critical time t_c), the second factor on the right-hand side is a dimensional factor representing the effects of quantum corrections, and the last factor contains in the exponential the bubble energy discussed in Section II.1. In the case of the QCD phase transition, we can solve these equations to obtain

$$\frac{E_B}{T} \simeq \ln(t_c T_c)^4 \tag{2.16}$$

where

$$\frac{E_B}{T} \simeq \frac{64\pi}{3} \frac{\sigma^3}{L^2 T_c} \frac{1}{(t/t_c - 1)^2} \tag{2.17}$$

with σ again denoting the surface tension, and L the latent heat of the phase transition. The factor depending on time in the denominator on the right-hand side comes from the fact that the difference between the pressures in the old and new phases must vanish when $t = t_c$, in which case the bubble radius and hence the bubble energy becomes infinite.

In order to estimate the temperature T_f and the time t_f when the QCD phase transition is completed, we need estimates of the surface tension σ and the latent heat L. Since QCD is a strongly-interacting theory in the neighbourhood of the critical temperature, it is difficult to calculate these quantities precisely. Some estimates have been made using phenomenological models and effective low-energy field theories [17]. However, the best estimates of σ and L are probably those from lattice gauge theory calculations, which suggest [18]

$$\frac{\sigma}{T_c^3} \simeq 0.1 \qquad \frac{L}{T_c^4} \simeq 10 \tag{2.18}$$

We see that the surface tension σ is rather small compared with the natural energy scale T_c. This combined with the relatively large latent heat L means that the amount of supercooling is expected to be very small [19]

$$1 - \frac{T_f}{T_c} \simeq 0.001 \tag{2.19}$$

Correspondingly, the bubble size is expected to be rather large [19]

$$r_f = \frac{2\sigma}{L(1 - T_f/T_c)} \sim \frac{\sigma}{L} \cdot \frac{L}{\sigma^{3/2}} \sim \frac{1}{\sigma^{1/2}} \simeq 20 \text{ to } 40 \text{ fm} \tag{2.20}$$

The basic reason why r_f is so large, much larger than an intrinsic strong interaction scale, is that the Hubble expansion rate $H \simeq T^2/m_P$ is so small. This is actually a necessary consistency condition for the reliability of the simple energetic analysis of bubble formation given in the previous section. We implicitly assumed there that the bubble wall was thin compared with the bubble radius. This thin-wall approximation is likely to be accurate for QCD, since the bubble-wall thickness is likely to be $\sim 1/\Lambda \sim 1$fm, which is much smaller than the critical bubble size $r_f \sim 30$fm. One can also calculate the typical distance between nucleation centres

$$\Delta r \sim 10^{-5} t_c \sim \text{few cm} \tag{2.21}$$

This gives the typical scale of inhomogeneities in the Universe after the quark/hadron phase transition. Although much larger than a natural strong-interaction scale, this scale of inhomogeneity is still much too small to have an appreciable effect on cosmological nucleosynthesis, as we discuss in the next two sections.

II.3 Inhomogeneous Big Bang nucleosynthesis

In recent years, this possibility has excited a lot of interest, mainly due to the possibility of a first-order QCD phase transition of sufficient strength to create inhomogeneities in the very early Universe that affect significantly primordial light element abundances [20]. The Universe could have been sufficiently lumpy at temperatures $T < 1.50$ MeV if the distance between nucleation sites of the new hadronic vacuum were large

$$\Delta r \gtrsim 0.5 \text{ m} \tag{2.22}$$

where the right-hand side represents the diffusion length of protons through the cosmological medium. In this case, neutrons would diffuse much further than protons, enabling the n/p ratio and η to vary in different parts of the Universe. As can be deduced from the discussion of the previous section, these possibilities would allow for different light element abundances. The question was then raised whether this flexibility would allow the Universe to be closed by baryons alone, thus obviating the need for non-baryonic dark matter.

The discussion of Section II.2 suggests that such a radical scenario is unlikely, given our knowledge of the quark/hadron phase transition. We recall that using the estimates [18]

$$\frac{\sigma}{T_c^3} \sim 0.1 \ , \quad \frac{L}{T_c^4} \sim 10 \tag{2.23}$$

of the QCD phase transition parameters given there, the typical separation between hadronic bubble nucleation sites was estimated as

$$\Delta r \sim \text{few cm} \tag{2.24}$$

which is much smaller than required above in Eq. (2.22). However, since the strong interactions are really strong in the neighbourhood of the critical temperature T_c, it is difficult to estimate the surface tension σ and the latent heat L very reliably. In particular, we cannot be absolutely sure that σ is not much larger than the estimate in Eq. (2.23), in which case the typical distance between nucleation sites could be in the interesting range given in Eq. (2.22). It should be reemphasized, though, that this possibility is disfavoured.

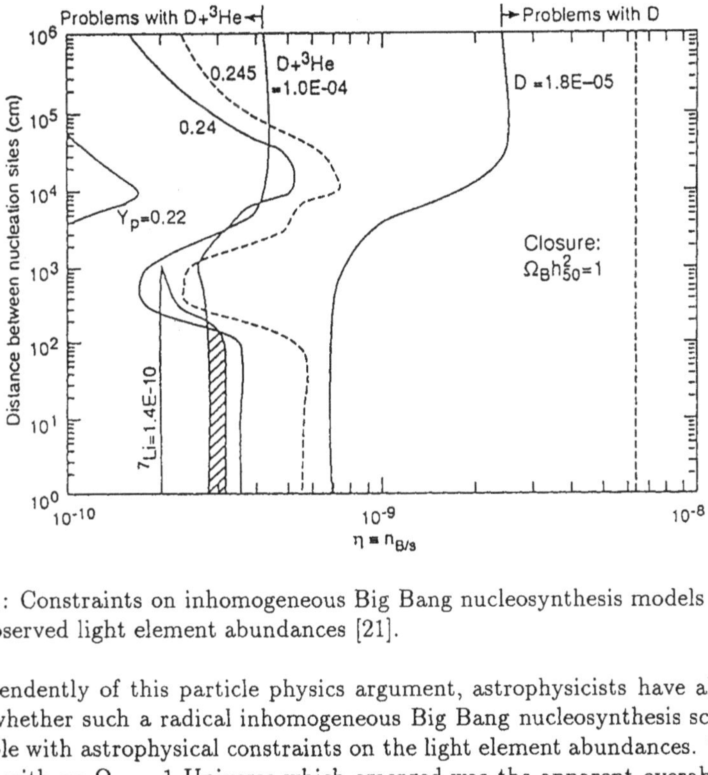

Figure 11: Constraints on inhomogeneous Big Bang nucleosynthesis models imposed by the observed light element abundances [21].

Independently of this particle physics argument, astrophysicists have also studied [21] whether such a radical inhomogeneous Big Bang nucleosynthesis scenario is compatible with astrophysical constraints on the light element abundances. The first difficulty with an $\Omega_B = 1$ Universe which emerged was the apparent overabundance of ^{7}Li. However, as discussed in the previous section, there is a question whether this might have been modified by a special depletion process in population II stars. We now know also that there is a problem with the abundance of ^{4}He in an inhomogeneous scenario. Regions with high baryon densities overproduce the heavier light elements, and this problem persists even if the ^{4}He bound is relaxed. Moreover, there is also a tendency to underproduce D in an inhomogeneous nucleosynthesis scenario, but this is a notoriously delicate element, as discussed in the previous section. However, if one takes all the astrophysical constraints at face value, then, as shown in Fig. 11, an $\Omega_B = 1$ Universe seems to be excluded [21]. There is very little room for deviation from the standard homogeneous Big Bang nucleosynthesis scenario, one still finds

$$\frac{n_B}{S} \simeq 3 \times 10^{-10} \tag{2.25}$$

and the scenario of a strongly first-order quark/hadron phase transition seems to be excluded by the weight of the astrophysical evidence as well.

II.4 Modelling the electroweak phase transition

There is a general effective description of the behaviour of an order parameter close to a critical temperature T_c, called the Landau mean-field expansion [22]. One introduces a parameter χ describing the order parameter, and writes down a general parametrization of the effective potential as a polynomial in χ

$$V_{eff}(\chi) = \frac{1}{2}\,\gamma(T^2 - T_0^2)\,\chi^2 - \frac{1}{3}\alpha T\chi^3 + \frac{1}{4}\lambda\chi^4 \tag{2.26}$$

In this expansion, α, γ and λ are model-dependent parameters. Later on we will discuss the determination of these parameters in the electroweak theory. The generic graph of pressure versus temperature is similar to that for the strong interactions, which we discussed earlier in Fig. 10. The pressures p_1 and p_2 of the low- and high-temperature phases become equal at the critical temperature T_c. Also shown and sketched in Fig. 10 as dashed lines are possible curves of superheating of the low-temperature phase to a maximum possible temperature T^+, and possible supercooling of the high-temperature phase to a minimum possible temperature T_0. The origins of T^+ and T_0 are shown in Fig. 12, which displays the likely behaviour of the effective potential as the temperature is increased. The zero-temperature potential has a well-defined global minimum at $\chi \neq 0$. At the temperature T_0, the curvature of the effective potential at the origin changes from being negative (the low-temperature case) to positive (the high-temperature case). It is only above this temperature that the $\chi = 0$ minimum of the effective potential is metastable, so T_0 represents the minimum possible supercooling temperature. At T_c, the minima at $\chi \neq 0$ and at $\chi = 0$ have equal heights, corresponding to the critical temperature. At T^+, the local minimum at $\chi \neq 0$ and the local maximum between there and $\chi = 0$ merge, giving a point of inflexion. Above this temperature, there is not even a metastable minimum with $\chi \neq 0$, so this represents the maximum possible superheating temperature. It is easy to calculate T^+, T_0 and T_c in the effective potential model of Eq. (2.26) . One finds

$$T_0 = T_c\sqrt{1 - \frac{2}{9}\,\frac{\alpha^2}{\lambda\gamma}} \qquad T^+ = T_0/\sqrt{1 - \frac{1}{4}\,\frac{\alpha^2}{\lambda\gamma}} \tag{2.27}$$

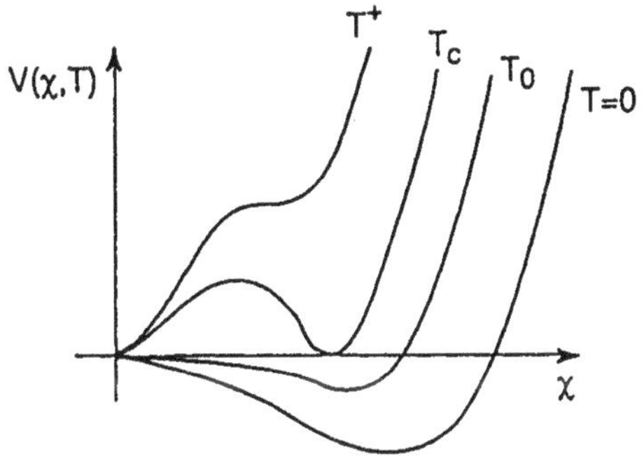

Figure 12: The finite-temperature effective potential, showing its form at $T = 0$, $T = T_0$ when the symmetric vacuum becomes classically unstable, $T = T_c$ the usual critical temperature, and $T = T^+$ above which the asymmetric vacuum does not exist.

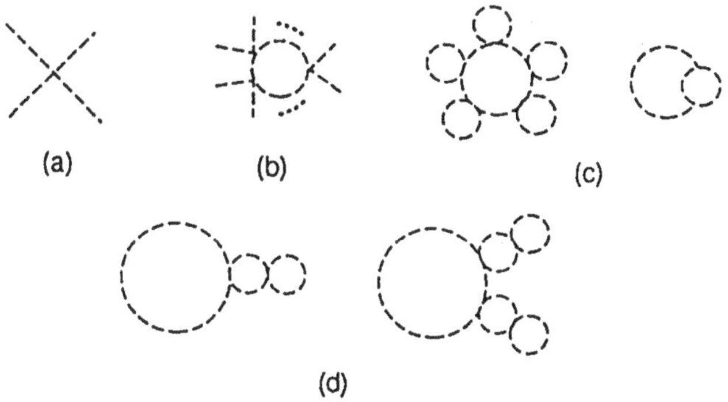

Figure 13: Contributions to the effective potential (a) at the tree level, (b) at the one-loop level, (c) due to daisy diagrams, and (d) due to superdaisy diagrams.

The effective potential for the electroweak Higgs boson has been calculated in successive approximations: tree-level, one-loop, daisies, superdaisies, illustrated in Fig. 13. We relate the calculations to the Landau mean-field expansion shown in Eq. (2.26), using as inputs the Standard Model parameters, principally m_H. To keep things simple, we will start working with a pure $SU(2)$ theory, i.e., take the limit $g' = 0$. The one-loop calculations in this approximate theory yield [23]

$$\begin{aligned} \alpha &= \frac{3}{16\pi} g^3 \simeq \frac{1}{18\pi} \\ \lambda &= \frac{g^2 m_H^2}{8 m_W^2} \simeq \left(\frac{m_H}{342 \text{ GeV}}\right)^2 \end{aligned}$$

$$\begin{aligned}\gamma &= \frac{3}{16}\,g^2 + \frac{\lambda}{2} \simeq \frac{1}{12} + \frac{1}{2}\left(\frac{m_H}{342\ \mathrm{GeV}}\right)^2 \\ T_0 &= \frac{m_H}{\sqrt{2\gamma}} \simeq m_H\sqrt{6}\end{aligned} \tag{2.28}$$

for the Landau mean-field expansion parameters, where we have shown approximate numerical values for $g \simeq 2/3$ as in the Standard Model. Now that we have these potential parameters, it is an easy matter to calculate quantities of interest concerning the nature of the electroweak phase transition

$$\begin{aligned}1 - \frac{T_0^2}{T_c^2} &= \frac{2}{9}\,\frac{\alpha^2}{\lambda\gamma} \simeq \left(\frac{10\ \mathrm{GeV}}{m_H}\right)^2 \ll 1 \\ \frac{v(T_c)}{v} &= \sqrt{\frac{4}{9}\,\frac{\alpha^2}{\lambda\gamma}\,\frac{T_c}{T_0}} \simeq \left(\frac{14\ \mathrm{GeV}}{m_H}\right) \ll 1 \\ \frac{L}{T_c^4} &= \frac{4}{9}\,\frac{\alpha^2\gamma}{\lambda^2}\,\frac{T_0^2}{T_c^2} \simeq \left(\frac{20\ \mathrm{GeV}}{m_H}\right)^4 \ll 1 \\ \frac{v(T_c)}{T_c} &= \frac{2}{3}\,\frac{\alpha}{\lambda} \simeq \left(\frac{37\ \mathrm{GeV}}{m_H}\right)^2 \ll 1\end{aligned} \tag{2.29}$$

The fact that all these quantities are much smaller than unity for physical values of the Higgs boson mass (we recall that direct searches at LEP force $m_H > 65$ GeV, as discussed in Lecture I) means that the electroweak phase transition is weakly first-order in this approximation. Note also that the scalar field correlation length, corresponding to the Compton wave-length and hence range of the Higgs field

$$\xi_H(T_c) = \frac{1}{m_H(T_c)} = \frac{3}{\sqrt{2}}\,\frac{\sqrt{\lambda}}{\alpha}\,\frac{1}{T_c} \simeq \frac{1}{10\ \mathrm{GeV}} \simeq 0.02\ \mathrm{fm} \tag{2.30}$$

is almost independent of m_H and T_c, and quite large by electroweak standards. Its large size also indicates that, although first-order, the electroweak transition may be "close" to being second order. The penetration depth corresponding to the range of the gauge field

$$\xi_W(T_c) = \frac{1}{m_W(T_c)} = \frac{2}{gv(T_c)} \simeq \left(\frac{m_H}{21\ \mathrm{GeV}}\right)^2 \frac{1}{T_c} \gg \frac{1}{T_c} \tag{2.31}$$

is also large at the critical temperature.

The first step beyond this one-loop approximation is to include daisy diagrams (see Fig. 13c), which give [24]

$$\begin{aligned}\delta m_H^2 &= \frac{1}{4}\left(\frac{3}{4}\,g^2 + h_t^2\right)T^2 \\ \delta m_{W_2}^2 &= \frac{11}{6}\,g^2T^2 \\ \delta m_{W_T}^2 &= 0\end{aligned} \tag{2.32}$$

where h_t denotes the Yukawa coupling of the top quark. From this we see that the Higgs and the longitudinal W have now acquired Debye screening masses, and are hence no longer in a position to contribute to α. Note, however, that the transverse W still has zero mass and hence contributes to α. The fact that the H and W_L no longer contribute to α means that the strength of the first-order transition is reduced by about 2/3:

$$T_c = 97 \text{ GeV} \;, \quad v(T_c) \simeq 40 \text{ GeV} \tag{2.33}$$

The next step is to calculate the superdaisy diagrams (see Fig. 13d), which yield [25]

$$\delta m_{W_T} = \frac{g^2 T}{3\pi} \tag{2.34}$$

However, this result is not reliable, because perturbation theory is not applicable to calculating the magnetic mass of a gauge boson. The normal assumption at this stage is a parametrization

$$\delta m_{W_T} = \tau \, \frac{g^2 T}{3\pi} \tag{2.35}$$

where τ is a parameter that is essentially non-perturbative. Studies show that the transition would become second-order if $\tau \gtrsim 2$.

This is as far as one can get with perturbation theory, and to get any further one needs a non-perturbative technique such as the lattice. Shown below are typical results of a high-temperature three-dimensional lattice simulation [26] of a pure $SU(2)$ electroweak theory:

m_H(GeV)	$\left.\frac{v(T_c)}{T_c}\right\vert_{1-\text{loop}}$	$\left.\frac{v(T_c)}{T_c}\right\vert_{\text{lattice}}$
35	1.1	1.1 to 2.0
80	0.2	0.7 to 0.9

(2.36)

These results and others indicate that the electroweak phase transition may be more strongly first order than suggested by the previous perturbative calcuations. If so, the origin is likely to be different from that in perturbation theory [27]. It remains to be seen whether these initial lattice calculations are reliable, and for the moment we take the view that the weight of evidence is that the electroweak phase transition is weakly first order, although probably not so weak as the QCD phase transition discussed in Section II.3.

III Supersymmetry and Grand Unification

III.1 Why supersymmetry?

The motivation for supersymmetry that I find most compelling is the hierarchy/ naturalness problem [28]. The general requirement of "naturalness" is that quantum corrections "should" be not much larger than their physical values. This requirement is respected by the quantum corrections to the mass of an elementary fermion such as the electron

$$\delta m_f \simeq m_f \left(\frac{\alpha}{\pi}\right) \ln\left(\frac{\Lambda}{m_f}\right) \tag{3.1}$$

Although formally infinite, as reflected in the logarithmic dependence on the cut-off Λ, this quantum correction is $\lesssim m_f$ for all values of the cut-off $\Lambda \lesssim m_P$. Many physicists consider that naturalness is a "reasonable" request for a "sensible" physical theory, but it is a requirement that goes beyond the pure mathematics of renormalization. The mechanics of renormalization theory, that is used to extract calculable results from field theory, does not care about the relative magnitudes of quantum corrections and physical values. However, if a quantum correction δX is much larger than the physical value of some quantity X_0, this means that the bare value must have been chosen in just such a way as to produce a large cancellation of the quantum correction

$$X = X_0 + \delta X \ : \ |X| \ll |X_0| \ , \ |\delta X| \tag{3.2}$$

which looks unnatural. It must be admitted, however, that naturalness is a subjective criterion that is in the mind of the beholder.

Naturalness is a problem for an elementary Higgs boson, as in the electroweak sector of the Standard Model , because it must have a mass in the range [29]

$$m_W = O\left(\sqrt{\frac{\alpha}{\pi}}\right)^{0\pm1} \times m_H \tag{3.3}$$

The reason why this is a problem is that quantum corrections like those in Fig. 14a are quadratically divergent and hence very large if the cut-off $\Lambda \gg 1$ TeV

$$\delta m_H^2 \simeq g_{f,W,H}^2 \int^{\Lambda} \frac{d^4k}{(2\pi)^4} \ \frac{1}{k^2} \simeq O\left(\frac{\alpha}{\pi}\right) \Lambda^2 \gg m_H^2 \tag{3.4}$$

for example if $\Lambda \sim m_P$ or m_{GUT}.

The hierarchy problem is why $m_W \ll m_P(m_{GUT})$. An alternative way of formulating this question is to ask why is $G_F \gg G_N$. Equivalently, one can ask why the Coulomb potential between electron and nucleus is much greater than the Newton gravitational potential

$$V_{\text{Coulomb}} = \frac{e^2}{r} \gg V_{\text{Newton}} = \frac{G_N m^2}{r} \simeq \left(\frac{m_W^2}{m_P^2}\right) \frac{1}{r} \tag{3.5}$$

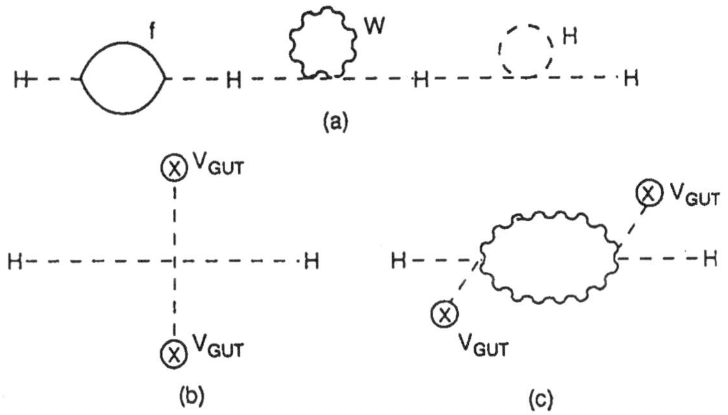

Figure 14: (a) Quadratically-divergent loop corrections to m_H^2, (b) a diagram which "leaks" the large mass-scale in a GUT into the small mass-scale, and (c) a quantum correction to this leakage.

One difficulty is to know why $m_W \ll m_P$ in some initial approximation, say the classical tree-level approximation, and the other difficulty is to understand how such a hierarchy can be maintained despite the depredations of the quantum corrections shown in Eq. (3.4). The danger is that they will drive $m_H \to m_P$, unless there is very clever fine-tuning of the parameters, and carry along m_W as suggested by Eq.(3.3).

Another aspect of this problem arises in theories with both large and small scales. There tends to be "leakage" of the large scale which pollutes the small scale [30]. For example, in a GUT there is almost inevitably a coupling λ between electroweak Higgses with a small v.e.v. v and Higgses with a large GUT v.e.v. V. Such a coupling as shown in Fig. 14b would give rise to a correction

$$\delta m_H^2 \simeq \lambda V_{GUT}^2 \tag{3.6}$$

to the "light" Higgs boson mass. Even if $\lambda = 0$ at the tree level (do not ask me why, that is the hierarchy problem), non-zero couplings are in general regenerated [31] by quantum corrections of the type shown in Fig. 14c

$$\delta m_H^2 \simeq O\left(\frac{\alpha}{\pi}\right)^2 V_{GUT}^2 \tag{3.7}$$

What one needs is a symmetry which suppresses not only λ but also all these quantum corrections through many, many orders of perturbation theory. As if this is were not enough, Hawking [32] has argued that a generic theory of quantum gravity will induce large corrections

$$\delta m_H^2 = O(m_P^2) \tag{3.8}$$

To be sure of this assertion, one needs a consistent quantum theory of gravity, and the only one available is string theory. As it is difficult if not impossible to formulate a consistent, stable string theory without invoking supersymmetry, let us now explore supersymmetry in more detail.

Supersymmetry works many miracles, for example in the loop diagrams shown in Fig. 14a. When I said a while back that each of these diagrams was quadratically divergent, I neglected to tell you that the fermion and boson diagrams had opposite signs, as a result of the difference between Fermi-Dirac and Bose-Einstein statistics

$$\delta m_{W,H}^2 \simeq -\left(\frac{g_F^2}{4\pi^2}\right)\left(\Lambda^2 + m_F^2\right) + \left(\frac{g_B^2}{4\pi^2}\right)\left(\Lambda^2 + m_B^2\right) \tag{3.9}$$

We see immediately that the quadratic divergences cancel if the couplings satisfy $g_F = g_B$ for pairs of bosons and fermions. Moreover, the residual corrections proportional to m_F^2 and m_B^2 would be small if $m_B \simeq m_F$. Equal couplings and almost equal masses mean approximate supersymmetry. Assuming this, we extract from Eq. (3.9)

$$\delta m_{W,H}^2 \simeq O\left(\frac{\alpha}{\pi}\right) |m_B^2 - m_F^2| \tag{3.10}$$

which is $\lesssim m_{W,H}^2$, in agreement with the physical requirement of naturalness, if

$$|m_B^2 - m_F^2| \lesssim 1 \text{ TeV}^2 \tag{3.11}$$

Historically, this was the first physical motivation for low-energy supersymmetry [33]. However, it must be emphasized again that this argument is a qualitative, physical argument that must be regarded as a matter of taste: an unnatural theory is still mathematically renormalizable, and hence acceptable to some people.

In addition to the above miracle, many logarithmic divergences are absent in supersymmetric theories [34]. Indeed, one finds that

$$\delta\lambda \propto \lambda \tag{3.12}$$

which vanishes if $\lambda = 0$. (Please do not ask me why $\lambda = 0$, that is again the hierarchy problem.) This means that if $\lambda = 0$ at the tree level, it vanishes to all orders of perturbation theory. As a result of these two miracles, if $m_W \ll m_P$ at the tree level, supersymmetry guarantees that it stays small through all orders of perturbation theory.

III.2 Model-building

The first question to ask when constructing a supersymmetric model is whether any of the known matter fermions q, ℓ can be related to any of the known bosons

γ, W, Z, H, g. Unfortunately, model-builders soon realized that this was impossible. For example, quarks q occur in **3** representations of colour, whereas bosons only appear in **1** (γ, W, Z, H) or **8** (g) representations of the colour group. Similarly, leptons carry a conserved lepton number $L = 1$, whereas all the known bosons have zero lepton number. In particular, if one tried to associate a lepton with a Higgs boson in a supermultiplet

$$\begin{pmatrix} \ell \\ H \end{pmatrix} \tag{3.13}$$

then the theory would have an uncontrollably large amount of lepton number violation [35]. For this reason, we are forced to introduce superpartners for all the known particles, as shown in Table 2. You may be appalled by the proliferation of new particles, but remember that this trick has served particle physicists well in the past. They have repeatedly found that introducing new particles was the only way out of their dilemmas. Remember the invention of the ν to rescue energy conservation in β-decay, the invention of the $W^{\pm}$ and Z^0 to make a sensible theory of the weak interactions, and the introduction of charm to suppress unwanted neutral current interactions. Even if supersymmetry is not economical in particles, it is economical in principles!

Table 2: Particles and Sparticles

Particle	Spin	Spartner	Spin
quark: q	$\frac{1}{2}$	squark: $\tilde{q}$	0
lepton: ℓ	$\frac{1}{2}$	slepton: $\tilde{\ell}$	0
photon: γ	1	photino: $\tilde{\gamma}$	$\frac{1}{2}$
W	1	wino: $\tilde{W}$	$\frac{1}{2}$
Z	1	zino: $\tilde{Z}$	$\frac{1}{2}$
Higgs: H	0	higgsino: $\tilde{H}$	$\frac{1}{2}$
gluon: g	1	gluino: $\tilde{g}$	$\frac{1}{2}$

Most supersymmetric theories contain one stable particle which should have been produced in the early Universe, and should be around today as a cosmological relic from the Big Bang [36]. This particle exists because most supersymmetric theories have a multiplicatively conserved quantum number called R parity [35], which takes the value +1 for all known particles, and −1 for all sparticles. It is easy to check using the forms of interaction given in the previous section that R parity is conserved by all the usual interaction vertices

$$\bar{f}\tilde{f}\tilde{g} \ , \quad \bar{f}\tilde{f}\tilde{H} \ , \quad |\tilde{f}^2|^2 \ , \quad |\tilde{f}H|^2 \ , \quad \ldots \tag{3.14}$$

The conservation of R parity is not mysterious, since it is linked to the conservation of baryon number B and lepton number L. Indeed, one can represent R parity by

$$R = (-1)^{3B+L+2S} \tag{3.15}$$

where S is the spin. This equation immediately tells us how it would be possible to violate R parity conservation, namely by breaking B and/or L in an appropriate way. This could be done in *vacuo* by a v.e.v. for some sneutrino field $\tilde{V}$; or by introducing a coupling between **H** and **L** supermultiplets. Both of these possibilities are constrained by laboratory limits on $\Delta L \neq 0$ interactions, and also by the cosmological constraints mentioned at the end of Lecture IV, but R violation cannot be completely excluded.

If we nevertheless assume R conservation, as occurs in most phenomenological models, we can draw three important conclusions.

1. Sparticles must always be produced in pairs, for example

$$e^+e^- \rightarrow \tilde{\mu}^+\tilde{\mu}^- \ , \quad \bar{p}p \rightarrow \tilde{q}\tilde{g}X \tag{3.16}$$

2. Heavier sparticles can only decay into final states containing lighter sparticles, for example

$$\tilde{e} \rightarrow e\tilde{\gamma} \ , \quad \tilde{q} \rightarrow q\tilde{g} \ , \quad \tilde{g} \rightarrow q\bar{q}\tilde{\gamma} \ , \quad \ldots \tag{3.17}$$

3. The lightest supersymmetric particle (LSP) is absolutely stable, since it has no available legal decay mode. It is this particle which is a candidate to be the supersymmetric relic from the Big Bang , discussed in more detail in Lecture V.

III.3 Grand unification

The simplest GUT model is based on the group SU(5) [37], [31], for which the 15 fermion helicity states of each generation

$$(ude\nu_e) \ , \quad (cs\mu\nu_\mu) \ , \quad (tb\tau\nu_\tau) \tag{3.18}$$

are each assigned to reducible $\bar{\mathbf{5}}$ and $\mathbf{10}$ representations of $SU(5)$

$$\bar{\mathbf{5}} = \begin{pmatrix} d_1^c \\ d_2^c \\ d_3^c \\ \cdots \\ -e^- \\ \nu_e \end{pmatrix} \qquad \mathbf{10} = \frac{1}{\sqrt{2}} \begin{pmatrix} 0 & u_3^c & -u_2^c & \vdots & u_1 & d_1 \\ -u_3^c & 0 & u_1^c & \vdots & u_2 & d_2 \\ u_2^c & -u_1^c & 0 & \vdots & u_3 & d_3 \\ \cdots & \cdots & \cdots & \cdots & \cdots & \cdots \\ -u_1 & -u_2 & -u_3 & \vdots & 0 & e^c \\ -d_1 & -d_2 & -d_3 & \vdots & -e^c & 0 \end{pmatrix} \tag{3.19}$$

Note that colour $SU(3)$ acts on the first three indices, weak $SU(2)$ on the last two indices, and the $U(1)$ of hypercharge is a diagonal generator of the group. The gauge bosons are represented by

$$\mathbf{24} = \begin{pmatrix} & & & \vdots & X_1 & Y_1 \\ & g_{1\ldots8} & & \vdots & X_2 & Y_2 \\ & & & \vdots & X_3 & Y_3 \\ \cdots & \cdots & \cdots & \cdots & \cdots & \cdots \\ \bar{X}_1 & \bar{X}_2 & \bar{X}_3 & \vdots & W & \\ \bar{Y}_1 & \bar{Y}_2 & \bar{Y}_3 & \vdots & & \end{pmatrix} \tag{3.20}$$

Symmetry breaking in $SU(5)$ is achieved via two sets of Higgs bosons, one of which yields the first stage of GUT symmetry breaking

$$SU(5) \rightarrow SU(3)_C \times SU(2)_L \times U(1)_Y \tag{3.21}$$

for which we need, at least, an adjoint $\mathbf{24}$ Higgs representation ϕ

$$\mathbf{24}\ \phi\ :\ \langle 0|\phi|0\rangle = V \begin{pmatrix} 1 & & & & \\ & 1 & & 0 & \\ & & 1 & & \\ & 0 & & -\frac{3}{2} & \\ & & & & -\frac{3}{2} \end{pmatrix} \tag{3.22}$$

The second stage of symmetry breaking

$$SU(2)_L \times U(1)_Y \rightarrow U(1)_{em} \tag{3.23}$$

is achieved by a $\mathbf{5}$ Higgs representation H

$$\mathbf{5}\ H\ :\ \langle 0|H|0\rangle = \begin{pmatrix} 0 \\ 0 \\ 0 \\ \cdots \\ 0 \\ 1 \end{pmatrix} v \tag{3.24}$$

which contains the doublet Higgs present in the Standard Model . The fermion assignments in the $SU(5)$ GUT are unique. The only place to put the colour singlet, $SU(2)$ doublet (ν, e) is in the $\bar{\mathbf{5}}$ representation, and for the total charge in that representation to be zero, one must put the three colours of d^c. As for the **10**, it contains a (1, 1) of $SU(3) \times SU(2)$ which is suitable for the e^c, a $(\mathbf{3}, \mathbf{2})$ that is suitable for the (u, d) doublet, and the remaining places are taken by the three colours of u^c. It is remarkable how simply and elegantly the quarks and leptons fit into this simplest GUT group.

The biggest apparent obstacle to Grand Unification is the inequality of the Standard Model gauge couplings

$$g_3 \gg g_2, g_1 \tag{3.25}$$

After all, the strong interactions are strong, and the weak weak, at present-day energies. This problem is resolved [38] by the logarithmic variations of g_3, g_2 and g_1, as exemplified by the asymptotic freedom of the strong coupling

$$\alpha_3(Q) \simeq \frac{12\pi}{(33 - 2N_q)\ln Q^2/\Lambda^2} + \ldots \tag{3.26}$$

discussed in Lecture I. We recall that in this leading-order formula, N_q stands for the number of light quarks, Q is the energy scale at which the coupling is measured, and Λ is the scale where the strong interactions become strong, which has an experimental value

$$\Lambda = 200 \text{ MeV} \times 2^{0\pm1} \tag{3.27}$$

The decrease Eq. (3.26) of the strong coupling, not forgetting the logarithmic variation of α_2 also, allows the possibility that $\alpha_3 = \alpha_2$ at some mass-energy scale m_X. This can be seen most conveniently in terms of the inverses of the gauge couplings

$$\frac{1}{\alpha_3(Q^2)} - \frac{1}{\alpha_2(Q^2)} = \frac{11 + N_H/2}{12\pi} \ln \frac{Q^2}{m_X^2} + \ldots \tag{3.28}$$

where N_H is the number of electroweak Higgs doublets, and the dots represent higher-order terms. Note that when $Q = m_X$, the first term on the right-hand side vanishes, and $\alpha_3(m_X) = \alpha_2(m_X)$. We also see from Eq. (3.28) that m_X must be exponentially high, because of the relatively slow logarithmic variation of the couplings. Indeed, since $\alpha_3 \simeq 1$ and $\alpha_2 \simeq \alpha_{em}$ in the neighbourhood of the strong interaction scale Λ, we find

$$\frac{m_X}{\Lambda} = \exp\left(\frac{O(1)}{\alpha_{em}} + O(\ln \alpha_{em}) + O(1) + \ldots\right) \tag{3.29}$$

The hypothesis of grand unification can be used to predict $\sin^2\theta_W$. The variations of couplings with energy Q are determined by renormalization group equations, which take the following form through two-loop order

$$Q\,\frac{\partial\alpha_i(Q)}{\partial Q} = -\frac{1}{2\pi}\left(b_i + \frac{b_{ij}}{4\pi}\,\alpha_j(Q)\right)\left[\alpha_i(Q)\right]^2 + \dots$$

$$b_i = \begin{pmatrix} 0 \\ -22/3 \\ -11 \end{pmatrix} + N_g \begin{pmatrix} 4/3 \\ 4/3 \\ 4/3 \end{pmatrix} + N_H \begin{pmatrix} 1/10 \\ 1/6 \\ 0 \end{pmatrix} \tag{3.30a}$$

$$b_{ij} = \begin{pmatrix} 0 & 0 & 0 \\ 0 & -136/3 & 0 \\ 0 & 0 & -102 \end{pmatrix} + N_g \begin{pmatrix} 19/5 & 3/5 & 44/15 \\ 1/5 & 49/3 & 4 \\ 4/30 & 3/2 & 76/3 \end{pmatrix} + N_H \begin{pmatrix} 9/50 & 9/10 & 0 \\ 3/10 & 13/6 & 0 \\ 0 & 0 & 0 \end{pmatrix} \tag{3.30b}$$

where the one-loop coefficients are given in (3.30a) and the two-loop coefficients in (3.30b). Note that in each of Eq. (3.30a) and Eq. (3.30b), the first terms are those associated with the gauge multiplets, the second are those associated with matter particles in $N_g = 3$ fermion generations, and the last terms are due to the Higgs fields, with N_H denoting the number of Higgs doublets. The renormalization coefficients are independent of the specific GUT model, and are due entirely to the conventional Standard Model particles.

One can use experimental values of α_{em} and α_3, together with the hypothesis that $\alpha_3 = \alpha_2 = \alpha_1$ at m_X to predict [39]

$$\sin^2\theta_W(m_Z)_{\overline{\rm MS}} = 0.208 + 0.004(N_H - 1) + 0.006\ln\left(\frac{400\ \text{MeV}}{\Lambda_{\overline{\rm MS}}\ (N_q = 4)}\right) \tag{3.31}$$

which yields

$$\sin^2\theta_W(m_Z)_{\overline{\rm MS}} = 0.214 \pm 0.002 \tag{3.32}$$

if $N_H = 1$ and Λ =100 - 200 MeV, to be contrasted with the latest LEP value [9]

$$\sin^2\theta_W(m_Z)_{\overline{\rm MS}} = 0.2324 \pm 0.0006 \tag{3.33}$$

Here, $\overline{MS}$ denotes a specific prescription for calculating higher-order diagrams. Can one resolve this discrepancy in a non-supersymmetric model? The answer is not by adding in more matter generations, which do not contribute to Eq. (3.31). You could get a better value of $\sin^2\theta_W$ by adding in more Higgs doublets, but this would simultaneously decrease the value of m_X and cause the proton lifetime to be catastrophically short. The discrepancy cannot be removed by varying m_t or α_3 within experimentally allowed ranges. Another way of stating the discrepancy between Eqs. (3.31) and (3.33) is that if one uses the experimental value Eq. (3.33) for $\sin^2\theta_W$ together with the experimental values of α_{em} and α_3, the extrapolated gauge couplings do not all become equal at a single energy [40], as seen in Fig. 15a.

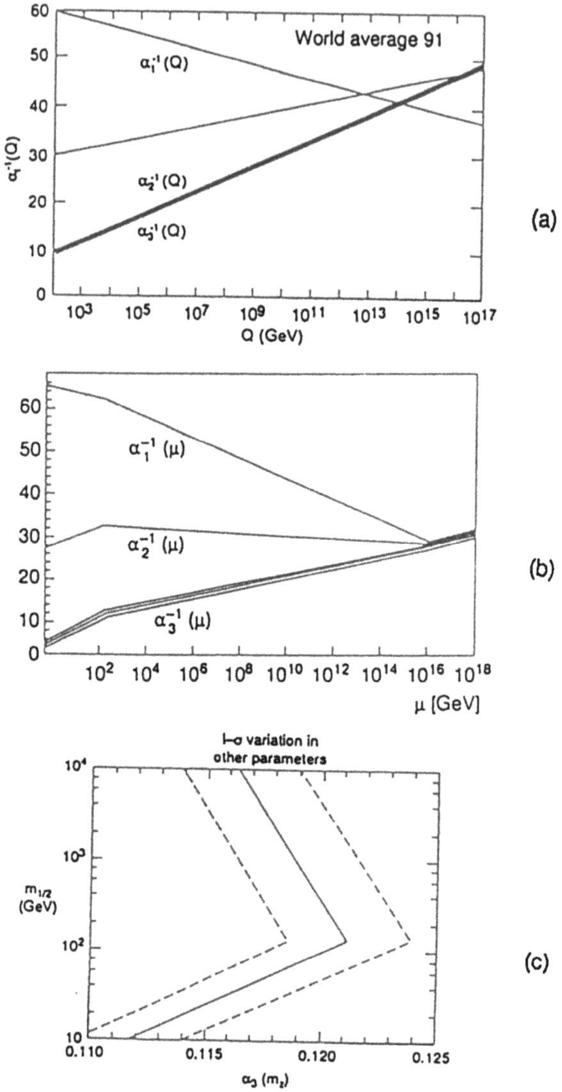

Figure 15: The low-energy couplings measured at LEP (a) do not extrapolate to grand unification without supersymmetry, but (b) do when supersymmetry is included [40], although (c) this agreement cannot be used to calculate reliably the supersymmetry-breaking scale [42].

The most economical way of resolving all these discrepancies and problems is to postulate low-energy supersymmetry [41]. In this case, the renormalization coefficients in Eq. (3.30) are modified as follows: at one loop

$$b_i = \begin{pmatrix} 0 \\ -6 \\ -9 \end{pmatrix} + N_g \begin{pmatrix} 2 \\ 2 \\ 2 \end{pmatrix} + N_H \begin{pmatrix} 3/10 \\ \frac{1}{2} \\ 0 \end{pmatrix} \tag{3.34a}$$

and two loops

$$b_{ij} = \begin{pmatrix} 0 & 0 & 0 \\ 0 & -24 & 0 \\ 0 & 0 & -54 \end{pmatrix} + N_g \begin{pmatrix} \frac{38}{15} & \frac{6}{5} & \frac{88}{15} \\ \frac{2}{5} & 14 & 8 \\ \frac{11}{5} & 3 & \frac{68}{3} \end{pmatrix} + N_H \begin{pmatrix} \frac{9}{50} & \frac{9}{10} & 0 \\ \frac{3}{10} & 7/2 & 0 \\ 0 & 0 & 0 \end{pmatrix} \tag{3.34b}$$

where the coefficients are once again independent of the specific GUT, as long as it incorporates supersymmetry. The gauge couplings measured at low energies appear to agree better with the predictions of a supersymmetric GUT [40], and Fig. 15b shows that the couplings measured at low energies extrapolated to high energies using Eqs. (3.34) are consistent with all being equal at the supersymmetric grand unification scale. However, it is not possible [42] yet to use this agreement to deduce the values of the supersymmetric particle masses, as shown in Fig. 15c. As one varies α_3 over the allowed range from 0.110 to 0.130, one can accomodate any value of the supersymmetry-breaking gaugino mass $m_{1/2}$ between 10 GeV and many orders of magnitude above the expected 1 TeV scale. This type of analysis of the gauge couplings is *consistent* with supersymmetric GUT unification, but does not *prove* it, nor *fix* the supersymmetry-breaking scale.

III.4 Neutrino masses, mixing and oscillations

Neutrino masses are much smaller than those of the corresponding leptons and quarks [43]

$$\begin{aligned} m_{\nu_e} &\lesssim 7.2 \text{ eV} \\ m_{\nu_\mu} &\lesssim 270 \text{ keV} \\ m_{\nu_\tau} &\lesssim 24 \text{ MeV} \end{aligned} \tag{3.34}$$

where the limit on m_{ν_e} comes from studies of the end-point in Tritium β-decay, the upper limit on m_{ν_μ} comes from studies of $\pi \to \mu\nu$ decay, and the upper limit on m_{ν_τ} comes from the end-point of $\tau \to \nu_\tau 5\pi$. The second distinctive feature of neutrinos is that only left-handed neutrinos have ever been observed in the laboratory. In the Standard Model , this is because only left-handed neutrinos appear in the electroweak charge current

$$J_\mu = \bar{e}\gamma_\mu(1-\gamma_5)\nu_e + \bar{\mu}\gamma_\mu(1-\gamma_5)\nu_\mu + \bar{\tau}\gamma_\mu(1-\gamma_5)\nu_\tau \tag{3.35}$$

Neither the Standard Model nor the minimal $SU(5)$ GUT contain right-handed neutrinos, whilst both quarks q and leptons ℓ have both left- and right-handed states. This means that quarks and charged leptons can have "Dirac" masses

$$g_{H\bar{f}f}\ H_{\substack{\Delta I=1/2\\ \Delta L=0}}\ \bar{f}_R f_L \Rightarrow m_f^D = g_{H\bar{f}f}\ \langle 0|H_{\substack{\Delta I=1/2\\ \Delta L=0}}|0\rangle \tag{3.36}$$

generated by a v.e.v. for the Standard Model Higgs boson with $\Delta I = 1/2, \Delta L = 0$. A puzzle then arises: if ν_R exists, why are the m_ν so much smaller than $m_{q,\ell}$? If the ν_R do not exist, can m_ν be non-zero?

The answer to the second question is affirmative, because one can replace the $\bar{f}_R$ in Eq. (3.36) by f_L^c. Thus there is a "Majorana" neutrino mass term

$$m^m \nu_L^T C \nu_L = m^m \bar{\nu}_L^c \nu_L \equiv m^m \text{“}\nu_L \nu_L\text{”} \tag{3.37}$$

which does not exist for either quarks or charged leptons. If we tried to couple two q^L fields together, we would have $\Delta Q_{em}, \Delta C \neq 0$, and if we tried to couple two ℓ_L together we would have $\Delta Q_{em} \neq 0$, which are not possible for mass terms. Note, however, that the Majorana mass term in Eq. (3.37) requires lepton number violation $\Delta L = 2$ and weak isospin $\Delta I = 1$. There is no Higgs field with such quantum numbers in either the Standard Model or minimal $SU(5)$. However, such a Majorana mass term

$$\tilde{g}_{H\bar{\nu}\nu}\ H_{\substack{\Delta I=1\\ \Delta L=2}}\ \text{“}\nu_L \nu_L\text{”} \Rightarrow m^M = \tilde{g}_{H\bar{\nu}\nu}\ \langle 0|H_{\substack{\Delta I=1\\ \Delta L=2}}|0\rangle \tag{3.38}$$

is possible and present in some more complicated GUT models, or could appear via higher-order non-renormalizable couplings [44] such as

$$\frac{1}{M}\ (H_{\Delta I=1/2}\nu_L)\ (H_{\Delta I=1/2}\nu_L) \tag{3.39}$$

Here M is a heavy mass parameter which could originate from the exchange of a massive ν_R

$$\frac{1}{M} = \frac{g^2}{M_{\nu_R}} \tag{3.40}$$

This mass could in turn arise from a Majorana mass term with $\Delta I = 0$.

Combining all the possible sources of neutrino masses discussed in the previous paragraph, we arrive at the following generic "see-saw" neutrino mass matrix [45]

$$(\nu_L\ ,\ \bar{\nu}_R) \begin{pmatrix} m^M & m^D \\ m^D & M^M \end{pmatrix} \begin{pmatrix} \nu_L \\ \bar{\nu}_R \end{pmatrix} \tag{3.41}$$

where m^M is a $\Delta I = 1$ Majorana mass, m^D is a Dirac mass which is $\cong m_{2/3}$ in many GUT models, and M^M is a $\Delta I = 0$ Majorana mass term. Generically, we expect in GUTs that

$$m_{\nu_e} \ll m_{\nu_\mu} \ll m_{\nu_\tau} \tag{3.42}$$

Diagonalizing the mass matrix in Eq. (3.41), we obtain the following mass eigenstates, where M should be understood as lying anywhere in the range $m_P \gtrsim M \gtrsim (\alpha/\pi)^2 \times m_X$

$$\begin{aligned} \nu_L + O\left(\frac{m_W}{M}\right)\bar{\nu}_R \ &: \ m = O\left(\frac{m_W^2}{M}\right) \\ \nu_R + O\left(\frac{m_W}{M}\right)\bar{\nu}_L \ &: \ m = O(M) \end{aligned} \tag{3.43}$$

If we extend the above picture to three generations, neglect possible neutrino mixing, and assume that the large Majorana masses M_i do not depend strongly on the generation index i, we arrive at the following likely hierarchy of light neutrino masses

$$m_{\nu_i} \sim \frac{m_{\ell_i}^2 \text{ or } m_{q_i}^2}{M_i} \Rightarrow m_{\nu_e} \ll m_{\nu_\mu} \ll m_{\nu_\tau} \tag{3.44}$$

In this picture, the most likely candidate to constitute Hot Dark Matter would be the ν_τ, which could have a mass around 10 eV if $M_3 \sim 10^{12}$ GeV.

In point of fact, we do expect the different flavours of neutrinos to mix analogously to the quarks, leading to neutrino oscillations [46]. Experiments searching for low-energy solar neutrinos ($E_\nu \lesssim 10$ MeV) see fewer than expected in the standard solar model. The explanation for this apparent deficit could be partly experimental and partly astrophysical. However, one favoured explanation is that the neutrinos produced by the sun mix and oscillate before reaching the Earth. The most likely neutrino oscillation scenario seems to be matter-enhanced mass oscillations of neutrinos within the Sun [47]. According to this hypothesis, Fig. 16 shows the preferred regions of the ($\sin^2\theta$, Δm^2) plane [48]. According to the generic GUT see-saw neutrino mass matrix shown in Eq. (3.41), the most likely hypothesis would be $\nu_e \to \nu_\mu$ oscillations, with

$$\sin^2\theta_{e\mu} \simeq 10^{-2} \tag{3.45}$$

If one scales up [49] the resulting $m_{\nu_\mu} \sim 3 \times 10^{-3}$ eV with the expected hierarchical ratio

$$m_\nu \sim m_u^2 \ : \ m_c^2 \ : \ m_t^2 \tag{3.46}$$

one easily finds $m_{\nu_\tau} \cong 10$ eV, which would be just right for the ν_τ to constitute Hot Dark Matter with $\Omega_\nu \cong 0.3$, as suggested by models [50] of structure formation in the aftermath of COBE [51]. Moreover, one is led [49] to the likelihood that

$$\sin^2\theta_{\mu\tau} \sim 10^{-3} \text{ to } 10^{-4} \tag{3.47}$$

which is within reach of a new round of neutrino oscillation experiments currently under preparation at CERN [52].

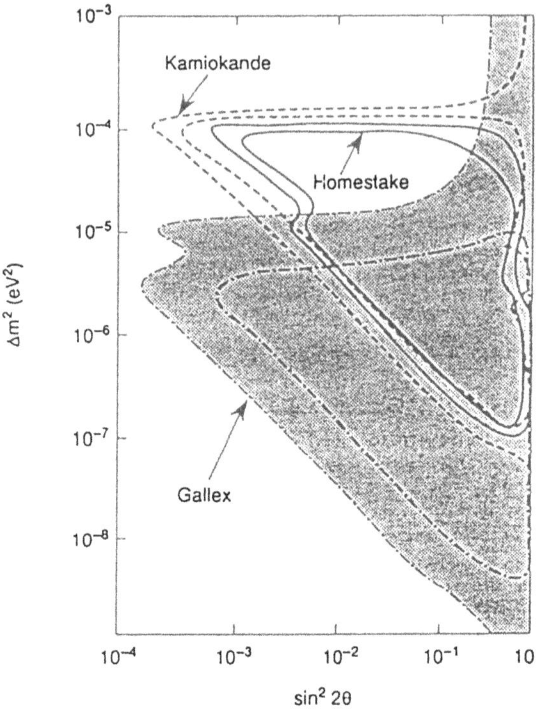

Figure 16: Matter-enhanced neutrino oscillation fits to the solar neutrino data [48].

IV Inflation

IV.1 The psychology of inflation

Just like the Standard Model of particle physics, the standard Big Bang cosmology leaves many unanswered puzzles, despite its manifold successes.

Foremost amongst these puzzles is the *horizon/smoothness problem.* Distant parts of the Universe that are becoming visible only just now, i.e., coming within our apparent horizon, are very similar. This is shown most strongly by the upper limit

$$\left(\frac{\delta T}{T}\right)_{\rm CMBR} \lesssim 10^{-5} \tag{4.1}$$

on the anisotropy of the microwave background radiation, as well as by the apparent homogeneity and isotropy of the matter distribution at sufficiently large distances.

The isotropy in Eq. (4.1) is a puzzle, because in standard Big Bang cosmology, those parts of the Universe would previously have been outside their relative event horizons. This is because [1]

$$a \sim t^{1/2} \text{ or } t^{2/3} \tag{4.2}$$

in a radiation- or matter-dominated cosmology, respectively, whereas the horizon $a_H \cong t$. The fact that distant parts of the Universe would have been beyond the event horizon is shown graphically in Fig. 17.

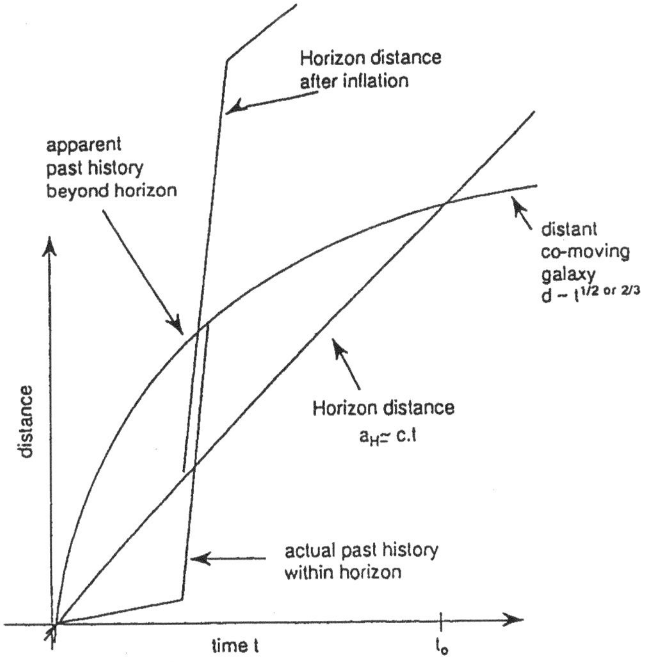

Figure 17: The horizon and the history of a distant galaxy at rest in a comoving reference frame, in conventional Big Bang cosmology and in inflationary cosmology.

Another difficulty for Big Bang cosmology is the *flatness/oldness problem.* If we consider the standard Friedman-Robertson-Walker cosmological equation

$$\left(\frac{\dot{a}}{a}\right)^2 = H^2 = \frac{8\pi G_N \rho}{3} - \frac{k}{a^2} \tag{4.3}$$

and, as in Lecture I, define

$$\Omega \equiv \frac{\rho}{\rho_c} \tag{4.4}$$

then we find that at different times

$$\Omega(t) = 1/(1 - \frac{k/a^2}{8\pi G_N \rho/3}) \tag{4.5}$$

where $\rho \sim a^{-4}$ for a radiation-dominated epoch and $\sim a^{-3}$ during a matter-dominated epoch. The fact that $|\Omega - 1| = O(1)$ today means that

$$|\Omega - 1| \lesssim 10^{-60} \tag{4.6}$$

at the initial Planck epoch $t \cong 10^{-43}$ s. In other words, the fact that the curvature term is not dominant in Eq. (4.3) today means that it must have been absolutely insignificant in the very early Universe, which requires fine tuning.

A third and, not unrelated, problem is that of *entropy*. Why does the visible Universe contain so many particles?

$$S \gtrsim 10^{90} \tag{4.7}$$

This is related to the size and age of the Universe, since a typical entropy density

$$S \sim T^3 \sim a^{-3} \tag{4.8}$$

The difficulty is that the only scale in the Friedman-Robertson-Walker Eq. (4.3) for Big Bang cosmology is $m_P \cong 10^{19}$ GeV. Therefore, a generic "typical" solution of the equation is expected to have size $a_P \sim 10^{-33}$ cm and age $t_P \sim 10^{-43}$ s. Moreover, according to Eq. (4.8), this "typical" solution, i.e., Universe, has $O(1)$ particles in it.

Finally, let us mention the *monopole problem* [53], which played an important historical role in the development of the inflationary idea. The problem is that GUTs predict the existence of magnetic monopoles with masses $m_M \sim 10^{17}$ GeV, and that standard Big Bang cosmology suggests

$$\frac{n_M}{n_\gamma} \sim \left(\frac{m_X}{m_P}\right)^3 \gtrsim 10^{-12} \tag{4.9}$$

which is much larger than allowed by cosmological nucleosynthesis and observational limits. This discrepancy provided the historical stimulus for inflation [54], but nowadays I worry less about it, since monopoles may not exist in any case. However, other unwanted relic particles such as gravitinos [55] may exist, and inflation provides a means for diluting their abundance until they are essentially undetectable.

The basic idea of inflation is illustrated in Fig. 17. At some early epoch, it is suggested [54] that the energy density of the Universe might have been dominated by an approximately constant term

$$\left(\frac{\dot{a}}{a}\right)^2 = \frac{8\pi\ G_N \rho}{3} - \frac{k}{a^2} \ : \ \rho = V \tag{4.10}$$

It is easy to check that, in this case, there would have been an epoch of exponential expansion

$$a = a_i \exp(Ht) \; : \quad H = \sqrt{\frac{8\pi\ G_N}{3} V} \tag{4.11}$$

with the Hubble constant related directly to the value of the energy density. This scenario has the following virtues.

The *horizon* would have expanded exponentially, meaning that all the observable Universe including the most distant galaxy and the points of emission of the microwave background radiation would have been within each other's event horizon before inflation. Therefore, homogeneity could have been established at an early epoch in the Universe. Another way of stating this is that the "real" horizon

$$a_i e^{H\tau} \tag{4.12}$$

is much larger than the naïve "apparent" horizon

$$a_H = ct_0 \tag{4.13}$$

as long as the Universe expanded by at least sixty e-foldings during the inflationary epoch.

The next benefit of inflation is the *flatness* of the Universe. It is clear that the exponential growth Eq. (4.11) in the scale factor means that the curvature term in the Friedman-Robertson-Walker equation (4.3) becomes negligible. Thus the Universe is indistinguishable from a critical Universe with $k = 0$, although, as we shall see in Section IV.3, perturbations lead one to expect

$$|\Omega - 1| \sim 10^{-5} \tag{4.14}$$

even today.

The *entropy* of the Universe is very large, just because the Universe is exponentially large and very old.

There is no *monopole* within our event horizon, since the closest one would have been inflated away beyond the microwave background radiation. A similar argument applies to any other unwanted particles such as gravitinos [55], ...

The approximately constant energy density postulated in Eq. (4.10) is very similar to a cosmological constant Λ

$$R_{\mu\nu} - \frac{1}{2} g_{\mu\nu} R = 8\pi\ G_N T_{\mu\nu} + \Lambda g_{\mu\nu} \tag{4.15}$$

If we compare this expression with the energy-momentum tensor for a standard fluid

$$T_{\mu\nu} = -p g_{\mu\nu} + (\rho + p) U_\mu U_\nu \tag{4.16}$$

where $U_\mu = (1,0,0,0)$ for a co-moving fluid, we see that we can write

$$R_{\mu\nu} - \frac{1}{2} g_{\mu\nu} R = 8\pi \ G_N T^\Lambda_{\mu\nu} \tag{4.17}$$

where

$$\rho_\Lambda = \frac{\Lambda}{8\pi \ G_N} = -p_\Lambda \tag{4.18}$$

This is the source of the common statement that inflation is driven by a negative pressure density.

There has been a lot of discussion at this school about the possibility that a cosmological constant could play a role in the present-day expansion of the Universe. Any such cosmological constant term would obey the constraint

$$\rho_\Lambda < \rho_c \Leftrightarrow \Lambda < 10^{-73} \ \mathrm{GeV}^2 \tag{4.19}$$

which is much smaller than any Λ suggested by any known microphysics source, as seen in Table 3. I see no reason within particle physics why one should expect a value of Λ which is in the range of present cosmological interest. Rather, the cosmological constant problem seen from the point of view of a particle physicist is: why is Λ not many, many orders of magnitude larger than the experimental upper bound of Eq. (4.19)? This, however, is a problem for the Theory of Everything including quantum gravity. Inflation is something we can achieve with quantum field theory, and we see in the next section various possible ways of doing this.

Table 3:

	ρ_Λ	$\div \rho_c$
QCD	10^{-4} GeV^4	10^{44}
Electroweak	10^{9} GeV^4	10^{57}
GUTs	10^{60} GeV^4	10^{110}
Q Gravity	10^{70} GeV^4	10^{120}

IV.2 Some inflationary models

Historically, the first particle physics model of inflation was the *old inflation* [54] illustrated in Fig. 18a. The basic idea was that the Universe could have become stuck in a false vacuum, i.e., a local, not global, minimum, for some time before making a first-order phase transition by tunnelling quantum-mechanically through the barrier shown in Fig. 18a. The little ball represents the classical value of the so-called "inflaton" field ϕ. We will discuss in Section IV.3 the important quantum fluctuations in the field ϕ. During inflation, there is an effective cosmological constant

$$\Lambda = 8\pi\ G_N V(\phi = 0) \tag{4.20}$$

resulting in an effective Hubble expansion rate

$$H = \sqrt{\frac{8\pi\ G_N}{3} V} \tag{4.21}$$

In order to carry out the inflationary programme, we need

$$H\tau \gtrsim 60 \tag{4.22}$$

where τ is the duration of the inflationary epoch. Being a first-order transition, we expect inflation to be terminated via bubble nucleation as discussed in Lecture III. The problem here is whether the transition would ever have been completed [56]. The question arises because the volume of the "cheese" in between the bubbles of the new vacuum is exponentially expanding

$$a^3(t) \sim e^{3Ht} \tag{4.23}$$

whilst the co-moving bubble sizes remain constant. The probability $p_{old}(t)$ that any piece of the Universe will remain in the old vacuum is determined by the nucleation transition rate Γ

$$p_{old}(t) \simeq \exp\left(-\frac{4\pi}{3}\left(\frac{\Gamma}{H^4}\right)Ht\right) \tag{4.24}$$

This means that the fraction of the Universe still in the old vacuum is

$$F_{old}(t) = p_{old}(t)a^3(t) \sim \exp\left(\frac{-4\pi}{3}\left(\frac{\Gamma}{H^4}\right)Ht\right) \times \exp(3Ht) \tag{4.25}$$

The dilemma is that completion of the transition requires $F_{old}(t) \to 0$, which in turn requires $\Gamma/H^4 \gg 1$. However, in this case the transition would be so rapid that the inflationary requirement of Eq. (4.22) would not be satisfied. Therefore, in old inflationary models, either the phase transition is complete or the inflationary epoch is long enough, but not both simultaneously.

Among the possible fixes of this problem is *new inflation* [57]. The idea here is illustrated in Fig. 18b, namely that the exponential expansion of the Universe would have occurred during a slow roll-over of the inflation field ϕ into the new vacuum. This roll-over might have started after a first-order transition, as illustrated by the dotted lines in Fig. 18b, but this is not essential. The key requirement in new inflation is that of a very flat potential, so that the roll-over may be long enough to obtain enough e-foldings, cf. Eq. (4.22). The idea of new inflation is that we live inside a bubble of the new vacuum that expanded exponentially with the exponentially-expanding cheese. In a generic new inflationary model, the bubble of new vacuum that we inhabit is larger than the observed Universe, although there could be regions beyond our bubble that are still expanding exponentially in the old vacuum. It should be noted that $V(\phi)$ is not quite constant during the roll-over illustrated in Fig. 18b, and hence that the Hubble expansion rate H is not quite constant, and hence as we shall see in Section IV.3, the quantum perturbations are not quite scale-invariant.

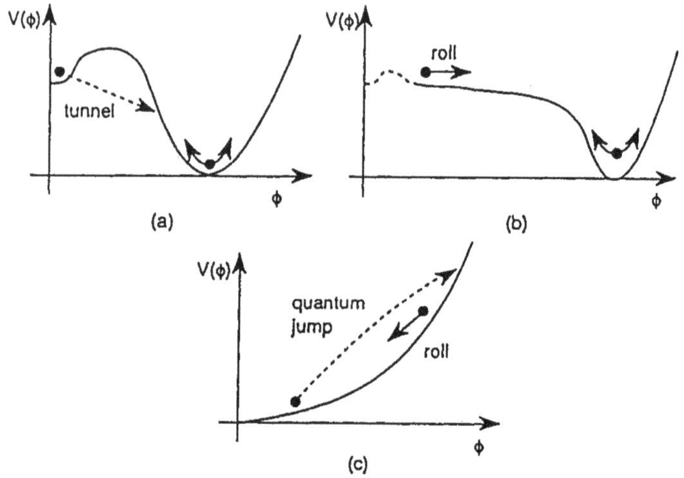

Figure 18: The behaviour of the Higgs field in (a) old inflation [54], (b) new inflation [57], and (c) a chaotic inflationary model [60].

An alternative fix to the dilemma of old inflation goes by the name of *extended inflation* [58]. Recall that the dilemma was that one requires $\Gamma/H^4 \leq 1$, corresponding to a slow transition rate with many e-foldings, but then the transition is never completed, as seen in Eq. (4.25). The solution proposed is that Γ/H^4 is variable, because of other parameters in the theory. Generically, one can represent this idea by introducing another scalar field

$$V(\phi) \to V(\phi, \chi) \; ; \quad \chi = \chi(t) \tag{4.26}$$

An example of extended inflation is provided by the Brans-Dicke [59] theory of gravity, according to which the gravitational constant is in fact variable

$$G_N \to G_{eff} = \Phi^{-2} \tag{4.27}$$

where Φ is the Brans-Dicke scalar field. As a result of the behaviour in Eq. (4.27), H decreases with time in such a way that

$$a(t) \sim t^{power} \tag{4.28}$$

whilst the efficiency of bubble nucleation

$$\frac{\Gamma}{H^4} \sim t^4 \tag{4.29}$$

This has the desired feature that it is small initially, leading to a large expansion, but eventually becomes large, ensuring completion of the inflationary phase transition.

A model of extended inflation with more than one scalar field, as shown in Eq. (4.26), strikes me personally as somewhat epicyclic. My preference is for the simplicity of *chaotic inflation* [60]. According to this idea, there is no need to postulate a metastable (false) vacuum, i.e., there is no secondary local minimum. Instead, one considers a sufficiently flat generic potential such as

$$V(\phi) \sim \lambda\phi^n \quad \text{or} \quad e^{\alpha\phi} \tag{4.30}$$

where favourite values are $n = 2$ or 4, higher values being disfavoured by the requirement of renormalizability, and the second form is motivated by some stringy considerations. The idea then is that the classical configuration of the inflaton field ϕ, indicated by the ball in Fig. 18c, rolls slowly classically down the potential. However, quantum fluctuations may drive the field ϕ locally up the slope in some parts of the Universe. If a quantum fluctuation is sufficiently strong, i.e., the ball is driven sufficiently high up the potential, then its subsequent classical roll-down will yield sufficient inflation to obey the constraint shown in Eq. (4.22). Clearly, the requirement that quantum fluctuations may plausibly be sufficiently strong imposes certain conditions on the parameters of the effective potential shown in Eq. (4.30).

This chaotic inflationary scenario suggests the possibility of a reincarnation [61] of the steady-state Universe, via a "stationary" Universe in which inflationary domains constantly reproduce themselves. Here "stationary" means that the distribution of probability P_p to find any given value of the inflaton field ϕ at any given time t in any given volume V is constant. In realistic models, it is claimed [61] that the expectation value of ϕ increases with time, in such a way that the energy density rapidly increases to order of m_P^4, the Planck density. One then finds that there is a fixed fraction of the volume of the Universe with any given properties, such as the particular set of field values and parameters that make human life possible. Since further speculations along these lines would rapidly become philosophical, let us return to experimental reality with a discussion of quantum fluctuations in a generic inflationary model.

IV.3 Inflationary perturbations

There are two important types of perturbations [62] which are generated during a generic inflationary epoch, namely density or scalar perturbations and gravity waves or tensor perturbations. The former arises in the density field $\rho(\mathbf{x})$, with the perturbations being defined by

$$\begin{aligned} \delta(\mathbf{x}) &\equiv \frac{\rho(\mathbf{x}) - \langle\rho\rangle}{\langle\rho\rangle} \\ &= \int d^3\mathbf{k}\delta_{\mathbf{k}}\, e^{-i\mathbf{k}\cdot\mathbf{x}} \end{aligned} \tag{4.31}$$

The magnitude of the density perturbation at a given length scale λ is defined by

$$\left(\frac{\delta\rho}{\rho}\right)^2_\lambda \equiv \left.\frac{k^3|\delta_{\mathbf{k}}|^2}{2\pi^2}\right|_{k^{-1}=\lambda} \tag{4.32}$$

The evolution of a perturbation depends on λ/a_H. When $\lambda/a_H < 1$, the evolution depends on the dynamics and hence on the astrophysics. It depends on the nature of matter, its equation of state, dissipation, and in particular the Jeans instability. The wave equation for a generic field with wave number k is given by

$$\ddot{\delta}_{\mathbf{k}} + 2H\dot{\delta}_{\mathbf{k}} + v_s^2\frac{k^2}{a^2}\,\delta_{\mathbf{k}} = 4\pi\; G_N\langle\rho\rangle\delta_{\mathbf{k}} \tag{4.33}$$

where $v_s^2 \equiv dp/d\rho$ is the square of the speed of sound. It is easy to see from this equation that if the wave number $k > k_J$, where the latter Jeans wave number is defined by

$$k_J^2 = \frac{4\pi\; G_N a^2\langle\rho\rangle}{v_s^2} \tag{4.34}$$

then the relevant mode oscillates. However, if the wave number $k < k_J$, then the relevant mode grows. For example, in a Universe dominated by Cold Dark Matter, $v_s \simeq 0$, and hence all perturbations within the horizon grow according to this theory of gravitational instability.

On the other hand, if $\lambda/a_H > 1$, so that the wavelength of the perturbation is larger than the horizon size, the above theory of gravitational instability does not apply. The evolution of these perturbations is discussed here in detail by Lukash [63]. For our purposes, I will just recall that the ratio

$$\frac{\delta\rho}{\rho + p} \tag{4.35}$$

is gravitational-gauge-invariant and remains constant as long as the perturbation is outside the horizon. Hence the value which it takes when inflated beyond the horizon during the original inflationary epoch is equal to the value of Eq. (4.35) when the perturbation is brought back inside the horizon by the subsequent expansion of the Universe.

As we have already seen, during inflation the scalar field energy (cosmological constant) contributions to ρ and p are equal and opposite [see Eq. (4.18)], which means that

$$\rho + p \sim \langle \dot{\phi}^2 \rangle \tag{4.36}$$

The numerator in Eq. (4.35) is given by

$$\delta\rho = \delta\phi \times \frac{\partial V}{\partial \phi} \equiv \delta\phi \times V'(\phi) \tag{4.37}$$

To estimate this we recall that during the roll-over

$$\ddot{\phi} + 3H\dot{\phi} + V'(\phi) = 0 \tag{4.38}$$

where the first term is negligible if the roll-over is sufficiently slow. In this case we may write

$$\dot{\phi} \simeq \frac{-V'(\phi)}{3H} \tag{4.39}$$

where

$$H^2 = \frac{8\pi}{3} G_N(V(\phi) + \frac{1}{2}\dot{\phi}^2) \simeq \frac{8\pi\ G_N}{3} V(\phi) \tag{4.40}$$

and in writing the second part of this equation we have assumed that the dominant contribution is that due to the potential $V(\phi)$. The quantum fluctuations of any scalar field in a de Sitter space have the magnitude [64]

$$\delta\phi \simeq \frac{H}{2\pi} \tag{4.41}$$

Inserting this and Eq. (4.39) into Eq. (4.35) we find that

$$\frac{\delta\rho}{\rho + p} \simeq \frac{\delta\phi \cdot V'}{\dot{\phi}^2} \simeq \frac{H^3 \cdot V'}{(V')^2} \simeq \frac{V^{3/2}}{V'} \tag{4.42}$$

during the inflationary epoch. Recalling that, as mentioned above, this quantity takes the same value when it returns within the event horizon, we find that

$$\left(\frac{\delta\rho}{\rho}\right)_{\lambda = a_H} \equiv A_S(\phi) = \frac{\sqrt{2}\kappa^2}{8\pi^{3/2}} \frac{H^2}{|H'|} \tag{4.43}$$

where we have defined $\kappa^2 \equiv 8\pi G_N$. Note that in a generic inflationary model neither factor on the right-hand side is constant, so we do not expect the scalar $\delta\rho/\rho$ to be exactly constant.

Now we turn to the gravity-wave perturbations [62],[63]. They are governed by a massless wave equation in curved space, similar to that discussed previously for the scalar field

$$\ddot{h}_{\mathbf{k}}^{1,2} + 3H\dot{h}_{\mathbf{k}}^{1,2} + \frac{k^2 h_{\mathbf{k}}^{1,2}}{a^2} = 0 \tag{4.44}$$

There are two such equations for the two physical polarization states of the graviton particle (gravitational field strength), where we have defined

$$g_{\mu\nu} = g_{\mu\nu}^{FRW} + h_{\mu\nu} \tag{4.45}$$

and rewritten Eq. (4.45) in terms of the Fourier modes of the gravitational perturbation $h_{\mu\nu}$. As in the case of the scalar modes discussed above, these also are unchanged as long as they remain outside the horizon, and hence have the same value when they return within the horizon as they had when inflation pushed them beyond the horizon. Again as for the scalar field the initial value was [64]

$$h_{\mathbf{k}} \simeq \frac{H}{2\pi} \tag{4.46}$$

Hence the strength of the gravity (tensor) perturbations is

$$|k^3 h_{\mathbf{k}}|_{\lambda=a_H} \equiv A_G(\phi) = \frac{\kappa}{4\pi^{3/2}}\, H \sim V^{1/2} \tag{4.47}$$

We see again that, in a generic inflationary model, the right-hand side is not constant, and hence gravitational waves also have a non-trivial scale dependence.

It is instructive to compare the relative magnitude of the density and gravitational (tensor) waves

$$\frac{A_S(\phi)}{A_G(\phi)} = \frac{\kappa}{\sqrt{2}}\,\frac{H}{|H'|} \qquad \kappa^2 \equiv 8\pi\, G_N \tag{4.48}$$

We see that this ratio is > 1 if $\ddot{a} > 0$, as we expect. However, the relative contributions of the scalar (density) and tensor (gravitational) waves to large-angle cosmic microwave background radiation anisotropies are given approximately by

$$\frac{25 A_G^2}{2 A_S^2} \tag{4.49}$$

Therefore, the gravitational waves can be important or even dominate, even though they are in some sense bounded by the scalar waves.

Equipped with these results, we now investigate what the COBE data tell us about the inflaton potential [65], and what possible future data could tell us. To interpret the COBE constraints on observations, we first need to understand the relation between a given value of the inflaton field ϕ and the cosmological scale parameter λ of a given perturbation, as measured in Mpc, for example. There is a one-to-one correspondence $\lambda(\phi)$ between any given cosmological scale and the value of the inflaton field ϕ when this scale crossed the Hubble radius $a(\phi)$ during the inflationary epoch. If we define the values

$$\phi \equiv \phi_e\ , \quad a \equiv a_e \tag{4.50}$$

at the end of inflation, the number of e-foldings is simply given by

$$a_e = a(\phi)\exp\left[N(\phi)\right] \tag{4.51}$$

where

$$N(\phi) \equiv \int_t^{t_e} H(t')dt' = \frac{\kappa^2}{2}\int_\phi^{\phi_e} \frac{H(\phi')}{|H'(\phi')|}\,d\phi' \tag{4.52}$$

Similarly, it is easy to verify that

$$\lambda(\phi) = \frac{\exp\left[N(\phi)\right]}{H(\phi)}\;\frac{a_0}{a_e} \tag{4.53}$$

which can in turn be related to the perturbation spectra of the tensor and gravitational waves defined in the previous section

$$\frac{1}{\lambda}\;\frac{d\lambda(\phi)}{d\phi} = \frac{\kappa}{\sqrt{2}}\left(\frac{A_S}{A_G} - \frac{A_G}{A_S}\right) \tag{4.54}$$

As mentioned there, we expect non-trivial spectral indices both for gravitational waves

$$1 - n_G \equiv \frac{d\ln A_G^2(\lambda)}{d\ln\lambda} = 2/\left(1 - \frac{A_G^2(\lambda)}{A_S^2(\lambda)}\right) \tag{4.55}$$

and for density perturbations

$$1 - n_s = \frac{d\ln A_S^2(\lambda)}{d\ln\lambda} \tag{4.56}$$

My personal prejudice is that the scalar spectral index is more likely to be close to unity than is the tensor spectral index.

The formulae of the previous section can be used to express the value of the inflaton potential at a given cosmological scale λ in terms of the magnitudes of the gravitational and scalar perturbations at the same scale

$$V\left[\phi(\lambda)\right] = \frac{16\pi^3 A_G^2(\lambda)}{\kappa^4}\;\left[3 - \frac{A_G^2(\lambda)}{A_S^2(\lambda)}\right] \tag{4.57}$$

where

$$\phi(\lambda) = \frac{\sqrt{2}}{\kappa}\int^\lambda \frac{d\lambda'}{\lambda'}\;\frac{A_S A_G}{A_S^2 - A_G^2} \tag{4.58}$$

COBE [51] tells us that the value of the potential Eq. (4.57) is small at the value of λ corresponding to about 10° in the microwave sky. This is because we can express Eq. (4.57) in the form [65]

$$V = \frac{16\pi^3}{\kappa^4}\;A_{\mathrm{COBE}}^2\;\left[\frac{3 - A_G^2/A_S^2}{1 + \frac{25}{2}\,A_G^2/A_S^2}\right]\;\frac{A_G^2}{A_S^2} \tag{4.59}$$

The quantity in parentheses has a maximum magnitude of 0.13, which is achieved if $A_G^2/{A_S}^2 \sim 0.10$. Here $A_{\rm COBE} \sim 10^{-5}$ is the effective perturbation strength measured by COBE, which has the weighting of scalar and gravitational modes given in Eq. (4.49). We immediately infer from Eq. (4.59) the striking fact that the fourth root of the value of the inflationary potential V is much smaller than the Planck scale $\sim 10^{18} - 10^{19}$ GeV, and indeed is very similar to the scale of supersymmetric Grand Unification discussed in Lecture III:

$$V \lesssim 10^{-8}\, m_P^4 \quad \text{i.e.} \quad \sqrt[4]{V} \lesssim 10^{16}\ \text{GeV} \tag{4.60}$$

For future reference, we note that the slope of the potential at a given scale λ can also be related to the magnitudes of the gravitational and scalar perturbations [65]

$$V' \equiv \frac{dV(\phi(\lambda))}{d\phi} = \frac{16\pi^3}{\sqrt{2}\kappa^3}\,\frac{A_G^3}{A_S}\left[7 - n_s - (5 - n_s)\,\frac{A_G^2}{A_S^2}\right] \tag{4.61}$$

where n_s is the scalar spectral index defined in Eq. (4.56). Likewise, the curvature of the effective potential is given by [65]

$$V'' \simeq \frac{4\pi^3}{\kappa^2}\,\frac{A_G^2}{A_S^2}\left[4(n_s - 4)^2 A_G^2 - (1 - n_s)\,(7 - n_s)\,A_S^2\right] \tag{4.62}$$

in the slow roll-over approximation. The curvature shown here can be regarded as the square of the effective mass of the inflaton field during the roll-over, and Eq. (4.62) suggests that this can be no more than about 10^{14} GeV. Since all the entropy of the Universe is generated by the oscillations of the inflaton field at the end of the inflationary epoch, everything in the Universe must be made out of particles coupled to the inflaton, either appearing in its decay or coupled by thermalized interactions. This means, in particular, that any particle whose out-of-equilibrium decay is supposed to be responsible for baryogenesis should weigh less than $m_{\rm inflaton}$, and hence no more than about 10^{13} GeV.

IV.4 Sphalerons and leptogenesis

Much of the interest in the electroweak phase transition has been triggered by the discovery that the electroweak interactions violate baryon and lepton number conservation non-perturbatively [12], and the possibility that this might provide scenarios for cosmological baryogenesis, if the electroweak phase transition were sufficiently strongly first order. In this section, I will try to give you some idea of the reasons how and why non-perturbative electroweak effects violate B and L. The first fact to recognize is that the vacuum structure in a gauge theory is very complicated, with many configurations whose energies are equal in perturbation theory, as seen in Fig. 19. The horizontal axis is labelled by the so-called Chern-Simons number.

The perturbative effects discussed up to now correspond to analyzing small quantum fluctuations in the gauge and Higgs fields A_μ, ϕ around one of these vacua. Non-perturbative effects include transitions between different vacua labelled by different values of the Chern-Simons number. In a pure (unbroken) gauge theory such as QCD, these transitions are generated by instantons [12], which are configurations of the gauge field A_μ

$$A_\mu \ : \ \frac{g^2}{32\pi^2}\int d^3x\ F_{\mu\nu}\tilde{F}^{\mu\nu} = N_I \ : \ \tilde{F}^{\mu\nu} \equiv \epsilon^{\mu\nu\rho\sigma}F_{\rho\sigma} \tag{4.63}$$

where N_I is an integer. Things are more complicated in the electroweak theory, which possesses classical non-perturbative field configurations of (A_μ, ϕ) called sphalerons [66], that correspond to bumps in the effective potential shown in Fig. 19. These are unstable saddle points of the effective potential, which have masses

$$E_{sph} \simeq \frac{m_W(T)}{\alpha} \simeq 10\ \mathrm{TeV} \tag{4.64}$$

These two classes of non-perturbative configurations mediate transitions with rates

$$e^{-4\pi/\alpha_W}\ ,\quad e^{-\beta E_{sph}} \approx e^{-\frac{m_W(T)}{\alpha_W \cdot T}} \tag{4.65}$$

The instantons at $T = 0$ are negligible, but the sphalerons at $T \neq 0$ can be very important [67]. Remember that $m_W(T) \to 0$ at high temperatures, so the exponential suppression factor vanishes.

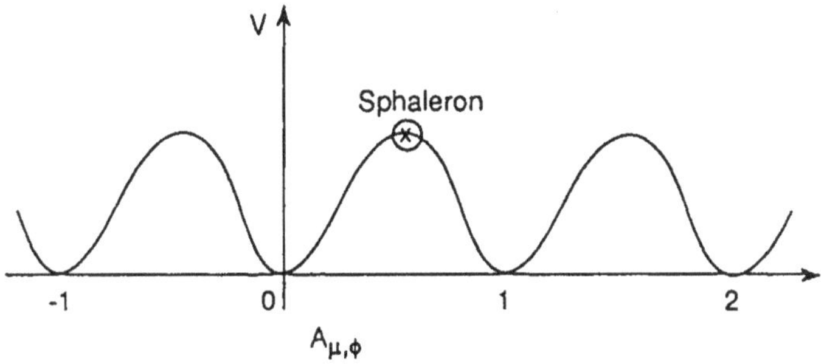

Figure 19: The effective potential in a spontaneously-broken gauge theory as a function of the Chern-Simons number.

The next point to recognize is that these non-perturbative transitions between vacua with different Chern-Simons number violate baryon number B and lepton number L, since these non-perturbative gauge field configurations contribute to the divergences of the corresponding baryon and lepton number currents

$$\partial^\mu J_\mu^{baryon} = N_g\,\frac{g^2}{32\pi^2}\int d^3x F_{\mu\nu}F^{\mu\nu} = \partial^\mu J_\mu^{lepton} \tag{4.66}$$

Here N_g is the number of matter generations, known from LEP to be three, and the remaining factor is an integer $N = 1, 2, \ldots$. The corresponding changes in B and L are

$$\Delta B = N_g = 3 = \Delta L \tag{4.67}$$

Note that although B and L are violated, the combination $(B - L)$ is conserved. The rate of baryon number violation is given by

$$(\dot{B} + \dot{L}) = -c(B + L)\, T \exp\left(-\frac{E_{sph}}{T}\right) \tag{4.68}$$

in the low-temperature phase where the electroweak gauge symmetry is broken. As already mentioned, the exponential suppression factor is expected to vanish in the high-temperature symmetric phase, since

$$E_{sph}(T) \simeq \frac{m_W(T)}{\alpha_W} \rightarrow 0 \tag{4.69}$$

The quantity c is a numerical coefficient that can only be estimated using non-perturbative techniques [68]. Lattice simulations of theories thought to be analogous to gauge theory indicate that

$$(\dot{B} + \dot{L}) = -c(B + L)T \tag{4.70}$$

Comparing the rate of $(B + L)$ violation with the Hubble expansion rate, we see that baryon and lepton number violating processes were presumably in equilibrium at temperatures above the electroweak critical temperature, implying that

$$(B + L) = 0 \tag{4.71}$$

This point should be taken into account in any scenario for cosmological baryogenesis.

Let us recall the three Sakharov [69] conditions for cosmological baryogenesis

1. Interactions that violate baryon number

   These are present non-perturbatively in the Standard Model , thanks to the sphalerons discussed in the previous section.

2. Interactions that distinguish matter and antimatter

   This means interactions that violate the following discrete symmetries. Charge conjugation C changes particles into antiparticles and vice versa

   $$f \leftrightarrow \bar{f}(f^c) \tag{4.72}$$

   The combination of charge conjugation with parity P, denoted CP, exchanges particles and antiparticles and simultaneously exchanges left with right

   $$(\overleftarrow{f} + \overrightarrow{f}) \leftrightarrow (\overrightarrow{f^c} + \overleftarrow{f^c}) \tag{4.73}$$

It has been known for a long time that the charged-current weak interactions violate C maximally. The violation of CP has been seen in a laboratory in the weak decays of kaons. In particular,

$$P(K \to \pi^+ e^- \nu) \neq P(K \to \pi^- e^+ \bar{\nu}) \tag{4.74}$$

This result is interpreted as evidence for CP violation in the mass matrix mixing K_0 and $\bar{K}_0$.

3. Interactions that are out of thermal equilibrium

   Thermodynamics in the form of Boltzmann's H-theorem tells us that the population of states in equilibrium depends only on their mass, and it is a sacred theorem of quantum field theory that particles and antiparticles have identical masses. Therefore, thermal equilibrium forbids any matter-antimatter asymmetry.

The non-perturbative electroweak baryon number violating interactions described in the previous section would wash out rapidly any previously-generated $(B + L)$ asymmetry, and hence tend to wash out any baryon asymmetry generated by GUTs, for example. However, if the GUT were able to generate a $(B - L)$ asymmetry, this would survive the electroweak sphaleron wash-out. Alternatively, one could try to generate the baryon asymmetry via the electroweak interactions, which would be possible if the electroweak phase transition were sufficiently strongly first order, so that sphaleron interactions were out of equilibrium below the electroweak critical temperature, as discussed above. A third possibility would be to use the equilibrium sphalerons at high temperatures to reprocess a pre-existing lepton asymmetry [70]. This works as follows: if there is an initial net lepton number $L_i \neq 0$, then the $(B + L)$-violating sphalerons ensure that

$$B_f + L_f = 0 \tag{4.75}$$

However, $B - L$ conservation ensures that

$$B_f - L_f = -L_i \tag{4.76}$$

The solution of these last two equations is to generate

$$B_f = -\frac{L_i}{2} \tag{4.77}$$

An example of such a scenario will now be given.

The suggestion [70] is that heavy neutrinos ν^c could have decayed out of equilibrium, their dominant decays being $\nu^c \to (\nu, \ell)H$ via a Dirac mass coupling. Higher-order diagrams could cause C and CP violation in ν^c decays, leading to a lepton asymmetry

$$\epsilon_1 = \frac{1}{2\pi(\lambda_L^{+}\lambda_L)_{ii}} \sum \mathrm{Im}\ \left[(\lambda_L^{+}\lambda_L)_{ij}\right]^2 f\left(\frac{m_j^2}{m_i^2}\right) \tag{4.78}$$

Subsequent reprocessing by electroweak sphalerons, which violate $(B+L)$ but conserve $(B-L)$, could turn this lepton asymmetry into a baryon asymmetry. For this mechanism to work, one needs

$$m_{\nu^c} < m_{inflaton} \lesssim 10^{13}\ \mathrm{GeV} \tag{4.79}$$

which is consistent with the above fit to the solar neutrino problem and the suggestion that the ν_τ could constitute the Hot Dark Matter in the Universe.

V Dark Matter

V.1 Running hot and cold dark matter?

Inflation tells us to expect that the density of the Universe as a whole is indistinguishable from the critical density: $\Omega = 1$. However, observations only provide $\Omega_{\mathrm{visible}} \lesssim 0.01$, whereas Big Bang nucleosynthesis apparently restricts $\Omega_B \lesssim 0.1$. Measurements of rotation curves suggest that galactic haloes contain $\Omega_{\mathrm{halo}} \sim 0.1$. Thus the halo could in principle be made out of baryons, although arguments have been given that these could not be in the form of gas, dust or snowballs. However, the possibility that they could be "Jupiters" or brown dwarfs, i.e., small, "failed" stars remains open. Some microlensing events caused by brown dwarfs have apparently been seen [71]. However, it is not clear whether they are due to objects in the galactic halo, or in the galactic disc, or the Large Magellanic Cloud. It still seems likely that there is a substantial contribution to the local halo density, estimated to be

$$\rho_{halo} \sim 0.3\,\mathrm{GeV\,cm^{-3}} \sim 0.01\, m_\odot\, pc^{-3} \tag{5.1}$$

from non-baryonic dark matter.

Theories of galaxy formation seem to require some non-baryonic dark matter to enhance initial density perturbations via gravitational instability. Theories of structure formation in the early Universe distinguish two categories of dark matter: hot, which was relativistic when galaxy-sized perturbations came within the event horizon and began to grow, and cold, which was non-relativistic at that epoch. The one you favour depends on your pet theory of galaxy formation. If you believe these formed from a Gaussian random field of perturbations laid down during inflation, then you should prefer cold dark matter. This is because it enables perturbations to grow on all scales from galaxies upwards, whereas hot dark matter escapes from small galaxy-sized perturbations, retarding their growth. Thus, in a pure hot dark matter scenario, galaxies only form late as a result of larger structures breaking up. This scenario seemed to be disfavoured by the observation of galaxies and quasars at high redshifts, as well as by other considerations. However, if galaxies formed from seeds, such as cosmic strings, hot and cold dark matter might be preferred.

Assuming some type of inflationary scenario, the comparison of COBE [51] and other large-scale data with other data on structures at smaller scales seems to require a modification of the pure cold dark matter scenario. One possibility is that there is a non-zero cosmological constant, but this seems very unlikely from the point of view of particle physics, since there is no known natural reason why the cosmological constant should be in the range of cosmological utility. The two prime candidates for modifying the standard cold dark matter scenario seem to be mixed dark matter with an admixture of hot and cold dark matter

$$\Omega_{\rm Cold} \sim 0.7, \; \Omega_{\rm Hot} \sim 0.3, \; \Omega_{\rm Baryons} \lesssim 0.1 \tag{5.2}$$

or a tilt in the spectrum of primordial density perturbations. The parameters of these two models can be fixed by galaxies and by the COBE data, which fix the overall normalization of $\delta\rho/\rho$: the question is then what the models predict at intermediate scales, in particular for the peculiar motions of clusters. At least some astrophysicists will tell you that the cold dark matter with tilt model predicts peculiar motions on the scale of clusters that are too small by comparison with the data. Moreover, at least in simple models one expects a very small amount of tilt. Therefore, my own preference at the moment is for a mixed dark matter model as in Eq. (5.2).

V.2 Massive neutrinos

Figure 20 shows a sketch of the relic "neutrino" density, or the density of some other neutral particle coupled to the Z^0, as a function of its possible mass. When the neutrino mass is very small (less than 1 MeV), the relic neutrino number density is independent of the mass, and hence the relic energy density

$$\rho_\nu \propto m_\nu \tag{5.3}$$

When the mass of the neutrino increases above about 30 eV, this neutrino density exceeds the critical density, and we enter a disallowed range of neutrino masses. We emerge from this disallowed region again when the neutrino mass is of order a few GeV [72], when the number density of neutrinos is suppressed by a Boltzmann factor $e^{-m_\nu/T}$. Since the neutrino energy density is inversely proportional to the $\bar{\nu}\nu$ annihilation cross-section, as we shall see in more detail in the next paragraph, the relic neutrino density is very small for neutrinos with mass about $m_Z/2$, whilst the relic neutrino density increases again for larger values of m_ν. We are therefore left with three interesting ranges of neutrino masses

$$m_\nu \sim 30 \text{ eV}\ ,\quad m_\nu \sim 3 \text{ GeV}\ ,\quad m_\nu \sim 1 \text{ TeV} \tag{5.4}$$

Neutrinos would constitute Hot Dark Matter in the first window, and Cold Dark Matter in either of the second or third windows. I know of no well-motivated alternatives to neutrinos as Hot Dark Matter, but there are other candidate Cold Dark Matter particles whose relic abundance is calculated in a similar way to that of neutrinos.

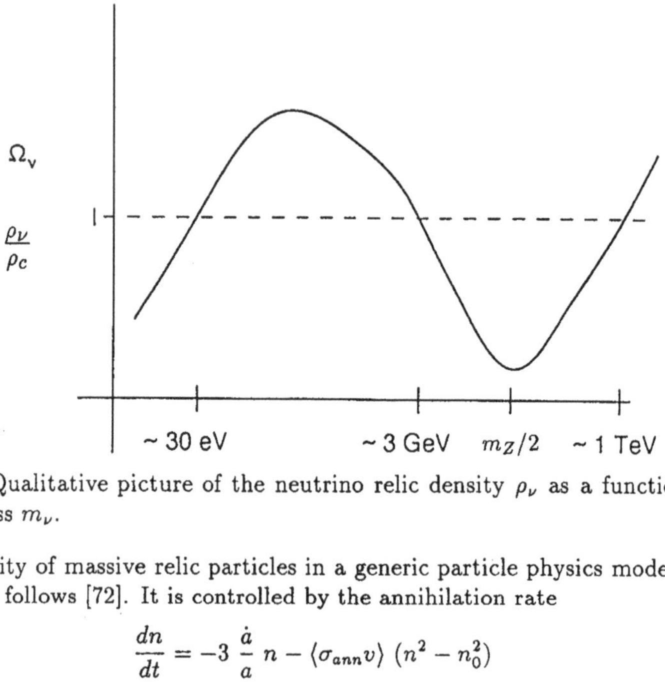

Figure 20: Qualitative picture of the neutrino relic density ρ_ν as a function of the neutrino mass m_ν.

The density of massive relic particles in a generic particle physics model may be estimated as follows [72]. It is controlled by the annihilation rate

$$\frac{dn}{dt} = -3\ \frac{\dot{a}}{a}\ n - \langle\sigma_{ann}v\rangle\ (n^2 - n_0^2) \tag{5.5}$$

where n is the particle number density, the first term on the right-hand side is the normal dilution due to the Hubble expansion, the last term is due to annihilation with a relative velocity v, and n_0 denotes the normal Bose-Einstein or Fermi-Dirac equilibrium number density. In most cases of interest, one can expand the low-energy annihilation cross-section in the general form

$$\langle\sigma_{ann}v\rangle \simeq a + bx\ :\quad x \equiv \frac{\langle v^2\rangle}{6} \tag{5.6}$$

There is a simple approximate solution for $n(t)$, called the freeze-out approximation. According to this, $n = n_0$ until

$$\dot{n}_0 \simeq \langle\sigma_{ann}v\rangle n_0^2 \tag{5.7}$$

which occurs at a characteristic temperature, that of freeze-out T_f. Thereafter, the evolution of the number density is supposed to be given by

$$\dot{n} \simeq \langle \sigma_{ann} v \rangle n^2 \tag{5.8}$$

Exact numerical calculations for a number of particle candidates confirm that this freeze-out approximation give results which are accurate to about 20 % in most cases, yielding

$$\rho \simeq 0.8 \; (T_\chi)^3 \; \frac{m_\chi}{a x_f + 1/2 b x_f^2} \tag{5.9}$$

where the first factor on the right-hand side is a correction for the naïve freeze-out approximation, and T_χ is the effective temperature of the relic particles, which is related to that of photons ($T_\gamma \simeq 2.73K$) by the number of effective degrees of freedom at the freeze-out temperature T_f

$$\left(\frac{T_\chi}{T_\gamma}\right)^3 \simeq \left(\frac{g_{eff}(T_f)}{g_{eff}(\text{low } E)}\right)^{-3/4} \tag{5.10}$$

One can represent the relic density as

$$\rho \simeq 4.0 \times 10^{-40} \; \left(\frac{T_\chi}{2.8^0\text{K}}\right)^3 \; g_{eff}^{1/2} \; \left(\frac{\text{GeV}^{-2}}{a x_f + 1/2 b x_f^2}\right) \; \text{g cm}^{-3} \tag{5.11}$$

in convenient units.

V.3 Lightest supersymmetric particle

We saw in Lecture III that in many supersymmetric theories the lightest supersymmetric particle or LSP is likely to be absolutely stable and hence a good candidate for a cosmological relic from the Big Bang [36]. The LSP is likely to be electromagnetically neutral and have only weak interactions with baryonic matter, and the possible candidates are: sneutrinos $\tilde{\nu}$ of spin 0, some gaugino/Higgsino mixture called generically a neutralino χ of spin 1/2, and the gravitino $\tilde{G}$ of spin 3/2. We will now discuss which of these candidates appears to be the most plausible.

Sneutrino $\tilde{\nu}$

Such a Cold Dark Matter particle would have coherent spin-independent interactions with heavy nuclei, which have not been seen in underground experiments using Si and Ge materials [73], yielding the constraints

$$m_{\tilde{\nu}} \lesssim 10 \text{ GeV} \quad \text{or} \quad \gtrsim 1 - 10 \text{ TeV} \tag{5.12}$$

The first inequality can be strengthened by considering possible sneutrino capture by the Sun, followed by annihilation with the production of high-energy neutrinos from the solar core. The fact that these have not been seen tells us that $\tilde{\nu}$ escapes from the Sun, and hence must be light [74]

$$m_{\tilde{\nu}_{e,\mu}} \lesssim 4 \text{ GeV} \tag{5.13}$$

Finally, as was pointed out in the previous section, the LEP neutrino counting experiments tell us that

$$m_{\tilde{\nu}} \gtrsim 40 \text{ GeV} \tag{5.14}$$

Combining these constraints we see that the $\tilde{\nu}$ is ruled out as the Cold Dark Matter particle unless it weighs more than 1-10 TeV, in which case it is unlikely to be the lightest supersymmetric particle.

Gravitino $\tilde{G}$

In the absence of cosmological inflation, the relic number density of gravitinos would be somewhat smaller than that of photons or neutrinos, since gravitinos would have decoupled from the Universe at a very early epoch [75]. This means that the gravitino mass required to give the critical density is about

$$m_{\tilde{G}} \sim 1 \text{ keV} \tag{5.15}$$

Such gravitinos would constitute "Warm" rather than Hot or Cold Dark Matter. If there was a cosmological inflationary epoch, the primordial density of gravitinos would be suppressed, and the above limit does not apply. Indeed, models suggest that

$$m_{\tilde{G}} \gtrsim m_{\tilde{e}} > 45 \text{ GeV} \tag{5.16}$$

in which case the gravitino would not be the lightest supersymmetric particle, and again the limit in Eq. (5.16) does not apply. In view of Eq. (5.16), the possibility that the gravitino is the LSP appears implausible, but it cannot be ruled out. We can only hope that the gravitino is indeed not the LSP, because its interactions with ordinary matter are so weak that it is unlikely it could ever be detected.

Neutralino χ

This is a generic name for a mixture of neutral spin-1/2 gauginos and higgsinos $\tilde{\gamma}/\tilde{Z}/\tilde{H}_1/\tilde{H}_2$, which is parametrized by the unmixed gaugino mass $m_{1/2}$, the Higgs mixing parameter μ, and $\tan\beta = v_2/v_1$, the ratio of Higgs v.e.v.'s in the MSSM. The full mass matrix for neutral gauginos and higgsinos is [36]

$$M_{\tilde{N}} = \begin{pmatrix} M_2 & 0 & \frac{-g_2 v_2}{v_2} & \frac{g_2 v_1}{v_2} \\ 0 & M_1 & \frac{g' v_2}{v_2} & \frac{-g' v_1}{v_2} \\ \frac{-g_2 v_2}{v_2} & \frac{g' v_2}{v_2} & 0 & \mu \\ \frac{g_2 v_1}{v_2} & \frac{-g' v_1}{v_2} & \mu & 0 \end{pmatrix} \tag{5.17}$$

The lightest mass eigenstate, which is the best candidate to be the LSP and hence Cold Dark Matter [36], simplifies in the limits $m_{1/2} \rightarrow 0$, in which case it is approximately a photino $\tilde{\gamma}$, and $\mu \rightarrow 0$, in which case it is a higgsino $\tilde{H}$. However, a generic LSP is a complicated mixture, as you can well imagine from the matrix in Eq. (5.17). Figure 21 shows the domains of parameter space allowed by searches at LEP and other accelerators for the LSP and other related supersymmetric particles. They tell us that [76]

$$m_\chi \gtrsim 10 \text{ to } 20 \text{ GeV} \tag{5.18}$$

The constraints used in drawing Fig. 21 include that from the invisible Z^0 decay width (neutrino counting), the total Z^0 decay width, the peak Z^0 peak cross-section, searches for $Z \rightarrow \chi\chi'$ followed by $\chi' \rightarrow \chi(\ldots)$ decays, called zen events, unsuccessful searches for charged partners of the neutralinos, and the unsuccessful search for gluinos at the FNAL $\bar{p}p$ collider. Also shown in Fig. 21 are regions of parameter space where an "interesting" amount of Cold Dark Matter can be found, namely

$$\Omega h^2 = 0.1 \text{ to } 1 \tag{5.19}$$

for some values of the other supersymmetric model parameters.

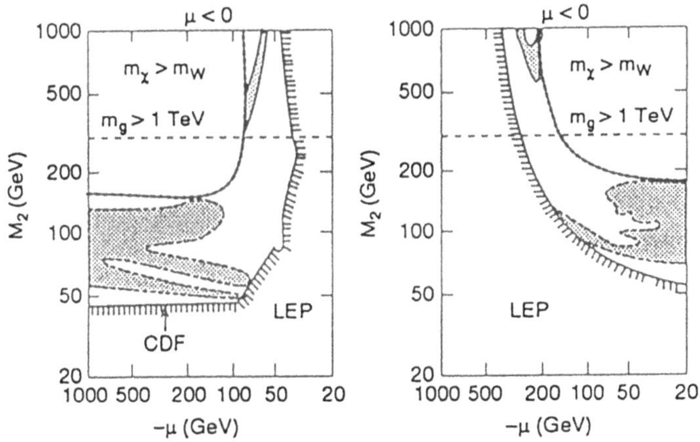

Figure 21: Regions of the $(\mu, m_{1/2})$ plane excluded by LEP and CDF searches, together with regions where the lightest neutralino mass $m_\chi < m_W$, and the relic neutralino density is in an "interesting" range $0.1 < \Omega_{LSP}h^2 < 1$.

We conclude that the spin-1/2 neutralino χ is a very good candidate to be the Cold Dark Matter beloved of astrophysicists, and we will devote the rest of this lecture to discussing strategies for experimental searches for this LSP.

V.4 Searches for cold dark matter particles

V.4.1 Annihilations in the galactic halo

Here the basic idea [77] is that massive Cold Dark Matter particles, such as the neutralinos discussed above, will occasionally meet in the galactic halo, annihilating into l^+l^- or $\bar{q}q$ pairs for the most part, which can then lead to stable particles such as antiprotons, positrons, photons or neutrinos in the cosmic rays. All such stable particles fluxes share the same dependences

$$\phi_{\bar{p},e^+,\gamma} \propto n^2 \sigma_{\rm Ann} \tau \tag{5.20}$$

on the local halo density of Cold Dark Matter particles n, corresponding to the expected halo density of around 0.3 GeV cm^{-3}, on the relic annihilation cross-section $\sigma_{\rm Ann}$, and on the length of time that the produced particle spends within the galaxy. In my view [78], the best way to fix the annihilation cross-section is to assume that the Cold Dark Matter particle has the closure density. The corrections in a mixed dark matter model, in which $\Omega_{cold} \simeq 0.7$, would not be very different, and are easy to take into account[1]. A much larger source of uncertainty is the time τ that the produced particle spends in the galaxy. This is easy to calculate in the case of a photon or neutrino, being given simply by the light travel time from the annihilation point, but is rather uncertain in the case of $\bar{p}$'s and e^+'s. The other factors in Eq. (5.20) have to be estimated from dynamical models of the differential distribution dn/dE expected from a given $\bar{f}f$ annihilation final state. The best guide [80] to the latter may be a Monte Carlo program of the type developed to model final-state particle spectra in e^+e^- annihilation at LEP and elsewhere.

The first stable annihilation product to be considered was the $\bar{p}$, particularly at low energies. The flux of $\bar{p}$'s at moderate energies above about 1 GeV is qualitatively consistent with cosmic-ray models assuming the secondary production of $\bar{p}$ by primary matter cosmic-ray particles. At one time, it was claimed that there might be an excess of low-energy $\bar{p}$'s with a typical energy around 200 MeV [81]. However, this apparent anomaly disappeared, and the present low-energy data [82] are also qualitatively consistent with the normal matter cosmic-ray model. The methods outlined in the previous paragraph can be used to derive from these experiments an upper bound on a number density of massive Cold Dark Matter relic particles, in particular on the number density of neutrinos. The cosmic ray e^+ flux can also be used to derive upper limits on the neutralino number density. Both of these limits are sensitive to the assumed time that the $\bar{p}$'s and e^+'s spend in a galactic magnetic field.

[1] If there is an admixture of Hot Dark Matter neutrinos: $\Omega_{\rm hot} \simeq 0.3$, these are not expected [79] to constitute a large fraction of the galactic halo.

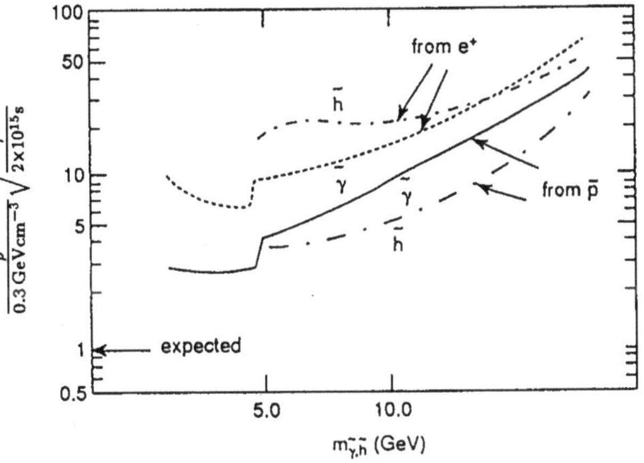

Figure 22: Upper limits on the photino or higgsino density in the galactic halo, from the absence of $\bar{p}$ or e^+ annihilation products [78].

Typical upper limits [78] on the neutralino density from $\bar{p}$ and e^+ fluxes are shown in Fig. 22. We see that the $\bar{p}$'s give slightly more stringent upper limits, and conclude that

$$\rho_{\tilde{\gamma},\tilde{h}} \lesssim (5 \text{ to } 10) \times \rho_{\text{Halo}} \tag{5.21}$$

is the conservative upper limit from these cosmic-ray studies. Since the stable particle fluxes depend quadratically on the neutralino density, as seen in Eq. (5.20), and the confinement time of antiprotons and positrons in the galactic magnetic field is rather uncertain, it seems to me unlikely that searches for these particles will give us useful information about the neutralino halo density.

V.4.2 Annihilations in the Sun or Earth

Here the basic idea is as follows. As a relic particle in a galactic halo passes through the solar system, it may pass through the Sun [83] (or Earth [84]) itself, and perhaps undergo there an elastic collision with one of the nuclei (mainly protons) beneath the solar surface. Such a collision would cause the relic particle to lose energy to the struck nucleus, and this energy loss might convert it from an initially hyperbolic orbit into an elliptic orbit with a perihelion below the solar surface. Then it would pass repeatedly through the Sun, losing more energy through collisions with nuclei inside the Sun from time to time, until eventually it settled into an approximately isothermal distribution, characterized by energy balance with the solar nuclei. It has been proposed [85] that such a process might actually cause the core of the Sun to cool, by transferring thermal energy to outer regions of the Sun. This would be of interest for low-energy solar neutrinos, since it would tend to reduce their calculated flux. Detailed calculations indicate, however, that this is not possible with supersymmetric particles. The population of relic particles within the Sun is controlled by annihilation and/or by evaporation [86], the particle analogues of civil war and emigration. Annihilations into $\bar{f}f$ final states could lead directly or indirectly to high-energy neutrinos detectable on Earth. Other annihilation products would of course be absorbed within the Sun. The high-energy neutrinos could be detected either directly in deep underground proton decay experiments, or indirectly via upward-going muons produced by neutrino collisions in the rock beneath the detector.

As in the case of annihilations in the galactic halo, one finds [87] that with conservative and reasonable assumptions, one can at best establish an upper limit

$$\begin{aligned} \rho_{\tilde{\gamma}} &\lesssim 10 \times \rho_{\mathrm{Halo}} \quad \text{for} \quad m_{\tilde{\gamma}} \lesssim 10 \text{ GeV} \\ \rho_{\tilde{h}} &\lesssim 10 \times \rho_{\mathrm{Halo}} \quad \text{for} \quad m_{\tilde{h}} \lesssim 20 \text{ GeV} \end{aligned} \tag{5.22}$$

No range of m_χ is yet excluded at a level of the expected halo density, if one assumes that the neutralino relic density is close to the closure dentity. You may sometimes see parts of the $(\mu, m_{1/2})$ plane that are apparently excluded by searches for high-energy solar neutrinos [88], but so far these are regions where the relic neutrino density is in any case much below the closure density, and probably unable to constitute the galactic halo.

Before leaving this method of searching for dark matter particles, it is worth commenting that matter-enhanced neutrino oscillations could conceivably cloud the interpretation of the experiments [89]. This is because even high-energy neutrinos produced in the core of the Sun may oscillate into other, less detectable, neutrino flavours before they reach the Earth. This effect is shown in Fig. 23 for a particular choice of oscillation parameters

$$\begin{aligned} \sin^2 2\theta &= 0.6 \\ \Delta m^2 &= 8 \times 10^{-6} \text{ eV}^2 \end{aligned} \tag{5.23}$$

that is consistent with the low-energy solar neutrino experiments. Here we see that the ν_e flux produced in the Sun is enhanced before arriving at the Earth, whilst the $\nu_{\mu,\tau}$ fluxes are relatively suppressed. It is also possible to find neutrino mass and mixing parameters such that the ν_e flux is suppressed, whilst the $\nu_{\mu,\tau}$ flux is enhanced. This possibility of neutrino flavour transmutation means that care must be exercised in interpreting upper limits from terrestrial searches for high-energy solar neutrinos. To be on the safe side, one should always assume [89] that by the time the neutrinos arrive at the Earth they have the flavour for which your detector efficiency is the smallest, generally the ν_τ.

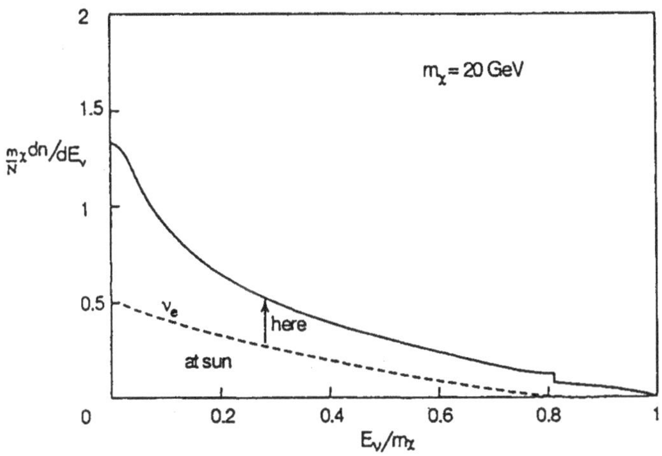

Figure 23: Possible matter-enhanced oscillation effects on the high-energy solar neutrino spectrum due to relic annihilations [89].

Another place to look for possible neutrino annihilations is in the core of the Earth [84], following trapping of the neutralinos there by a process similar to that in the Sun. Since the composition of the Earth is different from that of the Sun, the calculation of the trapping rate is different, and the trapping efficiency has a different dependence on the neutralino mass. First searches have been made for a high-energy neutrino flux emanating from the earth core, but these have also not yet attained the sensitivity required to see the galactic halo density of neutralinos.

V.4.3 Dark matter searches in the laboratory

Here the principal interest is in elastic scattering of Cold Dark Matter particles on heavy nuclei [90]. The typical recoil energy in such a collision is

$$\Delta E < m_\chi v^2 = 10 \left(\frac{m_\chi}{10 \text{ GeV}}\right) \text{ keV} \tag{5.24}$$

where we have assumed a typical galactic velocity $v \simeq 300$ km s^{-1}. The neutralino-nucleus scattering rate is built up in several steps, starting with the effective interaction Lagrangian between neutralinos and the quark constituents of a proton or neutron inside the nucleus. This takes the form

$$\mathcal{L}_{eff} = \sum_q [(\bar{\chi}\gamma_\mu\gamma_5\chi) \;\; (\bar{q}\gamma^\mu(A_q P_L + B_q P_R)q) + (\bar{\chi}\chi) \cdot C_q m_q \bar{q}q] \tag{5.25}$$

where $P_{L,R} \equiv 1/2(1 \pm \gamma_5)$, the first term is due to sfermion $\tilde{f}$ and Z_0 exchange and is spin-dependent, whilst the second term is due to $\tilde{f}$ and H exchange and is spin-independent.

The next step is to estimate the spin-dependent neutralino-nucleon scattering matrix elements, which may be written as

$$M_{SD}(\chi_{(p,n)} \to \chi_{(p,n)}) = 4\mathrm{S}_\chi \cdot \mathrm{S}_{(p,n)} \sum_{q\in(p,n)} (B_q - A_q)\, \Delta q \tag{5.26}$$

Initially, theorists used the naïve constituent quark model to estimate the Δq [90]. However, recent experiments at CERN and elsewhere [91] indicate that the naïve quark model does not give an adequate picture of the quark contributions to the proton and neutron spins. Instead, a better approximation seems to be given [92] by the Skyrme model, which treats protons and neutrons as rotating coherent clouds of very light u, d and s quarks. In what follows, we will assume [93] the following values for the quark contributions to the proton spin

$$\Delta u = 0.77 \pm 0.08\ ,\quad \Delta d = -0.49 \pm 0.08\ ,\quad \Delta s = -0.15 \pm 0.08 \tag{5.27}$$

which are extracted from the EMC experiment and close to the Skyrme model predictions. Recent experiments on polarized muon- and electron-neutron scattering give estimates of the Δq which are quite consistent with Eq. (5.27), but with smaller errors. For a recent review of experimental determinations of Δq, see Ref. [94].

The matrix element for the spin-independent part of the neutralino-nucleon scattering amplitude may be parametrized by [95]

$$M_{SI}\Big(\chi(p,n) \to \chi(p,n)\Big) = (\bar{\chi}\chi)\cdot(m_p\bar{p}p, m_n\bar{n}n)$$
$$\left[\hat{f}\ \frac{m_u C_u + m_d C_d}{m_u + m_d} + f C_s + \frac{2}{27}(1 - f - \hat{f})\left(C_u + C_b + C_t\right)\right] \tag{5.28}$$

Here $\hat{f} \simeq 0.05$ and $f \simeq 0.2$ have the interpretations of the fractions of the proton mass carried by $u+d$ and s quarks, respectively. These quantities are related to other experimental observables, in particular the cross-section for low-energy π-nucleon scattering. For a review of the relevant strong interaction uncertainties, see Ref. [96]. The coefficients A_q, B_q and C_q are given in [93] and references therein.

The next step is to apply the above scattering amplitudes in Eqs (5.26) and (5.28) to estimate the scattering amplitudes for neutralinos on heavy nuclei containing many protons and neutrons. This yields [97],[93],[98] the following expression for the elastic scattering rate of an LSP on a nucleus

$$R = (R_{SD} + R_{SI})\ Y\left(\frac{4m_\chi m_N}{(m_\chi + m_N)^2}\right)$$
$$\times\left(\frac{\rho_{LSP}}{0.3\ \mathrm{GeV\ cm^{-3}}}\right)\left(\frac{\langle|\mathbf{v}_E|\rangle}{320\ \mathrm{km\ s^{-1}}}\right)\ \text{events kg}^{-1}\ \text{day}^{-1} \tag{5.29}$$

where Y is the isotopic abundance of the nuclei considered, ρ_{LSP} is a local halo density of LSPs, and $\mathbf{v}_E$ is the mean LSP velocity relative to the Earth. The first two terms in Eq. (5.29) are the spin-dependent and spin-independent expressions shown below

$$R_{SD} = 580\ \lambda^2 J(J+1)\ \zeta(r_{spin})\ \left(10^4\ \mathrm{GeV}^2 \sum_q (B-A)\Delta q\right)^2 \tag{5.30}$$

where λ is given by one's nuclear model, $\zeta(r_{spin})$ is the spin form factor, and

$$\begin{aligned} R_{SI} &= 210 m_N^2 m_Z^4\ \zeta(r_{charge}) \\ &\times \left(\hat{f}\ \frac{m_u C_u + m_d C_d}{m_u + m_d} + f C_s + \frac{2}{27}\left(1 - f - \hat{f}\right)\left(C_c + C_b + C_t\right)\right)^2 \end{aligned} \tag{5.31}$$

where $\zeta(r_{charge})$ is the Fermi mass form factor. To normalize the scattering rate, we adjust the sparticle parameters so that $\Omega_{LSP} = 1$ for $H_0 = 50$ km s^{-1} Mpc^{-1}.

Typical results [98] are shown in Fig. 24. Comparing Figs. 24a and 24b, we see that the scattering rates are dependent on the Higgs sector of the theory, parametrized here by the pseudoscalar Higgs mass m_A. In the case of Fluorine, the rate is dominated by the spin-dependent interaction. We see that, at least in the case m_A =200 GeV, a large fraction of the $(\mu, m_{1/2})$ space in which $\Omega_{LSP} = 1$ is possible can be covered by realistic experiments. Figures 24c and 24d show the corresponding rates for a target containing equal amounts of ^{73}Ge and ^{76}Ge. In this case, the interaction is dominated by the spin-independent term, and there is little difference between the two isotopes of Germanium. We see that in this case, another large area of the $\Omega_{LSP} = 1$ space can be covered, and one which is somewhat complementary to that reachable using a Fluorine target. Finally, Figs. 24e and f show analogous rates for elastic scattering off Thallium, which seems to be the best choice if m_A is large. However, it should be emphasized that the elastic scattering rates for heavier nuclei are intrinsically less reliable. Figure 25 shows, for the combined Germanium target, the "probability" of LSP detection, defined to be the fraction of the parameter space where $\Omega_{LSP} = 1$ is possible for some choice of supersymmetric model parameters in which the elastic scattering rate is "observable", in the sense that the scattering rate exceeds 0.1 events kg^{-1}d^{-1}. We see that the elastic scattering experiment is likely to be successful if m_A is small, but is unlikely to be successful with this nucleus if m_A is very large. For comparison, Fig. 24 also shows the domains of this parameter space which can be explored using LEP and/or the LHC. We see that there is a fair amount of both redundancy and complementarity between the accelerator searches and the searches for Cold Dark Matter particles in the galactic halo.

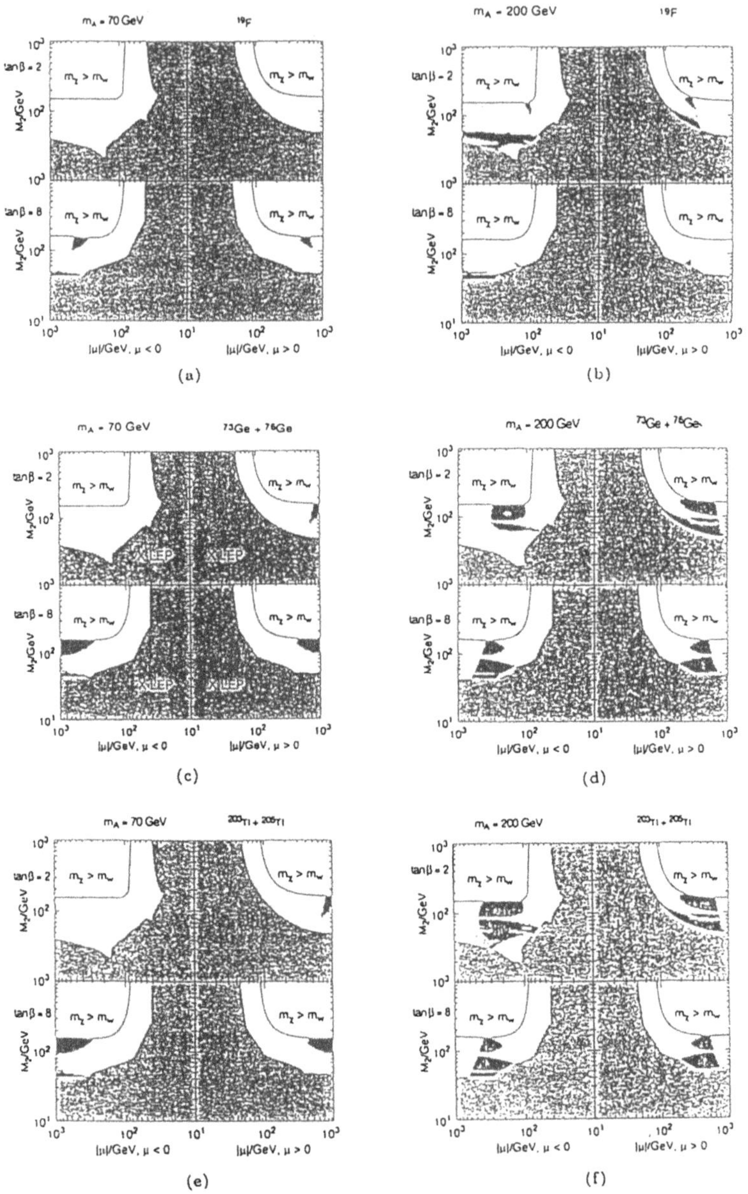

Figure 24: Elastic scattering rates for (a), (b) ^{19}F, (c), (d) ^{73}Ge + ^{76}Ge, (e) (f), Tl, for the pseudoscalar Higgs mass m_A =70 GeV in (a), (c), (e), 200 GeV in (b), (d), (f) [94], [99]. The rates are above 0.1 events kg^{-1}d^{-1} in the shaded regions.

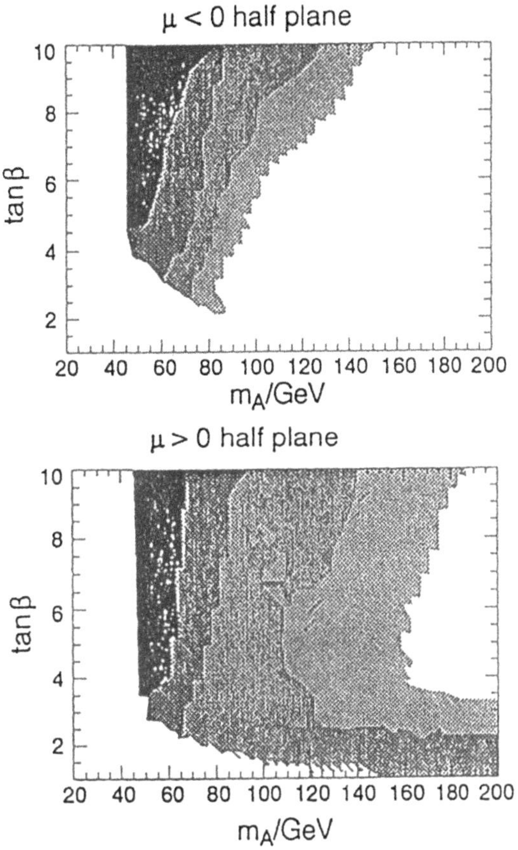

Figure 25: The "probability" of detecting neutralino dark matter, as defined by the fraction of the parameter space where the critical density $\Omega_{LSP} = 1$ can be attained in which the χ-Ge elastic scattering rate exceeds 0.1 events $kg^{-1}d^{-1}$ [94], [99].

Before leaving this subject, your attention is drawn to Ref. [99], in which inelastic LSP-nucleus scattering is considered. A priori, such a process is of experimental interest because it has a double signature, given by the combination of recoil energy and a de-excitation γ with an energy in the range 10-50 KeV, which might render the process observable at room temperature in a large detector. Unfortunately, as shown in Ref. [99], the rates found in a naïve shell model are discouragingly low, only a few exceeding 10^{-4} $kg^{-1}d^{-1}$. Therefore, I am unaware of any active experimental development programme to explore the inelastic scattering signature.

I conclude this lecture with the following Copernican challenge. Copernicus showed us that we were not at the centre of the Universe. Modern astrophysicists claim that we are not even made of the same stuff as most of the matter in the Universe. The challenge is to prove this latter assertion!

References

[1] S. Weinberg, "Gravitation and Cosmology" (Wiley, New York, 1972).

[2] E.W. Kolb and M.S. Turner, "The Early Universe" (Addison-Wesley, Redwood City, 1990).

[3] J. Ellis and G. Steigman, *Phys.Lett.* **89B** (1980) 186.

[4] C. Quigg, "Gauge Theories of the Strong, Weak and Electromagnetic Interactions" (Benjamin-Cummings, Reading, 1983).

[5] D. Ross and M. Veltman, *Nucl.Phys.* **B95** (1975) 135.

[6] E.S Abers and B.W. Lee, *Physics Reports* **9C** (1973) 1.

[7] M. Veltman, *Nucl.Phys.* **B123** (1977) 89.

[8] M.S. Chanowitz, M. Furman and I. Hinchliffe, *Phys.Lett.* **B78** (1978) 285.

[9] The LEP Electroweak Working Group, CERN Preprint LEPEWWG/95-01.

[10] J. Ellis, G.L. Fogli and E. Lisi, *Phys.Lett.* **B333** (1994) 118.

[11] C. Bouchiat, J. Iliopoulos and Ph. Meyer, *Phys.Lett.* **138B** (1972) 652 and references therein.

[12] G. 't Hooft, *Phys.Rev.* **D14** (1976) 3432;
R. Jackiw and C. Rebbi, *Phys.Rev.Lett.* **37** (1976) 172;
C.G. Callan, R.F. Dashen and D.J. Gross, *Phys.Lett.* **63B** (1976) 334.

[13] H.D. Politzer, *Phys.Rev.Lett.* **30** (1973) 1346;
D.J. Gross and F.A. Wilczek, *Phys.Rev.Lett.* **30** (1973) 1343.

[14] J. Gasser and H. Leutwyler, *Nucl.Phys.* **B250** (1985) 465.

[15] E. Witten, *Nucl.Phys.* **B223** (1983) 422,433;
G. Adkins, C.R. Nappi and E. Witten, *Nucl.Phys.* **B228** (1983) 552.

[16] A. Chodos et al., *Phys.Rev.* **D9** (1984) 3471.

[17] B.A. Campbell, J. Ellis and K.A. Olive, *Nucl.Phys.* **B345** (1990) 57.

[18] Y. Iwasaki et al., *Phys.Rev.* **D46** (1992) 4657;
B. Grossmann and M.L. Laursen, Jülich Preprint HLRZ-93-7 (1993);
Y. Iwasaki et al., CERN Preprint TH. 6798/93 (1993).

[19] J. Ignatius, K. Kajantie, H. Kurki-Suonio and M. Laine, *Phys.Rev.* **D49** (1994) 3854 and **D50** (1994) 3738.

[20] J.H. Applegate and C.J. Hogan, *Phys.Rev.* **D31** (1985) 3037;
J.H. Applegate, C.J. Hogan and R. Scherrer, *Phys.Rev.* **D35** (1987) 1151;
C. Alcock, G.M. Fuller and G.J. Mathews, *Ap.J.* **320** (1987) 439.

[21] D. Thomas, D.N. Schramm, K.A. Olive and B.D. Fields, *Ap.J.* **406** (1993) 569.

[22] A.D. Linde, *Nucl.Phys.* **B216** (1983) 421;
M. Dine et al., *Phys.Rev.* **D46** (1992) 550.

[23] G.W. Anderson and L.J. Hall, *Phys.Rev.* **D45** (1992) 2685.

[24] M.E. Carrington, *Phys.Rev.* **D45** (1992) 2933;
M. Dine et al., Ref. [45].

[25] J.R. Espinosa, M. Quirós and F. Zwirner, *Phys.Lett.* **B314** (1993) 206;
W. Buchmüller, Z. Fodor, T. Helbrig and D. Walliser, *Ann.Phys.* **234** (1994) 260.

[26] K. Kajantie, K. Rummukainen and M. Shaposhnikov, *Nucl.Phys.* **B407** (1993) 356.

[27] M. Shaposhnikov, CERN Preprint TH. 6918/93 (1993).

[28] G. 't Hooft, "Recent Developments in Field Theories", eds. G. 't Hooft et al. (Plenum Press, New York, 1980).

[29] M. Veltman, *Acta Phys.Pol.* **B8** (1977) 475;
C. Vayonakis, *Lett. Nuovo Cimento* **17** (1976) 383;
B.W. Lee, C. Quigg and H.B. Thacker, *Phys.Rev.* **D16** (1977) 1519.

[30] E. Gildener and S. Weinberg, *Phys.Rev.* **D13** (1976) 3333;
E. Gildener, *Phys.Rev.* **D14** (1976) 1667.

[31] A.J. Buras, J. Ellis, M.K. Gaillard and D.V. Nanopoulos, *Nucl.Phys.* **B135** (1978) 66.

[32] S. Hawking, D.N. Page and C. Pope, *Phys.Lett.* **86B** (1979) 175 and *Nucl.Phys.* **B170** (1980) 283.

[33] S. Dimopoulos and H. Georgi, *Nucl.Phys.* **B193** (1981) 150;
N. Sakai, *Z.Phys.* **C11** (1982) 153.

[34] J. Iliopoulos and B. Zumino, *Nucl.Phys.* **B76** (1974) 310;
S. Ferrara, J. Iliopoulos and B. Zumino, *Nucl.Phys.* **B77** (1974) 413.

[35] P. Fayet, *Phys.Lett.* **69B** (1977) 489 and **84B** (1979) 416.

[36] H. Goldberg, *Phys.Rev.Lett.* **50** (1983) 1419;
J. Ellis, J.S. Hagelin, D.V. Nanopoulos, K.A. Olive and M. Srednicki, *Nucl.Phys.* **B238** (1984) 453.

[37] H. Georgi and S.L. Glashow, *Phys.Rev.Lett.* **32** (1974) 438.

[38] H. Georgi, H. Quinn and S. Weinberg, *Phys.Rev.Lett.* **33** (1974) 451.

[39] W. Marciano and A. Sirlin, *Phys.Rev.Lett.* **46** (1981) 163.

[40] P. Langacker, University of Pennsylvania Preprint UPR 0238T (1990);
J. Ellis, S. Kelley and D.V. Nanopoulos, *Phys.Lett.* **B249** (1990) 441;
U. Amaldi, W. de Boer and H. Furstenau, *Phys.Lett.* **B260** (1991) 447.

[41] S. Dimopoulos and H. Georgi, Ref. [74];
S. Dimopoulos, S. Raby and F. Wilczek, *Phys.Rev.* **D24** (1981) 1681;
L. Ibáñez and G.G. Ross, *Phys.Lett.* **105B** (1981) 439.

[42] F. Anselmo, L. Cifarelli, A. Petermann and A. Zichichi, CERN Preprints PPE-91-123 (1991), TH. 6429/92 (1992), PPE-92-103 (1992);
J. Ellis, S. Kelley and D.V. Nanopoulos, *Nucl.Phys.* **B373** (1992) 55.

[43] Particle Data Group, K. Hikasa et al., *Phys.Rev.* **D45** (1992) 1.

[44] R. Barbieri, J. Ellis and M.K. Gaillard, *Phys.Lett.* **90B** (1980) 249.

[45] T. Yanagida, Proc. Workshop on the Unified Theory and the Baryon Number in the Universe (KEK, Japan, 1979);
R. Slansky, Talk at the Sanibel Symposium, Caltech Preprint CALT-68-709 (1979).

[46] B. Pontecorvo, *Zh.Eksp.Teor.Fiz.* **33** (1957) 549; **34** (1958) 247; **53** (1967) 1717.

[47] L. Wolfenstein, *Phys.Rev.* **D17** (1978) 2369;
S.P. Mikheyev and A.Yu. Smirnov, *Nuovo Cimento* **9C** (1986) 17.

[48] P.I. Krastev and S.T. Petcov, *Phys.Lett.* **B299** (1993) 99.

[49] J. Ellis, J. Lopez and D.V. Nanopoulos, *Phys.Lett.* **B292** (1992) 189.

[50] See, e.g.: G. Efstathion, J.R. Bond and S.D.M. White, *Mon.Not.R. Ast.Soc.* **258** (1992) 1P;
and M.S. Turner, FNAL Preprint Conf-92/313-A (1992).

[51] G.F. Smoot et al., *Ap.J.Lett.* **396** (1992) L1;
E.L. Wright et al., *Ap.J.Lett.* **396** (1992) L13.

[52] CHORUS Collaboration, N. Armenise et al., CERN-SPSC/90-42 (1990);
NOMAD Collaboration, P. Astier et al., CERN-SPSC/91-21 (1991).

[53] J. Preskill, *Phys.Rev.Lett.* **43** (1979) 1365.

[54] A.A. Starobinsky, *Phys.Lett.* **91B** (1980) 99;
A. Guth, *Phys.Rev.* **D23** (1981) 349.

[55] J. Ellis, J.E. Kim and D.V. Nanopoulos, *Phys.Lett.* **145B** (1984) 181.

[56] A. Guth and E. Weinberg, *Phys.Rev.* **D23** (1981) 876;
S.W. Hawking, I.G. Moss and J.M. Stewart, *Phys.Rev.* **D26** (1982) 2681.

[57] A.D. Linde, *Phys.Lett.* **108B** (1982) 389;
A. Albrecht and P.J. Steinhardt, *Phys.Rev.Lett.* **48** (1982) 1220.

[58] D. La and P.J. Steinhardt, *Phys.Rev.Lett.* **62** (1988) 376.

[59] C.H. Brans and R.H. Dicke, *Phys.Rev.* **124** (1961) 925.

[60] A.D. Linde, *Phys.Lett.* **129B** (1983) 177.

[61] A.D. Linde, D. Linde and A.Mezhlumian, Stanford Preprint SU-ITP-93-13 (1993).

[62] J. Bardeen, P.J. Steinhardt and M.S. Turner, *Phys.Rev.* **D28** (1983) 679;
A.H. Guth and S.-Y. Pi, *Phys.Rev.Lett.* **49** (1982) 1110;
A.A. Starobinsky, *Phys.Lett.* **117B** (1982) 175;
S.W. Hawking, *Phys.Lett.* **115B** (1982) 295.

[63] V. Lukash, Lectures at this School.

[64] N.D. Birrell and P. Davies, "Quantum Fields in Curved Space" (Cambridge University Press, Cambridge, 1982).

[65] E.J. Copeland, E.W. Kolb, A.R. Liddle and J.E. Lidsey, Sussex Preprints SUSSEX-AST 93/3,4-1 (1993);
M.S. Turner, FNAL Preprint Pub-93/026-A (1993);
J.R. Bond et al., CITA/University of Pennsylvania/Oxford Preprint (1993).

[66] F. Klinkhamer and N. Manton, *Phys.Rev.* **D30** (1984) 2212.

[67] V. Kuzmin, V. Rubakov and M. Shaposhnikov, *Phys.Lett.* **B155** (1985) 36.

[68] J. Ambjorn, T. Aksgaard, H. Porter and M. Shaposhnikov, *Nucl.Phys.* **B353** (1991) 346.

[69] A.D. Sakharov, *Pis'ma Zh.Eksp.Teor.Fiz.* **5** (1967) 32.

[70] M. Fukugita and T. Yanagida, *Phys.Lett.* **174B** (1986) 45.

[71] C. Alcock et al., *Nature* **365** (1993) 621;
E. Aubourg et al., *Nature* **365** (1993) 623.

[72] B.W. Lee and S. Weinberg, *Phys.Rev.Lett.* **39** (1977) 165;
P. Hut, *Phys.Lett.* **69B** (1977) 85.

[73] S.P. Ahlen et al., *Phys.Lett.* **B195** (1987) 603;
D.O. Caldwell et al., *Phys.Rev.Lett.* **61** (1988) 510;
F. Boehm et al., *Phys.Lett.* **B255** (1991) 143.

[74] J.S. Hagelin, K.W. Ng and K.A. Olive, *Phys.Lett.* **180B** (1986) 375.

[75] H. Pagels and J. Primack, *Phys.Rev.Lett.* **48** (1982) 223.

[76] L. Roszkowski, *Phys.Lett.* **B252** (1990) 471.

[77] J. Silk and M. Srednicki, *Phys.Rev.Lett.* **53** (1984) 624.

[78] J. Ellis et al., *Phys.Lett.* **B214** (1988) 403.

[79] J. Ellis and P. Sikivie,.*Phys.Lett.* **B321** (1994) 390.

[80] S. Ritz and D. Seckel, *Nucl.Phys.* **B304** (1988) 877.

[81] A. Buffington, S.M. Schindler and C. Pennypacker, *Ap.J.* **248** (1981) 179.

[82] S.P. Ahlen et al., *Phys.Rev.Lett.* **61** (1988) 145;
T. Bowen et al., Arizona Preprint LHEA-89-006 (1989).

[83] J. Silk, K.A. Olive and M. Srednicki, *Nucl.Phys.* **B279** (1987) 804.

[84] A. Gould, *Ap.J.* **321** (1987) 571.

[85] G.B. Gelmini, L.J. Hall and M.J. Lin, *Nucl.Phys.* **B281** (1987) 726.

[86] A. Gould, *Ap.J.* **321** (1987) 560.

[87] J. Ellis, R.A. Flores and S. Ritz, *Phys.Lett.* **B198** (1987) 493.

[88] A. Bottino et al., Torino Preprint DFTT 33/92 (1992) and references therein; M. Mori et al., *Phys.Lett.* **B270** (1991) 89.

[89] J. Ellis, R.A. Flores and S. Masood, *Nucl.Phys.* **B294** (1992) 229.

[90] M. Goodman and E. Witten, *Phys.Rev.* **D30** (1985) 3059.

[91] EMC Collaboration, J. Ashman et al., *Nucl.Phys.* **B328** (1989) 1.

[92] S.J. Brodsky, J. Ellis and M. Karliner, *Phys.Lett.* **B206** (1988) 309.

[93] J. Ellis and R.A. Flores, *Phys.Lett.* **B263** (1991) 259; *Nucl.Phys.* **B400** (1993) 25.

[94] J. Ellis and M. Karliner, *Phys.Lett.* **B313** (1993) 131.

[95] K. Griest, *Phys.Rev.* **D38** (1988) 2357, **E, D39** (1989) 3802.

[96] J. Gasser, H. Leutwyler and M.E. Sainio, *Phys.Lett.* **B253** (1991) 252.

[97] J. Ellis and R.A. Flores, *Nucl.Phys.* **B307** (1988) 883.

[98] J. Ellis and R.A. Flores, *Phys.Lett.* **B263** (1991) 259.

[99] J. Ellis, R.A. Flores and J.D. Lewin, *Phys.Lett.* **B212** (1988) 375.

DARK MATTER IN GALAXIES

K.C. FREEMAN
Mount Stromlo and Siding Spring Observatories
The Australian National University

Abstract

There is not much doubt now that most spiral and dwarf irregular galaxies have dark halos which are at least a few times more massive than their luminous components. The Milky Way is no exception: its ratio of dark to luminous matter is probably at least 10. Elliptical and dwarf spheroidal galaxies also appear to have massive dark halos, but the evidence is not so direct.

1 Spiral Galaxies

1.1 Structure of the Dark Halos

Spiral galaxies commonly have flat rotation curves, which cannot be produced by a mass distribution that follows the surface brightness distribution. From the observed light distribution, and assuming that mass follows light, we can compute the expected rotation curve. There is also a small contribution to the rotation curve from the gas in the galaxy, which can be computed directly. This procedure gives the shape of the rotation curve, but there is a scaling parameter (the mass-to-light ratio M/L) which determines the amplitude of the rotation curve. The usual situation is that the rotation curve computed from the light distribution fits the observed rotation curve well enough in the inner regions of the galaxy (out to about 3 scalelengths of its exponential surface brightness distribution), but beyond about 5 scale lengths the expected rotation curve lies far below the observed rotation curve.

This situation is typical of most spirals with measured rotation curves extending well beyond 3 scalelengths (these extended rotation curves are measured from the 21-cm line of neutral hydrogen, because optical rotation curves can rarely be measured beyond a few scale lengths). The usual interpretation is that luminous matter in these galaxies dominates the radial gradient of the gravitational potential in the inner regions, and that the galaxies have a massive dark halo which makes up the shortfall in the outer parts of the rotation curve. The usually flat shape of the rotation curves in the outer regions of spirals suggests that the radial density distribution of this dark halo goes like $\rho(r) \sim r^{-2}$ at large radius. A simple model that is often used has

$$\rho(r) = \rho_o/[1 + (r/a)^2] \tag{1}$$

This procedure, of fitting the inner parts of the rotation curve with the expected rotation curve from the light + gas, and then attributing the shortfall in the outer regions to the dark halo, maximises the contribution from the disk and minimises the contribution from the dark halo; models with less massive disks and more massive halos often fit just as well [1]. Even with the maximum disk model, the mass of the inferred dark halo is about five times the mass of the luminous disk and gas within the outermost point on the observed rotation curve.

The simple halo given by equation (1) has two scales: the core radius a and the central density ρ_o. The estimated values of these two scales from the rotation curve analysis depend on the adopted value for the disk's M/L ratio. The maximum disk models give the largest core radii and the smallest ρ_o. The typical values of ρ_o from maximum disk solutions are in the range $\rho_o = 0.001$ to 0.01 $M_\odot$ for giant spirals and 0.01 to 0.1 $M_\odot$ for the smaller spirals and irregulars. The core radii for the large spirals derived from maximum disk solutions are typically about 10 kpc, but could be much smaller if the maximum disk analysis is not the correct procedure [1].

The core radius, central density and shape of the dark halos are interesting because they relate to numerical simulations of galaxy formation . For example, Warren *et al.* [2] made simulations of a 400 Mpc^3 piece of the universe, using 10^6 particles and various perturbation spectra. In these simulations, the halos grow by dissipationless merging of smaller collapsed structures via dynamical friction and tidal stripping. They find that (i) halos tend to be more prolate than oblate, and (ii) the core radii are typically about 1 kpc for large halos with circular velocities of 250 km s^{-1}. This is much smaller than the halo core radii inferred from maximum disk analysis of rotation curves of large spirals. These results suggest that either the maximum disk solutions are not appropriate or some mechanism is needed to enlarge the cores of large spirals. This core growth may occur naturally as gas settles to a disk during the hierarchial merging: Katz and Gunn [3] showed that the settling gas can transfer more than 50% of its original angular momentum to the dark halo.

The shapes of dark halos are not yet well constrained by observation. The flaring of the gas layer observed in edge-on spirals gives some information about the shape of the potential in the outer parts of these systems, and dark matter distributions as flat as the disk can probably be excluded. In a rare class of galaxies, the polar ring systems, material is orbiting in two roughly orthogonal planes, and it is possible to estimate the shape of the dark matter distribution from the kinematics of gas and stars in the two perpendicular planes of the stellar disk and the polar ring. Recent results for the system NGC 4650A [4] are consistent with a fairly flat dark halo.

1.2 Maximum Disks

This question of the maximum disk analysis of rotation curves has been widely discussed (see the review by Sancisi and van Albada [5]), because it is important to know whether this is the right way to proceed. Our knowledge of the structure of dark halos depends on the assumptions about the M/L ratio for the stellar disk. It also affects our view of the dynamics of the inner disk: does the disk dominate the radial potential gradient in the inner two or three scalelengths, or is it dominated even in the inner regions by the dark halo.

Analysis of the rotation curve of our Galaxy does not give a clear answer to the maximum disk question. The rotation curve in the inner regions is not well determined, and the relative contributions to the potential field from the disk and the bulge are difficult to assess. Different studies indicate that the surface density of a maximum disk near the sun would lie somewhere in the range 70 to 130 $M_\odot$ pc^{-2}, depending on the adopted mass of the galactic bulge [6,7]. Recent estimates of the local surface density, from the kinematics and distribution of stars towards the galactic poles [8,9], are in the range 50 to 80 $M_\odot$ pc^{-2}.

Swing amplifier theory for spiral structure can give some useful limits on the density of the disk, by requiring that the disk should not be so dense as to allow azimuthal $m = 1$ modes. Athanassoula *et al.* [10] made maximum disk fits to a large sample of rotation curves, with the extra constraint of "no $m = 1$ amplification". They found that their fits (i) are consistent with almost all of their galaxies, (ii) are compatible with the $m > 2$ spirals seen in the outer parts of some of their galaxies, and (iii) give M/L ratios for the disks that are mostly within 20% of the unmodified maximum disk values.

Buchhorn [11] has analysed optical rotation curves of 552 spirals. The expected rotation curves were computed from I-band CCD surface photometry, which minimises the effects of internal absorption and regions of young population in the disks. For most of his sample (> 97%), the calculated rotation curves are an excellent fit to the data within the region of the optical rotation curves (typically 2 to 3 scalelengths). The good fits in these systems are particularly significant for the maximum disk question, because his galaxies show a very wide range of rotation curve morphology, extending from flat rotation curves to almost solid body rotation.

The success of the fits for this large and varied sample suggests that either

- the disk dominates the radial potential gradient in the inner few scalelengths of most disk galaxies, or
- the detailed density distribution of the dark halo mimics the surface density structure in the disk (but the structure in the halo must be more prominent than in the disk because the gravitational effects of the halo structure are diluted by its more spheroidal shape).

The sample includes galaxies with rotational velocities from about 40 to 350 km s^{-1}, but shows little trend of the disk's M/L_I ratio with rotational velocity. This indicates that, for most galaxies, the contribution that the halo makes to the inner rotation curve remains small as we go from the largest to the smallest spirals. However, there are certainly a few galaxies for which the maximum disk approach does not work. These are mostly smaller spirals and irregular galaxies, such as NGC 5585 [12], and in these systems it seems likely that the dark halo *is* making a significant contribution to the gravitational field in the inner regions of the disk.

1.3 The Dark Halo of our Galaxy

For our Galaxy, four methods are used to estimate the enclosed mass $M(R)$ within some radius R: (i) the rotation curve, (ii) the escape velocity of high velocity stars in the solar neighborhood, (iii) the random velocities of distant halo stars and satellite galaxies, and (iv) timing arguments. From the inner rotation curve and the adopted structure of the disk and bulge, the total mass of the luminous component (disk and bulge) is probably in the range (5 to 12)$\times 10^{10}$ $M_\odot$. For further references for this section, see Freeman [13].

The rotation curve of the Galaxy is roughly flat between about 3 kpc and 15 kpc from the galactic center. Beyond 15 kpc the velocity uncertainties become large and the rotation curve less secure, but it appears to rise slowly all the way from $R = 2$ kpc to the maximum extent of the measured rotation curve at 20 kpc. The galactic rotation curve out to a radius of 20 kpc gives a total $M(20\ kpc) = 22 \times 10^{10}$ $M_\odot$. This is at least double the estimated mass of the visible components of the Galaxy, and already provides a strong indication that the Galaxy has a dark corona. In some spirals it is possible to estimate the parameters of the dark corona (*e.g.* its core radius and scale density) from the shape of the rotation curve; for the Galaxy this is difficult, because the rotation curve itself remains uncertain in the inner few kiloparsecs of the Galaxy.

In the solar neighborhood, stars of the metal poor stellar halo have a velocity dispersion of about 140 km s^{-1} in the radial direction. If the halo stars in the high velocity tail of the velocity distribution are bound to the Galaxy, then their distribution of total space motions gives a lower limit on the local escape velocity V_{esc} which in turn sets a lower limit on the total galactic mass. The mass estimates from escape velocity arguments are lower limits on the total mass because the estimated V_{esc} may be smaller than the true escape velocity in the solar neighborhood. For example, the most energetic halo stars in the solar neighborhood may be firmly bound to the Galaxy if the stars of the halo come from accreted satellites or if the velocity distribution of the halo stars has been truncated in the close approaches of the Magellanic Clouds.

Leonard and Tremaine [14] used a maximum likelihood analysis to estimate V_{esc} from the velocity distribution of the highest velocity stars. Their analysis included the observational biases associated with the proper motion selection criteria for high velocity stars. They found that the 90% confidence limits on V_{esc} are 450 and 650 km s^{-1}. The radial velocity alone of one particular high velocity star gives a limit of $V_{esc} > 430$ km s^{-1}. For a simple galactic mass model with $M(R) \propto R$ out to some limiting radius R_{lim} and constant for $R > R_{lim}$, the total mass M_{total} is related to the local escape velocity V_{esc} by

$$\log M_{total}/M(R_{\circ}) = (1/2)(V_{esc}/V_c)^2 - 1$$

where $R_{\circ}$ and V_c are again the radius of the solar orbit and the circular velocity. With this model, the Leonard and Tremaine estimate of V_{esc} indicates that $M_{total} > 3 \times 10^{11}$ $M_{\odot}$, again well in excess of the mass inferred from the rotation curve

For the distant halo stars and satellites, the distribution of rv^2 provides a mass estimator [15], where r is the distance of the object and v its radial velocity. From the halo stars, $M(50$ kpc$) = (4 \times 10^{11}\ M_{\odot})$ [16], assuming that their orbits have an isotropic velocity distribution.

The timing arguments use the present separation and relative velocity of M31 and the Galaxy [17] to estimate the total mass of the M31/Galaxy system, assuming that their initial separation was small and adopting some age for the universe. M31 is about 710 kpc from the Galaxy and has a galactocentric radial velocity of -118 km s^{-1}. If the age of the universe is 18 Gyr, then the mass of the Galaxy is $(13 \pm 2) \times 10^{11}\ M_{\odot}$. A similar argument using the dwarf galaxy Leo I [18] at a distance of about 230 kpc gives $(12 \pm 2) \times 10^{11}\ M_{\odot}$. The two estimates from M31 and Leo I agree remarkably well, and are consistent with a flat rotation curve ($V_c = 220$ km s^{-1}) extending out to at least 100 kpc. We note that these estimates assume radial orbits, so are lower limits.

More elaborate studies which include the dynamics of the angular momentum of the Local Group and interactions with nearby galaxies give similar estimates for the total galactic mass: see for example Raychaudhury and Lynden-Bell [19]. Peebles [20] modelled the formation of the Local Group and derived a total mass of about 2×10^{12} $M_{\odot}$ for the Galaxy. Kroeker and Carlberg [21] checked the accuracy of the timing argument mass estimates by examining binary systems identified in $\Omega = 1$ CDM simulations. They find that timing arguments (assuming radial orbits) typically underestimate the total masses by a factor of about 1.7.

In summary, the data for our Galaxy at large R consistently indicate that the Galactic mass distribution follows

$$M(R) \approx 10^{10} \times R(\text{kpc})\ M_{\odot} \qquad (2)$$

with a total mass in excess of $10^{12}\ M_{\odot}$. This is consistent with a flat rotation curve extending out to about 100 kpc from the galactic center. The mass of the dark halo of our Galaxy is then about ten times the mass of the luminous matter.

We should mention here the microlensing experiments that are in progress now in Chile and Australia. These are aimed at detecting dark matter objects in the galactic halo. For example, the MACHO experiment (a collaboration between the Lawrence Livermore National Laboratory, the Center for Particle Astrophysics at the University of California, Berkeley, and Mount Stromlo and Siding Spring Observatories) uses a large CCD camera and the dedicated 50-inch telescope at Mt Stromlo to monitor about 10^7 stars in the LMC and SMC (and a similar number in the galactic bulge) for brightness fluctuations due to microlensing by intervening dark matter objects. The experiment is sensitive to dark matter objects in the mass range $2.10^{-8} \lesssim M \lesssim 200$ $M_\odot$, after a long enough run of photometry. At the time of writing, and after analysis of two years of data, the MACHO experiment has discoved several microlensing events which may be produced by objects in the galactic dark halo. These are consistent with about half of the dark halo mass being in the form of dark compact objects, and with the mass of these dark objects being about half the solar mass. As the experiment proceeds and the photometric time series become longer, the experiment becomes sensitive to microlensing events of longer duration. In the mean, these longer events will be associated with more massive lensing objects, which also make a larger contribution to the lensing optical depth. It therefore seems likely that the estimates of the fraction of the dark halo in dark compact objects, and also of the mass of these dark objects themselves, will gradually rise as the experiment goes on.

1.4 Conclusion for Spirals

There seems little doubt now that spirals have extended dark halos with masses of order 3 to 10 times the mass of their luminous components. The rotation curve studies described above support the maximum disk concept in most spirals; for the Milky Way the situation is still unclear. Before leaving the spirals, we should mention two particular problems.

- If the maximum disk concept is correct, then the disk dominates the radial potential gradient over the inner two to three scalelengths where the optical rotation curves can be measured. The optical rotation curves typically end (*i.e.* the HII region emission ends) close to the radius at which the disk's contribution to the rotation curve begins to decline and the halo begins to contribute significantly to the radial potential gradient. This observation requires some explanation. We note that, at this radius in most galaxies, the dark halo does not yet dominate the potential gradient in the radial direction and certainly not in the vertical direction.

- The dark halo is seen most clearly in galaxies for which the HI distribution extends out to many disk scalelengths. Several authors, most recently Carignan [22)] have pointed out that the ratio $\sigma_{HI}/\sigma_{dark}$ of the directly observed surface density of HI to the inferred surface density of the dark halo is roughly constant with radius in these galaxies. Is this an important clue to the nature of dark matter, or is it simply an artifact of the maximum disk decomposition procedure? For example, the maximum disk procedure forces a large value of the core radius for the dark halo. The HI is by selection very extended in galaxies for which rotation curves can be measured out to large radii. Hence we are comparing two extended and usually radially decreasing density distributions which may then be roughly similar even if there is no physical basis to the effect.

2 The Dwarf Irregular Galaxies

HI synthesis observations of dwarf irregular galaxies give virial mass estimates from the HI profile width, assuming that the HI is in equilibrium. These estimates [23)] show that the low luminosity dwarfs ($M_B > -12$) have M/L ratios $\gtrsim 12$ and are likely to have significant dark matter fractions. The brighter dwarfs ($-14.5 < M_B < -12$) have lower M/L ratios, around 3. These lower M/L values tell us little about the dark matter content, because the HI in these systems is often no more extended than the light distribution; the low M/L values just mean that the inner regions of these brighter dwarfs are not dark matter dominated. Two of the well studied dwarf irregulars are particularly interesting.

The first is DDO 154 [24)]. This dIrr galaxy has $M_B = -13.8$ and is immersed in an extended HI envelope. Its rotation curve extends out to 15 disk scalelengths and shows that the dark halo dominates the gravitational field at almost all radii; this system is an extreme example of no conspiracy (see §1.4).

The second is GR8 [25)]. In this very small system ($M_B = -10.6$), the gas is supported by random motions. Its rotation is small and disorderly (recall that the rotational velocity $\propto$ luminosity$^{1/4}$). However it is still possible to make a fairly straightforward estimate of the enclosed mass $M(R)$. The HI synthesis observations show that the HI velocity dispersion is constant with radius, at about 10 km s^{-1}, and the HI surface density distribution is very close to gaussian: $\mu_{HI} \propto \exp(-R^2/2h^2)$, where h is the scalelength. Assuming that the HI density distribution is spherical, then the volume density distribution of the HI is also gaussian and the equation of hydrostatic equilibrium takes the simple form $GM(R)/R^2 = c^2 R/h^2$, where c is the random velocity of the HI. Out to the edge of the observable HI, the gas mass is $2.10^6\ M_\odot$, the stellar mass is $3.10^6\ M_\odot$, the total mass is $4.10^7\ M_\odot$, and the M/L ratio is 15. The hydrostatic equilibrium equation for this gaussian HI distribution with its isothermal turbulent velocity shows that the enclosed mass $M(R) \propto R^3$; *i.e.* the HI component of GR8 lies in the approximately uniform density core of its dark halo. This is independent evidence for a core in a dark halo. The density in this core is about $0.07\ M_\odot$ pc^{-3}, which is much higher than the typical value for spirals ($\lesssim 0.01\ M_\odot$ pc^{-3}) and for DDO 154 ($0.016\ M_\odot$ pc^{-3}).

3 The Dwarf Spheroidal Galaxies

These systems have very little gas, very low surface brightness (typically 0.1 of the night sky) and masses of about 10^6 to 10^7 $M_\odot$. For the dwarf spheroidal companions of the Galaxy, the M/L ratios are estimated by using star counts to measure the surface brightness distribution $I(R)$ and radial velocities of individual stars to measure the velocity dispersion $\sigma(R)$. If the mass follows the light, and the system is spherical, isotropic and has a homogeneous core, then the central M/L ratio is given by $M/L = 9\eta\sigma^2_{obs}(0)/2\pi G I(0) R_c$, where $\sigma_{obs}(0)$ is the observed central velocity dispersion, R_c is the half-peak-brightness radius, $I(0)$ is the central surface brightness and $\eta \approx 1$ (see ref. 24). Several of these dwarf spheroidal systems have now been studied. Two of them (Carina and Sculptor) have $M/L \approx 5$ which provides no clear result on their dark matter content. For three others (Fornax, Draco and Ursa Minor) there is strong evidence now that dark matter contributes significantly to the potential gradient in the inner regions.

Fornax [26] has an absolute magnitude $M_V = -12.6$. There is some uncertainty in its $I(R)$ distribution from different sources, and also in its velocity dispersion (depending on whether marginal members are included). The velocity dispersion is in the range 10 to 14 km s^{-1} and the central M/L ratio is between 9 and 15. The central density is 0.07 ± 0.03 $M_\odot$ pc^{-3}.

Draco and Ursa Minor [27] are very small systems, with $M_V \approx -8.5$. Their velocity dispersions are 9 ± 2 and 11 ± 2 km s^{-1} respectively. If their mass follows the light, then their central densities have very high values, about 1.0 $M_\odot$ pc^{-3}, and their M/L ratios are between 60 and 80. These values are lower by about a factor 2 if the dark matter is much more extended than the light. In any case, the origin of these extremely dense low-mass dark halos is an interesting problem.

4 Dark Matter in Ellipticals

Elliptical galaxies usually have no HI at large radii (there are some exceptions: see below), so other dynamical tracers are used. These include the integrated starlight, the hot X-ray emitting gas, globular clusters and planetary nebulae, and we give a brief discussion of these different tracers.

4.1 Integrated Light

The radial equilibrium of elliptical galaxies is primarily between gravity and the pressure gradient associated with the velocity dispersion. In most ellipticals, rotation is less important. The momentum equation is

$$\frac{\partial\Phi}{\partial R} = \frac{V_\varphi}{R} - \sigma^2_R\left\{\frac{\partial \log \rho\sigma^2_R}{\partial R} + \frac{1}{R}\left(1 - \frac{\sigma^2_\varphi}{\sigma^2_R}\right)\right\}, \tag{3}$$

where σ_φ and σ_R are the azimuthal and radial components of the velocity dispersion and ρ is the stellar density. Equation (3) can be used to estimate the total enclosed mass within radius R from measurements of the line of sight components of the velocity dispersion σ and the mean rotation V_φ. However the observations do not yet reach into the outer regions ($r \gtrsim 2r_e$, where r_e is the half-light radius) where dark matter would dominate the potential. There is also uncertainty about the isotropy of σ, although there are indications that the velocity dispersion in the outer parts of giant ellipticals is close to isotropic [28].

4.2 Hot X-ray Gas

The extended X-ray emission from many giant ellipticals indicates the presence of hot gas ($\sim 10^7$ K). Assuming spherical symmetry, the equation of hydrostatic equilibrium

$$\frac{GM(r)}{r} = -\frac{kT(r)}{\mu m_p}\left\{\frac{d\log\rho}{d\log r} + \frac{d\log T}{d\log r}\right\} \tag{4}$$

gives an estimate of the enclosed mass $M(R)$, where T and ρ are the gas temperature and density. The temperature distributions in the X-ray emitting gas are not yet well determined. Using arguments about convective stability of the hot gas, Fabian *et al.* [29] showed that for 14 E/SO galaxies the M/L ratios within the X-ray emitting region are greater than about 70, compared to typical values of about 5 in the inner regions.

4.3 Centaurus A

Planetary nebulae [30] and globular clusters [31] in Cen A give independent estimates of the enclosed mass M(r) out to about 20 kpc or $4r_e$. The planetary nebulae belong to the diffuse stellar population. The kinematics of a sample of about 500 planetary nebulae show that the velocity dispersion σ decreases with increasing radius and the rotation velocity V_{rot} increases: V_{rot}/σ increases from about 0.25 at $r = r_e$ to about 1.4 at $r = 4r_e$. In contrast, the globular clusters show no systemic rotation, and in this sense are similar to the halo clusters in our Galaxy. In the inner regions of Cen A ($r < 5$ kpc), the velocity dispersions of planetary nebulae and globular clusters are similar: 141 ± 23 km s^{-1} and 130 ± 20 km s^{-1}, respectively.

To estimate the enclosed mass $M(r)$ from the planetary nebulae and globular cluster data, using the momentum equation, some further assumptions are made: (i) isotropy of σ, (ii) spherical potential, and (iii) spherical density distribution $\rho(r)$ derived from the light distribution. These assumptions are obviously a bit rough. However the estimates of $M(r)$ from the globular clusters and the planetary nebulae agree very well, despite the different dynamics of these two populations. The M/L_B ratio increases from 3 at the edge of the dust lane to 15 at a radius of 20 kpc, with $M(20\text{ kpc}) \approx 3.10^{11}\ M_\odot$.

4.4 Globular Clusters and Planetary Nebulae

From the kinematics of the globular clusters around the giant elliptical M87, Mould *et al.* [32] showed that the M/L_V ratio in M87 increases from 6.5 at $r = 72''$ to 31 at $r = 8'$. Grillmair *et al.* [33] measured velocities for 57 globular clusters around the giant elliptical NGC 1399 in the Fornax cluster and found again that the M/L ratio increased outwards to at least $M/L = 80$. This large value of M/L was independently confirmed for NGC 1399 by velocity measurements of planetary nebulae by Arnaboldi *et al.* [34]. However, in this case it is not certain whether the M/L value really pertains to the galaxy alone or partly also to the Fornax cluster of galaxies. The dynamical relationship of the globular cluster system around NGC 1399 to the Fornax cluster itself is not clear: the Fornax cluster has a relatively small velocity dispersion and the velocity dispersions of the globular clusters and planetary nebulae are similar to that of the Fornax cluster itself.

4.5 Ellipticals with Outer HI Rings

The galaxy IC 2006 is a good example of an elliptical with an outer HI ring [35]. This system has an absolute magnitude $M_B = -19.7$. From its stellar velocity dispersion, the central $M/L_B = 5 \pm 1$. From the kinematics of the outer HI ring, the M/L_B ratio within $6.5r_e$ is 16^{+4}_{-2}. Several examples of related objects all show moderate increases of M/L by 2 to 4 times within 1 to 2 Holmberg radii.

4.6 Summary for Ellipticals

All of the tracers which extend out to large enough radii in elliptical galaxies (X-ray emission, planetary nebulae in Cen A and NGC 1399, globular clusters around Cen A, M87 and NGC 1399, and the HI ring data for IC 2006) indicate that the M/L ratio increases from about 5 in the inner regions to values in excess of 15 in the outer regions.

5 Conclusion

Dark matter in galaxies is apparently ubiquitous. The ratio of dark to luminous matter out to the most distant tracer is in the range 3 to 10 for the ellipticals and large spirals, and up to about 80 for some dwarf irregulars and dwarf spirals.

References

1) Van Albada, T.S., Bahcall, J.N., Begeman, K., Sancisi, R. 1985 Astrophys.J., 295, 305.
2) Warren, M., Quinn, P., Salmon, J., Zurek, W. 1992. Astrophys.J., 399, 405.

3) Katz, N., Gunn, J. 1991. Astrophys.J., 377, 365.
4) Sackett, P.D., Rix, H-W., Jarvis, B.J., Freeman, K.C. 1994. Astrophys.J., 436, 629.
5) van Albada, T.S., Sancisi, R. 1986. Phil. Trans. R. Soc. Lond. A, 320, 447.
6) van der Kruit, P.C. 1987. in The Galaxy, ed G. Gilmore, R. Carswell, (Dordrecht: Reidel), p 27.
7) Sellwood, J.A., Sanders, R.H. 1988. Mon.Not.R.Astron.Soc., 233, 611.
8) Kuijken, K., Gilmore, G. 1989. Mon.Not.R.Astron.Soc., 239, 605.
9) Bahcall, J.N., Flynn, C., Gould, A. 1992. Astrophys.J., 389, 234.
10) Athanassoula, E., Bosma, A., Papaioannou, S. 1987. Astron.Astrophys., 179, 23.
11) Buchhorn, M. 1992. ANU thesis.
12) Cote, S., Carignan, C., Sancisi, R. 1991. Astron.J., 102 904.
13) Freeman, K.C. in IAU Symposium 169 "Unsolved Problems of the Milky Way", ed L. Blitz (Kluwer), in press.
14) Leonard, P.J.T., Tremaine, S. 1990. Astrophys.J., 353, 486.
15) Little, B., Tremaine, S. 1987. Astrophys.J., 320, 493.
16) Norris, J.E., Hawkins, M. 1991. Astrophys.J., 380, 104.
17) Kahn, F., Woltjer, L. 1959. Astrophys.J., 130, 105.
18) Zaritsky, D., Olszewski, E., Schommer, R., Peterson, R., Aaronson, M. 1989. Astrophys.J., 345, 759.
19) Raychaudhury, S., Lynden-Bell, D. 1989. Mon.Not.R.Astron.Soc., 240, 195.
20) Peebles, P.J.E. 1990. Astrophys.J., 362, 1.
21) Kroeker, T.L., Carlberg, R.G. 1991. Astrophys.J., 376, 1.
22) Carignan, C. 1991. Contribution to Space Telescope Workshop on Dark Matter.
23) Freeman, K.C. 1986. in Nearly Normal Galaxies, ed S. Faber (New York: Springer), p 317.
24) Carignan, C., Freeman, K.C. 1988. Astrophys.J., 332, L33.
25) Carignan, C., Beaulieu, S., Freeman, K.C. 1990. Astron.J., 99, 178.
26) Mateo, M., Olszewski, E., Welch, D., Fischer, P., Kunkel, W. 1991. Astron.J., 102, 914.
27) Pryor, C., Kormendy, J. 1990. Astron.J., 100, 127.
28) Winsall, M.L., Freeman, K.C. 1993. Astron.Astrophys., 268, 443.
29) Fabian, A.C., Thomas, P.A., Fall, S.M., White, R.E. 1986.
30) Hui, X. *et al.* 1992. Astrophys.J., 449, 592.
31) Harris, H.C., Harris, G.L., Hesser, J.E. 1988. in Globular Cluster Systems in Galaxies, ed J.E. Grindlay and A.G.D. Philip (Dordrecht: Kluwer), p 205.
32) Mould, J.R., Oke, J.B., de Zeeuw, P.T., Nemec, J.M. 1990. Astron.J., 99, 1823.
33) Grillmair, C. *et al.* 1994. Astrophys.J., 422, L9.
34) Arnaboldi *et al.* 1994. ESO Messenger, June 1994, p 40.
35) Franx, M., van Gorkom, J., de Zeeuw, P.T. 1994. Astrophys.J., 436, 642.

HIGH-REDSHIFT GALAXIES

R.W. HUNSTEAD
School of Physics, University of Sydney
NSW 2006, Australia

Much of our current understanding of high-redshift galaxies has come from detailed studies of the absorption signatures they impress on the spectra of background QSOs. The absorbers attracting the most attention are the damped Lyman α systems, because they contain a significant fraction of all the neutral hydrogen in the early universe. We have measured the chemical abundances in damped Lyman α galaxies using various techniques and show that they are typically 10 times below solar at redshifts ~2–3. Future studies of these systems, especially with the new generation of 8–10 m class telescopes, hold considerable promise for building a full picture of galactic evolution.

1 Introduction

The detection and study of high-redshift galaxies has been a major goal of observational cosmology over the past decade. We would obviously like to be able to answer the big questions such as: *When* did galaxies form? *How* did galaxies form? Was *star formation* a sudden or steady process? However, despite intense efforts and significant technical advances in the last few years, searches for the high-redshift counterparts of galaxies like our own have proven fruitless. Deep spectroscopic surveys of faint field galaxies have not found a population of objects at $z > 1$, but have revealed instead that at the faintest magnitudes sampled, $B \simeq 24$, the counts are dominated by low-luminosity galaxies at only modest redshifts, $z < 0.4$. [1,2] Similarly, searches in blank fields for Lyman α emission at $z = 2$–5 from primeval galaxies undergoing an initial burst of star-formation have yet to produce an unequivocal detection of a 'normal' young galaxy. [3]

In contrast, we routinely detect 'galaxies' (or at least gas which we suspect to reside in galaxies) out to redshifts $z > 4$ [4] via the heavy-element (or so-called 'metal') absorption lines they produce in the spectra of background quasi-stellar objects (QSOs). The technique of QSO absorption line spectroscopy is intrinsically very powerful in that it can measure directly a wealth of physical parameters — such as chemical composition, temperature, dust content, molecular fraction and star-formation rates — of the gas intersected by the line-of-sight to the QSO. Moreover, galaxies may be studied in absorption more or less independently of redshift or luminosity, whereas galaxies selected by their emission are subject to strong selection biases associated with their colour and surface brightness.

The only technique which so far has proven successful in locating high-z galaxies in emission has been the optical identification of steep-spectrum radio sources, [5,6,7,8] the most distant to date being at $z = 4.25$. [9] It is important at this stage to clarify the distinction between the types of galaxies sampled by the two methods. Whereas distant radio galaxies are highly luminous and probably powered by an active nucleus, most of the sightline to a high-z QSO samples passive material which is expected to be typical of *normal* galaxies and intergalactic gas at intervening redshifts. Radio galaxies are important because they can be imaged directly and we can obtain spatially resolved spectra. However, there are concerns that these galaxies may be atypical, even pathological, and therefore of limited use in constructing a global picture of galaxy formation and evolution.

In the following sections we will explore some of the properties of the galaxies detected by their absorption signatures, notably those in which the Lyman α line shows strong damping wings. Because of the limitations imposed by the ultraviolet cutoff of the earth's atmosphere, this restricts ground-based studies to $z > 1.7$. For an up-to-date account of high-redshift radio galaxies the reader is referred to the excellent review by McCarthy. [8]

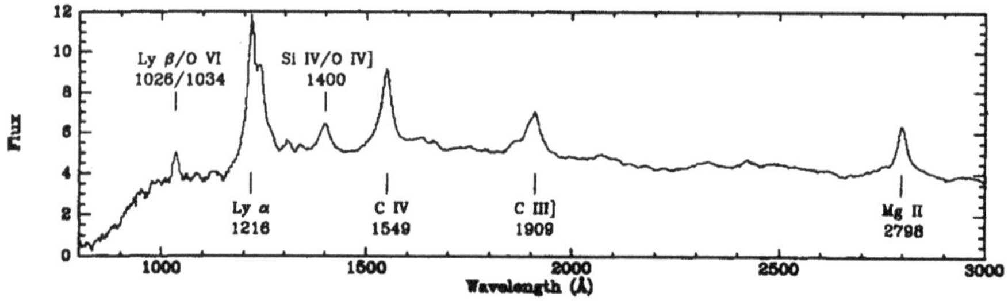

Figure 1: Composite QSO spectrum from the Large Bright Quasar Survey (LBQS). [10] Flux density per unit wavelength (arbitrary scale) is plotted against rest wavelength. The prominent broad emission lines seen at high redshift are marked, along with their rest wavelengths in Å. The drop in flux below Lyman α emission is due to the Lyman α absorption forest (see Fig. 2).

2 QSO absorption lines

The spectrum of a QSO[a] is characterised by strong broad emission lines — resonance lines of the most abundant elements — and a predominantly non-thermal continuum, as shown in Fig. 1.

The chance alignment of gas clouds and galaxies along the line of sight to a high-z QSO is shown schematically in the upper panel of Fig. 2 and a typical spectrum is shown in the lower panel. With a ten-fold increase in resolution over Fig. 1 the absorption lines stand out clearly. The prominent broad emission line near 4000 Å is Lyman α ($z_{em} = 2.25$) and the spectrum extends to Lyman β. Most of the narrow absorption lines are Lyman α lines arising in H I clouds spanning the range $z_{abs} = 1.76$–2.25 — the so-called Lyman α forest — but some are due to metals associated with the prominent damped Lyman α lines at 3510 and 3730 Å.

2.1 The Damped Lyman α (DLA) systems

Among the different classes of QSO absorption lines,[11] the damped Lyman-α systems are currently attracting the most attention. Wolfe and collaborators[12] first recognised them as a distinct class of absorbers and defined them, from practical considerations, as systems with a neutral hydrogen column density $N(\mathrm{HI}) \geq 2 \times 10^{20}$ cm^{-2}. This limit was chosen originally to match the H I $\lambda 21$ cm studies of nearby galaxies.[13] It is also approximately the minimum column density for which a damped Lyman α line can be recognised in low-resolution spectra.

Although relatively rare, the damped Lyman α systems account for most of the neutral hydrogen seen at high redshifts. This extraordinary result is what makes the study of DLA systems so important cosmologically. To determine the cosmological co-moving mass density of neutral gas[4] we need to know the number density $n(z)$ of DLA systems as a function of hydrogen column density N. In practice, this amounts to determining the column-density distribution function $f(N)$, which is well described by a power law of the form

$$f(N) = BN^{-\beta},$$

where the index β is in the range 1.2–1.8.[14] The comoving mass density of neutral gas Ω_g, in units of the critical density, is given by

$$\Omega_g = \frac{H_0}{c} \frac{\mu m_{\mathrm{H}}}{\rho_{\mathrm{crit}}} \int_{N_{\mathrm{min}}}^{N_{\mathrm{max}}} N f(N) dN,$$

[a]The abbreviations QSO and quasar are used interchangeably, regardless of whether the quasi-stellar object is radio-loud or radio-quiet

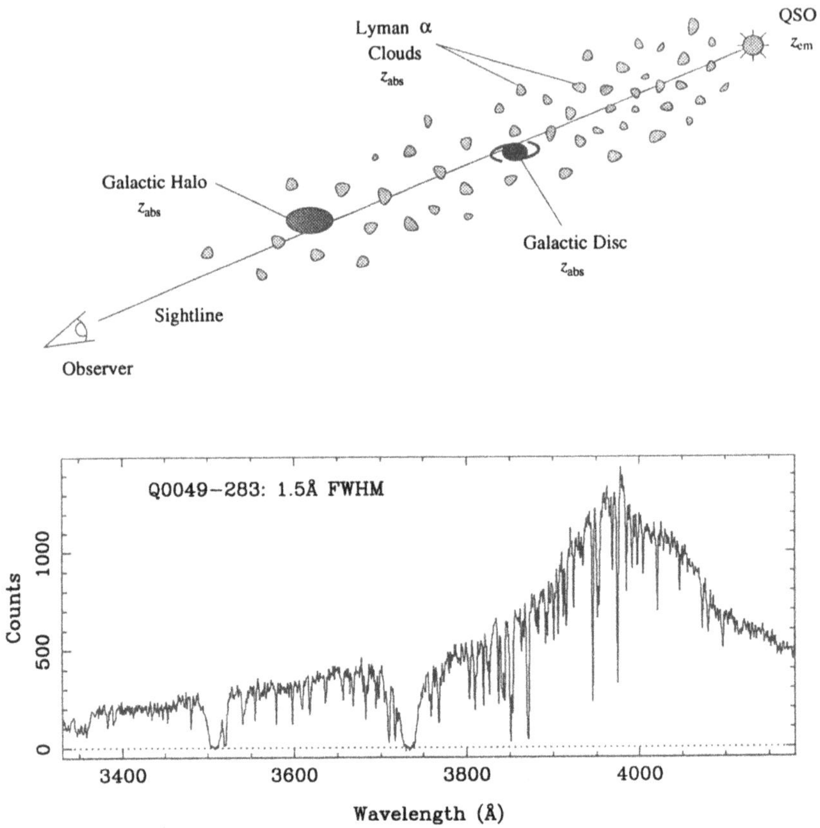

Figure 2: *Upper panel*: Schematic diagram showing how the line of sight to a high-z QSO can intersect gas clouds and galaxies over the entire path from $z = 0$ to z_{em}. *Lower panel*: Anglo-Australian telescope (AAT) spectrum of the $z_{em} = 2.25$ QSO Q0049−283 at a resolution of 1.5 Å (~100 km s^{-1}). The most prominent features are the broad Lyman α emission line near 4000 Å and two damped Lyman α absorption lines near 3510 and 3730 Å, probably the result of the sightline intersecting galactic disks at $z_{abs} = 1.89$ and 2.07. Most of the weak, narrow absorption lines are Lyman α lines which arise in hydrogen 'clouds' distributed along the line of sight to the quasar. Until recently it was assumed that these clouds were truly intergalactic and consisted of pristine material, unenriched by the products of stellar nucleosynthesis; both of these assumptions are now being thrown into question by new data from the *Hubble Space Telescope* and large ground-based telescopes.

where m_H is the mass of the hydrogen atom, ρ_{crit} is the critical density, and N_{min} and N_{max} define the range of column densities of interest. The surprising result is that Ω_g is independent of the sizes, shapes or covering factors of the absorbers. However, it does depend on the cosmological model and is dominated by N_{max}, the highest column density DLA systems, currently $\approx 6 \times 10^{21}$ cm^{-2}.

The compilation of DLA systems by Lanzetta *et al.* [14] shows that, with a mean number density $\langle n(z) \rangle = 0.16$–$0.25$ at $\langle z_{abs} \rangle \simeq 2.5$ and mean $\langle N(\mathrm{H\,I}) \rangle \simeq 1 \times 10^{21}$ cm^{-2}, the mass density contributed by the damped Lyman α systems (in units of the current closure density) is $\Omega_g \simeq (0.7$–$1.5) \times 10^{-3}\ h^{-1}$ (for $q_0 = 0$–0.5, and $H_0 = 100\,h$ km s^{-1} Mpc^{-1}). For comparison, stars in present-day galaxies[15] are thought to account for $\Omega_s \simeq 3 \times 10^{-3}\ h^{-1}$.

The similarity between Ω_g and Ω_s has led to the suggestion that the DLA systems may be the progenitors of present-day luminous galaxies — or at least the neutral cores of such galaxies — observed at a time prior to the bulk of star formation, when the gas fraction was high and the galaxies were still in the process of gravitational contraction. An alternative interpretation is that the lines may trace a population of dwarf galaxies which even at the present time are visible only during relatively brief episodes of enhanced star formation.[16,17] Direct information on the size of the damped Lyman α absorbers is very limited at present, but in at least two cases there are indications that we may be dealing with structures extending more than $\sim 10\ h^{-1}$ kpc across the line of sight.[18,19]

2.2 Chemical Evolution of Damped Lyman α Galaxies

A key property of the DLA systems is their chemical composition. If we could chart the progress of metal enrichment through stellar nucleosynthesis in these high-z galaxies this would be an important clue to their evolutionary status. Although the column density of neutral hydrogen is well determined by the profile of the damped Lyman α line (see Fig. 3), the measurement of metal abundances encounters two major difficulties which are well known from analogous work on the local interstellar medium.[20] Firstly, for $N(\mathrm{H\,I}) > 2 \times 10^{20}$ cm^{-2} the accessible absorption lines of the most abundant elements are heavily saturated, making the determination of column densities uncertain by at least an order of magnitude. Secondly, this is compounded by the fact that most elements are depleted from the gas phase onto interstellar dust grains by large and varying amounts.

The strategy adopted to circumvent these difficulties[21,22] draws on the experience of UV observations over the past two decades with the *Copernicus* and *International Ultraviolet Explorer* satellites. We target the resonance lines

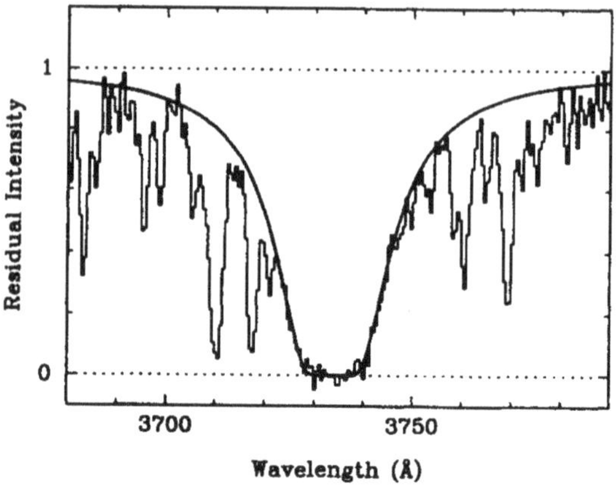

Figure 3: The region of spectrum encompassing the damped Lyman α line at $z_{abs} = 2.07$ in Q0049−283 (see Fig 2). As can be seen, this line is well fitted by a theoretical damped profile for a column density of 3.1×10^{20} cm^{-2}. The uncertainty in N(H I) is ∼10–15%.

of Zn II $\lambda\lambda$2025.5, 2062.0 and Cr II $\lambda\lambda$2055.6, 2061.6, 2065.5 for the following reasons:

1. In Galactic stars, and in LMC and SMC stars, both Zn and Cr track closely the abundance of Fe down to the lowest metallicities measured,[23,24] i.e. [Zn/Fe] and [Cr/Fe] $\simeq 0^b$.

2. In the local interstellar medium (LISM) Zn differs from most other iron-peak elements in that it is *not* readily incorporated into dust grains. Extensive surveys [25] have shown that generally Zn is present in the gas phase in near-solar proportions (relative to hydrogen) and that its depletion onto dust is at most by factors of 2–3 in the densest clouds sampled. Cr, on the other hand, is one of the most heavily depleted elements in the LISM with typically only ≈ 1% remaining in the gas phase.

3. In H I regions both Zn and Cr are predominantly singly ionised, thus avoiding the need to account for unobserved ion stages.

4. Zn and Cr are both trace constituents of the LISM and the absorption lines are expected to be weak, even at the high H I column densities of the DLA systems. This means that accurate values of column density

[b]This follows the normal convention for comparing abundances with solar values, $[\mathrm{Zn/Fe}] \equiv \log_{10}(\mathrm{Zn/Fe}) - \log_{10}(\mathrm{Zn/Fe})_\odot$.

can be obtained without line saturation becoming the problem it is for the more abundant elements.

5. Finally, for $z \simeq 2$–3 the wavelength interval near 2050 Å is shifted into a convenient region where detector efficiency is high and the sky background is tolerable. Moreover, this region falls redward of Lyman α emission and is therefore free of the confusion caused by Lyman α forest lines.

The combination of all these factors means that the N(Zn II)/N(H I) ratio measures directly and reliably the abundance of iron-peak elements in the DLA galaxies, while N(Cr II)/N(Zn II) reflects the degree of depletion of refractory elements onto dust. Taken together, the two abundance measurements give an approximate indication of the dust-to-gas ratio.

Pettini *et al.*,[22] in a program spanning several years, have carried out an extensive study of Zn and Cr abundances in 15 DLA systems based on spectra obtained at the Anglo-Australian Telescope (AAT) at Siding Spring Observatory, Australia, the William Herschel Telescope on La Palma, Canary Islands, and the Hale Telescope at Palomar Observatory, USA. Zn and Cr lines were detected in approximately half the DLA systems observed. The examples of two detections and one non-detection shown in Fig. 4 serve to emphasise the weakness of the absorption lines.

The derived abundances of Zn and Cr are plotted against redshift in Fig. 5. There are several trends evident in these plots:

1. At $z \simeq 2$–3 damped Lyman α galaxies are generally metal-poor; almost all the points in Fig. 5(*a*) fall below the line of solar abundance. For the nine Zn II detections the median metallicity is $Z_{\rm DLA} = 1/10\, Z_{\odot}$. As discussed by Lanzetta,[4] this can be taken as the characteristic metallicity of the neutral content of the universe at these redshifts, indicating that the process of metal enrichment through stellar nucleosynthesis had apparently not proceeded far in most galaxies by $z \simeq 2$.

2. At essentially the same epoch, $z \simeq 2$, there is a considerable spread in the degree of metal enrichment attained by different DLA galaxies. While some galaxies had apparently already evolved to abundances comparable with present-day spirals, in others the interstellar medium had hardly been enriched at all.

3. As shown in Fig. 5(*b*), Cr is consistently less abundant than Zn by factors ranging from 3.5 to > 13. By analogy with the local ISM, it is assumed that the missing fraction of Cr has been removed from the gas phase and

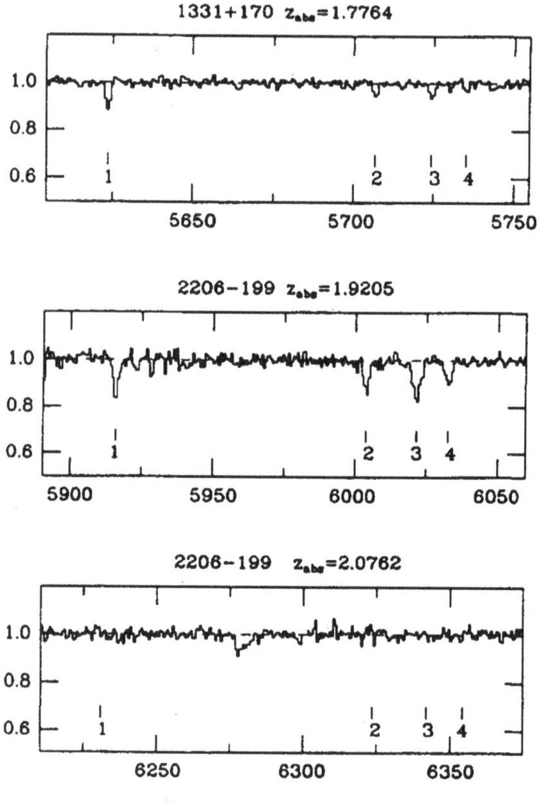

Figure 4: Portions of the QSO spectra[22] in the regions of Zn II and Cr II in three damped Lyman α systems; exposure times were typically ~30 000 s. The spectra have been normalised to the local continuum and residual intensity is shown as a function of wavelength in Å; note the expanded vertical scale. The numbered vertical tick marks indicate the expected positions of the absorption lines. Line 1: Zn II λ2025.484, 2: Cr II λ2055.596, 3: Cr II λ2061.575 + Zn II λ2062.003 (blended), 4: Cr II λ2065.501. The feature near 6276 Å in the bottom panel is due to atmospheric O_2.

presumably incorporated in dust grains. Pettini *et al.*[22] take this result as evidence for the existence of dust in the DLA systems, suggesting that dust is a constituent of the ISM of galaxies even at this early stage of chemical evolution. The presence of dust deduced from the [Cr/Zn] ratio is in good agreement with estimates of the dust-to-gas ratio from the mild reddening of the continua of QSOs containing a DLA system.[26]

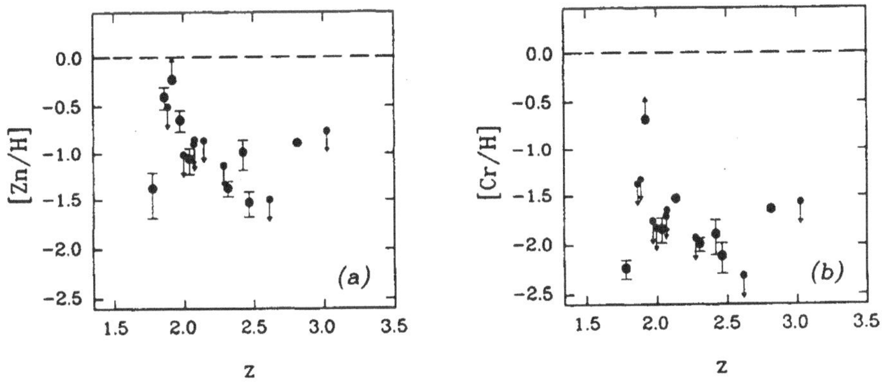

Figure 5: Abundances of *(a)* Zn and *(b)* Cr in the 17 DLA systems from the survey by Pettini *et al.*[22] plotted against redshift z. The abundances are measured on a log scale relative to solar values, indicated by the dashed lines; thus [Zn/H] = -1.0 denotes an underabundance of Zn by a factor of 10. Upper limits, corresponding to non-detections, are indicated with downward-pointing arrows and one lower limit with an upward-pointing arrow.

2.3 Extension to Higher Resolution and Higher Redshift

The results shown in Fig. 5 have been interpreted by some as evidence for a rapid build-up of metals in the universe between z = 3 and 2. Such an effect may go hand-in-hand — at least qualitatively — with the significant consumption of gas suggested by the evolution of the H I content of DLA lines over the same redshift interval.[27] If correct, this would be an important result signalling the epoch of galaxy formation. However, as yet the published data are insufficient to measure with confidence any trends in metallicity or H I content with redshift. Indeed, new data on the rate of occurrence of DLA systems in the spectra of very high redshift QSOs ($z_{em} > 4$)[28] suggests that the H I redshift evolution may have been overestimated.

It was noted above that about half of the Zn II observations produced only upper limits to the abundances. In some cases we *know* that the true abundances lie well below these limits. For metallicities $Z_{\rm DLA} < 1/50\, Z_\odot$, we enter a regime where some of the UV resonance lines of the more abundant elements (eg. Si, O, Fe) become accessible to abundance studies through high-resolution observations with echelle spectrographs (see Fig. 6). Abundance estimates obtained in this way may have greater uncertainties in comparison with [Zn/H], because of non-solar relative abundances, but they serve to highlight the fact that at $z \simeq 2$ DLA systems span more than two orders of magnitude in metallicity, and that systems with $Z_{\rm DLA} < 1/100\, Z_\odot$ are not uncommon. Such a wide range is unlikely to be due to radial abundance gradients similar to those seen in present-day spiral galaxies, but more likely reflects a spread in the starting times and rates of chemical enrichment in the galaxies sampled by the damped systems.

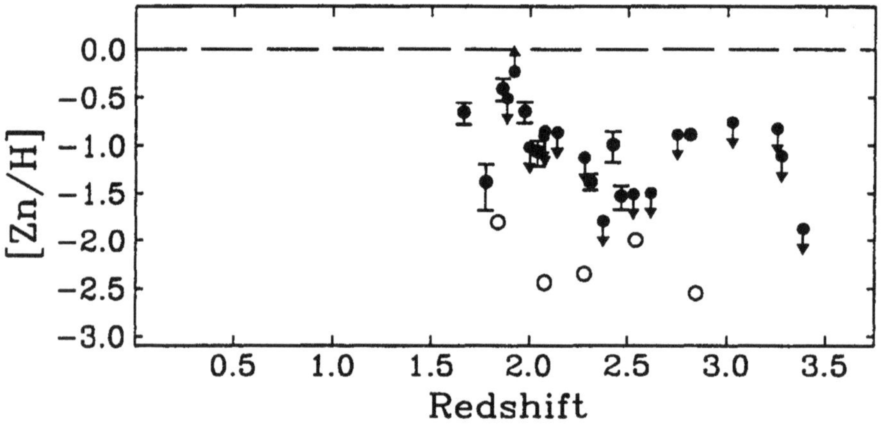

Figure 6: Abundances of Zn in the 17 DLA systems shown in Fig. 5 augmented by more recent data at higher redshift, mostly at $z_{\rm abs} \geq 2.5$. Also shown are metal abundances measured from very high resolution echelle observations of unsaturated lines of O, Si or Fe in DLA systems (open circles); two of these open circles have corresponding [Zn/H] upper limits.

The Zn II survey by Pettini *et al.*[22] has been extended to higher redshifts[29] with the aim of defining better any trends with redshift. In fact, *no* new detections were made at $z \geq 2.5$, as shown in Fig. 6, suggesting that *all* DLA systems sampled are metal poor (<1/10 solar), in contrast to $z \simeq 2$ where at least some galaxies have reached abundances comparable with spiral galaxies today. It is tempting, therefore, to interpret the interval from $z = 3$–2 as indicating the period over which the first episodes of star formation took place. In this sense, at least, we may indeed have identified the epoch of galaxy formation.

A further point of interest concerns the *lowest* abundances recorded in Fig. 6, $Z_{\rm DLA} < 1/300\, Z_\odot$, the same as that which apparently applies to clouds in the Lyman α forest, as recently discovered by Tytler.[30] One possible interpretation is that this value of $[Z] = -2.5$ may represent a universal base level of metallicity[c] on which subsequent galactic chemical evolution can build.

If the Lyman α forest does indeed trace an IGM which, by a redshift ~3, has already been contaminated with heavy elements, abundance measurements of a range of elements in DLA systems with $[Z] \sim -2.5$ may provide vital clues to the stellar populations which produced this initial enrichment. Such studies have now begun [31] with the DLA system at $z = 2.28$ towards Q2348−147. A portion of the AAT echelle spectrum of Q2348−147 is shown in Fig. 7 to emphasise the narrowness of the metal lines which can only be recorded adequately at this high spectral resolution; most of the other lines in this spectral region are due to Lyman α.

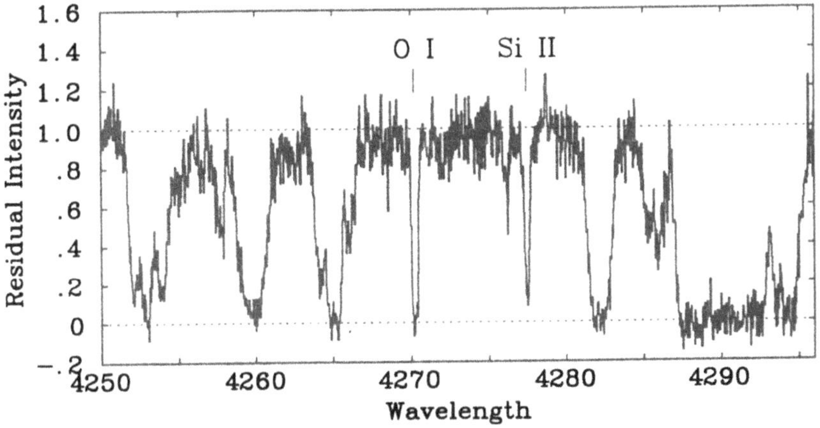

Figure 7: Portion of the AAT echelle spectrum of Q2348−147 encompassing the O I $\lambda1302$ and Si II $\lambda1304$ transitions in the $z_{\rm abs} = 2.27936$ DLA system. The resolution is $R \simeq 45\,000$, corresponding to 0.095 Å or 6.7 km s^{-1} FWHM.

The abundances of 10 elements in this DLA system are shown in Fig. 8. The absorption lines of C II and O I are saturated, so the abundances are poorly constrained. Most of the other elements are underabundant by about

[c]Some stars in our galaxy must have formed from more pristine gas, since there are stars known with abundances as low as $[Z] \simeq -4$

two orders of magnitude relative to the Sun. The uniformity of the relative abundances points to significantly reduced dust depletions compared with the local interstellar medium. A striking feature of Fig. 8 is the underabundance of N relative to the other elements, a result which has important consequences for the nucleosynthetic origin of nitrogen.[31]

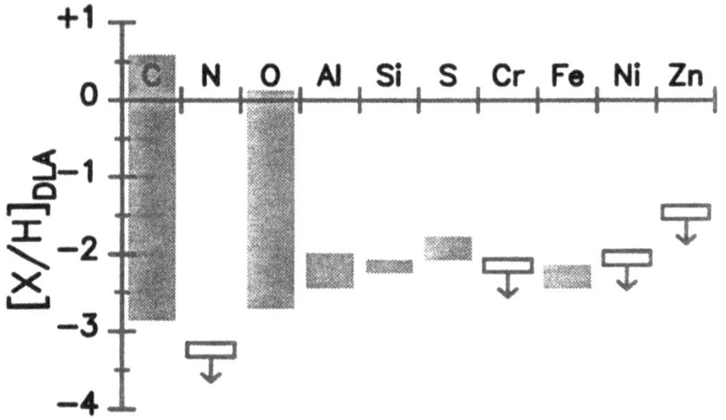

Figure 8: Element abundances in the z_{abs} = 2.27936 DLA system towards Q2348−147, plotted on a log scale relative to solar abundances. Grey vertical bars indicate positive detections, with their length defining the allowed range of values. Open boxes indicate upper limits, with the upper and lower edges representing 3σ and 2σ limits respectively.

2.4 Metal Abundances at Intermediate Redshifts

Ideally we would like to be able to fill out our picture of the chemical evolution of galaxies by tracking the build-up of metals to the present time. The difficulty here is first in identifying even a modest sample of DLA systems at redshifts down to $z_{abs} \simeq 0.6$ (corresponding to Zn II falling at the atmospheric cutoff) and then obtaining the necessary confirmation and measurement on N(H I) with *HST*. To date there are only three measurements of metallicity in damped systems at intermediate redshift, and in all three cases the abundances deduced are approximately one order of magnitude below solar and not significantly higher than the typical metallicity at $z = 2$. For two of the DLA systems the candidate absorbing galaxy has been identified from deep images of the field: one is a low-surface-brightness galaxy and the other a low-luminosity galaxy. In each case the low abundances are not surprising, since star formation probably proceeds at a slower pace in LSB and dwarf galaxies.

These results raise the possibility that we may be subject to serious selection effects introduced by dust as the metal content of the universe grows with time.[32] Our view of distant galaxies may become progressively more skewed in favour of metal-poor systems with decreasing redshift. QSOs which lie behind metal-rich — and presumably dusty — galaxies may be reddened by amounts sufficient to exclude them from the magnitude limited surveys from which the current DLA systems are drawn.

3 Lyman α Emission from High-redshift Galaxies

Since we have established that DLA systems at $z \simeq 2$–3 arise in galaxies which are mostly metal-poor, the fields of QSOs containing damped systems might be good places to search for evidence of on-going star formation. For instance, we might expect to detect Lyman α emission from high-redshift H II regions, particularly as the QSO light is completely extinguished by the DLA absorption line over a wavelength interval of several tens of Å, corresponding to $\sim$1000 km s^{-1} or more.

However, intensive searches using intermediate-resolution long-slit spectroscopy by our group and by others have been either fruitless or produced only marginal detections. A typical null result is shown in Fig. 3. The only published detection of emission in the base of a damped Lyman α line has been for the $z_{abs} = 2.4651$ system in Q0836+113.[33] This result has been disputed by some, but confirmed by new observations obtained in 1994. The spectrum is shown in Fig. 9.

The lack of significant Lyman α emission from DLA galaxies has motivated a considerable amount of theoretical work. The observations are consistent with three possibilities:

(i) Starbursts are short-lived and even massive bursts of star formation are probably bright in Lyman α for only a relatively brief period;[34]

(ii) Lyman α line radiation can be readily quenched by small amounts of dust,[35] due to the large optical depths involved; and

(iii) Lyman α photons probably escape from the parent galaxy along preferred directions, determined by the inhomogeneous nature of the ISM.[36]

The rapid improvements in infrared detectors suggest that searches for redshifted Balmer lines at $z > 2$ may be the most promising strategy for assessing the relative importance of these processes, and measuring the star-formation rate of high-z galaxies.

In conclusion, it is clear that the field of QSO absorption line spectroscopy has matured significantly over the past decade, to the point where quantitative measurements of chemical abundances can be probed over cosmologically long timescales. It provides a powerful new observational tool in an area of study which until now has been confined to studies of Galactic stellar populations and H II regions in nearby galaxies. We can confidently expect that, with the new generation of 8–10 m class telescopes, further progress will be made towards understanding the processes regulating the early stages of galactic chemical enrichment and, thereby, building a more complete picture of galaxy evolution.

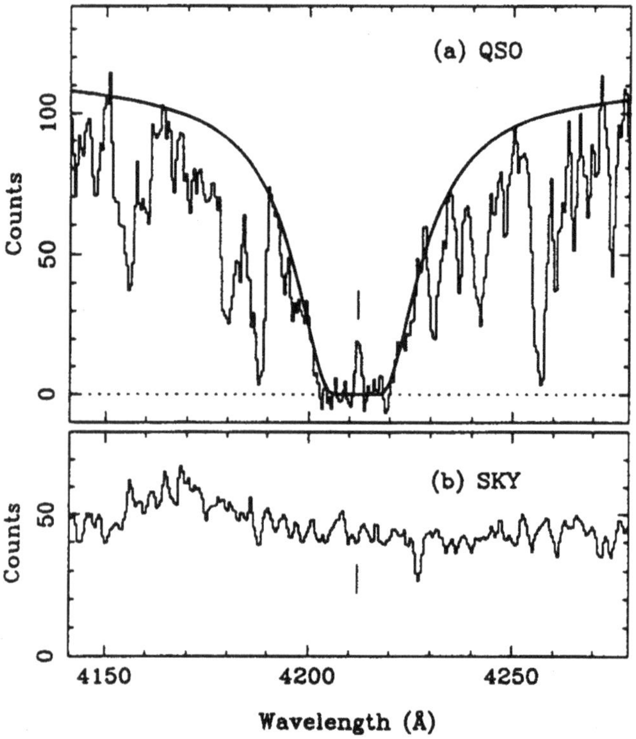

Figure 9: *(a)* AAT spectrum[33] of Q0836+113 in the region of the damped Lyman α line at $z_{abs} = 2.4651$, together with the fitted profile for $N(\mathrm{H\,I}) = 4.2 \times 10^{20}$ cm^{-2}. The emission line in the base of the damped absorption line is marked. *(b)* Sky spectrum spanning the same wavelength interval.

Acknowledgments

It is a pleasure to acknowledge my collaborators in the QSO absorption line program: Max Pettini, Linda Smith, Dave King, Keith Lipman and David Mar. I also thank David Mar for creating the upper panel in Fig. 2. My AAT observing program is supported by a grant from the Australian Research Council.

References

1. L.L. Cowie, A. Songaila & E.M. Hu, *Nature* **354**, 460 (1991)
2. M. Colless, R.S. Ellis, T.J. Broadhurst, K. Taylor & B.A. Peterson, *MNRAS* **261**, 19 (1993)
3. S. Djorgovski, D. Thompson & J.D. Smith, in *First Light in the Universe: Stars or QSOs?*, ed. B. Rocca-Volmerange, M. Dennefeld, B. Guideroni & J. Tran Thanh Van (Editions Frontières, Gif-sur-Yvette, 1993), 67
4. K.M. Lanzetta, in *The Environment and Evolution of Galaxies*, ed. J.M. Shull & H.A. Thronson (Kluwer, Dordrecht, 1993), 237
5. K.C. Chambers, G.K. Miley & R.R. Joyce, *Ap. J.* **329**, L79 (1988)
6. S.J. Lilly, *Ap. J.* **340**, 77 (1989)
7. H.S. Spinrad, in *The Epoch of Galaxy Formation*, ed. C.S. Frenk, R.S. Ellis, T. Shanks, A.F. Heavens & J.A. Peacock (Kluwer, Dordrecht, 1989), 39
8. P.J. McCarthy, *Ann. Rev. Astron. Astrophys.* **31**, 639 (1993)
9. M. Lacy *et al.*, *MNRAS* **271**, 504 (1994)
10. P.J. Francis, P.C. Hewett, C.B. Foltz & F.H. Chaffee, *Ap. J.* **398**, 476 (1992)
11. J.C. Blades, D.A. Turnshek & C.A. Norman (eds), *QSO Absorption Lines* (Cambridge University Press, Cambridge, 1988)
12. A.M. Wolfe, D.A. Turnshek, H.E. Smith & R.D. Cohen, *Ap. J. Supp.* **61**, 249 (1986)
13. A. Bosma, *AJ* **86**, 1825 (1981)
14. K.M. Lanzetta, A.M. Wolfe, D.A. Turnshek, L. Lu, R.G. McMahon & C. Hazard, *Ap. J. Supp.* **77**, 1 (1991)
15. N.Y. Gnedin & J.P. Ostriker, *Ap. J.* **400**, 1 (1992)
16. N.D. Tyson, *Ap. J.* **329**, L57 (1988)
17. D.G. York, in *QSO Absorption Lines*, ed. J.C. Blades, D.A. Turnshek & C.A. Norman (Cambridge University Press, Cambridge, 1988), 227
18. F.H. Briggs, A.M. Wolfe, H.S. Liszt, M.M. Davis & K.L. Turner, *Ap. J.* **341**, 650 (1989)

19. C.C. Steidel & D. Hamilton, *AJ* **104**, 941 (1992)
20. E.B. Jenkins, in *Interstellar Processes*, ed. D.J. Hollenbach & H.A. Thronson (Reidel, Dordrecht, 1987), 533
21. M. Pettini, A. Boksenberg & R.W. Hunstead, *Ap. J.* **348**, 48 (1990)
22. M. Pettini, L.J. Smith, R.W. Hunstead & D.L. King, *Ap. J.* **426**, 79 (1994)
23. J.C. Wheeler, C. Sneden & J.W. Truran, *Ann. Rev. Astron. Astrophys.* **27**, 279 (1989)
24. S.C Russell & M.A. Dopita, *Ap. J.* **384**, 508 (1992)
25. M.E. van Steenberg & J.M. Shull, *Ap. J.* **330**, 942 (1988)
26. S.M. Fall & Y.C. Pei, *Ap. J.* **402**, 479 (1993) (and references therein)
27. K.M. Lanzetta, A.M. Wolfe & D.A. Turnshek, *Ap. J.* **440**, 435 (1995)
28. L.J. Storrie-Lombardi, R.G. McMahon, M.J. Irwin & C. Hazard, in *QSO Absorption Lines*, ed. G. Meylan (Springer, Berlin, 1995), in press
29. M. Pettini, D.L. King, L.J. Smith & R.W. Hunstead, in *QSO Absorption Lines*, ed. G. Meylan (Springer, Berlin, 1995), in press
30. D. Tytler, in *QSO Absorption Lines*, ed. G. Meylan (Springer, Berlin, 1995), in press
31. M. Pettini, K. Lipman & R.W. Hunstead, *Ap. J.* (1995), in press
32. S.M. Fall & Pei, Y.C., in *QSO Absorption Lines*, ed. G. Meylan (Springer, Berlin, 1995), in press
33. R.W. Hunstead, M. Pettini & A.B. Fletcher, *Ap. J.* **356**, 23 (1990)
34. S. Charlot & S.M. Fall, *Ap. J.* **415**, 580 (1993)
35. W.L. Chen & D.A. Neufeld, *Ap. J.* **432**, 567 (1994)
36. D.A. Neufeld, *Ap. J.* **370**, L85 (1991)

THE VERY EARLY UNIVERSE

V.N.Lukash

Astro Space Center of the Lebedev Physical Institute
Profsoyuznaya, 84/32, 117810 Moscow, Russia

In these lectures we dwell upon the cosmological corner-stones of Very Early Universe (VEU) theory: the Parametric Amplification Effect (PAE) responsible for the generation of Primordial Cosmological Perturbations (PCPs), Chaotic and Stochastic Inflation, and the Principal Tests of VEU theory.

1. INTRODUCTION

A great success of VEU theory as the theory of the beginning of the Big Bang is related to its semiclassical nature, allowing us to operate productively in terms of classical space-time filled with quantum physical fields (including the gravitational perturbations). It (VEU theory) connects like a bridge the theory of our Universe based on the Friedmann model (FU) with theories of Everything (TOE) essentially employing quantized gravity (still very ambiguous). This connection is already realized itself in the important understanding that the quasi-homogeneous and isotropic state of the Universe on the horizon scale (on one hand) and the primordial cosmological perturbations which gave a birth to the Universe structure on smaller scales (on the other hand) are just two features in the low-energy limit of some theory of the VEU based on a model of the inflationary Universe (IU). Up to now we have no alternative to Inflation for producing the cold remnants which we observe today as the micro and macro worlds; it is commonly considered as a necessary element and a key test of high energy physics and any TOE.

When discussing VEU theory we need to decide whether to start from cosmological or particle physics foundations. The particle physics is not yet fixed well at high energies: to follow this direction today means to start from N particle and modified gravity theories, where N is a big number, and then to build up N inflationary models based on them. For this reason, it seems rather preferable to investigate cosmologically standard VEU theories based on three standard points:

(i) setting some cosmological postulates within General Relativity (GR);

(ii) deriving the theory from these postulates, and

(iii) confronting the predictions of the theory with observations.

Along this way, which I will follow in these lectures, we are independent of future particle physics and, thus, may try to find some basic properties and principal features of the VEU theory (tested by observations!) which stay independent of the particle physics uncertainties as well. We know two good examples of such theories. The first is the Friedmann-Robertson-Walker model which has ensured the great success of observational and theoretical cosmology, just based on the cosmological postulate (the homogeneity and isotropy of 3-space). The second is the PAE, the theory of the generation of the PCPs which gave birth to the Large Scale Structure (LSS) of the Universe, just based on linear perturbation theory in the FU.

Regarding the VEU our goal would be to find the basics for the IU in GR which could create the Friedmannian region we live in now (which, in its turn, would provide the initial conditions for both our examples, the FU and PAE theories). The answer which we know today is the model of Chaotic Inflation based on the assumptions of the existence of the inflation (the scalar field weakly coupled to all other physical fields) and of the initial-inflation-condition postulating a quasi-homogeneous spatial distribution of the initial inflaton in a Compton-wavelength region (i.e., in the finite-scale region ! A great step in comparison with the FU cosmological postulate !).

Below, our lectures will be grouped in four parts. The first is devoted to the PAE of scalar perturbations in the FU (Lukash 1980). Next, we consider the cosmological applications to the *scattering* problem and chaotic inflation (Lukash & Novikov 1992). In the third part we present some recent developments of the theory of chaotic and stochastic inflation proposed by Linde (1983, 1986). The last Chapter deals with the problem of testing and confronting these theories with observations.

Certainly, we do not pretend to cover all the relevant references and give a review of all recent ideas and speculations in VEU theories. Our main goal here is to present some basic properties of the VEU which seem today more or less settled and independent of future theoretical constructions. We try to consider the simplest mathematical models, paying particular attention to the physical meaning of the effects considered. Some necessary mathematical calculations are given in three Appendices. Hereafter, our units imply $c = 8\pi G = \hbar = 1$ and $H_o = 100h^{-1}\ \ kms^{-1}Mpc^{-1}$.

2. PARAMETRIC AMPLIFICATION EFFECT

The formation of the structure of the Universe is one of the fundamental problems of modern Cosmology. The two following properties of the present Universe are very important in understanding the physics of its early expansion.

(i) A high degree of homogeneity and isotropy on large scales ($\delta\rho/\rho \lesssim 10^{-4}$ on the cosmological horizon), along with a well-developed structure on scales less than 0.01 of the horizon.

(ii) A high specific entropy ($N_\gamma/N_b \simeq 10^9$), along with the mean baryon density being less than $\sim 10\%$ of the critical value ($\Omega_b \lesssim 0.1$).

Property (i) proves that the large scale structure of the Universe (galaxies, clusters and superclusters) stemmed from initially small amplitude perturbations of a homogeneous and isotropic cosmic medium, since it is small perturbations that may grow up to the order of unity (and then form gravitationally bound objects) when the horizon becomes many times larger than their linear scale. We do not know yet whether PCPs formed together with the cosmological Friedmann model at Planckian curvatures or whether they originated in the process of the homogeneous and isotropic expansion which is described by Classical General Relativity (CGR). In the first case we have no theory. However, an important point is that the quasi-Friedmannian Cauchy-hypersurface is already a classical object after Planckian time. So, if the PCPs are made by an evolutionary process and their scales are less than the Friedmann-hypersurface scale then they are likely to form at a semiclassical stage when the large-scale gravity was governed by the CGR equations. We know one example of this kind - it is inflation: galaxy-scale PCPs form at the very late stages of the inflationary expansion when the Friedmannian Cauchy-hypersurface (which forms the background of our local Universe today) has been already prepared by the inflation. Below, we will develop the theory of small potential (scalar) PCPs, assuming the existence of the Friedmann background model.

Also, it is quite evident that by no process could the Universe be born strictly homogeneous and isotropic: there always exist quantum fluctuations of the metric and the physical fields. The seed fluctuations could be of a statistical, random character or they might be thermal, etc. An important point here is as follows: the inevitable minimum level of seed initial perturbations is always maintained by the quantum zero-point fluctuations of a quasi-homogeneous gravitating medium which is the basis of a spatial slice of the FU.

One of the basic implications of property (ii) is that in the past the Universe was hot and its expansion was governed by the gravitational field of intensively interacting relativistic particles. At this stage the matter represented to a high degree of accuracy a hydrodynamic perfect fluid with equation of state $p = \epsilon/3$ (the Hot Friedmann Universe, HFU).

Small perturbations of such a radiationally dominated gravitating fluid are sound waves propagating through the matter with a constant amplitude (adiabatic decrease of the wave amplitude due to the Universe expansion is exactly compensated in this situation by the increase of the amplitude gained due to the pressure gradient in the comoving (with the wave front) reference system). Mathematically, such perturbations in the expanding Universe are governed by the same equation as that for the acoustic perturbations of a non-gravitating homogeneous static thermal bath in flat Minkowski space-time (conformal invariance). Deep physics is behind it: no new phonons, cosmological potential perturbations, can be produced during the expansion of the hot Universe.

For the real generation of PCPs to occur in the early Universe, one has to reject the hot equation of state ($p = \epsilon/3$) at some expansion epoch, which can be done in principle in the VEU only before the primordial heating of cosmic matter to high relativistic temperatures. In this case, as we shall see below, the number of phonons is not conserved and new phonons (the density perturbations) can be created in the course of the expansion. The mechanism is purely classical and works as a parametric amplification: the energy of a large-scale non-stationary gravitational potential is pumped to the energy of small-scale perturbations (as new photons are created in an electromagnetic resonator when its size changes in a non-adiabatic way). This effect, which we will generally call the "parametric amplification effect", has nothing to do with Jeans instability: actually it is a CGR-effect since the typical scales to be amplified are just the cosmological horizon (and the effect involves the velocity of light and gravitational fundamental constants). Before we present the mathematical formalism the physical meaning of the parametric amplification is discussed in the next Section.

2.1. Physical Meaning of the Parametric Amplification

The theory of small perturbations in the FU was constructed by Lifshitz in 1946. According to this theory there are three types of perturbations of the homogeneous isotropic model: density (potential) perturbations, vortex perturbations and gravitational waves. We are interested now in the first type of perturbations, potential perturbations, because they are related to galaxy formation.

Let us consider the spatially Euclidean background model:

$$ds^2 = dt^2 - a^2 d\vec{x}^2 = a^2(d\eta^2 - d\vec{x}^2), \tag{1}$$

where a is the scale factor which is a function of time, and t and η are universal and conformal times respectively,

$$\eta = \int \frac{dt}{a} \tag{2}$$

It is usual to present the density perturbations $\delta\rho$ in a Fourier expansion

$$\delta = (2\pi)^{-3/2} \int d^3\vec{k}\delta_{\vec{k}} e^{i\vec{k}\vec{x}}, \tag{3}$$

where $\delta = \delta\rho/(\rho + p)$ (a more rigorous definition which makes $\delta = \delta(t, \vec{x})$ a gauge invariant function is given later on). Here, the wavenumber $k = |\vec{k}|$ and the physical wavelength $\lambda = 2\pi a/k$.

Also we shall use the perturbation scale

$$l_k = \lambda/2 = \frac{\pi a}{k}, \tag{4}$$

and the horizon scale [1]

$$l_H = H^{-1} = \frac{a}{\dot{a}} \tag{5}$$

The latter is a typical scale of the causally connected region during the evolution time scale of the a-function. (The dot denotes the derivative with respect to the universal time $\dot{()} = \frac{d}{dt}$).

We consider here only homogeneous and isotropic states of the perturbation fields (random spatial phase fields). Their important characteristic is the power spectrum of the density perturbations Δ_k^2:

$$\frac{k^3}{2\pi^2}\langle \delta_{\vec{k}} \delta^*_{\vec{k}'} \rangle = \Delta_k^2 \, \delta(\vec{k} - \vec{k}'), \tag{6}$$

where brackets $\langle ... \rangle$ mean an average over the field state, $\delta(\vec{k} - \vec{k}')$ is the 3-dimensional δ-function and (*) is the complex conjugate. This spectrum determines the second correlation of the density perturbations

$$\xi(r) = \langle \delta(t, \vec{x})\delta(t, \vec{x} + \vec{r}) \rangle = \int_0^\infty \frac{dk}{k} \Delta_k^2 \frac{\sin(kr)}{kr}, \tag{7}$$

[1] For brevity, we refer to the Hubble scale H^{-1} as the horizon, although this is not technically correct.

where $r = |\vec{r}|$. All the odd correlations are identically zero. Gaussian fields, which are particular cases of the random fields, can be totally described by this correlation function only (all higher order correlations are negligible in linear approximation). Eq. (7) also clarifies the physical sense of the power spectrum. The amplitudes Δ_k are just the corresponding density perturbations in the scale interval $\Delta k \sim k$ around k; they are additive (the total power is a sum of the partial ones over all scales):

$$\langle \delta^2 \rangle = \sum_{\Delta k \sim k} \Delta_k^2$$

Let us now consider the hot Universe (HFU) when the equation of state was [2] $p = \epsilon/3$. The solution for Δ_k is

$$\Delta_k^2 = (c_1(k) f_1(k))^2 + (c_2(k) f_2(k))^2, \tag{8}$$

where

$$\kappa = \omega\eta \sim l_H/l_k \sim t/l_k \sim a \sim t^{1/2}, \ \omega = \frac{k}{\sqrt{3}},$$

$$f_1(\kappa) = -\cos\kappa + 2\left(\frac{\sin\kappa}{\kappa} + \frac{\cos\kappa - 1}{\kappa^2}\right),$$

$$f_2(\kappa) = \sin\kappa + 2\left(\frac{\cos\kappa}{\kappa} - \frac{\sin\kappa}{\kappa^2}\right),$$

and $c_{1,2}(k)$ are the amplitudes of the growing and decaying modes respectively . The sound velocity here is $\beta = 1/\sqrt{3} \simeq 1$, so the function κ is a ratio of the distance travelled by sound during the cosmological time t to the perturbation scale l_k, which is about the horizon to perturbation scale ratio.

The perturbations of the gravitational potential or metric perturbations are as follows (exact definitions are given below):

$$h_1 = c_1 \frac{1 - \cos\kappa}{\kappa^2} \quad \text{for the growing mode,}$$

$$h_2 = c_2 \frac{\sin\kappa}{\kappa^2} \quad \text{for the decaying mode} \tag{9}$$

An important feature of these perturbations is the following. Neither growing nor decaying modes increase catastrophically in time. Both of them are described by sin and cos functions, so, if c_1 and c_2 are less than unity (and it should be this way,

[2] In our units the energy density and matter density are the same function, $\epsilon = \rho c^2 = \rho$.

otherwise $h_{1,2}$ would be large at smaller times) then both these modes are just sound waves with constant (in time) amplitudes c_1 and c_2 and with different time phases.

First of all, this result means that the HFU is absolutely gravitationally stable against small perturbations: if initial perturbations are less than unity then they remain small while the equation $p = \epsilon/3$ holds.

A more elegant proof of this important conclusion may be given with the help of the q-scalar (Lukash, 1980) which is, generally, a gauge invariant combination of matter velocity and gravitational perturbation potentials. Below, we shall see that potential perturbations in the FU are totally described by this scalar (and conversely: all matter and metric perturbations can be expressed as functions of q). The physical meaning of the q-scalar follows easily from its definition: for large scales, $l_k > l_H$, q is mainly the gravitational potential (matter effects are not important), while inside the horizon, $l_k < l_H$, gravitational perturbations are negligible and q is just the matter (velocity) potential.

In the HFU the q-field obeys the following equation:

$$\ddot{q} + 3H\dot{q} - \frac{1}{3a^2}\Delta q = 0, \tag{10}$$

where $H = \dot{a}/a$ is the Hubble function and $\Delta = \partial^2/\partial\vec{x}^2$ is the spatial Laplacian. Transformations $\bar{q} = aq$, $(') = d/d\eta = ad/dt$ reduce eq. (10) to (note that $a \sim \eta$ for the HFU)

$$\bar{q}'' - \frac{1}{3}\Delta\bar{q} = 0, \tag{11}$$

which is just the non-gravitating acoustic wave equation in the flat spacetime (η, x):

$$\bar{q}_k \sim c_1 \sin\kappa + c_2 \cos\kappa. \tag{12}$$

Eqs. (1,10,11,12) indicate the conformal invariance of potential perturbations in the HFU and, as a result, the conservation of the adiabatic invariant (the total number of phonons, the sound wave quanta) which proves the stability of the HFU expansion against small matter perturbations [3]. Note in this connection that gravitational waves

[3]We do not go into further detail about this stability effect since we have emphasized it many times in our previous lectures. We mention only that the increase in time of the density contrast at $\kappa < 1$ (see eq. (8)), which some people interpret as an instability period, simply corresponds to a period of time during which the oscillatory functions change monotonically. Thus, to speak of instability in this case is as incorrect as to speak of the instability of a mathematical pendulum when it moves, say, out of its stable point for a time which is less than the oscillatory period. Returning to our case, note that the potential energy of such a pendulum at $\kappa < 1$ is all in the gravity (see eqs. (9)). The Jeans thinking fails here because the $\kappa < 1$ region is a purely relativistic one.

have a similar invariance property (see Grichshuk 1974) but we do not discuss them here.

The lumps of matter in these sound waves start growing only after the equality epoch ($l_H \sim 10^4$ yrs), when the non-relativistic particles come to dominate in the expansion and the pressure falls in comparison with the total density. This process develops due to the Jeans gravitational instability causing the fragmentation of the medium into separate bodies at the late stages of the expansion ($l_H \sim 10^9 - 10^{10}$ yrs).

We shall not discuss here these late processes of galaxy formation. For us the following is important: for the formation of large structures (superclusters and clusters), we need a definite amplitude of the sound waves $\sim 10^{-4} - 10^{-5}$ at the linear scales which encompass a number of baryons big enough for these structures to form. So c_1 and/or c_2 must be of the order of 10^{-4} on these scales.

This is a very serious demand on the initial perturbations. Indeed, when t is small, $\kappa \ll 1$, we have (see eqs. (9,12)):

$$\begin{aligned} &c_1 \ll 1, \\ &c_2 \ll \kappa. \end{aligned} \tag{13}$$

From these expressions we can see that c_2 must be extremely small and cannot be of the order of 10^{-4}. So, we need in fact the following conditions to be met for $\kappa \ll 1$:

$$\begin{aligned} &(i)\ c_1 \gg c_2, \\ &(ii)\ c_1 \sim 10^{-4}. \end{aligned} \tag{14}$$

But both of them look very strange.

Indeed, any general natural initial conditions assume a random time phase state for the seed fluctuations

$$c_1 = c_2 \ll\ll 1. \tag{15}$$

The first eq. in (15) holds for example for vacuum or thermal fluctuations. More than that, any natural fluctuations in a hot gravitating medium imply that c_1 and c_2 are dozens orders of magnitude less than 10^{-4}.

The last point is demonstrated with the help of the following example. Let us suppose that the time of origin of the fluctuations is the Planckian one and let us put $k = 1$ for l_{pl}. Then on the galactic scale $k_{gal} \sim 10^{-26}$. Now, let us take a thermal fluctuation spectrum at this moment with the Planckian temperature and, thus, the maximum at $l \sim l_{pl}$. Then the amplitude of the perturbations for $k < 1$ would be proportional to $k^{3/2}$ and, on the galactic scale, it would be $\sim 10^{-40}$. So, c_1 has to be $\sim 10^{-40}$ and it is 35 orders of magnitude less than we need.

Our results are the following:

1) The classical cosmology of the hot VEU has serious difficulties in the explanation of origin of the PCPs. Both the requirements given by eqs. (14) for the large scale structure formation cannot be naturally explained within the framework of the HFU.

2) To account for the appearance of the PCPs at the hot Universe expansion period we need, as a necessary condition, to reject the $p = \epsilon/3$ equation of state at the VEU stage. Modern cosmology provides for a variety of possibilities of this type: from quantum-gravity effects to vacuum phase transitions, cosmic strings, textures, etc. Here we consider the most general conditions for the parametric amplification effect appearing in theories with one scalar field φ coupled to gravity in the minimal way.

Parametric amplification means the production of the gravitating potential inhomogeneities (PCPs) in a non-stationary gravitational background of the expanding Universe: the large scale dynamic gravitational field parametrically creates (amplifies) the small scale perturbation fields. Mathematically, potential perturbations of the FU with a general expansion law are governed by the q-scalar which, after the conformal transformations ($q \to \bar{q}$, $t \to \eta$), satisfies the following equation:

$$\Box_\beta \bar{q} = U\bar{q}, \tag{16}$$

where $\Box_\beta = \partial^2/\partial\eta^2 - \beta^2\Delta$ is the light ($\beta = 1$) or sound ($\beta < 1$) type d'Alembertian operator in the conformal spacetime and $U = U(\eta)$ is the effective potential of the q-field, which is a function of the expansion rate of the FU. Eq. (16) is a type of parametric equation in mathematical analysis capable of amplifying the fields with scales $k \lesssim U^{1/2}$ which are usually outside or about the horizon size (i.e., in the purely relativistic region).

For example, for the massless scalar field φ with minimal coupling the effective potential is $U = a''/a$, and so the typical frequency is just the horizon one (see eq. (16)):

$$\frac{U^{1/2}}{a} = \frac{(a^2H)^{\cdot 1/2}}{a} = H(2 - \frac{dl_H}{dt})^{1/2} \sim H. \tag{17}$$

For the HFU, $a \sim \eta$ and the effective potential is identically zero ($U = 0$), which reduces eq. (16) to eqs. (11,12) considered before. In this case we can define the vacuum state of the q-field for all spatial frequencies and introduce, for instance, a standard technique for the scattering problem with $| \, in\rangle$ and $| \, out\rangle$ states, vacua, and so on, to see how many phonons are spontaneously created during expansion, what their spectrum is, etc.

Further applications depend on the sign of the second derivative of the initial scale factor. The point is that this sign can give us an idea about which scale expands faster, the perturbation scale l_k or the horizon l_H (see eqs. (4,5)). Indeed, the first derivative of their ratio is just proportional to the second derivative of the scale factor:

$$(\frac{l_k}{l_H})^{\cdot} \sim \ddot{a}. \tag{18}$$

So, if $\ddot{a} < 0$ at the beginning, then the galactic scales are found initially outside the horizon, and the Cauchy initial data should be set up outside the horizon as well. On the contrary, if $\ddot{a} > 0$, then the initial conditions for the scales of interest can be set up inside the horizon.

We shall investigate both cases. The qualitative result is as follows: under natural initial conditions for eq. (15) (e.g., $| \, in\rangle$ vacuum state for the q-field) it is the growing mode of perturbations that is finally created in the $| \, out\rangle$ state due to the parametric effect. So, the resulting perturbation field is described by the first line of eq.(14) with the $c_1(k)$ spectrum depending on the expansion factor behaviour at the time period when the parametric amplification condition was met ($k \leq U^{1/2}$).

The Lagrangian theory and the quantization of potential perturbations in the FU are considered below. The next chapter deals with some cosmological applications.

2.2. Lagrangian Theory and Perturbations

Let us consider a scalar field $\varphi = \varphi(x^i)$, with the Lagrangian density depending on φ and its first derivatives in the following general form:

$$L = L(w, \varphi), \quad w^2 = \varphi_{,i}\varphi^{,i} = \varphi_{,i}\varphi_{,k}g^{ik}, \tag{19}$$

The action of the gravitating φ field is as follows:

$$W[\varphi, g^{ik}] = \int (L - \frac{1}{2}R)\sqrt{-g}\ d^4x, \tag{20}$$

where g_{ik} and R_{ik} are the metric and Ricci tensors respectively, $R = R^i_i$, $g = det(g_{ik})$. Variations of eq. (20) over φ and g^{ik} at an extremum give the classical equations of motion of the φ field

$$(\frac{n}{w}\varphi^{,i})_{;i} + n\dot{\nu} = 0, \tag{21}$$

and of the gravitational field created by the φ-field-source

$$G_{ik} = T_{ik}, \quad T_{ik} = \frac{n}{w}\varphi_{,i}\varphi_{,k} - g_{ik}L, \tag{22}$$

where $n = \partial L/\partial w$, $n\nu = -\partial L/\partial\varphi$, $G_{ik} = R_{ik} - Rg_{ik}/2$ and (;) is the covariant derivative in the metric g_{ik}. Note that eq. (21) can be obtained from the Bianchi identities $T^k_{i;k} = 0$ as well.

Useful constructions are the comoving (with the φ-field) energy density and the total pressure of the φ field

$$\epsilon = T_{ik}\varphi^{,i}\varphi^{,k}/w^2 = nw - L,$$
$$p = \frac{1}{3}(\epsilon - T) = L, \tag{23}$$

where $T = T^i_i$. Also, the following equations are useful:

$$\epsilon + p = nw, \quad \beta^{-2} = \frac{w}{n}\frac{\partial^2 L}{\partial w^2} = \frac{\partial \ln n}{\partial \ln w},$$

$$m^2 = -\frac{w}{n}\frac{\partial^2 L}{\partial\varphi^2}, \quad \Gamma = \frac{w}{n}\frac{\partial^2 L}{\partial w\partial\varphi} = w\frac{\partial \ln n}{\partial\varphi}. \tag{24}$$

(The functions m^2 and β^2 can be negative.)

When considering linear perturbation theory φ and g^{ik} are presented as sums of some known functions (the background (^{o}) solution) and small perturbations ϕ and h_{ik}:

$$\varphi = \varphi^{(o)} + \phi, \quad g^{ik} = g^{ik(o)} - h^{ik}. \tag{25}$$

Below, we consider classical backgrounds (eqs. (21,22) are satisfied automatically in (^{o}) order); however, the perturbations can be quantum ones.

The Lagrangian of the perturbation field is obtained by expanding the integrand of eq. (20) up to the second order in ϕ and h^{ik} with all divergent terms excluded (see Appendix A):

$$W^{(2)}[\phi, h_i^k] = \int L^{(2)}\sqrt{-g^o}\, d^4x,$$

$$L^{(2)} = L^{(2)}(v, \psi_i^k) = \frac{\epsilon + p}{2}[v_i v^i + \chi^2(\beta^{-2} - 1) - 2v_i\psi_k^i u^k + v(\nu\psi - m^2 v + 2\Gamma\chi)] +$$
$$+\frac{\epsilon - p}{2}(\psi_{ik}\psi^{ik} - \frac{1}{2}\psi^2) + \frac{1}{8}(\psi_{ik;l}\psi^{ik;l} - 2\psi_{ik;l}\psi^{il;k} - \frac{1}{2}\psi_{,l}\psi^{,l}), \tag{26}$$

where $v = \phi/w$, $v_i = v_{,i} + v(w_{,i}/w)$, $u_i = \varphi^{(o)}_{,i}/w^{(o)}$, $\chi = v_i u^i - h_{ik}u^i u^k/2 = \delta w/w$, $\psi_i^k = h_i^k - h\delta_i^k/2$, $h = h_i^i = -\psi$. (Hereafter all manipulations with indices are carried out with the help of the background metric tensors $g_{ik}^{(o)}$ and $g^{ik(o)}$, and the background index $(^o)$ is omitted where possible). Obviously,

$$\frac{\delta p}{\epsilon + p} = \chi - \nu v, \quad \frac{\delta\epsilon}{\epsilon + \rho} = \beta^{-2}\chi + (\nu + \Gamma)v. \tag{27}$$

The classical field equations which couple the metric and scalar perturbations can be obtained either by taking the first variations of the action (26) equal to zero or, directly, by expanding eqs. (21,22) to the linear order terms. Generally, these equations describe three oscillators coupled to each other through the background shear and vorticity (the system of equations is 6th-order in time): one oscillator is the scalar potential perturbations and the other two are just two polarizations of the gravitational waves. [4]

To find the physical degrees of freedom of the perturbation fields and to approach the problem of the origin of the PCPs, the following steps have to be developed.

(i) Gauge invariant functions must be introduced instead of ϕ and h_{ik}.

The point is that, although the original fields, scalar φ and tensor g_{ik}, are fixed by definition, their decomposition into background and perturbation parts is not unambiguous at all. Indeed, if we transform infinitesimally the reference system,

[4]The vortex perturbations (if any) are found in the standard way as the first integrals of the 6th-order equation system.

$$x^i = \tilde{x}^i + \xi^i, \tag{28}$$

where $\xi^i = \xi^i(x^k)$ are small arbitrary functions, then the new separation in the coordinates $\tilde{x}^i$ will take the following form:

$$\varphi = \varphi^{(o)}(x^i) + \phi = \varphi^{(o)}(\tilde{x}^i) + \tilde{\phi}, \quad \tilde{\phi} = \phi + w\xi_i u^i,$$

$$g_{ik}dx^i dx^k = \tilde{g}_{ik}d\tilde{x}^i d\tilde{x}^k, \quad \tilde{h}_{ik} = h_{ik} + \xi_{i;k} + \xi_{k;i}, \tag{29}$$

where the background metric $g_{ik}^{(o)}$ has the same functional dependence in the new coordinates $g_{ik}^{(o)}(\tilde{x}^l)$ as that in the old ones $g_{ik}^{(o)}(x^l)$.

To develop the gauge invariant theory one has, first, to expand the perturbation tensor h_{ik} over the irreducible representations of the background geometry to mark off the scalar and gravitational wave polarizations and, second, to find the appropriate gauge invariant (i.e., independent of the transformations (29)) linear superpositions of the perturbation functions.

(ii) The Lagrangian and Hamiltonian formalisms of the perturbation fields should be developed on the basis of the gauge invariant theory.

An important point here is to obtain the canonical field variables which account for the physical degrees of freedom.

(iii) Secondary quantization of the perturbations and cosmological applications need to be considered in connection with the PCP problem.

Here, we are going to analyze all these points for the FU background metric. In this case

$$g_{oi} = u_i = \delta_i^o, \quad g_{\alpha\beta} = -a^2\gamma_{\alpha\beta},$$

$$H^2 = \frac{1}{3}\epsilon - \frac{K}{a^2}, \quad \dot{H} = -\frac{\epsilon + p}{2} + \frac{K}{a^2},$$

$$\frac{\dot{n}}{n} + 3H + \nu = 0, \quad v_i = v_{,i} - \beta^2 v(3H + \nu)u_i, \tag{30}$$

$$\chi = \dot{v} - \beta^2 v(3H + \nu + \Gamma), \quad \frac{\delta\epsilon}{\epsilon + p} = \beta^{-2}\dot{v} - 3Hv,$$

where $\gamma_{\alpha\beta} = \gamma_{\alpha\beta}(x^\gamma)$ is the metric tensor of the homogeneous isotropic 3-space with the spatial curvature $K = 0, \pm 1$ (manipulations with the Greek indices are done with the help of $\gamma_{\alpha\beta}$). All the perturbation types evolve independently of each other in the linear approximation since the background shear and vorticity are identically zero. Below, only potential perturbations are considered.

To avoid formal mathematical constructions we use here the spatially flat FU ($\gamma_{\alpha\beta} = \delta_{\alpha\beta}$, $K = 0$, see eq. (1)) and the synchronous reference system for its perturbations ($h_{oi} = 0$) unless explicitly stated otherwise. The necessary general formulae are given in Appendix B.

2.3. Potential Perturbations in Friedmann Cosmology

Let us obtain equations of motion of the gauge invariant potential perturbations directly from the linearized Einstein eqs. (22), and their Lagrangian from eqs. (26).

General metric perturbations in the synchronous reference system are presented in terms of two gravitational potentials, A and B:

$$ds^2 = dt^2 - a^2(\delta_{\alpha\beta} + h_{\alpha\beta})dx^\alpha dx^\beta,$$

$$h_{\alpha\beta} = A\delta_{\alpha\beta} + B_{,\alpha\beta}. \tag{31}$$

Since this metric is governed by the scalar field (19), we have three perturbation potentials v, A and B entering the field equations. In fact, only one of them is independent.

We will consider as the independent one a gauge invariant scalar q which is a linear combination of the perturbation potentials. Let us first define the q-scalar and then, using the low order equations, express v, A and B in terms of this scalar. The gauge freedom in the choice of these potentials follows from eqs. (29):

$$\tilde{v} = v + \frac{F}{2}, \quad \tilde{A} = A + HF, \quad \tilde{B} = B + F\int\frac{dt}{a^2} + G, \tag{32}$$

where F and G are small arbitrary functions of the space coordinates, and

$$2\xi_i = Fu_i - a^2(F_{,k}\int\frac{dt}{a^2} + G_{,k})P_i^k \tag{33}$$

is the most general form of the ξ_i-vector in the synchronous gauge (Landau & Lifshitz 1967), where $P_i^k = \delta_i^k - u_i u^k$ is the projection tensor.

Eqs. (32) show that the following function is gauge invariant:

$$q = A - 2Hv. \tag{34}$$

In fact, this function is independent of any other gauge as well (see Appendix B), which proves that $q = q(x^i)$ is a 4-scalar in the unperturbed FU.

To derive the inverse transformations and the equation of motion for the q-field we will need only the low-order (in time) Einstein equations:

$$\delta G_o^o = H\dot{h} - \frac{\Delta A}{a^2} = \delta\epsilon, \tag{35a}$$

$$G_{o\alpha} = -\dot{A}_{,\alpha} = (\epsilon + p)v_{,\alpha}, \tag{35b}$$

$$\delta(G_\alpha^\beta - \frac{1}{3}G_\gamma^\gamma\delta_\alpha^\beta) = \frac{1}{2a^2}(C_{,\alpha}^{,\beta} - \frac{1}{3}\Delta C\delta_\alpha^\beta) = 0, \tag{35c}$$

where $h = 3A + \Delta B$ and $C = A - (\dot{B}a^3)\dot{}/a$. The first two eqs. (35a,b) are just the conservation of energy and momentum respectively, whereas the last one (35c) states the Pascalian condition (the absence of pressure anisotropies for the φ-field). In the class of functions of interest eq. (35c) has only the trivial solution [5]:

$$C = 0 \tag{36}$$

which relates the potentials A and B straightforwardly.

Making use of eq. (35b), we can now express the functions v, A and B in terms of the q-scalar:

$$v = \frac{1}{2}(Q - \frac{q}{H}), \quad A = HQ, \quad \dot{B} = a^{-2}Q - a^{-3}P,$$

$$Q = \int \gamma q dt, \quad P = \int a\gamma q dt,$$

[5]The formal solution is $C = f\vec{x}^2 + \vec{g}\vec{x} + e$, where $f, \vec{g}$ and e are functions of time. The last two terms on the r.h.s. can be excluded because they do not enter the original $h_{\alpha\beta}$ functions. The quadratic term can be ascribed only to the B potential, which results in $h_{\alpha\beta} \sim \tilde{f}(t)\delta_{\alpha\beta}$ and the latter is excluded by redefining the scale factor.

$$\gamma = -\frac{\dot{H}}{H^2} = \frac{\epsilon + p}{2H^2} = \frac{3}{2}\frac{\epsilon + p}{\epsilon}. \tag{37}$$

The functions $Q = Q(x^i)$ and $P = P(x^i)$ are q-integrals over the Friedmannian world line $dt = u_i dx^i$. They are determined up to additive functions of the space coordinates. This freedom for the Q-function is just the gauge one (see (32)). The P scalar is gauge invariant, so its "ambiguity" is physically meaningful and related to a certain class of perturbations of the Bianchi type I model. To prove this we need another relation between the functions q and P which we are going to get from eq. (35a).

Let us recover the energy perturbation using eqs. (30, 37):

$$\frac{\delta\epsilon}{\epsilon + p} = -\beta^{-2}\dot{q}/2H - 3Hv. \tag{38}$$

Now, substituting this into eq.(35a), we have the key equation for the q-scalar:

$$\gamma a^3 \dot{q} = \beta^2 \Delta P \tag{39}$$

which is obviously the GR-analogue of the Poisson equation.

Let us first assume that $\beta \neq 0$. Comparison of eqs. (37, 39) shows that $P(t, \vec{x})$ is specified by the q-scalar up to an additive harmonic function of the spatial coordinates $P(\vec{x})$:

$$\Delta P(\vec{x}) = 0. \tag{40}$$

In the class of uniformly bounded (in 3-space) functions $h_{\alpha\beta} \sim P(\vec{x})_{,\alpha\beta}$ the solution is a bilinear form with zero trace:

$$P(\vec{x}) = a_{\alpha\beta} x^\alpha x^\beta, \quad a_{\alpha\beta} = const, \quad a^\alpha_\alpha = 0. \tag{41}$$

Thus the potentials v, A and B have been reconstructed from the given q-scalar but there is a partial solution that does not vanish under gauge transformations:

$$v = A = 0, \quad B = -P(\vec{x}) \int \frac{dt}{a^3}. \tag{42}$$

Appendix C demonstrates that these perturbations (42) are homogeneous and belong to the Bianchi type I cosmological model.

For $\beta = 0$, the function $P(\vec{x})$ is arbitrary and eqs. (42) describe the decaying mode of perturbations. The growing mode is determined by another arbitrary function of the space coordinates, $q(\vec{x})$. The general solution in this case is

$$v = 0, \quad Q = q(\vec{x})/H$$
$$P = q(\vec{x}) \int a\gamma dt + P(\vec{x}), \quad \delta\epsilon = \frac{H\Delta P}{a^3}. \tag{43}$$

In view of the above we may conclude that, for $\beta \neq 0$, the q-scalar is totally responsible for the evolution of the physical potential perturbations in the spatially flat Friedmann model. Division by β^2 and differention of eq. (39) (with eq. (37) for the P-scalar taken into account) gives the second-order equation for the q-scalar:

$$\ddot{q} + (3H + 2\frac{\dot{\alpha}}{\alpha})\dot{q} - (\frac{\beta}{a})^2 \Delta q = 0, \tag{44}$$

where $\alpha^2 = \gamma/2\beta^2$.

Now we can derive the Lagrangian density for the q-field. Substituting eqs. (37) into eq. (26) and leaving out all divergent terms, we have after rather lengthy calculations:

$$W^{(2)}[\phi, h_i^k] = W[q] = \int L(q) a^3 dt d^3x$$
$$L(q) = \frac{1}{2}\alpha^2(\dot{q}^2 - (\frac{\beta}{a})^2 q_{,\alpha}q^{,\alpha}), \tag{45}$$

where $L(q)$ and $L^{(2)}$ differ from each other only in the divergent terms.

Eqs. (45) show that the q-scalar is the unique single canonical variable for the physical degrees of freedom of potential perturbations in the FU driven by a scalar field of type (19). The Lagrangian density depends only on the first derivatives of the q-field. However, if $\alpha \neq const$ then q acquires a mass. Indeed, introducing the following transformation

$$\tilde{q} = \alpha q \tag{46}$$

we may rewrite the Lagrangian in the form of that for a standard scalar field with squared-mass $\mu^2 = -\ddot{\alpha}/\alpha$:

$$L(q) = \frac{1}{2}(\dot{\tilde{q}} - (\frac{\beta}{a})^2 \tilde{q}_{,\alpha}\tilde{q}^{,\alpha} - \mu^2 \tilde{q}^2) \tag{47}$$

Appendix B confirms eqs. (44,45) for the general case. The covariant generalization is as follows:

$$(D^{ik}q_{,i})_{;k}=0,\quad D_{ik}=\frac{1}{2}\alpha^2(u_iu_k+\beta^2P_{ik}),\tag{48}$$

$$L(q)=\frac{1}{2}D^{ik}q_{,i}q_{,k},\quad W[q]=\int L(q)\sqrt{-g}\,d^4x,\tag{49}$$

where $\alpha=(\epsilon+p)^{1/2}/2\beta H,\ \ P_{ik}=g_{ik}-u_iu_k,\ \ g=det(g_{ik})$, all Friedmann functions.

Before going to the next point, we give some relations for density perturbations in the different systems most frequently used in literature.

For the synchronous gauge we have from eq. (38):

$$\frac{\delta\epsilon}{\epsilon+p}=-\frac{\dot{q}}{2H\beta^2}+\frac{3}{2}(q-HQ).\tag{50}$$

For the comoving reference system with the synchronized time ($\tilde{v}_i\sim u_i$):

$$\frac{\delta\tilde{\epsilon}}{\epsilon+p}=-\frac{\dot{q}}{2H\beta^2},\tag{51}$$

where

$$d\tilde{s}^2=(1+\dot{q}/H)d\tilde{t}^2-a^2(\delta_{\alpha\beta}(1+q)+\tilde{B}_{,\alpha\beta})d\tilde{x}^\alpha d\tilde{x}^\beta,$$

$$\tilde{t}=t+v,\quad \tilde{x}^\alpha=x^\alpha+\int\frac{v^{,\alpha}}{a^2}dt,\tag{52}$$

$$\dot{\tilde{B}}=\frac{q}{Ha^2}-P/a^3.$$

For the Newtonian gauge (zero shear reference system, $\tilde{\tilde{B}}=0$):

$$\frac{\delta\tilde{\tilde{\epsilon}}}{\epsilon+p}=-\frac{\dot{q}}{2H\beta^2}+\frac{3}{2}(q-HP/a),\tag{53}$$

where

$$d\tilde{\tilde{s}}^2=(1-HP/a)d\tilde{\tilde{t}}^2-a^2(1+HP/a)\delta_{\alpha\beta}d\tilde{\tilde{x}}^\alpha d\tilde{\tilde{x}}^\beta,$$

$$\tilde{\tilde{t}}=t+\frac{1}{2}a^2\dot{B},\quad \tilde{\tilde{x}}^\alpha+\frac{1}{2}B^{,\alpha}.$$

For scales within the horizon ($k\eta\gg1$) all the three expressions for $\delta\epsilon$ coincide since the leading term is the first one ($\delta\epsilon\sim\dot{q}$). In the relativistic region ($k\eta\leq1$), $\delta\epsilon$ depends explicitly on the spatial slice taken.

2.4. Quantization and Conformal Non-Invariance

Bearing in mind eqs. (48,49) one can formally consider the q-field as a test scalar field in the Friedmann models. It allows for a standard development of the Hamiltonian formalism.

First, we can construct the Hilbert space of all complex solutions of eq. (48) with the scalar product

$$(q_1, q_2) = \int_{\Sigma} J^i_{12} d\Sigma_i,$$

$$J^i_{12} = iD^{ik}(q^*_1 q_{2,k} - q^*_{1,k} q_2), \tag{54}$$

where $d\Sigma_i$ is the invariant measure on a Cauchy-hypersurface Σ. The integral in eq. (54) does not depend on the choice of Σ because of the 4-flux conservation law $J^i_{12;i} = 0$.

Next, the canonically conjugate gauge invariant scalar is introduced:

$$\sigma = \sigma(x^i) = \frac{\partial L(q)}{\partial \dot{q}} = \alpha^2 \dot{q}. \tag{55}$$

Further steps in constructing the field Hamiltonian and canonical quantization are as simple as in the case of any other scalar field. We would like to emphasize here two points.

The quantization is based on the equal-time commutation relation for the canonically conjugate operators q and σ:

$$[q(t, \vec{x}), \sigma(t, \vec{x}')] = q\sigma - \sigma q = i\sqrt{-g}\ \delta(\vec{x} - \vec{x}'). \tag{56}$$

This equation can be compared with the commutator between the velocity potential and density perturbation operators of sound waves in nongravitating static matter, $[v, \delta\epsilon] = i\delta(\vec{x} - \vec{x}')$ (see Lifshitz & Pitaevski 1978).

It is worthwhile rewriting eq. (48) in terms of the conformal coordinates (η, x) for the conformal field $\bar{q}$:

$$\Box_\beta \bar{q} = U\bar{q}, \quad \bar{q} = \alpha a q, \tag{57}$$

where $\Box_\beta = \frac{\partial^2}{\partial \eta^2} - \beta^2 \Delta$ is the d'Alembertian operator in the Minkowski metric $d\bar{s}^2 =$

$ds^2/a^2 = d\eta^2 - d\vec{x}^2$. The function $U = U(\eta) = (\alpha a)''/(\alpha a)$ plays the role of an effective potential of scalar perturbations in the FU. It is an unambiguous function of the background expansion or, more precisely, of the scale factor and its time derivatives up to the fourth order. Note that the Lagrangian for eq.(57)

$$\bar{L}(\bar{q}) = \frac{1}{2a^4}(\bar{q}'^2 - \beta^2\bar{q}_{,\alpha}\bar{q}^{,\alpha} + U\bar{q}^2) \qquad (58)$$

coincides with $L(q)$ up to divergent terms.

The total energy of potential perturbations in the Friedmannian space $t = const$ — the field Hamiltonian — can be also presented in terms of the conformal field:

$$H = \frac{1}{a}\bar{H} = a^3 \int E d^3\vec{x},$$
$$E = \frac{1}{2a^4}(\bar{q}'^2 + \beta^2\bar{q}_{,\alpha}\bar{q}^{,\alpha} - U\bar{q}^2), \qquad (59)$$

where $E = E(\eta, \vec{x})$ is the local energy density of the q-field. Note that for non-gravitating matter (or for short wavelengths $k\eta \gg 1$) E is analogous to the sound wave energy density:

$$E \simeq \frac{(\epsilon + p)\vec{v}^2}{2} + \frac{(\beta\delta\epsilon)^2}{2(\epsilon + p)},$$

where $\vec{v} = (v^{,\alpha}/a)$ is the matter velocity.

Because of this formal analogy with sound waves and the fact that PCPs, which are just the resulting (after amplification) q-field, are usually found at the beginning of the HFU expansion stage, we will call below the q-field quanta phonons. These cosmological phonons remind us of the standard physics phonons only when the phonon wavelength is inside the horizon ($k\eta \gg 1$) [6]; in this case gravity is negligible and $q \sim v$ (see eq. (34)). For large scales ($k\eta \leq 1$), matter effects are not important and q is mainly the gravitational field potential.

When the scale factor is proportional to the conformal time, the q-scalar appears conformally coupled to the FU (see eq. (57)):

$$a \sim \eta/\alpha : \quad U(\eta) = 0. \qquad (60)$$

[6]Rigorously speaking, inside the sound horizon, $k\eta \gg |\beta|^{-1}$ (see the d'Alembertian in eq. (57)). For estimates, we assume in the main text that $\beta \sim 1$.

In all the other cases $U \neq 0$ and the q-field is conformally non-invariant. This means that q interacts with the background non-stationary metric, which provides for the spontaneous and induced production of phonons in the process of cosmological expansion.

The secondary quantization of the q-scalar results in the following expansion:

$$q = \int d^3\vec{k}(a_{\vec{k}} q_{\vec{k}} + a^{\dagger}_{\vec{k}} q^*_{\vec{k}}), \tag{61}$$

where

$$(q_{\vec{k}}, q_{\vec{k}'}) = [a_{\vec{k}}, a^{\dagger}_{\vec{k}'}] = \delta(\vec{k} - \vec{k}'),$$

$$(q_{\vec{k}}, q^*_{\vec{k}'}) = [a_{\vec{k}}, a_{\vec{k}'}] = 0,$$

$a_{\vec{k}}$ and $a^{\dagger}_{\vec{k}}$ are the annihilation and creation operators respectively,

$$q_{\vec{k}} = q_{\vec{k}}(x^i) = \frac{\nu_k}{(2\pi)^{3/2}\alpha a} e^{i\vec{k}\vec{x}},$$

and $\nu_k = \nu_k(\eta)$ satisfies the following equations:

$$\nu_k'' + (\beta^2 k^2 - U)\nu_k = 0, \quad \nu_k \nu_k^{*'} - \nu_k^* \nu_k' = i.$$

Below, we apply the theory of the q-field for the VEU. We shall assume that the $| \, in >$ vacuum is the initial state for the q-field. To define it explicity, we will consider in the next Chapter two cases for $\eta \to 0$: $\ddot{a} < 0$ and $\ddot{a} > 0$.

3.ORIGIN OF PRIMORDIAL COSMOLOGICAL PERTURBATIONS

Here, we consider some cosmological applications of the theory of the q-field: the scattering problem and the problem of the generation of PCPs in chaotic inflation.

Let us separate explicitly the kinetic and potential terms in the Lagrangian:

$$L = p(w) - V(\varphi) \tag{62}$$

From eqs. (23), we have

$$6a\ddot{a} = -(\epsilon + 3p) = 2V(\varphi) - (\epsilon(w) + 3p(w)),$$

where $\epsilon(w) = nw - p(w)$. It is seen that, if $V(\varphi) > 0$ and $\epsilon(\omega) > 0$, then the case $\ddot{a} < 0$ can be realized only when the potential term is negligible, whereas for the opposite case ($\ddot{a} > 0$) the potential term may play a dominant role. For this reason, we will consider two interesting asymptotic behaviours.

First, let us suppose that the kinetic term dominates the potential term in the general Lagrangian. The following *theorem* can be easily proved in this connection:

The theory of a real scalar field with Lagrangian depending only on the kinetic term,

$$L = p(w), \quad w^2 = \varphi_{,i}\varphi^{,i}, \tag{63}$$

is mathematically equivalent to the theory of potential motions of the ideal fluid with arbitrary equation of state

$$p = p(w), \quad \epsilon = \epsilon(w) = w\frac{dp(w)}{dw} - p(w). \tag{64}$$

The 4-velocity of the ideal fluid is a time-like vector ($w > 0$):

$$u^i = \frac{dx^i}{ds} = \varphi^{,i}/w. \tag{65}$$

Thus the φ-field acts here as the velocity potential.

In the other case $\ddot{a} > 0$, the potential term becomes important and we may expand the L-function in the small parameter w^2:

$$L(w, \varphi) = -V(\varphi) + W^2(\varphi)w^2/2 + O(w^4). \tag{66a}$$

A simple redefinition of the φ-field

$$\varphi \to \int W(\varphi)d\varphi$$

reduces eq. (66a) to the following case with standard kinetic term :

$$L = \frac{1}{2}\varphi_{,i}\varphi^{,i} - V(\varphi). \tag{66b}$$

The next Section deals with a general scattering approach for the q-field (the first case can be solved only in this approximation). In the last Section of this Chapter we consider the Lagrangian (66b) with the initial condition $V(\varphi) \gg |\varphi_{,i}\varphi^{,i}|$.

3.1. Scattering Problem for the q-Field

First, let us suppose that $V(\varphi) = 0$.

Applying the *theorem* to the flat FU, we see that for the equation of state $p = \epsilon/3$ the q-field is conformally coupled ($U = 0$). Eqs. (59) give the integral of motion

$$a \sim \eta : \quad \bar{H} = const, \tag{67}$$

which means the conservation of the total number of phonons during the process of the cosmological expansion.

In fact, the phonon number is conserved for any frequency mode. Let us dwell on this in a bit more detail.

At the relativistic stage (67) phonons are represented by the following choice of functions [7]:

$$\nu_k = (2\omega)^{-1/2} e^{-i\omega\tau}, \quad \omega = k/\sqrt{3}, \tag{68}$$

where $a' = const$, $a/a' = \tau = \eta + const$, and the prime ($'$) is the derivative with respect to conformal time. The field Hamiltonian is a sum over the energies of all quanta:

$$H_{Reg} = \int d^3\vec{k} E_k N_{\vec{k}}, \tag{69}$$

where $E_k = \omega/a$ is the phonon energy and $N_{\vec{k}} = a^{\dagger}_{\vec{k}} a_{\vec{k}}$ is the operator of the number of phonons with physical momentum $\vec{k}/a$. The eigenvalues of the mean energy density operator

$$E = \frac{H}{a^3 V}, \quad V = \int d^3\vec{x} \tag{70}$$

are

$$(2\pi a)^{-3} \int d^3\vec{k} E_k (n_{\vec{k}} + \frac{1}{2}), \tag{71}$$

where the $n_{\vec{k}}$ are the occupation numbers of the phonon states.

---

[7] We do not go into detail about such standard things for any quantized theory as separation of the Hilbert space into the positive and negative frequency subspaces, the Fock space of states, Bogoliubov transformations, etc.

Eq.(67) also allows for the introduction of the growing and decaying mode operators:

$$C_1 = C_1(\vec{k}) = (\omega/2)^{1/2}(\frac{a_{\vec{k}} - a^{\dagger}_{-\vec{k}}}{ia'}),$$

$$C_2 = C_2(\vec{k}) = (\omega/2)^{1/2}(\frac{a_{\vec{k}} + a^{\dagger}_{-\vec{k}}}{a'}). \quad (72)$$

In terms of these operators the field expansions have the following form (cf. eq. (12)):

$$q = (2\pi)^{-3/2} \int d^3\vec{k} e^{i\vec{k}\vec{x}} (C_1 \frac{\sin \omega\tau}{\omega\tau} + C_2 \frac{\cos \omega\tau}{\omega\tau}),$$

$$H = \frac{1}{6} \epsilon a^3 \int d^3\vec{k} (\mid C_1 \mid^2 + \mid C_2 \mid^2), \quad (73)$$

where $\mid C \mid^2 = CC^{\dagger} = C^{\dagger}C$.

Now, let us calculate the number of phonons created during a period, with some arbitrary expansion law governed by the general Lagrangian (19):

$$a = a(\eta), \quad \eta_1 < \eta < \eta_2. \quad (74)$$

We can do this directly by taking into account the phonon numbers before ($\eta \leq \eta_1$) and after ($\eta \geq \eta_2$) this period and just comparing them (the scattering problem).

As we have already seen, it is possible to calculate the occupation numbers for any wavelength during the linear expansion stage (see eq. (67)). Thus, to solve our problem, we should match the a-function (and its first derivative $\dot{a}$ [8]) to the linear behaviour $a \sim \eta$ at the beginning ($\eta = \eta_1$) and at the end ($\eta = \eta_2$) of the considered period.

So, from the resulting normalization (the Planck scale at the Planck moment of time is $k^{-1} = 1$) we have:

$$a = \begin{cases} \tau = \eta = (2t)^{1/2}, & \eta \leq \eta_1 \\ a(\eta), & \eta_1 < \eta < \eta_2 \\ A_1\tau = A_1(\eta + \eta_o), & \eta > \eta_2, \ \eta_o = const \end{cases} \quad (75)$$

where $\tau = \tau(\eta)$ is the conformal horizon (Hubble) time:

[8] We need this condition in order to avoid the unwanted non-physical effects of creation at the matching points. For further details about the scattering problem, which are standard for any test field theory, see Grib et al. (1980), Birrell & Davies (1981), and others.

$$\tau = \frac{a}{a'} = \eta - \int_{\eta_1} \frac{aa''}{a'^2} d\eta = \frac{1}{2} \int (1 + \frac{3p}{\epsilon}) d\eta. \tag{76}$$

The value of the physical constant

$$A_1 = \frac{a'(\eta = \eta_2)}{a'(\eta = \eta_1)} \tag{77}$$

depends on the average expansion rate during the period (74). Locally, the factor A_1 can be related to the conformal acceleration:

$$\frac{a''}{a'} = \lim_{\eta_2 \to \eta_1} \left(\frac{A_1 - 1}{\eta_2 - \eta_1} \right). \tag{78}$$

Let $a_{\vec{k}}$ and $b_{\vec{k}}$ be the phonon representations (68) diagonalizing the Hamiltonian at the stages $\eta \leq \eta_1$ and $\eta \geq \eta_2$ respectively. Then

$$b_{\vec{k}} = \alpha_k a_{\vec{k}} + \beta_k^* a^\dagger_{-\vec{k}}, \quad |\alpha_k|^2 - |\beta_k|^2 = 1, \tag{79}$$

where α_k and β_k are Bogoliubov coefficients. The $|\, in\rangle$ and $|\, out\rangle$ vacua are defined accordingly:

$$a_k \,|\, in\rangle = 0, \quad b_k \,|\, out\rangle = 0. \tag{80}$$

The Heisenberg state of the q-field is assumed to coincide with the $|\, in\rangle$ vacuum. This means that there are no phonons (i.e., potential perturbations) initially.

Taking an average over the $|\, in\rangle$ vacuum state, we derive the mean occupation numbers of phonons spontaneously created during the period $\eta_1 < \eta < \eta_2$,

$$\langle b^\dagger_{\vec{k}} b_{\vec{k}} \rangle = n_k \delta(\vec{k} - \vec{k}'), \quad n_k = |\beta_k|^2, \tag{81}$$

and the ratio of the energy density of the field perturbations to that of the homogeneous cosmological field (see eqs. (70,71)):

$$E_{Reg}/\epsilon = (24\pi^3 A_1^2)^{-1} \int d^3\vec{k}\ \omega \,|\, \beta_k |^2 . \tag{82}$$

The factor A_1^{-2} takes into account phonon energy cooling during the expansion period (74).

Calculation of the spectrum produced is also straightforward:

$$\langle q^2\rangle = \int_0^\infty \frac{dk}{k} q_k^2$$
$$q_k^2 = \left(\frac{k}{\pi\tau A_1}\right)^2 (\mid \beta_k \mid^2 + Re(\alpha_k \beta_k^* e^{-2i\omega\tau})). \tag{83}$$

The dependence of the spectrum on the oscillatory exponent means that growing and decaying modes are created unequally (see eq. (73)). Now, we are going to prove that under the general condition $a'' > 0$ ($A_1 > 1$, see eq. (78)), which is most frequently met in the applications, it is the growing mode of perturbations that is created preferentially by the parametric mechanism.

Indeed, our initial conditions generally imply ($\eta \leq \eta_1$, cf. eq. (15)):

$$C^2 = \langle \mid C_1 \mid^2 \rangle = \langle \mid C_2 \mid^2 \rangle \lll 1. \tag{84}$$

The situation is trivial in the case $\ddot{a} > 0$, when the perturbation scales inflate from inside to outside the horizon. The decaying mode, appearing originally at the horizon with the same amplitude as the growing one, decays quickly for larger times while the growing mode is frozen. This can be demonstrated with the help of the general solution of the dynamical equation (46) at large scales (the Laplacian term is negligible):

$$q = q_1(\vec{x}) + q_2(\vec{x}) \int \frac{dt}{\alpha^2 a^3}, \tag{85}$$

where $q_{1,2}(\vec{x})$ are arbitrary functions of the spatial coordinates. The integral rapidly converges to a function constant in time

$$q = q(\vec{x}), \tag{86}$$

which describes the growing perturbation mode for $\eta \mid \nabla q \mid \ll q$ for any expansion law.

Let us consider in more detail the case when the initial conditions (84) are set up outside the horizon. The general solution (85) helps again. Comparing it with the functions $\nu_k = \nu_k(\eta)$ in the a-representation, which have the form (68) for $\eta \leq \eta_1$, and

$$\nu_k = (2\omega)^{-1/2}(\alpha_k e^{-i\omega\tau} + \beta_k e^{i\omega\tau})$$

for $\eta \geq \eta_2$, we obtain for the Bogoliubov coefficients at $k\tau_1 \ll 1$

$$\alpha_k = \frac{1}{2}(A_1^{-1} + ig_k A_1), \quad \beta_k = \frac{1}{2}(A_1^{-1} - ig_k A_1), \tag{87}$$

where $g_k = \chi_1/k - i$ is the amplification coefficient and

$$\chi_1\tau_1/\sqrt{3} = 1 - A_1^{-2}\tau_1/\tau_2 - \int_{\eta_1}^{\eta_2}(\alpha a)^{-2}d\eta = const.$$

Obviously, $g_k = \chi_1/k$ for $g_k \gg 1$ and $g_k \sim (\omega\tau_1)^{-1}$ for $A_1 \gg 1$.

Substitution of eqs. (87) into eq. (83) gives the following spectrum for $A_1 \gg 1$:

$$q_k = (2\pi)^{-1}k^2 \mid g_k \mid \frac{\sin\omega\tau}{\omega\tau}. \tag{88}$$

The comparison with eqs. (73) reveals easily that only the growing mode is created.

To clarify the physical meaning of this effect, let us relate directly the $C_{1,2}$ operators in the a and b representations (see eqs. (72,79,87), $k\tau_1 \ll 1$):

$$C_1^{(b)} = C_1^{(a)} + \frac{\chi_1}{k}C_2^{(a)}, \quad C_2^{(b)} = A_1^{-2}C_2^{(a)}. \tag{89}$$

So, if one begins with eq. (84) then, in the end, eqs. (89) give for $A_1 \gg 1$ and $g_k \gg 1$

$$\langle\mid C_1 \mid^2\rangle = g_k^2 C^2 \gg C^2 \gg \langle\mid C_2 \mid^2\rangle, \tag{90}$$

just what is required by the galaxy formation theories (cf. eq.(14)). In fact, the effect is simply a game of mode mixing; it is not the q-field itself that is created but rather the q-momentum (the time derivative $\dot{q}$). In other words one can say that the parametric amplification effect brings about the creation of a squeezed state from an initially random state.

So, as we can see, it is not a problem to produce the growing mode of perturbations with the necessary amplitude. The amplification coefficient is the larger the earlier the HFU expansion is violated. The typical spectra (88) go like

$$q_k = Mk, \quad k < M, \tag{91}$$

with the maximum amplitude corresponding to the mass scale $M \sim \tau_1^{-1}$, when the linear expansion law is broken for the first time. (Note that the spectrum (91) decreases at large wavelengths in comparison with the Harrison-Zeldovich scale-free spectrum $q_{HZ} \sim const.$) For $k \gg M$, the amplification coefficient is exponentially small.

There are two points concerning eq. (91).

(i) The initial (vacuum) conditions are set up outside the horizon, which requires a physical explanation.

(ii) To get the required expansion factor for typical scales $k \leq M$ to be of the order of the galactic scales, it is necessary to ensure the acceleration (inflationary) condition $\ddot{a} > 0$ during the period (74).

In the latter case, the initial vacuum condition must be set up in the adiabatic zone (within the horizon), which can be done independently of the expansion law at the beginning.

3.2. Generation of Perturbations during Inflation

Let us consider the Lagrangian (66) with a potential $V = V(\varphi) \in C_3$ (at least the first three derivatives defined). We shall generally assume it to be a monotonically increasing function of φ for $\varphi > \varphi_o$. Without loss of generality we can put $\varphi_o = 0$ and $V = dV/d\varphi = 0$ at $\varphi = 0$ which, with the symmetric condition $V(\varphi) = V(-\varphi)$, gives a stable minimum of the potential $V(\varphi)$ at the point $\varphi = 0$. With such a normalization, our main assumption takes the following form:

$$V > 0, \quad dV/d\varphi > 0 \;\; for \;\; \varphi > 0. \tag{92}$$

Also, we define three auxiliary functions of φ related to the potential derivatives:

$$c = \frac{d\ln V}{d\ln\varphi}, \quad e = \frac{d\ln c}{d\ln\varphi}, \quad f = \frac{d\ln e}{d\ln\varphi}.$$

The simplest examples are $V_1 = m^2\varphi^2/2$ and $V_2 = \lambda\varphi^4/4$, where the constants m, λ are the field mass and dimensionless parameter respectively. Evidently, for the power-law potentials $V_n \sim \varphi^{2n}$, $c = 2n = const$ and $e = f = 0$.

The background Friedmann quantities and eqs. (30) are

$$2\alpha^2/3 = (1 + 2V/\dot{\varphi}^2)^{-1}, \tag{93}$$

$$\beta^2 = 1, \quad w = n = -\dot{\varphi}, \quad p = -V + \dot{\varphi}^2/2, \tag{94}$$

$$\epsilon = 3H^2 = V + \dot{\varphi}^2/2, \quad \dot{H} = -\dot{\varphi}^2/2, \tag{95}$$

$$\ddot{\varphi} + 3H\dot{\varphi} + dV/d\varphi = 0. \tag{96}$$

Inflation occurs when

$$\ddot{a} = a(\dot{H} + H^2) = a(V - \dot{\varphi}^2)/3 > 0. \tag{97}$$

For $V > \dot{\varphi}^2$ the potential energy density makes the dominant contribution to the Hubble parameter. For functions satisfying (92) this may happen at large φ, and the inflationary solution can be obtained as an expansion in the inverse powers of $|\varphi| \gg 1$:

$$H = (V/3)^{1/2}(1 + \frac{1}{12}c^2\varphi^{-2} + O(\varphi^{-4})), \tag{98}$$

$$\dot{\varphi} = -(cH/\varphi)(1 + \frac{1}{3}(e-1)c\varphi^{-2} + O(\varphi^{-4})), \tag{99}$$

$$a = exp\left[-\int d\varphi \;\; \varphi(c^{-1} + \frac{1}{3}(1-e)\varphi^{-2} + O(\varphi^{-4}))\right], \tag{100}$$

$$\alpha = (c/2\varphi)(1 + \frac{1}{3}(e-1)c\varphi^{-2} + O(\varphi^{-4})). \tag{101}$$

Then the inflationary condition can be rewritten in terms of the function c:

$$|c/\varphi| < \sqrt{3}, \tag{102}$$

or, in the terms of potential V,

$$|dV/d\varphi| < \sqrt{3}V. \tag{103}$$

Eqs. (98,99,100,101) are obviously true in a so-called slow-roll approximation which assumes a friction-dominated equation of motion for the φ-field (the second-derivative term in eq. (96) is unimportant). There are two conditions for this approximation, which follow directly from the expansions (98 – 101):

$$|c/\varphi| \leq 2, \quad |(e-1)c/\varphi^2| \leq 2. \tag{104}$$

The first condition almost coincides with eq. (102). From the second condition we have

$$| d^2V/d\varphi^2 | \leq 2V. \tag{105}$$

So the inflationary solution (98, 99, 100, 101) requires certain analytical properties of the potential $V(\varphi)$ (see eqs. (103,105)). We can put it in other terms: eqs. (98, 99, 100, 101) are valid for large φ,

$$| \varphi | \geq \varphi_I = max\left(| c | /2, \ | (e-1)c/2 |^{1/2}\right). \tag{106}$$

One can show that, within the class of functions (92) satisfying the inequality (106) for

$$| \varphi | \geq \varphi_1 = const, \tag{107}$$

where φ_1 is a positive root of the equation $\varphi_1 = \varphi_I(\varphi_1)$ [9], eqs. (98, 99, 100, 101) describe an attractive (for growing time) separatrix towards which all the other dynamical trajectories [10] with initial $| \varphi | > \varphi_1$ approach rapidly during their dynamical evolution to the stable point $\varphi = \dot{\varphi} = 0$. On the inflationary separatrix (98, 99, 100, 101) φ, H and α vary slowly, while the scale factor increases nearly exponentially in time, $a \sim exp(Ht)$.

Eqs. (98 – 101) fail when the field reaches the point $| \varphi | \sim \varphi_1 \sim 1$ and the further evolution proceeds via damping oscillations around $\varphi = \dot{\varphi} = 0$. If $d^2V/d\varphi^2 > 0$ at $\varphi = 0$, then

$$H = \frac{2}{3}t^{-2}, \quad a \sim t^{2/3}, \quad \varphi = (8/3)^{1/2}(mt)^{-1}\sin(m(t+t_o)) \leq 1, \tag{108}$$

where $V \simeq m^2\varphi^2/2$ for $| \varphi | \leq 1$; here m and t_o are constants. At this stage the Universe expands like a pressureless medium since the average cosmological pressure is exponentially small. The medium — coherent oscillations of the spatially homogeneous field — is unstable in this situation and will decay into particles. The result is reheating and the beginning of the HFU expansion.

[9] For making estimates, we take $\varphi_1 \sim 1$. If there are few roots of the equation $\varphi = \varphi_I(\varphi)$ for $\varphi \geq 1$, then the solution (98, 99, 100, 101) can fail for some large φ. We do not analyse here such a possibility.

[10] In the phase space $(\varphi, \dot{\varphi})$, the initial conditions for the classical trajectories of eqs. (95, 96) are set up on the quantum boundary $\epsilon = 1$, which represents a kind of ellipse (circle in the case of V_1) around the central point $\varphi = \dot{\varphi} = 0$. The radii of this ellipse along the φ and $\dot{\varphi}$ axes are $V(1)^{-\frac{1}{2}}$ and $\sqrt{2}$ respectively.

This reheating process, although producing some inhomogeneities on the horizon scale $\sim k_1$, cannot damage the large scale perturbations created already during the inflationary epoch ($k < k_1$).

To find the postinflationary PCP spectrum we must solve Eqs. (61) with the parameters (98, 99, 100, 101) for $\varphi > \varphi_1$:

$$\nu_k'' + (k^2 - U)\nu_k = 0, \tag{109}$$

$$\begin{aligned} U = \frac{(\alpha a)''}{\alpha a} &= 2(aH)^2\left(1 + \frac{\dot{H}}{2H^2} + \frac{(a^3\dot{\alpha})^{\dot{}}}{22a^3\alpha H^2}\right) = \\ &= 2(aH)^2\left(1 - \frac{1}{4}c_2\varphi^{-2} + O(\varphi^{-4})\right) = \\ &= \frac{2}{\eta^2}\left(1 + \frac{3}{4}c_3\varphi^{-2} + O(\varphi^{-4})\right), \end{aligned} \tag{110}$$

$$\eta = \int\frac{dt}{a} = -\frac{1}{aH}\left(1 + aH\int\frac{\dot{H}dt}{aH^2}\right) = -\frac{1}{aH}\left(1 + \frac{1}{2}c^2\varphi^{-2} + O(\varphi^{-4})\right), \tag{111}$$

where $c_2/c = c + 6(1-e)$ and $c_3/c = c + 2(e-1)$. The conformal time $\eta < 0$ and the initial conditions are

$$\nu_k = (2k)^{-1/2}exp(-ik\eta) \;\; for \;\; k \mid \eta \mid \gg 1. \tag{112}$$

Eqs. (109,110,111,112) can be solved explicitly by matching the two following solutions in the overlapping region

$$c_4\varphi^{-4} < k \mid \eta \mid < 1, \tag{113}$$

where $c_4/c^2 = (c+e-2)(1-e) - ef$. For the first solution, assuming the left inequality (113), the U-potential is approximately $U = const/\eta^2$ near $k \mid \eta \mid \sim 1$ (cf. eq. (110)). Since $U \ll k^2$ for $k \mid \eta \mid \gg 1$, we have for $k \mid \eta \mid > c_4\varphi^{-4}$:

$$\nu_k = \frac{1}{2}\sqrt{\pi \mid \eta \mid}H_\nu^{(1)}(k \mid \eta \mid) = (2k)^{-1/2}(1 - \frac{i}{k\eta})\left(e^{-ik\eta} + O(\varphi^{-2})\right), \tag{114}$$

where $H_\nu^{(1)}(x)$ is the Hankel function and $\nu = (3 + c_3\varphi^{-2})/2$.

Under the right inequality (113), eq.(85) describes the general solution for any $U(\eta)$. Since the integral in eq. (85) converges rapidly in time, we have for $k \mid \eta \mid < 1$:

$$\nu_k = i\pi\sqrt{2}k^{-3/2}\alpha a q_k = -\frac{i\pi}{\sqrt{2}\eta}k^{-3/2}(\frac{c}{\varphi H})q_k, \tag{115}$$

where the constants q_k are the PCP spectrum (see eq. (83)).

Matching eqs. (114,115) in the region (113) gives the following spectrum of the perturbations parametrically created outside the horizon:

$$q_k = \frac{1}{\pi}(\frac{\varphi H}{c})_k = \frac{1}{\pi}(\frac{H^2}{\mid \dot{\varphi} \mid})_k = \frac{1}{\pi\sqrt{3}}(\frac{V^{3/2}}{dV/d\varphi})_k, \tag{116}$$

where $\varphi = \varphi_k$ at the horizon crossing ($k\eta = -1$) can be expressed directly in terms of the wave number

$$k = aH = a(\varphi)H(\varphi). \tag{117}$$

The resulting spectrum (116,117) obviously belongs to the growing perturbation mode since only for this mode is the q-field constant in time outside the horizon. We have already emphasized that this property of the growing mode is independent of any expansion law or equation of state of matter. In particular, the spectrum (116,117) holds at large scales for any microphysical processes after inflation like phase transitions or reheating.

Sometimes, people prefer to deal with the power spectrum of the density perturbations. Below, the relation between the q-scalar and the coupled density perturbation field is obtained in a general form.

Indeed, eq. (44) has the growing mode general solution outside the horizon:

$$q = q(\vec{x}) + Q(t)\Delta q(\vec{x}),$$
$$\dot{Q}(t) = -(\beta/a)^2(H/\dot{H})(1 - \frac{H}{a}\int a dt). \tag{118}$$

Thus the comoving density perturbations are (see eq. (52)):

$$\delta = \frac{\delta\tilde{\epsilon}}{e} = -\frac{2\alpha^2\dot{q}}{3H} = \frac{1}{3}(aH)^{-2}(1 - \frac{H}{a}\int a dt)\Delta q(\vec{x}). \tag{119}$$

Now it is not a problem to get the desired relation between the power spectra:

$$\delta_k = \frac{1}{3} q_k (\frac{k}{aH})^2 (1 - \frac{H}{a} \int a dt) = q_k O((l_H/l_k)^2) \ll q_k. \tag{120}$$

We return now to eq. (117) which gives us the connection between k and φ on the inflationary separatrix (98,99,100,101). This equation can be solved explicitly for a class of so-called smooth potentials $V(\varphi)$.

Let us introduce smooth potential functions $V(\varphi)$ for which $c(\varphi)$ varies even more slowly than φ [11]:

$$| e(\varphi) | \ll 1. \tag{121}$$

For such a potential eq. (100) yields

$$a \simeq \varphi^{-1/3} exp(-\frac{\varphi^2}{2c}), \quad \varphi \geq \varphi_1, \tag{122}$$

which shows that φ varies logarithmically with the conformal time (cf. eq. (111)). If $c_1 = c(\varphi_1) \geq 2$, then $\varphi_1 = c_1/2 \geq 1$ (see eqs. (106,107)) and

$$\varphi^2 = (c/2) \left(c_1 + 4 \ln(\frac{\eta H \varphi_1^{1/3}}{\eta_1 H_1 \varphi^{1/3}}) \right) \tag{123}$$

for $\varphi \geq \varphi_1$ ($| \eta | \geq | \eta_1 |$). Substitution into eq. (117) gives for $k \leq k_1$:

$$\varphi_k^2 = (c/2) \left(c_1 + 4 \ln(\frac{k_1 H \varphi_1^{1/3}}{k H_1 \varphi^{1/3}}) \right) \simeq 2c \ln(k_1/k), \tag{124}$$

where the second equality assumes that $k \ll k_1$.

Thus smooth potentials generate spectra of the Harrison-Zeldovich type (see eq. (116)), growing only logarithmically at large scales.

A typical example of a smooth potential is the power-law potential

$$V_n = \frac{1}{2n} a_n^2 \varphi^{2n}$$

which generates the following spectrum:

$$q_k = (4\pi n^{3/2})^{-1} a_n \varphi^{n+1} =$$

[11] Physically, the characteristic scales of smooth potentials are not much shorter than φ.

$$= (2/\pi)(4n)^{(n-2)/2}a_n \left[n/2 + \ln\left(\frac{k_1}{k}(\ln\frac{k_1}{k})^{(3n-1)/6}\right)\right]^{(n+1)/2}, \tag{125}$$

where a_n and $c = c_1 = 2n \geq 2$ are constants.

Important cases are the massive field ($n = 1,\ a_1 = m$):

$$q_k = (m/\pi)\ln\left(\frac{k_1}{k}(\ln\frac{k_1}{k})^{1/3}\right), \tag{126}$$

and the λ-field ($n = 2,\ a_2 = \sqrt{\lambda}$):

$$q_k = (2\sqrt{\lambda}/\pi)\left[\ln\left(\frac{k_1}{k}(\ln\frac{k_1}{k})^{5/6}\right)\right]^{3/2}. \tag{127}$$

For non-smooth $V(\varphi)$ the spectrum q_k can have, in principle, any form depending on the shapes of the given potential and its first derivative. Moreover, it is possible to invert the problem and to find the potential $V(\varphi)$ for any [12] given postinflationary PCP spectrum (Hodges & Blumenthal 1989). True, some potentials appear to be rather exotic ones, but the result is very important: PCP spectra are very sensitive to the potential forms [13].

We shall return to the latter problem in Chapter 5. But now, let us stress two more points to conclude this Chapter.

Postinflationary perturbations (116,117) are Gaussian with random spatial phases, since the seed zero-point vacuum fluctuations (of the q-field) from which they were parametrically created are Gaussian by definition. Here, we have no problem with initial conditions for the q-scalar because they are determined by microphysics inside the horizon.

Another interesting point is that most spectra grow with increasing scale. This means that there exists some critical field (and, thus, a critical scale) for which the corresponding amplitude $q_k \sim 1$:

$$(\varphi H/c)_{cr.} \sim 1, \quad k_{cr.} \sim (aH)_{cr.} \ll k_1. \tag{128}$$

[12] Some spectra may violate the slow-roll conditions, which makes the inverse problem self-inconsistent.

[13] Physically, non-Harrison-Zeldovich spectra appear from potentials which have characteristic scales less then φ.

For the potentials (125,126,127) we have for example

$$
\begin{array}{lll}
n \geq 1: & (\varphi H/c)_{cr.} \sim 1, & k_1/k_{cr.} \sim \exp(a_n^{-\frac{2}{n+1}}), \\
n = 1: & \varphi_{cr.} \simeq 2\sqrt{\pi} m^{-1/2}, & k_1/k_{cr.} \simeq (m/\pi)^{1/3} \exp(\pi/m), \\
n = 2: & \varphi_{cr.} \simeq 3\lambda^{-1/6}, & k_1/k_{cr.} \sim \exp(\lambda^{-1/3}).
\end{array}
$$

We shall see in the next Chapter that the Universe on large scales, $k \leq k_{cr.}$, is globally non-linear and that it is stochastic (dominated by quantum fluctuations) for $\varphi > \varphi_{cr.}$.

4. INFLATION

It is no secret that Inflation is a corner stone of VEU theories. This is not a surprise since up to now we have no alternative to the Inflationary Paradigm.

We are not going to discuss the Paradigm here. There are a lot of reviews and courses devoted to the subject (e.g. see Kolb & Turner 1989 and references therein). Instead, we would like to dwell on chaotic inflation which, in our view, is the first theory of this kind that can be called a cosmologically standard theory. At least, it has the same status as the standard Friedmann cosmology or the parametric amplification theory. All of them, based on simple cosmological postulates which are not directly related to any particular particle physics, can explain and predict a lot of observational results (see the Introduction).

The goal of inflationary theory is to prepare the initial conditions for the standard FU. There are the following five items among them.

(i) Homogeneity and isotropy, along with the Euclidean geometry, of the spatial slice on scales near the contemporary horizon.

(ii) The amplitude $\sim 10^{-4}$ of PCPs on this slice on galactic to supercluster scales.

(iii) Reheating sufficient for the primordial entropy production, nucleosynthesis and baryogenesis met in Friedmann cosmology.

(iv) Small particle numbers ($\Omega \leq 1$) of the unwanted massive relics created by the Big Bang and primordial reheating.

(v) Small density of the Friedmann vacuum which is the Λ-term ($\Omega_\Lambda < 0.7$).

The first inflationary models were connected to specific physical theories and hypotheses like GUTs, phase transitions, quantum-gravity effects, etc. However, in view of the absence of a reliable understanding of high energy physics and, what is more important, taking into account the purely cosmological status of the first three items above, there was a recognition of the need to construct a cosmologically standard inflationary theory which could be independent of any current speculations about future fundamental physics, on one hand, and could account for the first three puzzles of the FU on the other. Certainly, such a theory would not solve the fourth and fifth problems, which are much more related to particle physics.

The first theory of such a type was proposed by Linde (1983). A basic assumption is that the potential energy of the inflaton φ-field grows with increasing φ (see eq. (92)). The word 'chaotic' comes from the requirement for a large value of the initial φ-field ($\varphi \gg 1$), which could be realized somewhere in spacetime under the hypothesis of chaotic initial conditions. However, we do not think that the latter requirement is a problem for the theory at all.

Below, we dwell on the necessary and sufficient conditions for chaotic inflation and then discuss some implications related to the subject.

4.1. Chaotic Inflation

Let us dwell on the Lagrangian (66) with the potential term of type (92). To start inflation, the latter must predominate at $\varphi \gg 1$:

$$V(\varphi) > \mid \varphi_{,i}\varphi^{,i} \mid . \tag{129}$$

Let us estimate the size of the region where eq. (129) is initially satisfied.

As we have seen in the previous Chapter the time derivative of the initial φ is not a problem for the inequality (129) if spatial homogeneity is postulated, since the inflationary solution is an attractive separatrix for $\varphi \gg 1$. So, of principal importance is the spatial gradient condition following from eq. (129):

$$\mid \nabla\varphi \mid \leq H. \tag{130}$$

Eq. (130) can be read as follows: to start inflation one has to prepare a quasi-homogeneous distribution of φ on the scale $\sim L = \varphi / \mid \nabla\varphi \mid$ which is much larger than the horizon scale:

$$L \geq l_{SI} = \varphi H^{-1} \gg l_H. \tag{131}$$

The initial inflation scale has the physical meaning of the Compton scale of the inflaton; this becomes explicitly clear in the case of the massive field ($V = V_1$):

$$l_{SI} = m^{-1}.$$

Taking this into account for estimates, the initial inflation condition (131) is probably not a great surprise in the general case as well.

Eq. (131) deals with the initial distribution of the φ-field. In principle, one can raise a question about another (additional to eq. (131)) initial inflation condition, namely, about the initial spatial distribution of the metric (curvature) on scales less than l_{SI}. However, we will not discuss this problem here by formal reasoning. From the beginning, we decided to restrict ourselves to the case when the φ-field is the only source of the metric g_{ik}. The point is that the small-scale nonlinear curvature perturbations (if any) are assumed to arise from a source unrelated to the φ-field since the latter is homogeneous on the scale l_{SI}. So, the curvature produced by the φ-field is supposed to be homogeneous on the scale l_{SI}, as well as the φ-field itself.

Eq. (131) can be interpreted in another way: the initial φ-field should be large enough for l_H to be small. However, more stringent constraints for the potential come from the slow-roll conditions (103), (105). Let us explicitly rewrite these conditions in terms of the PCP spectrum (106).

From eq. (103) we have the potential restrictions:

$$V^{1/2} \leq 10 q_k. \tag{132}$$

Eqs (105,116) restrict the range of the spectrum index:

$$\left(\frac{d \ln q_k}{d \ln k}\right) \leq 2. \tag{133}$$

Thus q_k cannot vary from the Harrison-Zeldovich spectrum faster than $k^{\pm 2}$ (which is quite compatible with the market of galaxy formation theories considered today).

Eq. (132) puts direct observational limits on the potential amplitude $V(\varphi)$ for $\varphi = \varphi_k$ within the structure scale range $k^{-1} \sim (10 - 10^4)h^{-1}$ Mpc. Eq. (132) with $q_k \sim 10^{-4}$ is evidence for the weak coupling of the φ-field to the potential $V(\varphi)$ in this region.

Indeed, let us demonstrate this when the c-function variation along the scale range is negligible. In this case $V = \lambda \varphi^c$ and we obtain a very small value for the coupling parameter (in Planck units):

$$\lambda \sim 3c^2 10^{-8}\varphi^{-(2+c)} \leq 10^{-8}. \tag{134}$$

Remember that a small coupling parameter is also required for the large size of the FU-bubble (more than the horizon today, see below).

Eq. (132) can be used for some other important constraints, e.g., on the reheating temperature. The radiation energy after reheating cannot actually exceed the inflation energy near the end of inflation:

$$\rho_{rad} = \frac{\pi^2}{30} g^* T_{RH}^4 \leq V(\varphi_1) \leq V(\varphi) \leq 10^2 q_k^2, \tag{135}$$

where $\varphi = \varphi_k \geq \varphi_1$ $(k < k_1)$ and g^* is the total number of massless degrees of freedom of the thermal bath particles. Eq. (135) gives the following upper limit for the reheating temperature:

$$T_{RH} < (\frac{10}{g^*})^{1/4} H^{1/2} \leq (\frac{300}{g^*})^{1/4} q_k^{1/2}. \tag{136}$$

Making use of the microwave quadrupole anisotropy $q_k \sim \Delta T/T \sim 10^{-5}$, we have for the standard model $(g^* \sim 100)$: $T_{RH} \leq 10^{16}$ GeV. The latter inequality can be confirmed with the help of a similar estimate for the gravitational waves produced during inflation (we do not discuss this problem here).

The next important parameter is the Friedmann slice scale l_F, i.e., a typical scale of the part of the Universe, created by inflation, which can be approximated by the Friedmann model. Any scale in such a quasi-homogeneous region is described by eq. (117):

$$k = H(\varphi)a(\varphi) = H(\varphi)e^{-N(\varphi)} \ [Mpc^{-1}], \tag{137}$$

where $N(\varphi) = \int H dt = N_I + N_F$ is the number of e-foldings of the expansion of the Universe from the moment when the perturbation was at the inflationary horizon up to now:

$$N_I = N_I(\varphi) = \int_{\varphi_1}^{\varphi} \frac{\varphi d\varphi}{c}, \quad N_F \simeq 60. \tag{138}$$

After substituting $\varphi = \varphi_{cr}$ from eq.(128) we have

$$l_F = k_F^{-1} = \exp(\varphi_{cr}^2/c) \sim (\lambda^{-\frac{1}{1+c/2}}) \sim exp(10^4) \ggg 10^{28} \ [\mathrm{cm}]. \tag{139}$$

The non-linear global Cauchy-hypersurface which develops as a result of the chaotic inflation dynamics has not yet built up. Nevertheless, we see no difficuties of principle in solving this problem. The point is that the global spatial Cauchy-hypersurface cannot exist everywhere in spacetime: it collapses in the spacetime regions where the φ-field reaches Planckian densities ($\varphi_{Pl} \sim \lambda^{-1/c}$), so that the semiclassical approach becomes self-inconsistent. We show in the next Section that the latter regions occupy the major part of the physical volume of the Universe produced by the chaotic inflation.

4.2. Stochastic Theory of the $\dot{q}$-Field

Let us return again to small scales $l_k < l_F$, where q can be treated as a linear quantum operator against the Friedmann background. The equation of motion of the q-field is

$$\ddot{q} + 3nH\dot{q} - a^{-2}\Delta q = 0, \tag{140}$$

where $n = 1 + 2\dot{\alpha}/3\alpha H$, and α and H are the classical background functions (see eqs. (93, 94, 95, 96, 98, 99, 100, 101)). As we know from the previous Chapters, the large scale perturbations are classical for $l_k \gg l_H$ while the quantum perturbations affect only small scales, $l_k \leq l_H$. Let us separate these two parts of the q-field at the inflation period, assuming that q is generated by quantum perturbations:

$$q = \Phi + F, \tag{141}$$

where Φ is the classical large scale part of the q-field operator.

To make this separation explicit let us introduce the notion of a miniuniverse (MU) as a part of the actual space-time with size proportional to the horizon:

$$l_{MU} = \zeta H^{-1} \geq l_H, \tag{142}$$

where $\zeta = const \geq 1$. Evidently, MUs do not expand with the comoving volume.

Now we can define the classical Φ-field as the mean value of q in a MU:

$$\Phi = \Phi(t, \vec{x}) = \int K_\sigma(\vec{x} - \vec{x}')q(t, \vec{x}')d^3\vec{x}', \tag{143}$$

where $K_\sigma(\vec{r}) = (2\pi)^{-3/2}\sigma^{-3}exp(-r^2/2\sigma^2)$ is the Gaussian MU-window,

$$\sigma = l_{MU}/a = \zeta/(aH) = -n_1\zeta\eta$$

is the MU-dimension in the comoving $\vec{x}$-space, and $n_1 = (-a\eta H)^{-1}$.

Similarly to Φ we can define the classical part of the q-field momentum:

$$V = V(t,\vec{x}) = \int K_\sigma(\vec{x} - \vec{x}')\dot{q}(t,\vec{x}')d^3\vec{x}. \tag{144}$$

The evolution of the coarse grained fields Φ and V is governed by the quantum perturbations represented by the F-operator: in the inflationary process successive perturbations created inside the horizon inflate one after another outside the horizon and start contributing to the classical fields Φ and V when their scales become about (and then larger than) l_{MU}. So F plays the role of a stochastic generator for Φ, the latter moving like a Brownian particle in the gas. More than this, the dynamical equations are similar as well.

Let us present eqs. (143,144) as Fourier integrals (see eq. (61)):

$$\Phi = \int d^3\vec{k}\Theta(a_{\vec{k}}q_{\vec{k}} + a^\dagger_{\vec{k}}q^*_{\vec{k}}),$$

$$V = \int d^3\vec{k}\Theta(a_{\vec{k}}\dot{q}_{\vec{k}} + a^\dagger_{\vec{k}}\dot{q}^*_{\vec{k}}), \tag{145}$$

where

$$\Theta = \Theta(k\sigma) = e^{-\frac{1}{2}k^2\sigma^2} = \int K_\sigma(\vec{r})e^{i\vec{k}\vec{r}}d^3\vec{r}.$$

Then the original eq. (140) can be rewritten in terms of the classical fields:

$$\begin{aligned}\dot{\Phi} - V &= f,\\ \dot{V} + 3nHV &= Hg/\zeta^2,\end{aligned} \tag{146}$$

where f and g can be called the noise functions (or generators) driven by the quantum fluctuations:

$$\begin{aligned}f &= mH\sigma^2\int d^3\vec{k}(k^2\Theta)(a_{\vec{k}}q_{\vec{k}} + a^\dagger_{\vec{k}}q^*_{\vec{k}}),\\ g &= H\sigma^2\int d^3\vec{k}(k^2\Theta)(a_{\vec{k}}p_{\vec{k}} + a^\dagger_{\vec{k}}p^*_{\vec{k}}),\end{aligned} \tag{147}$$

with

$$p_{\vec{k}} = -q_{\vec{k}} + m\zeta^2 \dot{q}_{\vec{k}}/H, \quad m = -\dot{\sigma}/(\sigma H).$$

Before we calculate correlators of the noise functions let us introduce the normal modes of the classical fields,

$$\tilde{\Phi} = \Phi + \frac{m_1}{3H}V, \quad \tilde{V} = V,$$

and separate the Eqs. (146):

$$\dot{\tilde{\Phi}} = \tilde{f}. \tag{148}$$

(Eq. (146) for the V-function does not change.) Here $\tilde{f} = f + m_1 g/\zeta^2$. The background functions n_1 and m_1 are easily derived on the inflationary separatrix ($\varphi \gg 1$):

$$\begin{aligned} n &= 1 + c(1-e)\varphi^{-2} + O(\varphi^{-4}), \\ n_1 &= 1 - \frac{1}{2}c^2\varphi^{-2} + O(\varphi^{-4}), \\ m &= 1 - \frac{1}{2}c^2\varphi^{-2} + O(\varphi^{-4}), \\ m_1 &= 1 - \frac{1}{6}c^2\varphi^{-2} + O(\varphi^{-4}). \end{aligned} \tag{149}$$

Eq. (148) coincides with the Langevin equation describing the drift of a Brownian particle if the particle coordinate is understood instead of Φ.

Comparing the windows Θ and $k^2\Theta$ for the classical and noise functions, we can see that in the latter case the main contribution comes from scales $l \sim l_{MU}$. It is clear that the field averaged over the mini-universe cannot change its value before the new perturbation reaches the MU-scale, which happens in a characteristic time (step-time) $\Delta t \sim l_{MU}$. Since the perturbation phases appearing on the MU-scale are random, the process of the classical field change is stochastic. To calculate the characteristic values of this process we must know the correlators of the noise functions on the Friedmann hypersurface $t = const$.

If the q-field is in the vacuum state then

$$\langle a^{\dagger}_{\vec{k}} a_{\vec{k}'} \rangle = 0, \quad \langle a_{\vec{k}} a^{\dagger}_{\vec{k}'} \rangle = \delta(\vec{k} - \vec{k}'),$$

and the process is the Gaussian one, so that it is sufficient to know the second correlators. In this case the perturbation amplitudes for $\varphi > 1$ and $\zeta > 1$ are as follows (see eqs. (61,115,118)):

$$q_{\vec{k}} = i\sqrt{2}(H\varphi/c)_k exp(i\vec{k}\vec{x}), \quad \zeta^2 \dot{q}_{\vec{k}}/H = k^2\sigma^2 q_{\vec{k}}.$$

Finally, after a straightforward calculation we have in the leading approximation for φ and ζ:

$$\langle \tilde{f}\tilde{f}' \rangle = \langle ff' \rangle = \frac{1}{2}(\langle fg' \rangle + \langle gf' \rangle) = \langle gg' \rangle = 2D\delta(t - t'), \tag{150}$$

where $D = D(t) = H^3(\varphi/2\pi c)^2$ is the diffusion coefficient. The δ-function in eq. (150) is used instead of each of the following expressions,

$$2H(\frac{\sigma}{\sigma'} + \frac{\sigma'}{\sigma})^{-2} \quad and \quad 12H(\frac{\sigma}{\sigma'} + \frac{\sigma'}{\sigma})^{-4},$$

because the halfwidths of these bell-functions are about the same as the cosmological horizon ($\Delta t \sim H^{-1}$), which is less (by the ζ-factor) than the MU-scale.

Now we can introduce the probability distribution $P = P(t, \tilde{\Phi})$ for the field $\tilde{\Phi}$ at time t. By definition,

$$\int P d\tilde{\Phi} = 1.$$

Following the standard methods, we can derive the Fokker-Planck equation for this function:

$$\frac{\partial P}{\partial t} = D\frac{\partial^2 P}{\partial \tilde{\Phi}^2}. \tag{151}$$

Obviously, the field dispersion grows in time in this stochastic process.

Let us take some arbitrary MU at time t_o with the classical field $\tilde{\Phi}_o$, which we call the mother. During inflation the physical spatial volume which belonged to the mother MU at $t \sim t_o$ expands to larger and larger scales. For $t > t_o$, this volume can be covered by other MUs (daughters) [14]. The $\tilde{\Phi}$-field varies from one daughter to another, and the r.m.s. deviation (from $\tilde{\Phi}_o$) $\tilde{\sigma} = \tilde{\sigma}(t)$ can be calculated by making the average either over the quantum q-state in one MU or over the assembly of daughter MUs:

$$\tilde{\sigma}^2 = \langle(\tilde{\Phi} - \tilde{\Phi}_o)^2\rangle = \int (\tilde{\Phi} - \tilde{\Phi}_o)^2 P d\tilde{\Phi} =$$

$$= 2\int D dt = -\frac{1}{2\pi^2}\int \frac{V^4 d\varphi}{(dV/d\varphi)^3} \sim t. \tag{152}$$

This can be confirmed also by the exact solution of eq. (151),

$$P(t, \tilde{\Phi}) = \frac{1}{\sqrt{2\pi}\tilde{\sigma}} exp\left(-\frac{(\tilde{\Phi} - \tilde{\Phi}_o)^2}{2\tilde{\sigma}^2}\right). \tag{153}$$

At the beginning,

$$\tilde{\sigma}(t_o) = 0, \quad P(t_o, \tilde{\Phi}) = \delta(\tilde{\Phi} - \tilde{\Phi}_o),$$

while, during time, the distribution (153) is broadened around $\tilde{\Phi}_o$, with equal probability for both signs of the deviation $(\tilde{\Phi} - \tilde{\Phi}_o)$. The typical one-step change of the $\tilde{\Phi}$-field is

$$\Delta(\tilde{\Phi}) = \tilde{\sigma}(\Delta t) = (\zeta D/H)^{1/2} \sim \zeta^{1/2} H. \tag{154}$$

Before we discuss some implications of this stochastic process, let us consider the necessary conditions for the diffusion approach to be valid.

Eqs. (148,150) are generally true if the D-function varies more slowly than the characteristic step-time $\Delta t \sim \zeta H^{-1}$. This requirement is commonly satisfied on the inflationary separatrix.

The Fokker-Planck approach (151) is less reliable here. Indeed, the notion of the P-function assumes that its characteristic change-time should be no larger than the step-time. However, taking note of eq. (153), we see that this can be marginally so if ζ is not too high. Below, we will assume that $\zeta \sim 1$.

[14] At each step $\Delta t \sim \zeta H^{-1}$ the mother volume expands by the factor $N \sim \exp(3\zeta)$, so that there are about N daughters of the first generation, N^2 of the second, and so forth, inside the volume.

4.3. Non-Linear Inflation

The mini-universes of size $H^{-1}(\zeta \sim 1)$ introduced in Section (4.2) are just the causally connected regions of the inflating space. If any two points with constant comoving $\vec{x}$-coordinates (i.e., expanding with the Universe) belong initially to the same MU then they will manage to exchange light signals at least once. But if they belong to two different MUs then the light signal sent from one point will never get to the other one. An important consequence is that each MU expands in time independently of any other.

This means that any MU can be chosen as the mother of the next generations of daughter MUs it produces. In its turn, any daughter at a moment $t \sim t_o$, although created by some mother at $t < t_o$, is a mother itself for $t > t_o$. This picture has neither beginning nor end. Actually, this assembly of MUs is an eternally self-reproducing inflationary Universe.

Let us consider some physical volume expanding with the comoving space. New MUs are constantly created inside the volume during the evolution. We can connect by time-like tracks causally related MUs (mother-daughter, mother-daughter, etc.). We saw in the previous Section that there is an equal probability for both signs of the field deviation $\Delta\tilde{\Phi}$ to be found along any track from the past to the future. Since any MU develops independently of its previous history and of its neighbours, we can forget about the seed mother field $\varphi^{(o)}$ and try to find the current classical quasi-Friedmannian field φ_{MU} which drives the given MU along its track and applies only to this MU. Certainly, φ_{MU} is the local φ-field renormalized each time by the classical part of the q-field (see eq. (141)).

Technically, we can use the Newtonian gauge to find φ_{MU}, since this particular frame most closely imitates the local Friedmannian expansion (see eqs. (53)):

$$ds^2_{MU} = (1 - HP_{MU}/a)dt^2 - a^2(1 + HP_{MU}/a)dx^2, \tag{155}$$

where

$$P_{MU} = \int a\gamma\Phi dt.$$

The local time is

$$t_{MU} = t - \int (HP_{MU}/2a)dt$$

and the local expansion factor is

$$a_{MU} = a(1 + HP_{MU}/2a).$$

In both equations we have disregarded the dependence of the P_{MU} function on the $\vec{x}$-coordinates within the MU. Then we can easily recover the φ_{MU}-field from the background eqs. (95). In the leading approximation for $\varphi \gg 1$ we have

$$\varphi_{MU}/\varphi^{(o)} = 1 + (\gamma\Phi - HP_{MU}/a)/c. \tag{156}$$

Below, we give only qualitative ideas about some of the results of this investigation.

The most interesting question to arise is: to which densities of the φ-field do most tracks lead during stochastic inflation? The answer is a combination of two factors: the probability of finding a certain φ-field on one track and the number of tracks carrying a given φ-field. As far as the first factor is concerned, the classical monotonic decrease of φ when sliding down the potential $V(\varphi)$ is the smaller the larger the field is: $\Delta\varphi_{cl} \sim -c/\varphi$ for one step-time $\Delta t \sim H^{-1}$. On the other hand, the stochastic change in the field due to the quantum perturbations proceeds in both directions of φ, with the amplitude $|\ \Delta\varphi_{st}\ | \sim H$ growing with higher field values. So, for large enough $\varphi(\varphi > (c^2/\lambda)^{1/(2+c)})$ the quantum stochastic process predominates, so that at each step half of the created daughters have higher φ than their mother. But the total number of tracks created per unit time grows with higher fields as well, $dN/dt \simeq 20H$. So, the majority of tracks lead to high densities; thus, the largest part of the global physical volume is occupied by the Planckian field density, i.e., by the space-time foam.

We can only guess what is happenning in these most typical high density states of the global Universe — the notions of space-time and inflaton break down and transitions to different physics, signatures, dimensions and other conceivable and inconceivable worlds may occur. In fact, we can only say that the inflation states with densities less than the Planckian one are non-typical and very improbable ones in this really chaotic Universe dominated by the sea of quantum fluctuations. The stochastic regime considered above is just a part, the semiclassical part, of this sea, where the space-time is already classical while the inflaton is still dominated by quantum fluctuations. Evidently, such *regions* decouple occasionally from the space-time foam and exist independently during some period of the *classical time.*

A very important conclusion is that inside these semi-classical regions there exist some very few tracks which lead occasionally (through the random stochastic process) to lower and lower densities of the φ-field. When the latter falls below the critical quantity ($\varphi \leq (c^2/\lambda)^{1/(2+c)}$), quantum fluctuations are not able any more to increase φ in the created daughter- MUs and the subsequent inflation continues with a monotonic decrease of the inflaton. Now the quantum fluctuations are responsible only for small density perturbations, varying slightly from one daughter to other. So, the result will be adiabatic Gaussian perturbations (with the amplitude decreasing at smaller scales) against the Friedmannian background patch surrounded by the non-linear chaotic Universe.

Let us emphasize two other points in the conclusion.

The global Cauchy-hypersurface does not rigorously exist. It can be constructed near the Friedmannian patch in a space-time region restricted by the Planckian densities. In fact, such a non-linear solution describes just a temporal island formed with a very small probability in the chaotic space-time foam of the Universe and suitable for life. In this connection, we can mention that we do not think it is worth while putting seriously the question as to how general are the initial inflation conditions for chaotic or other inflation theories from the point of view of the general solutions of the CGR equations. In our opinion, it is quite enough that the probability for creating a low-density world where life can appear is non-zero. We have no time to discuss this subject in more detail here.

The next important point is as follows. It is the inflaton quantum fluctuations in the chaotic inflation theory that are the reason for both the non-linear global structure of the Universe as a whole and the adiabatic density perturbations responsible for the large scale structure formation in the Friedmannian patch of such a Universe. Therefore, we can test inflationary theories just by investigating the spectrum of PCPs and then reconstructing the global structure within the framework of the theory. This is probably the only informative cosmological way to learn anything about features beyond the optical horizon, as well as the fundamental physics parameters beyond direct experiment. We can see that the problem of testing inflation becomes part and parcel of the VEU theories. We are going to discuss this briefly in the next Chapter.

5. TESTING VEU THEORIES

There are few important cosmological predictions coming from the very early inflationary epoch. Among them are the total energy density in the Universe and the spectrum of adiabatic PCPs.

The former quantity must be equal to the critical density up to the accuracy of the PCP amplitude on the contemporary horizon:

$$\Omega_{tot} = 1 \pm \delta_H, \qquad |\, K \,| \leq \delta_H. \tag{157}$$

This amplitude $\delta_H = \delta_k(k = H_o)$ can be easily estimated by the quadrupole anisotropy of the microwave background radiation:

$$\delta_H \sim 10^{-4}.$$

The real dynamical density is close to the critical one with an accuracy $\sim$ 30%. However, even if eq. (157) were confirmed with a much higher accurary by future observations, it could not, unfortunately, give us much information about the principal parameters of inflation. In this respect, more informative is the PCP spectrum generated by the last stages of the inflationary epoch before the beginning of the FU.

As far as chaotic inflation is concerned, the postinflationary PCP spectrum is very sensitive to the potential form (see eqs. (116, 117)). On the other hand, the shape of the potential energy of the inflaton found in a given particle physics model must be unambiguously fixed by the fundamental Lagrangian of all the particle fields and interactions.

Thus the principal test of inflation is the large scale structure of the Universe. Its analysis permits the reconstruction of the postrecombination PCP spectrum on galactic to horizon scales. This part of the spectrum is obviously related to the postinflationary PCPs which, in their turn, depend directly on the inflaton potential within the field interval responsible for the scales mentioned above (see eq. (117)).

The two following questions arise from this consideration:

* How are the postinflationary and postrecombination spectra related?

** Which are the basic cosmological observations now shedding light on the PCP spectrum?

Below, both topics are discussed very briefly.

5.1. Transfer Functions

The point is that any real confrontation of theory with observations can be done only within the framework of some cosmological model allowing the transfer of PCPs from times when they were produced up to the moment when they entered the non-linear evolution to form the hierarchy of the objects observed. The principal parameters of the model are those of the dark matter components which affect the gravitational evolution of PCPs.

Regarding the gravitational impact on PCPs there are two components of the Dark Matter, Cold (CDM) and Hot (HDM), that are important. The cold particles may be heavy relics ($m_c \gg 10$ eV) or coherent axions which behave like a non-relativistic medium. The hot particles are those like a massive neutrino with the equilibrium particle density and corresponding rest mass ($m_h \leq 10$ eV). These two components, which evolved very differently in the past, are both non-relativistic and maintain the critical density now:

$$\Omega_c + \Omega_h + \Omega_\Lambda = \Omega_{tot} = 1. \tag{158}$$

Baryons can be included in Ω_c. The third term, which is the energy of the physical vacuum, cannot be totally excluded today ($\Omega_\Lambda < 0.7$).

The rest of the dark matter, which we call the ν-component, does not contribute crucially to eq.(158) and consists of relativistic and semirelativistic weakly interacting particles ($m_\nu \ll 1$ eV). We can characterize ν-particles by their total number N_ν. If N_γ is the number of relic photons, we define

$$\nu = \frac{N_\nu}{N_\nu + N_\gamma} \in (0, 1). \tag{159}$$

For the standard CDM and HDM models, $\nu = 0.4$ and $\nu = 0.3$ respectively, counting three or two types of massless neutrino. Generally, ν-particles include gravitons and light SUSY and other hypothetical ...inos probably existing in the Universe.

So, in the simplest case ($\Omega_{tot} = 1$, $\Omega_\Lambda = 0$, stable particles) we have two free parameters, Ω_h and ν, both ranging from zero to one, which determine the past history of PCPs, beginning from Inflation. The goal is to find the ratio of the final to the initial

PCP spectrum as a function of these two parameters. This ratio is generally called the transfer function,

$$T(k) = q_k^{(f)}/q_k^{(i)}, \tag{160}$$

where the postinflationary spectrum $q_k^{(i)}$ coincides with the function q_k from eqs. (116, 117) and $q_k^{(f)}$ is the postrecombination PCP spectrum responsible for LSS formation in the Universe. Evidently, $T(k)$ does not depend on the inflationary period; it is a functional of the Friedmann model dynamics from the beginning to the present.

$T(k)$ is equal to unity for very large scales ($T(0) = 1$) and then decreases monotonically with increasing k. It still remains about unity up to some characteristic scale l_{eq} which coincides with the horizon at the period when equality of all relativistic and non-relativistic component densities occurs. The shape of the further fall in $T(k)$ at shorter wavelengths is an intrinsic property of the model (this subject is considered in other courses at this school).

For the standard models $l_{eq} \sim 30h^{-2}$ Mpc but, for arbitrary ν, l_{eq} grows with ν like $l_{eq} \sim (1-\nu)^{-1/2}$.

As we see, the resulting spectrum $q_k^{(f)}$ is a sensitive function of both the fundamental properties of the early Universe, the postinflation PCPs and the dark matter composition. So, the investigation of their direct consequence — the large scale structure of the Universe — is of great importance. Another critical test — the laboratory detection of dark matter particles — is not discussed here.

5.2. Observations

Since there are other special courses devoted to this subject we only briefly outline the highlights of the confrontation between observations and theory which are important for us. We see the tremendous importance for modern cosmology of two groups of experiments: $\Delta T/T$ observations on large scales ($\Theta \geq 1^o$) and direct observations of the distribution and evolution of the hierarchies of Large Scale Structures (LSS).

The point is that these experiments confront and complement each other.

While the upper limits on $\Delta T/T$ which are just now becoming available make us lower the primordial perturbation amplitudes on scales larger than $l > 10h^{-1}$ Mpc, at the same time LSS needs for its existence sufficiently high cosmological perturbation amplitudes on scales $l \sim 10 - 100h^{-1}$ Mpc. For most theories of galaxy formation the gap between these two requirements is quite negligible. For example, within the Gaussian perturbation theories any reasonable asumption that large superclusters and voids are more or less standard phenomena in the visible Universe leads inevitably to predictions of $\Delta T/T$ on degrees of arc which are capable of detection using current techniques. This brings a very great optimism that we can obtain the large scale primordial perturbation spectrum directly from observations with a high degree of accuracy.

The current situation with regard to $\Delta T/T$ is well known. For our case of Gaussian PCPs we may directly relate our scalar q on the last scattering surface at recombination with the map of the temperature anisotropies on the celestial sphere $\vec{n} = \vec{n}(\theta, \varphi)$. Indeed, let us decompose the latter into spherical functions:

$$\frac{\Delta T}{T}(\vec{n}) = \sum_{l,m} a_{lm} Y_{lm}(\theta, \varphi). \tag{161}$$

Then, after simple calculations, the temperature correlation function takes the following form:

$$C(\alpha) \equiv \langle \frac{\Delta T}{T}(\vec{n}_1) \frac{\Delta T}{T}(\vec{n}_2) \rangle = \sum_l (\Delta T/T)_l^2 P_l(\cos\alpha), \tag{162}$$

where $\vec{n}_1 \vec{n}_2 = \cos\alpha$, $P_l(\cos\alpha)$ are the Legendre polynomials, $\langle ... \rangle$ is the average over the field state, and

$$\left(\frac{\Delta T}{T}\right)_l^2 = \frac{2l+1}{4\pi} a_l^2 = \frac{1}{4\pi} \sum_m \langle a_{lm}^2 \rangle. \tag{163}$$

Generally, there are three main sources of the primordial temperature anisotropies: fluctuations of the gravitational potential and matter density and velocity perturbations. For vanishing pressure they may be reduced to the following expression taken at recombination:

$$\frac{\Delta T}{T}(\vec{n}) = \frac{1}{3} q(\vec{x}_r) + \frac{1}{4} \delta_{\gamma,r} - \vec{n}\vec{v}_r. \tag{164}$$

The first term (Sachs-Wolf effect) dominates for $\alpha > 1^\circ (l > 100\ h^{-1}$ Mpc), which yields a direct connection between $C(\alpha)$ and the PCP spectrum q_k:

$$a_l^2 = \frac{2}{\pi} \int \frac{dk}{k} q_k^2 J_l^2(k), \quad C(\alpha) = \frac{1}{2\pi^2} \int \frac{dk}{k} q_k^2 J_0(k\beta), \tag{165}$$

where J_l are the Bessel functions and $\beta = 2\sin\frac{\alpha}{2}$.

The most important observation of $\Delta T/T$ has been made by COBE for $\alpha \sim 10^o$ which shows consistency with the HZ-spectrum on very large scales, $l \sim 1000\ h^{-1}$ Mpc:

$$\Delta_k^2 \sim k^4 q_k^2 \sim k^\alpha, \quad \alpha = 4 \pm 0.5, \tag{166}$$

where $\delta^2 = (1+z)^{-2} \int \frac{dk}{k} \Delta_k^2$ is the mean square density perturbation.

As for the LSS data, we have many independent indications today for the existence of large scale structures up to a typical scale $l_{LS} \cong 100 - 150h^{-1}Mpc$. The most important data come from the Great Attractor ($z \leq 0.03$), distributions of clusters of galaxies ($z \leq 0.1$) and pencil beam galactic surveys ($z \leq 0.3$). It is now planned to test even deeper samples, e.g., radiogalaxies and active galactic nuclei (AGNs), quasars (QSOs), absorption (L_α, CIV) and emission (21 cm) clouds, X-ray clusters and others, to see how far these large structures extend into the past. I would like to dwell upon the evolutionary aspects of LSS formation and, in particular, on some hints about the spectrum coming from quasars.

5.3. Formation of Large Scale Structure and Dark Matter Models

Over the years we have observed the tragic demise of two fundamental models, consisting of hot or cold dark matter, for the evolution of the gravitational potential in the Universe. The crucial test which resulted in rejecting both of them was the LSS: HDM proved to be helpless to account for galaxy origin (because of lack of power on the megaparsec scale), while CDM could not account for the existence of large voids and superclusters (because of the lack of power on scales $\in (10, 100)h^{-1}$ Mpc when normalized by the galaxy spatial distribution).

The other interesting discrepancy between these different scale ranges came from observations of the total matter density Ω_o governing their dynamics: there we have two groups of experiments resulting in different conclusions, but for which consensus is still possible.

The first experiment deals with megaparsec scales – galaxy halos, groups and X-ray clusters – $l < l_D$ (the dynamical scale in the Universe $l_D \sim 10h^{-1}Mpc$, which is the transit pass of a galaxy for the current Hubble time, is just the scale of the richest clusters). The assumption of virial equilibrium yields a low dynamical mass forming the structure on small scales: $\Omega_o \sim 0.1 - 0.3$. Another important observation is the large fraction of baryons in X-ray clusters which somehow reaches $\sim 25\%$ in $l \leq 3Mpc$:

$$\frac{M_b}{M_o} \sim 0.2, \tag{167}$$

which also confirms the low matter density involved dynamically at megaparsec scales (since $\Omega_b \leq 0.1$ due to the primordial nucleosynthesis and $M_b/M_o = \Omega_b/\Omega_o$ for the dynamical scale).

The other experiment deals with LSS and is consistent with $\Omega_o \simeq 1$. There are two principal arguments in this case:

* the existence of subclusters in the majority of galaxy clusters (which confirms that the clusters are just forming today, which is possible only in a Universe with dynamical density close to the critical one); and

* the large bulk coherence velocities obviously of cosmological origin, thus allowing for the reconstruction of the total density contrast (and as a consequence, after comparing it with the distribution of galaxies, consistency with the "standard" model $\Omega_o \simeq 1$ and the biasing factor $b \sim 1$).

Finally, a possible reconciliation between the experiments on small and large scales is as follows: some fraction of the dark matter in the Universe is distributed on large scales and does not enter the galaxy halos and groups. What are the possible models? They are purely theoretical today. The most frequently discussed now are the model with mixed dark matter (hot+cold, with hot particles like massive neutrinos with a few eV rest mass and the corresponding density parameter $\Omega_\nu \in (0.2 - 0.4)$), and that with non-zero Λ-term ($\Omega_\Lambda \in (0.5 - 0.7)$). For both cases, the cold particles form the dynamical structure beginning from small scales, while on the large scale there is an additional contribution coming from light neutrinos or the vacuum density (the Λ-term) respectively. A very doubtful point concerning these and other cosmological models considered today as possible candidates for the real Universe is as follows: they all are multi-parameter and thus non-fundamental models, so they differ drastically from the purely hot or purely cold dark matter models. Is there something very important which we are missing in our discussion of the formation of the Universe structure? Maybe. I can only conclude here by saying that none of the models under discussion meets all the requirements of observations. For example, regarding the two previous examples, for $\Lambda \neq 0$ models one can expect a large fraction of old (relaxed) galaxy clusters and lensed quasars as well as relatively small bulk velocities, whereas the hot+cold models require $H_o < 60\, km\, s^{-1} Mpc^{-1}$ and a too small abundance of X-ray clusters and high-redshift quasars.

In such a situation, observational verifications become extremely important. The principal test is the LSS of the Universe.

All currently discussed models of cosmological structure formation are aimed at fitting the observational data at $z = 0$. So, one cannot in principle separate them without going to the evolution at medium and high redshifts where the models begin demonstrating their essential difference. Let me show briefly recent progress in our understanding of how and what quasars — the most dramatic evolutionary systems known today — may tell us about the history of LSS formation and how they may help in building up the true cosmological model (Komberg & Lukash 1994).

There are two basic "objects" on large scales ($l > l_D$) which are essentially quasilinear: enhanced and de-enhanced density regions. As for the local Great Attractor (GA), by a distant GA we understand a patch of enhanced total density, scaled in any dimension from larger than the richest cluster size up to $\sim l_{LS}$. The opposite construction is a void (Great Repulsor), a patch of decreased total density. Both objects - GAs and voids - are still expanding like the Universe (the density variation may be $\sim 10-30\%$ on the scale $\sim 100h^{-1}Mpc$), which makes them differ drastically from objects just collapsed or collapsing at least in one direction (the latter direction size being $\leq 10h^{-1}Mpc$) - galactic clusters, filaments and walls.

There are two steps on the way to relating quasar groups (QGs) and GAs:

* Clusters, where they are concentrated (the superclusters), trace the mass density enhancements;

* Clusters of distant clusters contain (at least some) QGs.

The first point is confirmed today in the nearby region where peculiar velocity measurements are done, and by the alignment of the cluster dipole with the microwave background dipole . As for the second point, there are a few observational indications:

1) Constructing the QSO correlation function and comparison with the cluster-cluster correlation function;

2) Similarity between distributions of cluster pairs and wide QSO pairs;

3) Direct investigations of QSO environments.

The QSO correlation function, displaying strong QSO clustering at $r \leq 10h^{-1}Mpc$, a weaker clumping at $\sim 20 - 50h^{-1}Mpc$, and the absence of correlations for $r > 150h^{-1}Mpc$, evolves by growing at smaller redshifts: for $z \rightarrow 0$ it looks like a clumped distribution of galaxies whereas for $z > 2$ the QSOs seem to be randomly distributed: $\xi_{qq} \sim (1+z)^{-a}$ with $a \sim 2$, consistent with the linear perturbation law. On taking this into account, the QSO correlation radius close to $z = 0$ approaches that of clusters, $r_c \sim 16-20h^{-1}Mpc$. As far as the wide QSO pairs are concerned ($\Delta\Theta \leq 1.5^o, \Delta v/c \sim 10^{-2}$) the QSOs there are mostly associated with different clusters, which is supported by the similarity of their distribution with that of the cluster pairs.

Today's observations show that QSOs at $z > 0.5$ are frequently associated with young clusters while those at $z < 0.5$ seem to be found in poor systems like galactic groups. This gives us a guess that distant quasars may form in merging and interacting galaxies. These merging effects, supplying accreting material (to massive black holes) just sufficient for QSO burning, can really exist in young protoclusters which are still in the process of their first collapse and first contraflows, i.e., well before cluster virialization and X-ray gas appearing. Taking into account that the first violent crossings (caustics) of the cosmic primordial medium must certainly occur in the central regions of preclusters, we may relate the majority of quasars at $z \sim 2.5$ with the epoch of subcluster formation. Of course, we do not associate every QSO with young clusters. There are a few generations of QSOs depending on the physical reason for the formation of the accreting gas disk. It is the dense caustics with a high merging activity of the host QSO galaxies that we associate with a typical cluster mass. The collapse of the lesser masses was not sufficiently powerful to ensure high gas densities, while the larger masses collapsed later, in a medium with a much smaller fraction of the gas (transferring to stars) already totally ionized .

If so, then two important topics are on the agenda:

* It is possible to find distant GAs in proximity to groups and pairs of QSOs on a scale of a few tens of Mpc ($l_{GA} \leq 100h^{-1}Mpc$).

* The relationship between the epochs of cluster formation ($z \sim 2-3$) and the appearance of GAs ($z < 2$) suggests the true model of large scale structure formation.

Today we have information on a dozen QGs. One of the most famous is the group at $z = 1.1$ (Crampton, Cowley & Hartwick, 1989) consisting of 23 QSOs within $\sim 60h^{-1}Mpc$. All QGs are found to have $z < 2$ and they are not seen at larger redshifts though the spatial number density of QSOs grows sharply up to $z \sim 2.5$. It is interesting to note that the number of observable wide QSO pairs seems to decay beyond $z \geq 2$ as well.

If GA structures really disappear at $z > 1$ and the cluster formation epoch lies in a broad interval centred at $z^*_{CL} \sim 2.5$, we have the following straightforward estimate for the spectrum of the Gaussian primordial density perturbations $\delta \equiv \delta\rho/\rho$ within the dynamical scale range $l \in 10 - 100h^{-1}Mpc$:

$$\Delta_k^2 \sim k^\gamma, \quad \gamma = 1^{+0.6}_{-0.4}. \tag{168}$$

This predicts that the correlation function of the dynamical mass is $\xi(r) \sim r^{-\gamma}$, which means that clusters are clumped in regions overdense in GA-scale perturbations.

Such a flat spectrum leads to the following conclusions.

(i) Clusters today should be far developed and concentrated where GAs are – clusters trace the mass in dense regions – while sparse young clusters might be found in nearby voids. (The formation time of galaxy clusters is strongly modulated by GA-scale perturbations.)

(ii) If the majority of the first clusters give birth to bright QSOs broadly peaked at $z \sim 2 - 3$ and appear in GA-peaks, then the mean separation between distant quasars should correlate with l_{LS} at $z \sim 2.5$, which is actually in good agreement with observations.

(iii) The Gaussian flat spectrum (168) creates by $z < 1$ a great variety of coherent structures (the hierarchies of walls, filaments, voids and GAs), produced by the broadscale perturbations with nearly equal amplitudes in $l \in 10 - 100h^{-1}Mpc$. Note that such structures form in the gravitating matter and do not require biasing type hypotheses.

(iv) Since galaxies form before clusters, their first generation is not modulated by GA-perturbations. However, merging and generating processes for galaxies, which are most active in dense regions, lead to successive generations of bright galaxies at GA locations. A test of this could be a search for dwarf galaxies in voids.

(v) Obviously, the spectrum (168) is more flat than that in the standard CDM model (for the latter we expect $\Delta_k^2 \sim k^{2-3}$ in the scale range). It can be realized in hybrid hot+cold dark matter models with $\sim 30\%$ of the total mass in the form of neutrino-like particles with rest mass $\sim 2 - 7$ eV.

(vi) Since COBE indicates consistency with the HZ spectrum at very large scales (see eq. (166)), the change from the flat part (168) to the HZ asymptotic form should happen at the supercluster scale $\sim 150h^{-1}Mpc$, which is obviously a real feature of the primordial spectrum (contrary to the case of galaxy clusters, which are the consequence of current dynamical time).

Summarising, we can say that the available data support the following two conjectures.

* QGs indicate the patches of enhanced matter density (distant GAs); they occur at $l_{GA} \sim 10 - 100h^{-1}Mpc$ and extend for $z < z_{LS} \leq 2$.

* The majority of distant QSOs are associated with the epoch of cluster formation, $z < 3$.

These two points are consistent if and only if QSOs belonging to QGs were created not long before the time at which the given QG is observed, which is also consistent with a relatively short lifetime of medium and bright QSOs. This means that the first young clusters and their associated QSOs at $z > z_{LS}$ appear more or less randomly in space (GA perturbations are still small). On the contrary, the next clusters and their related QSOs originating at $z < z_{LS}$ are born already in groups (i.e. they are clumped at the GA density peaks).

Certainly, both points lead to the important conclusion that the part of the primordial spectrum between the dynamical scale ($\sim 10h^{-1}Mpc$) and the coherence length ($\sim 100 - 150h^{-1}Mpc$) is nearly flat (see eq. (168)). So, if the HZ perturbation spectrum or anything close to it really exists on larger scales, then the change in the spectrum shape must take place just at the scale $l_{LS} \sim 100 - 150h^{-1}Mpc$, which is obviously a "signature of God" requiring its explanation in the physics of the very early Universe.

6. CONCLUSION

As we have seen, even in the absence of well established high energy physics, we may successfully develop the theory of the VEU on purely cosmological grounds and come to important conclusions about the spectrum of PCPs capable of being tested by current observations. On the other hand, we are very close to recovering the postrecombination PCP spectrum directly from the observations of both $\Delta T/T$ and LSS (especially its evolutionary aspects), and thus to reconstructing the true cosmological model and making the exciting link to VEU physics. Both the confronting branches – theory and observations – are developing fruitfully and make us hope that we can find the main answers concerning the evolutionary model of the Universe during our time.

This work was supported by the Russian Foundation for Fundamental Research (the code 93 02 29 29) and partly by the ISF grant MEZ300.

APPPENDIX A

Here we obtain the Lagrangian $L^{(2)}$ (see eqs. (26)).

Asuming that eqs. (25) are exact ones, all the auxiliary quantities are expanded to the second order in ϕ and h^{ik}:

$$\delta w = \frac{1}{2w^{(o)}}(\delta(\varphi_{,i}\varphi_{,k}g^{ik}) - (\delta w)^2) = w^{(o)}(\chi - \frac{1}{2}\chi^2 + \frac{1}{2}v_i v^i - u_i v_k h^{ik}),$$

$$\delta L = n^{(o)}(\delta w - \nu^{(o)}\phi + \Gamma\frac{\delta w}{w}\phi + \frac{1}{2w}((\frac{\delta w}{\beta})^2 - m^2\phi^2)) =$$

$$= n^{(o)}w^{(o)}(\chi - \nu^{(o)}v + \Gamma\chi v + \frac{\chi^2}{2}(\beta^{-2} - 1) + \frac{1}{2}v_i v^i - u_i v_k h^{ik} - \frac{1}{2}m^2 v^2)\ ,$$

$$\delta g_{ik} = h_{ik} + h_{il}h^l_k, \quad \ln(\frac{g}{g^{(o)}}) = h + \frac{1}{2}h^k_l h^l_k,$$

$$\delta\sqrt{-g} = \frac{1}{2}\sqrt{-g^{(o)}}\ (h + \frac{1}{2}h^k_l h^l_k + \frac{1}{4}h^2),$$

$$\delta\Gamma^l_{ik} = \frac{1}{2}(h^l_{i;k} + h^l_{k;i} - h^{;l}_{ik} + h^m_i h^l_{m;k} + h^m_k h^l_{m;i} + h^l_m h^{;m}_{ik} - (h_{im}h^m_k)^{;l}),$$

$$\delta\Gamma^l_{il} = \frac{1}{2}(h + \frac{1}{2}h^k_l h^l_k)_{,i},$$

$$\delta R = g^{ik}\delta R_{ik} - h^{ik}R^{(o)}_{ik},$$

$$\delta R_{ik} = (\delta\Gamma^l_{ik})_{;l} - (\delta\Gamma^l_{il})_{,k} + (\delta\Gamma^l_{ik})(\delta\Gamma^m_{lm} - (\delta\Gamma^m_{il})(\delta\Gamma^l_{km}),$$

where $\delta f = f - f^{(o)}$, $v = \phi/w^{(o)}$, $v_i = \phi_{,i}/w^{(o)}$ and Γ^l_{ik} are the Christoffel symbols.

Substitution into eq. (20) yields:

$$(L - \frac{1}{2}R)(g/g^{(o)})^{1/2} = L^{(o)} - \frac{1}{2}R^{(o)} + \frac{1}{2}(1 + \frac{1}{2}h)(G^{(o)}_{ik} -$$

$$-T^{(o)}_{ik} - \phi((n^{(o)}u^l)_{;l} + n^{(o)}\nu^{(o)})) + S^l_{;l} + L^{(2)},$$

$$S^l = \frac{1}{2}(1 + \frac{1}{2}h)(g^{il}\delta\Gamma^k_{ik} - g^{ik}\delta\Gamma^l_{ik}) + n^{(o)}u^l\phi,$$

$$L^{(2)} = \frac{1}{2}nw(v_iv^i + \chi^2(\beta^{-2} - 1) - 2v_i\psi^i_k u^k + \nu v\psi -$$

$$-m^2v^2 + 2\Gamma v\chi) + \frac{1}{4}(L^o - \frac{1}{2}R^{(o)})(\psi_{ik}\psi^{ik} - \frac{1}{2}\psi^2) +$$

$$+\frac{1}{8}(\psi_{ik;l}\psi^{ik;l} - 2\psi_{ik;l}\psi^{il;k} - \frac{1}{2}\psi_{,l}\psi^{,l}).$$

If the background metric satisfies the classical equations,

$$G^{(o)}_{ik} - T^{(o)}_{ik} = 0, \quad (n^{(o)}u^l)_{;l} + n^{(o)}\nu^{(o)} = 0,$$

then substitution into $L^{(2)}$ gives eq. (26). Note that the linear terms in ϕ and h^{ik} (see eq. (20)) prove to be zero since the background equations are satisfied.

APPENDIX B

Here we derive general relations between the q-scalar and the perturbations in an arbitrary reference frame.

Let us decompose the perturbations (25) over the irreducible representations of potential type in the general Friedmann model with the metric g_{ik} and 4-velocity u^i:

$$v = X + \frac{1}{2}(C + D_{,i}u^i),$$
$$h_{ik} = Ye_{ik} + Zg_{ik} + (Cu_{(i})_{;k)} + D_{;ik}, \tag{169}$$

where the 4-tensor $e_{ik} = 2u_iu_k - g_{ik}$ has Euclidean signature and subbrackets imply symmetrization. The functions X, Y, Z, C, D are coefficients of the linear decomposition. X, Y and Z are gauge invariant 4-scalars. C and D are arbitrary functions specifying the gauge freedom of the perturbations in eqs. (25) (see eqs. (28, 29)):

$$\xi_i = \frac{1}{2}(Cu_i + D_{,i}).$$

The Einstein linear equations can be decomposed as a 4-tensor over the irreducible representations as well:

$$h_{ik;l}^{;l} - 2h_{(i;k)l}^{l} + h_{;ik} + (p-\epsilon)h_{ik} +$$
$$+(\epsilon+p)(\tilde{\chi}(\beta^{-2}-1)e_{ik} + 4v_{(i}u_{k)} - 2\nu v g_{ik}) =$$
$$= Ee_{ik} + Fg_{ik} + (Iu_{(i})_{;k)} + J_{;ik} = 0,$$

where

$$E = Y_{;l}^{;l} - 4HY_{,l}u^l + (p - \epsilon + 4(\dot{H} + H^2))Y +$$
$$+(\epsilon+p)(\tilde{\chi}(\beta^{-2}-1) + 2(4H+\nu)X),$$

$$F = Z_{;l}^{;l} + (p-\epsilon)Z - 4H(Y_{,l}u^l + HY - (\epsilon+p)X),$$

$$I = -2(Y_{,l}u^l + HY - (\epsilon+p)X),$$

$$J = 2Z,$$

and $\tilde{\chi} = w^{-1}(wX)_{,l}u^l - (Y+Z)/2$. Here, the auxiliary relations

$$u_{i;k} = HP_{ik} = \frac{1}{2}H(g_{ik} - e_{ik}), \quad H = \frac{1}{3}u^l_{,l},$$

and some other formulae for the background were used.

From $J = 0$ we have [15]

$$Z = 0 \tag{170}$$

From $I = F = 0$ the following relation between the scalars X and Y is obtained:

$$\dot{Y} + HY = (\epsilon+p)X. \tag{171}$$

For the spatially flat model ($K = 0$) the q-scalar is given by the linear superposition of the gauge invariant functions:

$$q = -(Y + 2HX) \tag{172}$$

[15] The scalar J is generally connected with the anisotropic pressure, which is zero for the φ-field.

Obviously, it is a real 4-scalar (as are X and Y) independent of any reference frame. The inverse transformations follow from eqs. (171, 172):

$$X = \frac{1}{2}\left(\frac{P}{a} - \frac{q}{H}\right), \quad Y = -\frac{H}{a}P, \tag{173}$$

where $P = \int a\gamma q dt, \ \gamma = -\dot{H}/H^2$.

From $E = 0$ we have the key equation [16]:

$$a^3\gamma\beta^{-2}\dot{q} = \Delta P. \tag{174}$$

It remains to prescribe the potentials C and D for different gauges.

Projecting eqs. (169) onto the Friedmann reference system we have for an arbitrary gauge:

$$v = X + \frac{1}{2}F, \quad h_{oo} = Y + \dot{F},$$

$$h_{o\alpha} = \frac{1}{2}\psi_{,\alpha}, \quad h_\alpha^\beta = A\delta_\alpha^\beta + B_{,\alpha}^{,\beta}, \tag{175}$$

where $F = C + \dot{D}$, $\psi = a^2(D/a^2)^{\dot{}} + F$, $A = HF - Y$, $B = -D/a^2$. Thus the most general definition of the q-scalar in terms of the 3-potentials is the following (cf. eq. (34)):

$$q = A - 2Hv. \tag{176}$$

For the orthogonal gauge ($\psi = 0$)

$$C = 2a(aB)^{\dot{}}, \quad D = -a^2B, \quad F = a^2\dot{B}. \tag{177}$$

The next examples specify the function B in eqs. (177).

For the synchronous gauge ($h_{oo} = \psi = 0$)

$$\dot{B} = a^{-2}Q - a^{-3}P, \quad Q = \int \gamma q dt. \tag{178a}$$

---

[16]The functions $(\epsilon + p)\tilde{\chi}$ and $(\epsilon + p)X$ should be excluded from E with the help of eq. (171).

For the comoving gauge ($v = \psi = 0$)

$$\dot{B} = \frac{q}{Ha^2} - a^{-3}P. \tag{178b}$$

For the Newtonian gauge $B = \psi = 0$ (cf. eqs.(51, 52, 53)).

APPENDIX C

One can easily verify that eqs. (42) are just the linear expansion terms of the exact solution

$$ds^2 = dt^2 - a^2 exp(2\tau a_{\alpha\beta})dx^\alpha dx^\beta, \quad \tau = \int \frac{dt}{a^3}, \tag{179}$$

where the function $a = a(t)$ can be found as follows:

$$H^2 = \frac{1}{3}\epsilon - \frac{1}{9}\Lambda^2/a^6, \quad \Lambda^2 = \frac{2}{3}a^\beta_\alpha a^\alpha_\beta,$$
$$\dot{H} = -\frac{1}{2}(\epsilon + p) + \frac{1}{3}\Lambda^2/a^6.$$

For $t \longrightarrow 0$, we have the Kazner asymptotic behaviour:

$$a^3 = \Lambda t, \quad g_{\alpha\beta} \sim diag(t^{2p_a}), \tag{180}$$

where the Kazner exponents ($p_1 + p_2 + p_3 = p_1^2 + p_2^2 + p_3^2 = 1$) are obviously related to the eigenvalues of the matrix $a_{\alpha\beta}$:

$$det(a_{\alpha\beta} - \lambda\delta_{\alpha\beta}) = 0, \quad \lambda_a = \Lambda(p_a - 1/3).$$

Eqs. (170) describe the solution for the Bianchi type I model in comoving space. The infinite scale vortex and gravitational-wave perturbations lead also to eqs. (179). Note that the infinite scale perturbations, although causing the expansion anisotropy (shear), do not perturb the spatial curvature and density perturbations ($\delta\epsilon = u_\alpha = 0$). This happens only in spatially flat Friedmann models.

REFERENCES

Birrell N.D. & Davies P.C.W.: 1981, *Quantum Fields In Curved Space* , Cambridge Univ. Press.

Grib A.A., Mamaev S.G. & Mostepanenko A.: 1980, *Quantum Effects In Intensive External Fields*, Atomizdat, Moscow.

Grichshuk L.P.: 1974, *Zh. Eksp. Teor. Fiz. (JETP)* **67**, 825.

Hodges H.M. & Blumenthal G.R.: 1989, *Inflation and Primordial Fluctuation Spectrum*, Preprint SCIPP 89/56.

Kolb E.W. & Turner M.S.: 1989, *The Early Universe*, Addison-Wesley Publ. Co.

Komberg B.V. & Lukash V.N.: 1994, *MNRAS* **269**, 277.

Landau L.D. & Lifshitz E.M.: 1967, *Field Theory*, Nauka, Moscow.

Lifshitz E.M.: 1946, *Zh. Eksp. Teor. Fiz. (JETP)* **16**, 587. Lifshitz E.M. & Pitaevski L.P.: 1978, *Statistical Physics*, Nauka, Moscow.

Linde A.D.: 1983, *Phys, Lett. B* **129**, 177.

Linde A.D.: 1986, *Phys. Lett. B* **175**, 395.

Lukash V.N.: 1980, *Zh. Eksp. Teor. Fis. (JETP)* **79**, 1601.

Lukash V.N. & Novikov I.D.: 1992, in *Observational and Physical Cosmology*, eds. F.Sanchez et al., Cambridge Univ. Press, 3.

THE HUBBLE CONSTANT AND THE AGE OF THE UNIVERSE

Jeremy Mould

Mount Stromlo & Siding Spring Observatories
Institute of Advanced Studies
Australian National University

1 The Expanding Universe

In these lectures I shall draw heavily on Peebles' (1992) work, *Principles of Physical Cosmology*, where physical cosmology is defined as "the attempt to make sense of the large scale nature of the material world by the methods of the natural sciences". The methods of astronomy are constrained by the necessity to "look, but don't touch" to quote Peebles again.

1.1 Elements of the standard world picture

- The mass distribution is close to homogeneous

  On the largest observable scales, which correspond to a Hubble length c/H_0, deviations from uniformity are very small:

  $$\delta M/M < 10^{-4}$$

  (where M is the mass enclosed within a given radius), according to our observations of the cosmic microwave background.

  On scales 1% of the Hubble length, however,

  $$\delta M/M \sim 1$$

  The details of this transition are discussed at length by Szalay in this volume.

- The Universe is expanding

  The mean distance l between conserved particles is given by:

  $$\frac{dl}{dt} = H_0 l \qquad\qquad H_0 l < c$$

- The dynamics of the expanding universe are general relativity

  This means that local physics is the same for all positions and times, although this is clearly not the case as t approaches zero.

- The Universe expanded from a hot, dense state

  At that time the mass in the Universe was dominated by thermal blackbody radiation. Note that inflation is not taken to be part of this standard model, for the reason that it has not been tested.

Let us examine the evidence that the first two of these statements describe the Universe we live in.

1.2 Hubble's test for the space distribution of galaxies

Suppose that space is static, the geometry of space is Euclidean and all galaxies have luminosity L. The flux

$$f = \frac{L}{4\pi r^2}$$

The volume in 1 sr out to radius r is $V = r^3/3$. For a mean density n, a uniform distribution implies that the number of galaxies with flux exceeding f

$$N(> f) = nV = nr^3/3 = \frac{n}{3}(\frac{L}{4\pi f})^{\frac{3}{2}}$$

If we allow a broad spread in luminosity L_i

$$N(> f) = \frac{1}{3}\ \Sigma_i\ n_i\ (\frac{L_i}{4\pi f})^{\frac{3}{2}} \propto f^{-\frac{3}{2}}$$

To introduce some astronomical terms now, the apparent magnitude

$$m = -2.5 \log f + constant \qquad\qquad f \propto 10^{-0.4m}$$

and the absolute magnitude

$$M = -2.5 \log L + constant$$

define a distance modulus

$$m - M = 5 \log r_{Mpc} + 25$$

where r_{Mpc} is the galaxy distance in Megaparsecs.
This yields a relation between galaxy counts and magnitude

$$N(< m) \propto 10^{0.6m}$$

with differential counts given by:

$$\frac{dN}{dm} \propto 10^{0.6m}$$

The evidence for this relation is shown in Figure 1.

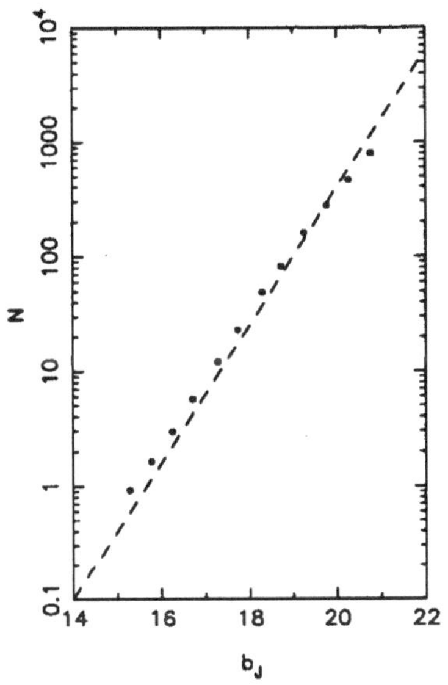

Figure 1: *Galaxy counts where N is 10 × count per 0.1 mag per square degree from Maddox et al. (1990). The blue b_J photometric system is used.*

Detailed discussion of galaxy counts is given by Koo & Kron (1992).

1.3 The expansion law

The expansion of the Universe means that the proper physical distance l between a pair of galaxies is increasing with time.

$$l(t) = l_0 a(t)$$

where a is the scale factor.

$$v = \dot{l} = l_0 \dot{a} = l\frac{\dot{a}}{a} = Hl$$

Since the wavelength of light λ expands in the expanding universe, the redshift of an observed galaxy, which is related to the ratio of observed and emitted radiation,

$$z = \frac{\lambda_0}{\lambda_{em}} - 1 = \frac{v}{c} = \frac{Hl}{c}$$

If v were time independent,

$$T_H = \frac{l}{v} = H^{-1}$$

would define the Hubble time, and,

$$l_H = \frac{c}{H}$$

would define the Hubble radius.

1.4 Dynamical effects of inhomogeneities

The expansion is neither time independent nor spatially invariant, however, because of inhomogeneities.

In the "linear approximation" velocity perturbations are taken to be proportional to a constant acceleration acting over a Hubble time.

$$\Delta V = \frac{GM}{r^2}\frac{1}{H_0}$$

at a distance r from mass M.

For the nearest spiral galaxy, M31, at a distance of 0.7 Mpc, assuming a mass for the Milky Way of $10^{12} M_\odot$,

$$\Delta V \sim \frac{6 \times 10^{-8} \times 10^{12} \times 2 \times 10^{33} \times 3 \times 10^{17}}{(2 \times 10^{24})^2} \quad cm\ s^{-1}$$

$$= 100 \quad km/sec$$

which is dynamically significant. Indeed, M31 is moving towards the Galaxy, rather than expanding away from it.

Similarly, we may consider the effect of the Virgo cluster on the local expansion, employing an estimate of $10^{15} M_\odot$ for its mass, and knowing that it is 17 Mpc away,

$$\Delta V = \frac{10^{15}}{10^{12}}\left(\frac{0.7}{17}\right)^2 \quad \Delta V_{M31}$$

$$= 170 \quad km/sec$$

1.5 Evidence for uniform expansion on large scales

Measurement of the distances of clusters of galaxies using the Tully-Fisher relation permits a demonstration of the uniformity of the Hubble expansion to 10^4 km/sec (Aaronson and Mould 1986).

Evidence that the expansion is uniform on the largest scales is to be seen in the Hubble diagram for 'standard candle' (constant luminosity) sources. For such objects,

$$m = 5 \log r + constant$$

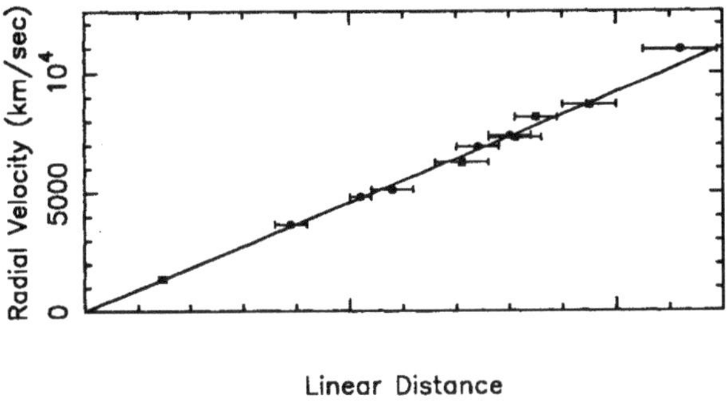

Figure 2: *Nearby clusters of galaxies follow the Hubble law.*

The Hubble diagram for classical radiogalaxies has been extended to $z = 3$ by McCarthy (1992) and Djorgovski (1995), measuring luminosity at 2.2μ wavelengths in the K band.

$$K = 5 \log \frac{cz}{H} + constant$$

The slope of this relation (see Figure 3) is consistent with uniform expansion for $z \lesssim 1$. Curvature of space and the evolution of sources make for uncertain expectations for $z > 1$.

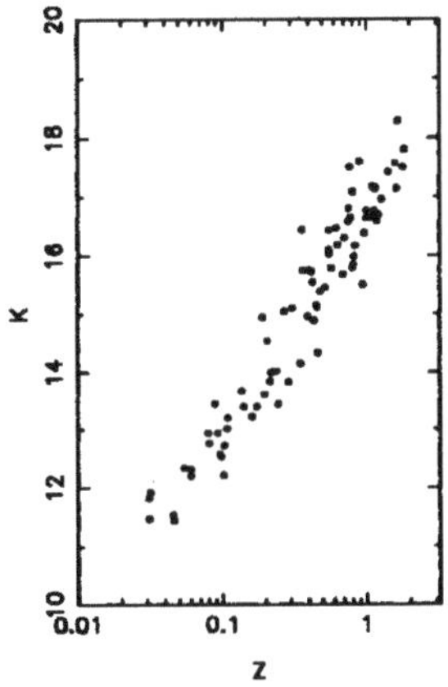

Figure 3: *Hubble diagram for classical radiogalaxies*

2 The Distance Scale Ladder

There are two classical ways of measuring astronomical distances

- geometrical: one measures the angle subtended by an object of known linear size. The problem here is that these angles are small at the distances of galaxies, frequently of the order of microarcseconds.
- standard candles: if flux and luminosity are known, then distance can be inferred.

$$f = \frac{L}{4\pi r^2}$$

The problem here is that luminosities are usually inferred astrophysically and there is an inherent dependence on the astrophysical model.

2.1 Trigonometric Parallaxes

One parsec is the distance at which 1 astronomical unit subtends an angle of one second of arc.

$$1pc = 2.06 \times 10^5 \frac{radians}{arcsec} 1.50 \times 10^{13} \quad cm$$

$$= 3.09 \times 10^{18} \quad cm$$

The typical ground based image size for a star is 1″. The corresponding state-of-the-art parallax measuring error is 1 milliarcsec. This yields 100% error at a distance of 1 kpc.

The current best work is that of the Hipparchos satellite (Perryman *et al.* 1992).

Optical interferometry can be expected to yield improvements in trigonometric parallaxes. For a 10 m baseline instrument operating at 500 nm, the fringe spacing is 10 mas. Relative measurements essentially consist of counting the fringes between two stars. If one can determine phase to 1%, then the measuring error can be brought down to 0.1 mas, corresponding to 10% error at 1 kpc.

2.2 The main sequence

The set of stars for which trigonometric parallaxes can be measured is predominantly main sequence (core hydrogen burning) stars. This technique, then, yields the distances of star clusters. Stellar evolutionary models, whose composition is parameterized by the mass fraction of hydrogen, helium and heavier elements (X,Y,Z), are useful for interpolating stars of different composition on to a common sequence.

The location of the main sequence in the luminosity-temperature plane is shown in Figure 4 for nearby stars of Population II, i.e. those very old stars of low Z. Bolometric magnitudes are defined by the relation:

$$M_{bol} = -2.5 \log \frac{L}{L_\odot} + 4.75$$

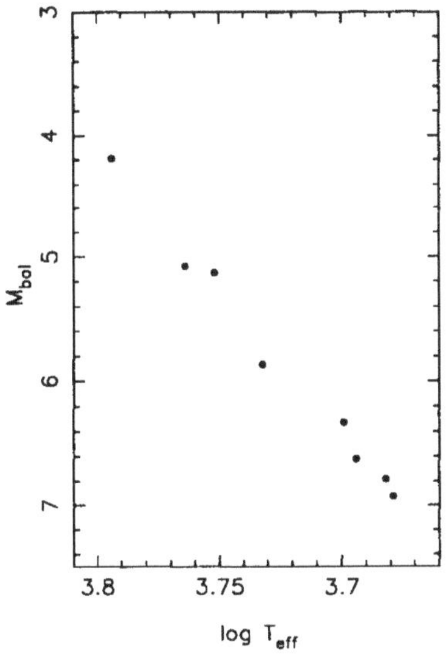

Figure 4: *The population II main sequence*

The corresponding diagram for Population I is shown in its empirical form, a colour-magnitude diagram of the young star cluster, the Pleiades, in Figure 5, where B–V is the flux ratio of the blue and visible bandpasses expressed in magnitudes. The distance modulus of the Pleiades is 5.6 mag from trigonometric parallax (Vandenberg & Poll 1989).

2.3 RR Lyrae

Main sequence fitting allows the distances of globular clusters to be determined (following Carney 1979). The quality of CCD colour-magnitude diagrams in the last decade (Hesser 1995) has made this step in the extragalactic distance scale a rather secure one. In turn, the distances of the globular clusters calibrate the luminosities of

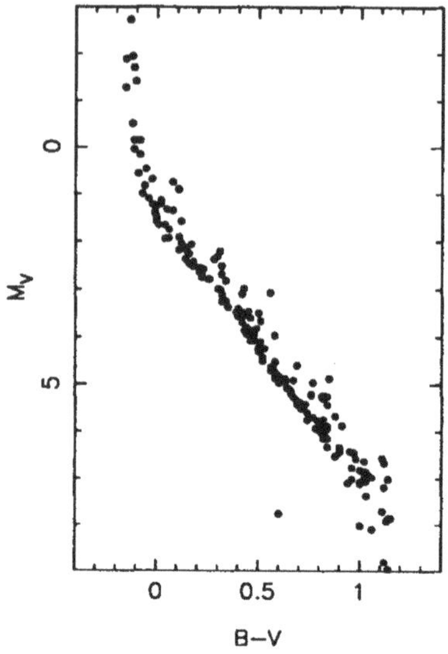

Figure 5: *The population I main sequence. The data are those of Johnson & Mitchell (1958).*

the core helium burning stars of Population II. Those "horizontal branch stars" which are unstable against radial pulsation, the RR Lyrae variable stars, can be detected in all the galaxies of the Local Group (Saha & Hoessel 1994).

A second, and quite fundamental, technique for calibrating the RR Lyrae stars is that pioneered by Baade & Wesselink, in which variations in the angular size of RR Lyrae stars are monitored by means of the associated flux and temperature variations, and variations in the linear size are monitored by integrating the radial velocity variations. For two phases, 1 and 2,

$$f_1/f_2 = (\theta_1/\theta_2)^2(T_{e1}/T_{e2})^4$$

where θ is the angular diameter of the star and T_e is the effective temperature of the star (determined from its colour temperature). At the same phases,

$$R_2 - R_1 = p\int_1^2 v\, dt$$

where R and v are the radius and radial velocity respectively, and p is the projection factor between the pulsation velocity and the line of sight integrated over the stellar disk.

Results have been summarized recently by Clementini *et al.* (1995) and are reproduced in Figure 6. They are fit by the relation:

$$M_v = 0.140[Fe/H] + 0.914$$

where

$$[Fe/H] = \log(Z/Z_{\odot})$$

provided that the ratios of the abundances of the heavy elements in Population II are approximately the scaled ratios observed in the Sun.

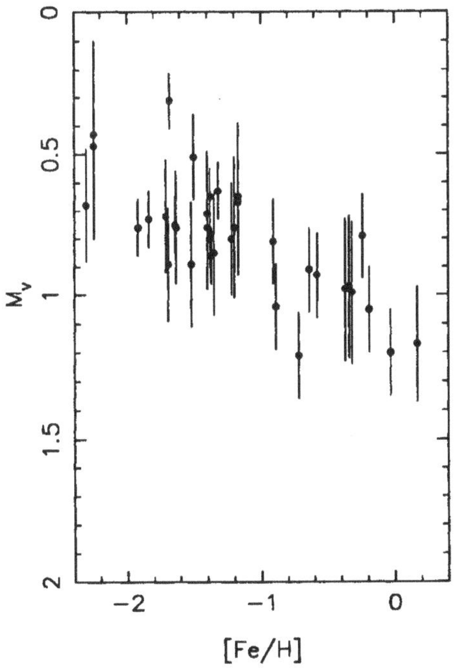

Figure 6: *The luminosities of RR Lyrae stars as a function of metallicity according to Clementini et al 1995*

2.4 Cepheids

Cepheid variable stars are the core helium burning radial pulsators of Population I. Their evolutionary status is explored in some detail by Becker *et al.* (1977). The existence of a relation between the period of pulsation and the stellar luminosity can be understood at a simple level.

The physical basis for the Cepheid PL relation is the very general relation for object oscillating in a gravitational field:

$$P = Q/\sqrt{\rho}$$

where P is the period and ρ is the density of the object supplying the gravitational restoring force. Dimensionally we can see that the basis of the above equation is the condition that the density of the star is just high enough to prevent the outer shell from escaping, for which energy considerations tell us that

$$\frac{1}{2}\left(\frac{R}{t}\right)^2 \lesssim \frac{GM}{R}$$

or

$$\frac{M}{R^3} \propto P^{-2}$$

where M and R are the stellar mass and radius respectively, and P is the relevant timescale, the period.

For core hydrogen and helium burning stars, power law relations exist between stellar mass and luminosity:

$$L \propto M^n$$

Eliminating the radius through its relation to the luminosity, we have a relation

$$L = AP^{1.3}$$

which is comparable to that which is observed (Feast & Walker 1987). (A is a constant.)

The same two techniques, main sequence fitting and the Baade-Wesselink method serve to calibrate the period luminosity relation for Cepheids, where calibration may be considered to be the evaluation of the constant A. An example of the sensitivity of standard candle distances to astrophysics is the question of whether A is sensitive to the composition of the star. According to the results of Chiosi *et al.* (1993) d ln A/d ln Z $\lesssim$ 0.03.

2.5 The Large Magellanic Cloud

The LMC is an anchor-point of the extragalactic distance scale. From the LMC it is possible to reach out into the Hubble flow using the Cepheid PL relation. The PL relation for the LMC itself is shown in Figure 7. Figure 7 shows H-band magnitudes (1.6μ effective wavelength) from the summary by Madore & Freedman (1990).

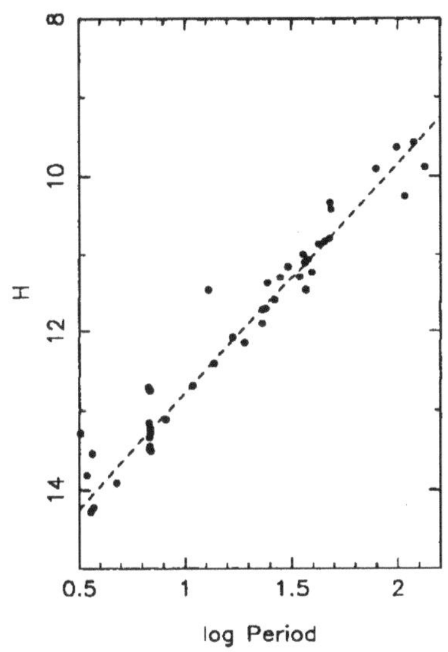

Figure 7: *The period luminosity relation for Cepheids in the LMC.*

The concordance of the Population I and II distance scales regarding the distance of the LMC is an important check on our uncertainty at the outset of the extragalactic distance scale. A third measure of the LMC distance is available from the angular and light travel time measurement of the ring around SN1987A (Plait *et al.* 1994).

Distance indicator	m-M	Reference
RR Lyrae	18.5	Walker (1993)
Cepheids	18.47 ± 0.15	Feast & Walker (1987)
SN1987A	18.5	Plait *et al.* (1994)

3 The Key Project for Hubble Space Telescope

In the previous section we saw that the absolute distance modulus of the LMC is 18.5 ± 0.1 mag (a distance of 50 ± 2.5 kpc). HST allows us to extend the reach of the distance scale based on Cepheid to the Virgo cluster at 17 Mpc. The aim of the key project for HST (Aaronson & Mould 1986, Mould 1989, Kennicutt, Freedman & Mould 1995) is to measure H_0 to 10% accuracy. Results to date are summarized by Mould *et al.* (1995).

In §1.4 we estimated the dynamical effects of local inhomogeneities. Szalay (this volume) has discussed the physical situation in greater depth. We can make a simple parameterized model of the effects of inhomogeneities by considering a spherical overdensity centred a distance r from the observer.

$$\Delta\rho \propto r^{-\gamma}$$

Integrating the overdensity, we can calculate the gravitational acceleration on the observer:

$$g \propto r^{1-\gamma}$$

In the linear approximation velocity perturbations are proportional to accelerations, which are considered to act steadily over a Hubble time [1]

$$\Delta V \propto g$$

This is the basis of a model which illustrates what has been termed (in comoving coordinates) the Virgocentric flow (Schechter 1980, Aaronson *et al.* 1982, Davis & Peebles 1983). The perturbation of a $\Delta V = 300$ km/sec infall velocity of the Local Group towards Virgo is illustrated in Figure 8.

3.1 Secondary Distance Indicators

Because the Hubble flow is significantly perturbed on the 17 Mpc scale of the Virgo cluster distance, redshift space as far as 10^4 km/sec must be investigated for a value of the Hubble Constant which is unaffected by "peculiar velocities" as large as the 600 km/sec dipole anisotropy of the microwave background.

This requires that Cepheid distances be used to calibrate secondary distance indicators which are capable of delivering accurate results within such a large volume.

[1] It is easy to show (Peebles 1992, Lightman & Schechter 1990) the relationship of this perturbation to the density of the Universe:

$$\frac{\Delta V}{V} \approx \frac{1}{3}\Omega^{0.6}\frac{\Delta\rho}{\rho}$$

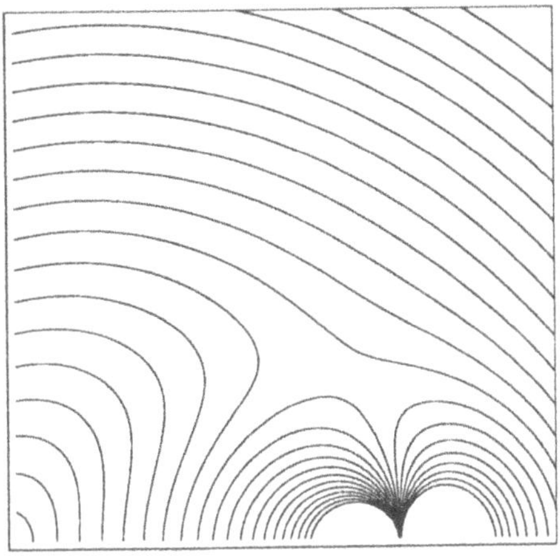

Figure 8: *Contours of constant velocity of recession from the origin in a linear Virgo-centric flow model. In this coordinate system the Local Group is at the origin and the Virgo cluster lies at the cusp on the x-axis. The infall velocity of the Local Group in comoving coordinates has been taken to be three tenths the recession velocity of Virgo. The contour spacing is one tenth of the Virgo redshift. The asymptotic Hubble flow is seen at large distances from the origin.*

The global dynamical relations (Tully-Fisher and (D_n, σ)), the luminosity function relations (surface brightness fluctuations, planetary nebulae, globular clusters), and the properties of supernovae are the secondary distance indicators which have been shown to be most robust (Jacoby *et al.* 1992).

The range of these secondary distance indicators is illustrated [2] in Figure 9.

[2] *References* : Arecibo: Aaronson *et al.* (1986), du Pont: Dressler & Faber (1990) EGALS: Burstein (1987), T93: Tonry (1993), (HST) (planned) Tonry *et al.* (1990), KPNO: Jacoby, Ciardullo, & Ford (1990), CFHT: Harris & Thompson (1987), KPNO: Harris (1988), SN1970J*: van den Bergh & Pazder (1992), SN1992am: Schmidt & Kirshner (1994) *Notes* : Planned, as opposed to published, observations are shown in parentheses. *Van den Bergh & Pazder list SN1970J as the most distant SNIa with a well-observed light curve, but see Hamuy *et al.* 1995) for SN1992aq at z = 0.1.

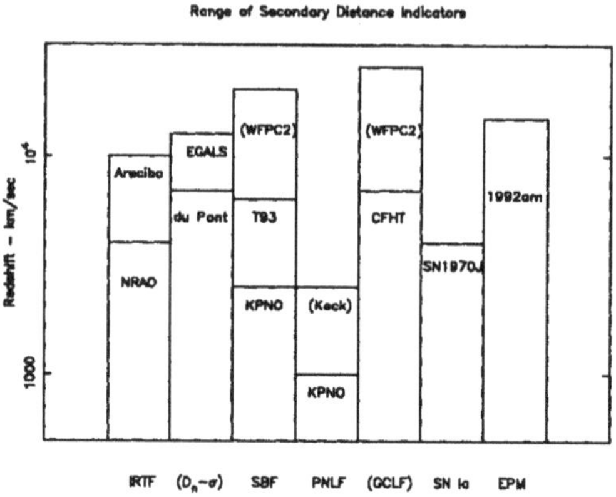

Figure 9: *For seven leading secondary distance indicators the maximum redshifts attained in different real or (planned) programs are indicated by bars of a histogram. Note that the vertical scale is logarithmic in redshift or distance.*

3.2 The Tully-Fisher Relation

This relation for dynamically supported disk galaxies was discovered by Fisher & Tully (1977). Aaronson, Mould & Huchra (1979) demonstrated the physical basis for the relation, firstly, centrifugal equilibrium:

$$M \propto v^2 R$$

where M is the mass of the galaxy, R its radius and v the rotational velocity.

Freeman (1970) showed that disk galaxies have constant central surface brightness, which is equivalent to constant surface density:

$$M \propto R^2$$

Spiral galaxies also tend to exhibit a constant mass to light ratio:

$$M/L = constant$$

These three relations yield:

$$L \propto v^4$$

The observables, L, the galaxy luminosity, and v are readily measured photometrically and spectroscopically, with 21 cm observations providing an excellent determination of rotational velocity. The Tully-Fisher relation can be applied much as the Cepheid PL relation is applied: galaxies of identical rotational velocity are standard candles. An equivalent relation governs pressure supported elliptical galaxies (Dressler *et al.* 1987).

The Tully-Fisher relation for galaxies in the Virgo cluster (together with some local galaxies) is shown in Figure 10.

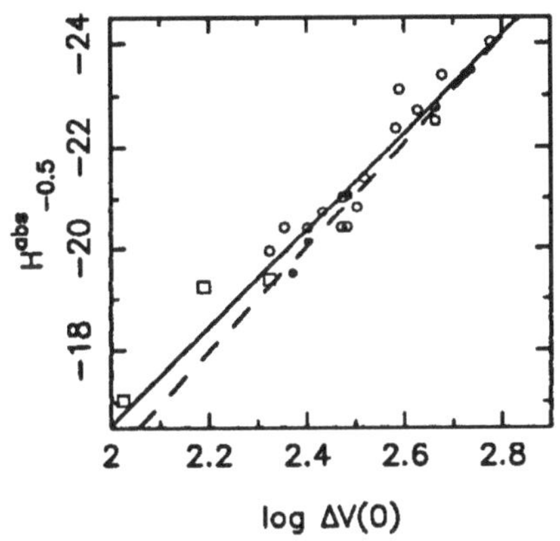

Figure 10: *The Tully Fisher relation for galaxies in the Virgo cluster and the Local Group (Mould et al. 1995). Magnitudes at 1.6μ effective wavelength are plotted versus the 21 cm velocity width of the galaxy corrected to edge-on.*

The scatter in the Tully-Fisher relation is typically 0.4 mag, or 20% in inferred distance. The distances of clusters of galaxies are thus better defined than those of individuals. Recession velocity is plotted as a function of cluster distance in Figure 2.

3.3 Surface Brightness Fluctuations

Surface brightness fluctuations (SBF) in early-type galaxies result from the counting statistics of the number of stars in a galaxy encompassed by a CCD pixel (Tonry & Schneider 1988). If N is the average number of stars in a pixel, and $\bar{f}$ is the average flux from a star, the average brightness in a pixel is

$$\mu = N\bar{f}$$

Since the number in a pixel of fixed angular size scales with distance as r^2 and the flux scales as r^{-2}, this mean surface brightness is independent of distance. The local *rms* from pixel to pixel, however, is

$$\sigma = \sqrt{N\bar{f}}$$

and this scales as r^{-1}: the smoothness of otherwise identical galaxies is proportional to distance.

We can derive a distance from a measurement of both μ and σ, provided that we know the average luminosity $\bar{L}$ corresponding to $\bar{f}$:

$$\bar{f} = \frac{\sigma^2}{\mu} = \frac{\bar{L}}{4\pi r^2}$$

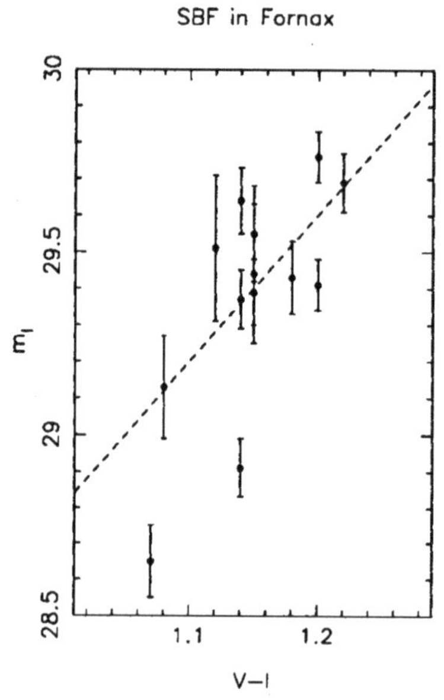

Figure 11: *Surface brightness fluctuation magnitudes in the I bandpass for elliptical galaxies in the Fornax cluster (Tonry 1991). The dashed line is the current calibration* $\bar{m}_I = -1.5 + 4.0(V - I - 1.15)$ *with an apparent distance modulus for Fornax of 30.9 mag*

The radial power spectrum of $\sigma/\sqrt{\mu}$ yields $\bar{f}$ for early type galaxies in a relatively straightforward manner, although the presence of globular clusters in these galaxies adds a source of noise and potential systematic error (Blakeslee & Tonry 1995). The quantity $\bar{f}$ has been measured at a number of wavelengths, and its behaviour as a function of galaxy colour is reasonably well understood (Worthey 1993; Pahre & Mould 1994; Ajhar 1993) in terms of giant branch systematics in stellar populations. Figure 11 shows $\bar{m}_I$ in the Fornax cluster (Tonry 1991).

3.4 The planetary nebula luminosity function

The wide range of stellar masses whose remnants yield 0.6 $M_\odot$ cores makes for a planetary nebula luminosity function (PNLF) which is largely invariant for galaxies of a wide range of star formation histories. Relative distances of galaxies can be derived by requiring that the PNLF is in fact invariant, and Figure 12 shows a highly linear relation between this secondary distance indicator and primary indicators.

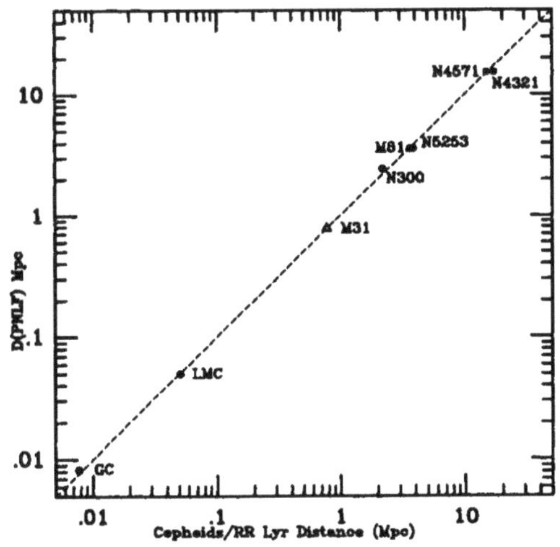

Figure 12: *Planetary nebula luminosity function distances versus variable star distances for galaxies from the Milky Way to the Virgo cluster (Jacoby 1995). The Cepheid distances of NGC 4571 and 4321 are discussed in §4.*

Similar mutual confirmation exists between the PNLF and SBF scales (Ciardullo *et al.* 1994) and other secondary distance indicators (Jacoby *et al.* 1992). The common link between PNLF and SBF is the behaviour of stars in their asymptotic giant branch and post-AGB phases of evolution (Dopita *et al.* 1990).

3.5 The globular cluster luminosity function

Tammann (1988) has used measurements of the globular cluster luminosity function (GCLF) to derive a distance modulus of 31.7 mag for the Virgo cluster. Secker and Harris (1993) argue that the data are compatible with a modulus of 30.9 mag. The M100-based Virgo distance modulus by Freedman et al. (1994a) of 31.15 ± 0.36 mag lies between these extremes. Study of a larger and more distant sample of galaxies to investigate the GCLF more fully will benefit from the high resolution of WFPC2. Since the method has not yet been applied beyond the Virgo cluster (but see Baum *et al.* 1995), this method does not offer us a significant constraint on the global value of H_0 at this stage.

3.6 Type I supernovae as standard candles

Type I supernovae are recognizable to $z \lesssim 0.5$ (Perlmutter *et al.* 1995, Schmidt 1995). When calibrated they will allow us to map the Hubble flow to large distances and probe the geometry of space. A controversy exists regarding the *rms* dispersion in the standard candle magnitude of Type Ia supernovae. Three recent estimates include <0.2 mag (Sandage & Tammann 1993), 0.36 mag (Branch & Miller 1993), and >0.6 mag (Phillips 1993). The standard model of an SN Ia is of a carbon deflagration event in a 1.4 $M_\odot$ degenerate object. Extinction (della Valle & Panagia 1992), variable decline rate, possible sub-Chandrasekhar-mass explosions (Woosley & Weaver 1993), and possible differences between supernovae in late-type calibrating galaxies and early-type galaxies (van den Bergh & Pazder 1992) increasingly suggest that the standard model is a naive one.

Phillips has noted a significant correlation between the maximum luminosity of SN Ia's and their decline rate. The decline rate is measured by the magnitude change in 15 days, and the correlation is shown in Figure 13. Tammann & Sandage (1995) argue that a larger sample, including new data, imply a slope four times smaller than that shown in the figure.

3.7 The Expanding Photospheres of Type II Supernovae

Although not blackbodies, as Baade & Wesselink assumed, Type II supernovae have photospheres, whose emissivity can be calculated. Distances are derived from a comparison of photospheric angular diameter and the time integrated expansion velocity (Kirshner & Kwan 1974). The approach is essentially that of §2.3. EPM has been employed to cz = 14,600 km/sec. Sophisticated radiative transfer calculations can make EPM an *ab initio* measurement of distance. A more empirical approach is to test the models with SNIIe like SN1987A and others in galaxies with Cepheid distances.

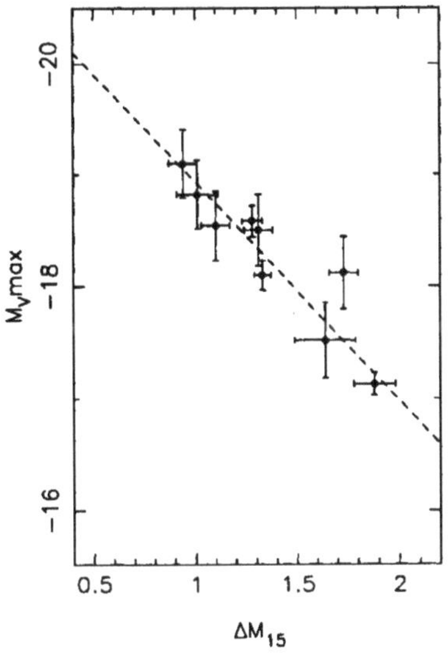

Figure 13: *Decline rate – peak luminosity relation for the nine best observed SNIa's (Phillips 1993).*

3.8 The need for calibration

What all the secondary distance indicators have in common is a need of calibration. The Key Project for HST aims to provide a calibration for all of the secondary distance indicators outlined above by measuring Cepheid distances to galaxies which will do this job. In many cases distances to important groups and clusters will provide multiple calibrators.

The number of galaxies to be observed is set by the need to be able to recognize systematic differences between distance indicators at the 0.1 mag level. In Table 1 σ is the intrinsic dispersion of the distance indicator and n is the number of calibrators in the project which contribute to determining the zeropoint of that distance indicator. For the two distance indicators whose dispersion is controversial (IRTF and SNIa) values are shown which are in the middle of the arguable range.

Table 1: Resultant Uncertainty in Zeropoint by Distance Indicator

		IRTF	SBF	PNLF	SNIa	SNII
σ	(mag)	0.40	0.15	0.15	0.36	0.32
n		30	12	11	11	5
$\sigma/\sqrt{n}$	(mag)	0.07	0.05	0.05	0.11	0.14

The error budget for the Key Project to measure H_0 is summarized in Table 2. Perhaps surprisingly, velocity field uncertainties contribute rather little to the budget. An all-sky survey of Tully-Fisher distances to clusters of galaxies provides empirical evidence (Mould *et al.* 1993), and models fitted to these data show only modest variations in the global value of H_0. A larger volume still has been explored: Lauer & Postman (1992) limit any variation in the apparent H_0 measured from our location to $\Delta H_0/H_0 < 0.07$ within $z < 0.05$.

Table 2: Error Budget – 1σ uncertainties

Source of Uncertainty	Distance Modulus Error	Section
LMC distance	±0.08 mag	§2
PL relation	±0.10 mag	§2
Systematics	±0.10 mag	*below*
Velocity field	±0.05 mag	*above*

The crucial category in Table 2 is discrepancies in H_0 from the five different distance indicators listed in Table 1, which contribute a residual uncertainty we have termed 'systematics'. In the absence of systematic error the five columns of Table 2, treated as five separate experiments to estimate the same quantity with no weighting of the results, would yield a random error in distance modulus of 0.04 mag. But Jacoby *et al.* (1992) find typical systematic differences between different distance indicators of 10% in distance. These systematic differences will contribute 0.09 mag to the error budget, again with no weighting of the results. This is combined in quadrature with the estimated random errors to arrive at the entry 'systematics' in Table 2. Combining the four entries in quadrature, we reach a 0.18 mag final uncertainty in the Hubble modulus of an object at the limits of our secondary distance indicators, or a 9% uncertainty in H_0.

4 Results to date from the Key Project

In the first two and half years of the operation of HST Cepheids were discovered in two of the program galaxies, M81 and M101 (Freedman *et al.* 1994b, Kelson *et al.* 1995). Following the first servicing mission, the improved resolution of the telescope enabled the discovery of Cepheids in M100, yielding a distance of 17 ± 3 Mpc for the Virgo cluster. Light curves for four of these Cepheids are shown in Figure 14.

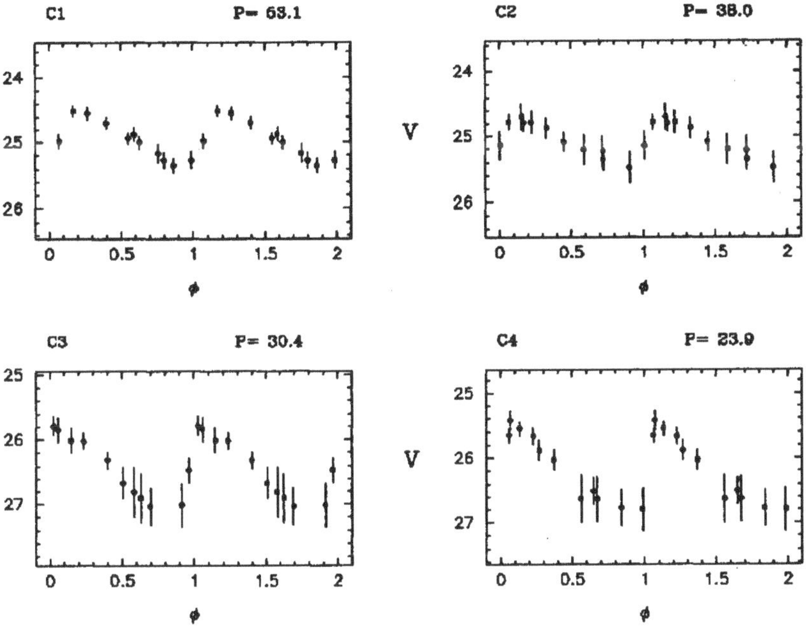

Figure 14: *Light curves for 4 Cepheids in M100 (Freedman et al. 1994a). Periods are given in days.*

The LMC and M100 PL relations are compared in Figure 15.

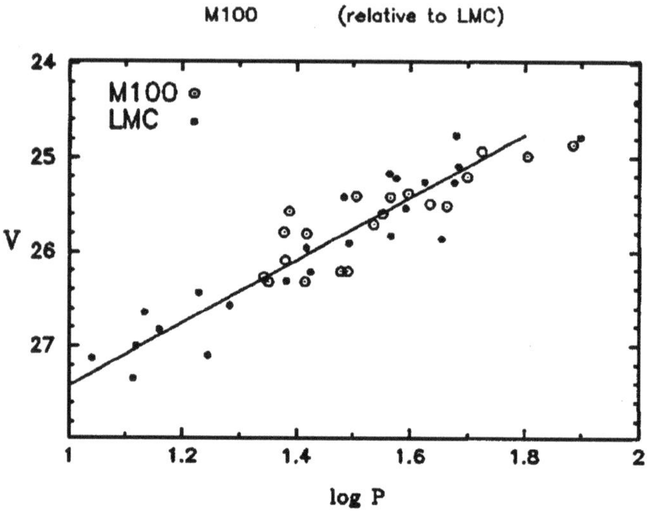

Figure 15: *Period luminosity relations for M100 and the LMC (Freedman et al. 1994a). The LMC data have been shifted in magnitude so as to superimpose them.*

4.1 The significance of a Virgo cluster distance

Because M100 is in the Virgo cluster, this measurement of its distance has an impact on the calibration of all of the extragalactic secondary distance indicators which reach beyond Virgo and define the expansion rate.

The distance of M100 provides two separate constraints on the Hubble Constant. First, it verifies the emissivity calculations for type II supernovae. Second, it constrains the distance of the Virgo cluster, which in spite of its apparent complex structure, provides an effective calibration for a set of reliable and well-used secondary distance indicators.

In its way the Virgo cluster is as significant a junction in the distance scale as the Magellanic Clouds, permitting a consistency check on many secondary distance indicators.

4.2 Membership of the Virgo cluster

Like many clusters of galaxies Virgo has a complex structure, which has been elucidated gradually, starting with the observations of de Vaucouleurs (1961) of distinct elliptical and spiral concentrations. M100 lies within the classical 6° circle of de Vaucouleurs, and just outside the last x-ray contour in the ROSAT map of Böhringer (1994) (Figure 16).

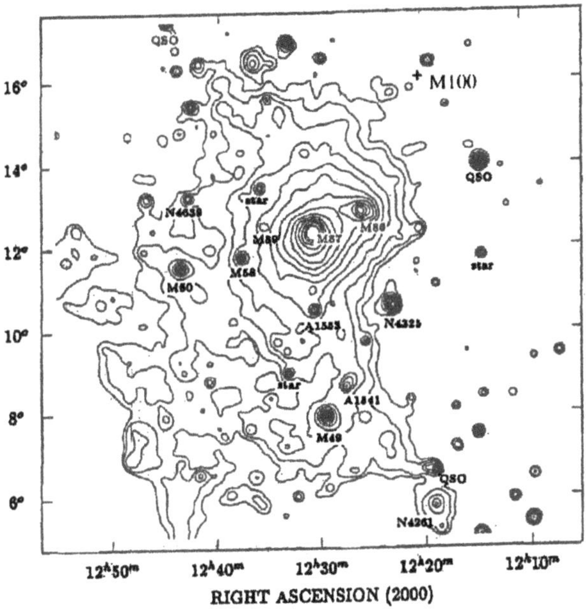

Figure 16: *X-ray map of the Virgo cluster Böhringer (1994).*

Marked substructure exists in the Virgo cluster (Huchra 1985, Bingelli, Sandage & Tammann 1985) with member galaxies exhibiting a spread in surface brightness fluctuation distances from 11 to 19 Mpc (Tonry et al. 1990). Mould *et al.* (1995) consider whether M100 is associated with the central concentration or the fringes of the cluster. A number of constraints are consistent with the conclusion that M100 is a member of the Virgo cluster, whose centroid is at 17 Mpc. Parameterized models of the cluster associate an uncertainty of ± 3 Mpc with that centroid, given the distances of M100 and NGC 4571 (Pierce *et al.* 1994).

4.3 The Redshift of the Virgo Cluster

The simplest estimate of H_0 can be derived from the Virgo distance measurement and its cosmological redshift. The cosmological velocity is estimated by determining the cluster's heliocentric velocity, correcting that to the centroid of the Local Group, and determining the peculiar velocity of the Local Group with respect to the Hubble flow. Modern determinations of the cluster velocity are based on Huchra's catalogue of Virgo galaxy velocities (Huchra 1985, 1988; Binggeli, Sandage and Tammann 1987).

A model fit to the TF relation for 308 spiral galaxies in the Local Supercluster by Aaronson et al. (1982) yielded 331 ± 41 km/s for the infall of the Local Group into the mass concentration centered on Virgo. The model is analogous to that for Galactic rotation in that there is a local standard of rest, which is a pattern velocity constant on shells of constant radius from the Virgo cluster (250 km/s locally), and a local random velocity (80 km/s in the Virgo direction). Subsequent work by Tonry et al. (1992) with more accurate distances from surface brightness fluctuations confirm that result, yielding 340 ± 80 km/s (90% confidence limits). Jerjen & Tammann (1993) obtained 240 ± 40 km/sec for the infall velocity from an analysis of more distant clusters. Averaging the first three estimates and combining this with the mean observed velocity of 1073 ± 75 km/sec, Mould *et al.* (1995) find that at the distance of Virgo, the expansion rate is 1380 ± 90 km/s.

The Hubble constant derived by Mould *et al.* (1995) from the cosmological redshift of Virgo is $H_0 = 81 \pm 11$ km/s/Mpc. The uncertainty is derived from the distance and velocity uncertainties taken in quadrature. There is an additional uncertainty arising from the extended nature of the Virgo cluster.

4.4 Supernovae

Galaxy distances can be inferred from the expansion of type II supernovae measured photometrically and kinematically (§3.6). There are now four galaxies with Cepheid distances, whose type II SNe yield EPM distances (Schmidt et al. 1994). The distance ratio PL/EPM is 1.02 ± 0.08 for the LMC, $1.01\,^{+0.23}_{-0.17}$ for M101 and 1.13 ± 0.28 for M100. These constraints, taken together without any weighting, amount to 90% confidence that empirical recalibration of EPM distances will be limited to multiplicative factors in the interval (0.88, 1.26). The relation of the data to this interval is shown in Figure 17.

Thus there is no evidence that any empirical recalibration of EPM is required at present by the Cepheid data. EPM provides independent and consistent constraints on the Hubble constant, currently yielding $H_0 = 73 \pm 11$ km/s/Mpc (see Figure 18 from Schmidt *et al.* 1994).

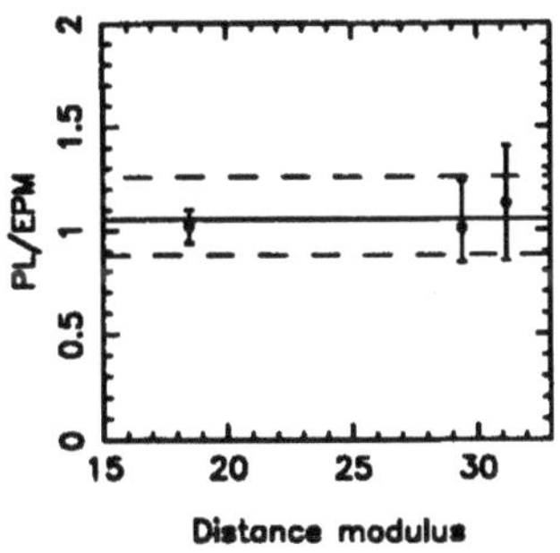

Figure 17: *Consistency of Cepheid PL and SNII EPM distances.*

The Hubble diagram for type Ia SNe at maximum light is discussed by Sandage and Tammann (1993). The mean value of V(max) for 6 well observed SNeIa in the Virgo cluster is 12.13 ± 0.14 mag. Scaling the prototypical SN1937C from 4.8 ± 0.3 Mpc in IC 4182 to our Virgo distance of 17 ± 3 Mpc, implies a visual magnitude of 12.01 ± 0.25 mag (Jacoby and Pierce 1994). The same calculation for SN1972E in NGC5253 (Sandage et al. 1994) yields 11.69 ± 0.25 mag, which would therefore appear to have been an unusually luminous event. Hamuy et al. (1995) assert that its high luminosity is related to its slow decline rate (Figure 13).

Like the type II SNe, the type Ia SNe offer a way to extend our knowledge of the expansion rate from the Local Supercluster, where the high density decelerates the expansion, to the freely expanding region outside it. The new estimate of type Ia absolute magnitudes based on the six Virgo supernovae permits a determination of H_0 from the 21 galaxies at redshift exceeding 3000 km/s collated by Sandage and Tammann (1993) and Tammann & Sandage (1994) . This yields $H_0 = 71 \pm 7$ km/s/Mpc plus an extra uncertainty arising from the extended nature of the Virgo cluster.

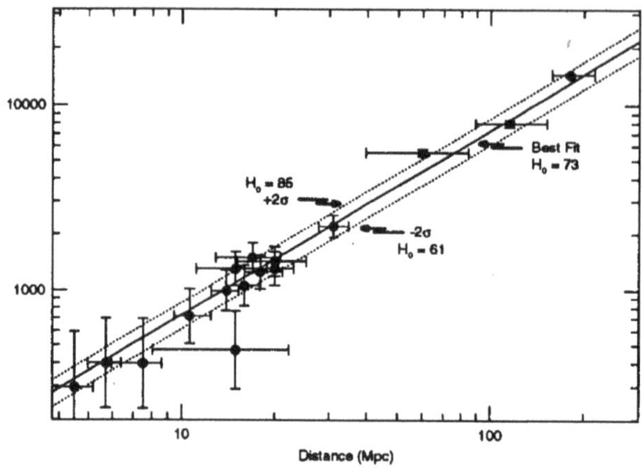

Figure 18: *The Hubble flow from EPM distances (Schmidt et al. 1994)* .

4.5 The Tully Fisher Relation for Clusters

The value of the Hubble Constant that follows from applying the calibration shown in Figure 10 to clusters beyond 4000 km/s, neglecting the effect of perturbations to the Hubble flow and the Virgo depth uncertainty, is $H_0 = 86 \pm 11$ km/s. If we exclude clusters within 45° of the Great Attractor (Lynden-Bell et al. 1988) (and its antipode), noting the evidence for large scale flows (Mathewson et al. 1993, 1994), we obtain $H_0 = 82 \pm 11$ km/s/Mpc. Again, there is an additional uncertainty arising from the extended nature of the Virgo cluster.

4.6 Surface brightness fluctuations

Tonry et al. (1990) and Ciardullo, Jacoby & Tonry (1993) have measured surface brightness fluctuations in 11 Virgo galaxies. Employing their current calibration, whose zeropoint is based on early type galaxies in the Local Group, one obtains a distance of 15.3 ± 1.1 Mpc. Adjusting the zeropoint to the 11% larger Virgo distance found from Cepheids, one can then examine the expansion rate of the most distant galaxies in the available sample. Tonry (1994, priv. comm.) has three galaxies with SBF measurements with $v_{CMB} > 4000$ km/s. These yield $H_0 = 84 \pm 16$ km/s.

4.7 Summary to date

Employing the Virgo cluster as calibrator, Mould *et al.* (1995) obtain measurements of the Hubble Constant from extant surface brightness fluctuation measurements, elliptical galaxy velocity dispersion measurements, the Tully Fisher relation, and the type Ia supernova standard candle. These yield $H_0 = 84 \pm 16$, 76 ± 10, 82 ± 11, and 71 ± 10 km/sec/Mpc respectively. All of these are consistent, but they are all subject to the additional uncertainty from Virgo's line-of-sight depth. Type II supernova distances for three galaxies with Cepheid distances support the recent SN result $H_0 = 73 \pm 11$ km/sec/Mpc. This support is independent of where M100 lies in the Virgo cluster. Reviewing the Hubble recession velocity (cosmological redshift) of the Virgo cluster, Mould *et al.* (1995) find $H_0 = 81 \pm 11$ km/sec/Mpc, plus an additional uncertainty arising from the extended nature of the Virgo cluster.

In combination these results strengthen the conclusion of Freedman et al. (1994a) that $H_0 = 80 \pm 17$ km/sec/Mpc and put rather useful limits on the Hubble Constant. Mould et al (1995) show that $50 < H < 100$ with 95% confidence. Sandage & Tammann (1995) find $H_0 = 55 \pm 2$ km/s/Mpc, which lies just within these bounds.

5 The age of the Universe

According to Carroll, Press & Turner (1992) The relationship between the age of the Universe and H_0 is:

$$t_0 \approx \frac{2}{3} H_0^{-1} \frac{sinh^{-1}[\sqrt{|1 - \Omega_a|/\Omega_a}]}{\sqrt{|1 - \Omega_a|}}$$

where

$$\Omega_a = \Omega_M - 0.3\Omega_{tot} + 0.3$$

Ω_M being the matter density of the Universe and Ω_{tot} being the total density, including the vacuum density associated with a cosmological constant, which we can denote as Ω_Λ. These quantities are expressed in units of the critical density.

A range of models is consistent with the new results. We assume $t_0 = 15 \pm 2$ Gyr based on the ages of globular clusters (VandenBerg 1991).

1. For $\Omega_{tot} = 1$, $\Omega_M = 0.1 \pm 0.1$ and $\Omega_\Lambda = 0.9 \pm 0.1$. Such a Universe is vacuum energy dominated with a trace of matter.

2. An open Universe $\Omega_{tot} \approx \Omega_{baryon} = 0.02$ (Steigman & Tosi 1992) is not excluded with the present 20% uncertainty in H_0 and and 13% uncertainty in t_0 (see Figure 19).

3. Since $H_0 > 50$ km/sec/Mpc

$$H_0 t_0 > 15/20 > 2/3$$

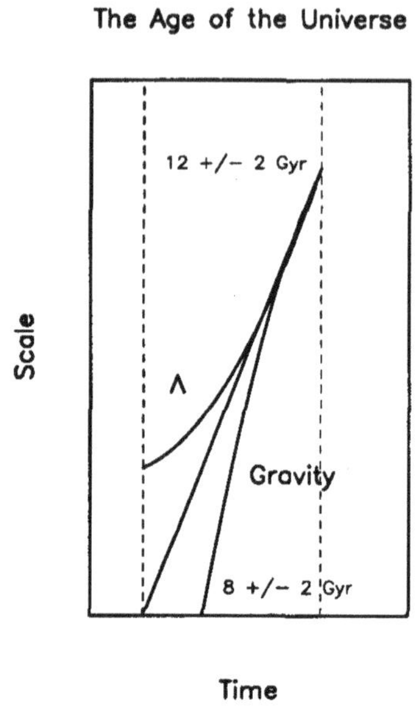

Figure 19: *Schematic of the evolution of the scale factor with time. The linear relation corresponds to an open Universe with* $\Lambda = \Omega = 0$.

For $\Omega_{tot} = 1$, $\Lambda = 0$ is excluded at some level which, because of the t_0 uncertainty, falls short of the 95% confidence level. But the elegant Einstein-de Sitter model is now an unlikely cosmological solution (see also Nakamura & Suto 1995).

6 The future

Assuming that the HST Key Project succeeds in reducing the uncertainty in H_0 to $\delta H/H \lesssim 0.1$, what is the next step ? The Hubble Constant appears to be a quantity which is well defined on large scales. Homogeneity and isotropy suggest that we are measuring a property of the Universe at the current epoch and not some local *ad hoc* parameter. A number of projects therefore suggest themselves.

- Microarcsecond astrometry from space will allows us to measure the trigonometric parallax of the LMC (Bahcall 1990). A parallax based Cepheid PL relation would be a nearer term goal.

- Interferometric measurement of Cepheid angular diameter changes will yield geometrical distances to these standard candles (see Figure 20).

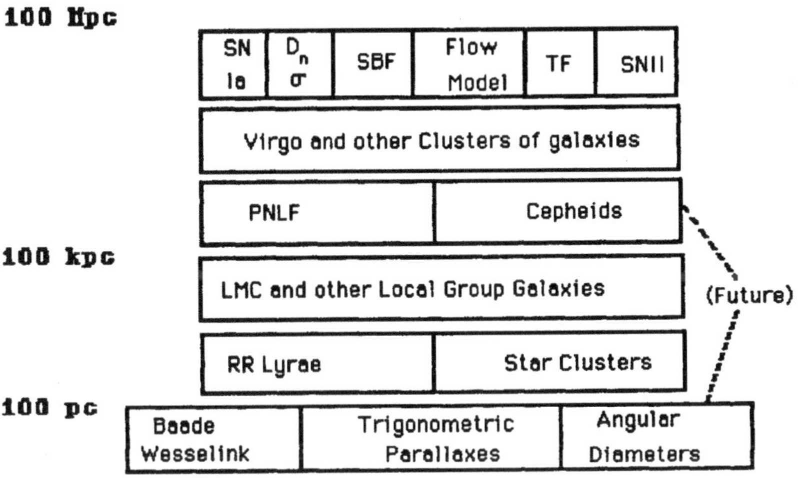

Figure 20: *The extragalactic distance scale from the solar neighbourhood to 100 Mpc.*

- The Key Project distances to galaxies in that program will be capable of improvement with NIC after the second HST servicing mission. This will essentially remove remaining uncertainties arising from extinction.
- Distance measurements which avoid the classical extragalactic distance scale will be developed. Examples include distances from gravitational lensing and from the Sunyaev-Zeldovich effect.
- Detection of Cepheids in the Coma cluster may be possible with Advanced Camera on HST (Madore 1995).

6.1 Gravitational Lenses

Unequal path lengths to multiply imaged QSOs lead to phase lags between flux variations between different images:

$$\delta l = c \quad \delta t$$

In some cases it may be possible to model the lens geometry so that:

$$\delta l = f(l)$$

In principle, one can then solve for the distance of the source. At present the model dependence of this approach has not yielded a strong constraint on H_0 (Bernstein, Tyson & Kochanek 1993).

6.2 Sunyaev Zeldovich effect

The thermal bremstrallung luminosity of the hot gas in x-ray emitting clusters of galaxies is given by:

$$L_x = 1.4 \times 10^{42} n_c^2 r_c^3 T_x^{1/2} \quad ergs/sec$$

where n_c is the number density in the cluster core, r_c is the core radius and T_x is the temperature of the gas.

The Compton effect of electron interactions with the microwave background radiation yields a temperature diminution of the radiation:

$$\delta T/T = -0.0128 n_c r_c \frac{kT_x}{m_e c^2}$$

One may then solve for r_c, whose angular diameter is also observable. A recent example of this work is that of Jones *et al.* (1993). X-ray and radio data of higher angular resolution will provide an interesting test of the appropriateness of the simple model employed here.

6.3 H_2O Megamasers

Miyoshi *et al.* (1993) have shown that H_2O masers orbiting a massive compact object and following a Keplerian rotation curve provide a remarkable good fit to the radial and transverse motions of the components in NGC4258. A model fit to the molecular disk in the centre of the galaxy yields a distance of 6.4 ± 0.9 Mpc. It will be interesting to compare that with a distance from the Cepheid PL relation. There is potential in this technique to achieve higher precision and greater distances.

I wish to acknowledge that the H_0 Key Project is a team effort. Team members are Robert C. Kennicutt, Wendy L. Freedman, Shaun M. Hughes, Barry F. Madore, Robert Hill, Laura Ferrarese, Holland C. Ford, Garth D. Illingworth, Sandra M. Faber, and Daniel Kelson, James E. Gunn, John A. Graham, John G. Hoessel and Mingsheng Han, John P. Huchra, Anne Turner, Paul Harding, Daya Rawson, Abhijit Saha, and Peter B. Stetson. I am grateful to Don Faulkner for a stimulating discussion of this subject at the Harley Wood Winter School of the Astronomical Society of Australia, and to George Jacoby and Brian Schmidt for providing figures reproduced here.

7 References

Aaronson, M. & Mould, J. 1986, ApJ 303, 1.
Aaronson, M., Bothun, G., Mould, J., Huchra, J., Schommer, R., & Cornell, M. 1986, ApJ, 302, 536
Aaronson, M., Huchra, J., Mould, J., Schechter, P., & Tully, R.B. 1982, ApJ, 258, 64
Aaronson, M., Mould, J., and Huchra, J. 1979, ApJ, 229, 1
Ajhar, E. 1993, PhD thesis, MIT.
Bahcall, J. 1990, in Decade of Discovery, Washington: National Academy Press
Baum, W. *et al.* 1995, BAAS, 27, 000
Becker, S., Iben, I. & Tuggle 1977, ApJ, 218, 633
Bernstein, G., Tyson, J.A., Kochanek, C. 1993, AJ, 105, 816
Bingelli, B., Sandage, A., & Tammann, G. 1985, AJ, 90, 1681
Bingelli, B., Sandage, A., & Tammann, G. 1987, AJ, 94, 251
Böhringer, H., Briel, U., Schwarz, R., Voges, W., Hartner, G. & Trümper, J. 1994, Nature 368, 828.
Blakeslee & Tonry, J. 1995, ApJ, 442, 579
Branch, D. & Miller, G. 1993, ApJ, 405, L5
Burstein, D. *et al.* 1987, ApJS, 64, 601
Carroll, S., Press, W. & Turner, E. 1992, Ann. Rev. Astr. Ap. 30, 499
Carney, B. 1979, ApJS, 42, 431
Chiosi, C., Wood, P. & Capitanio 1993, ApJS, 86, 541
Ciardullo, R., Jacoby, G., & Tonry, J. 1993, ApJ, 419, 479
Clementini, G. *et al.* 1995, A&A, in press
Davis, M. & Peebles, P.J. 1983, ARAA, 21, 109
de Vaucouleurs, G. 1961, ApJS, 6, 213.
Della Valle, M. & Panagia, N. 1992, AJ, 104, 696
Djorgovski, S.G. 1995, private communication.
Dopita, M., Jacoby, G. & Vassiliades, E. 1992, ApJ, 389, 27
Dressler, A. & Faber, S. 1990, ApJ, 354, 13
Dressler, A. *et al.* 1987, ApJ, 313, 42
Feast, M. & Walker, A. 1987, ARAA, 25, 345
Fisher, R. & Tully, R.B. 1977, A& A, 54, 661
Freeman, K. 1970, ApJ, 160, 811
Freedman, W. et al. 1994a, Nature, 371, 757
Freedman, W. et al. 1994b, ApJ, 427, 628
Hamuy, M., Phillips, M., Maza, J., Suntzeff, N., Schommer, R., & Aviles, R. 1995, AJ, 109, 1
Harris, W. 1988, in The Extragalactic Distance Scale,eds. S. van den Bergh and C. Pritchet, A.S.P. Conference Series # 4,
Harris, W. and Thompson, L. 1987, ApJL, 315, L29
Hesser, J. 1995, in IAU Symposium 164, in press, ed. P. van der Kruit.

Huchra, J. 1985 in The Virgo Cluster, eds. O. Richter & B. Bingelli, (Garching: ESO), p.181.
Huchra, J. 1988, in The Extragalactic Distance Scale, eds. S. van den Bergh and C. Pritchet, A.S.P. Conference Series # 4, p. 257
Jacoby, G. 1995, private communication.
Jacoby, G. et al. 1992, PASP, 104, 599
Jacoby, G., Ciardullo, R., & Ford, H. 1990, ApJ, 356, 332
Jerjen, H. & Tammann, G. 1993, A& A, 276, 1
Johnson, H. & Mitchell, R. 1958, ApJ, 128, 31
Jones, M. *et al.* 1993, Nature, 365, 320
Kelson, D. et al. 1995, ApJ, submitted.
Kelson *et al.* 1994, BAAS, 26, 963
Kennicutt, R., Freedman, W., & Mould, J. 1995, AJ, in press
Kirshner, R. and Kwan, J. 1974, ApJ, 193, 27
Koo, D. & Kron, R. 1992, ARAA, 30, 613
Lauer, T. and Postman, M. 1994, ApJ, 425, 418
Lightman, & Schechter, P. 1990, ApJS, 74, 831
Lynden-Bell, D. *et al.* 1988, ApJ, 326, 19
Maddox, S. *et al.* 1990, MNRAS, 247, 1P
Madore, B. 1995, private communication.
Madore, B. & Freedman, W. 1990, PASP, 103, 933
Mathewson, D., Buchhorn, M. & Ford, V. 1993, ApJ, 389, L5
Mathewson, D. and Ford, V. 1994, ApJL 434, L39
McCarthy, P. 1992, private communication
Miyoshi, M. *et al.* 1995, Nature, 373, 127
Mould, J. et al. , 1993, ApJ, 409, 14
Mould, J., 1989, in The Use of Pulsating Stars in Fundamental Problems in Astronomy, p.169
Mould, J. et al. 1995, ApJ, in press, August 20 issue.
Nakamuro, T. & Suto, Y. 1995, ApJL, in press.
Pahre, M. & Mould, J. 1994, ApJ, 433, 567
Peebles, J. 1993, Principles of Physical Cosmology, Princeton Series in Physics, (New Jersey: Princeton University Press).
Perlmutter, S. *et al.* 1995, ApJL, in press.
Perryman, *et al.* 1992, A&A, 258, 1
Phillips, M. 1993, ApJL, 380, L23
Pierce, M. Welch, D., van den Bergh, S., McClure, R., Racine, R. and Stetson, P. 1994, Nature, 371, 385
Plait *et al.* 1994, ApJ, 439, 730
Saha, A. & Hoessel J. 1994, AJ, 108, 645
Sandage, A. & Tammann, G. 1993, ApJ, 415, 1

Sandage, A. & Tammann, G. 1995, in Current Topics in Astrofundamental Physics, ed. N. Sanchez, in press.
Sandage, A. et al. 1994, ApJ, 423, L14
Schechter, P. 1980, AJ, 85, 801
Schmidt, B., Kirschner, R., & Eastman, R. 1992, ApJ, 395, 366
Schmidt, B. & Kirshner, R. 1994, AJ, 107, 1444
Schmidt, B. 1995, private communication.
Schmidt, B. et al. 1994, ApJ, 432, 42
Secker, J. & Harris, W. 1993, AJ, 105 1358
Steigman, G. & Tosi, M. 1992, ApJ, 401, 150
Tammann, G.. 1988, in The Extragalactic Distance Scale, eds. S. van den Bergh and C. Pritchet, A.S.P. Conference Series # 4
Tammann, G. & Sandage, A. 1995, submitted to ApJ.
Tonry, J. 1991, ApJ, 373, L1
Tonry, J. & Schneider, D. 1988, A.J. ,96, 807
Tonry, J., Ajhar, E., & Luppino, G. 1990, AJ, 100, 1416
van den Bergh, S. & Pazder, J. 1992, ApJ, 390, 34
VandenBerg, D. 1991, in *The Formation & Evolution of Globular Clusters*, Proc. A.S.P. 13, 183
Vandenberg, D. & Poll, H. 1989, AJ, 98, 1451
van den Bergh, S. & Pazder, J. 1992, ApJ, 390, 34
Walker, A. 1993, AJ, 106, 999
Woosley, S. & Weaver, T. 1994, ApJ, 423, 371
Worthey, G. 1993, ApJ, 409, 530

TOPOLOGICAL DEFECTS

STEPHEN POLETTI
Department of Physics and Mathematical Physics,
University of Adelaide, S.A. 5005, Australia.

CONTENTS

1. Introduction

A fascinating and wonderful feature of gauge theories is the topological objects (defects) associated with spontaneous symmetry breaking. In these lectures I hope to convey to the reader a flavour of the lovely interplay between mathematics and physics that the study of these strange objects entails. Surprisingly it is quite possible that much of the structure of the known universe may have resulted from the formation of a type of topological defect known as a cosmic string. In the later lectures I will explain the cosmic string scenario for the generation of the primordial perturbations, which are responsible for the large scale structure of the universe. There are a number of other possible experimental imprints which could be generated by cosmic strings and other topological defects which we will also consider.

Loosely speaking we can think of topological defects as regions of false vacuum trapped in the true vacuum due to the non trivial topology of the vacuum manifold. It is expected that in the early universe there were several phase transitions each of which could have resulted in the formation of topological defects, which may be all quite different in character. The different kinds of topological defects that can form include point like defects known as monopoles, string like defects (cosmic strings), and domain walls which are sheet like defects. However it is possible to have even more exotic objects such as textures[1] (which are like three dimensional knots), superconducting cosmic strings[2], cosmic springs[3], cosmic strings attached to monopoles[4] , semi-local strings[5], cosmic strings attached to black holes[6] and many more weird and wonderful things dreamed up by theorists with too much time on their hands! Unfortunately we won't have time to discuss all these strange objects, however hopefully, after reading these notes the reader will be inspired to go away and investigate them in more detail.

In preparing these lectures I have relied heavily on the excellent reviews by Vilenkin[7], and Hindmarsh and Kibble[8] I am also told that the recent book by Shellard and Vilenkin[9] is well worth looking at. Other useful references include Ryder's[10] book on Quantum Field Theory and the collection of Coleman's[11] lectures as well as the book by Kolb and Turner[12].

2. What is a Topological Defect?

Consider a real scalar field ϕ which has Hamiltonian density

$$\mathcal{H} = \frac{1}{2}(\nabla\phi)^2 + V(\phi) \tag{2.1}$$

with

$$V(\phi) = \frac{1}{8}\lambda(\phi^2 - a^2)^2 \tag{2.2}$$

where λ and a are real parameters. The important thing to notice about the potential (plotted in fig. 1) is that the vacuum state is degenerate. The minimum energy states are those for which the scalar field ϕ is constant with the value $\phi = \pm a$. As we shall see later, it is the degeneracy of the vacuum which signals the possible existence of topological defects.

Suppose that initial conditions are such that in one region of space the field ϕ is the the 'vacuum' state $\phi = -a$ and in another region of space the field is the 'vacuum' state $\phi = +a$. We shall see later that configurations such as this arise quite naturally in the early universe. In figure 2 the setup is sketched schematically. What happens in between the two regions?

Well, we know that along any path between the two regions such as the one indicated on the diagram the field ϕ must go through zero at least once, as indicated in figure 3. Although we don't know the exact shape of the field profile along the chosen path in figure 2, it should qualitatively look like the configuration sketched in figure 3.

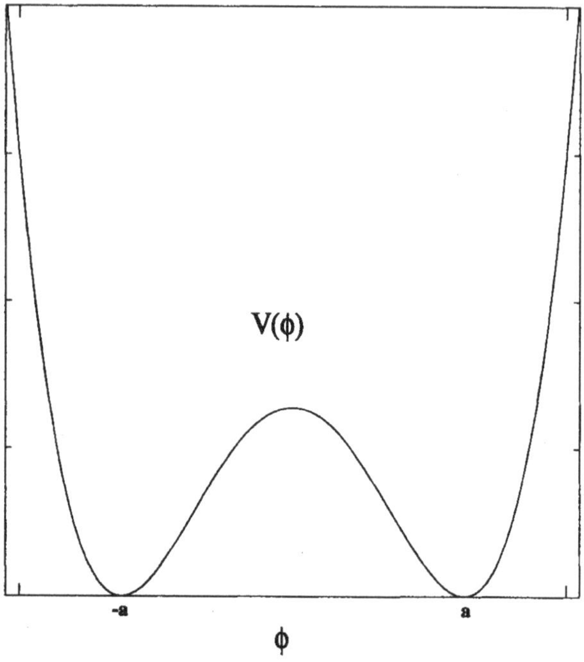

Fig. 1: Potential $V(\phi)$ for the real scalar field ϕ.

In the transition region the field will not be constant which means that the kinetic energy $(\nabla\phi)^2 > 0$ and the potential energy $V(\phi) > 0$. Hence the energy density is greater than zero in the transition region. One can imagine extending this simple 2-D example to 3-D using continuity arguments. The picture then would be a 2-D sheet where the energy density was greater then zero which would be a domain wall. The important point here is that the field cannot relax into the lower energy true vacuum state without the input of a large amount of energy to lift the field in one of the minima of the potential over the potential barrier into the other minimum. If both of the vacuum regions have infinite area then we would need an infinite amount of energy!

Problem: If we restrict ourselves to one space dimension find the shape of the domain wall. (Answer $\phi = a \tanh\left[a\sqrt{\frac{\lambda}{8}}(x - x_0)\right]$, where x_0 is the position of the wall on the x axis.)

Problem: Calculate the energy per unit area of a domain wall.

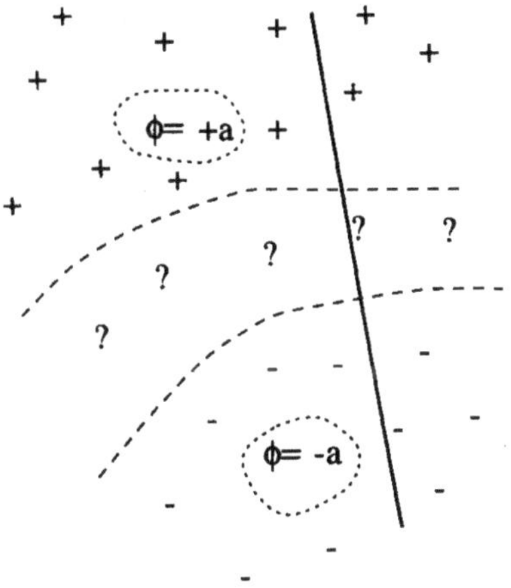

Fig. 2: Two dimensional slice with the scalar field ϕ varying between the $\phi = +a$ vacuum state and the $\phi = -a$ vacuum state. We would like to know what happens in the transition region specified by question marks.

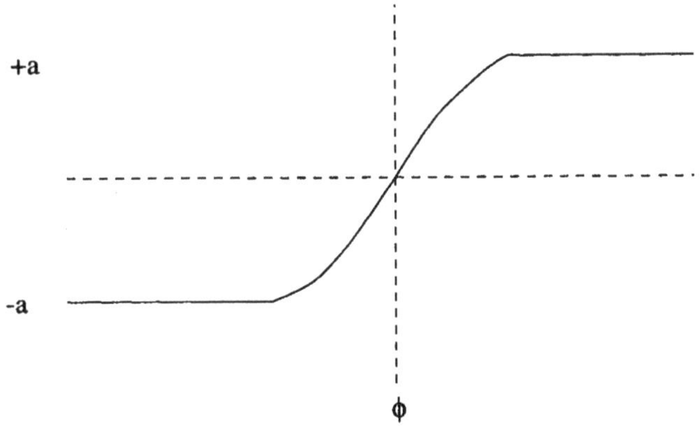

Fig. 3: Profile of the scalar field between the two regions illustrated in figure 2.

3. Nielsen-Olesen Vortex

As well as wall type structures the presence of degenerate vacuum states can also give rise to string like defects. Consider now a *complex* scalar field with Lagrangian

$$\mathcal{L} = \frac{-1}{2}\nabla_\mu\phi^*\nabla^\mu\phi - V(\phi) \tag{3.1}$$

with a "mexican hat" potential (see figure 4)

$$V(\phi) = \frac{1}{2}\lambda(|\phi|^2 - a^2)^2 \tag{3.2}$$

The vacuum state is again degenerate, but this time the minimum potential energy configurations lie on a circle,

$$\phi = ae^{i\alpha} \tag{3.3}$$

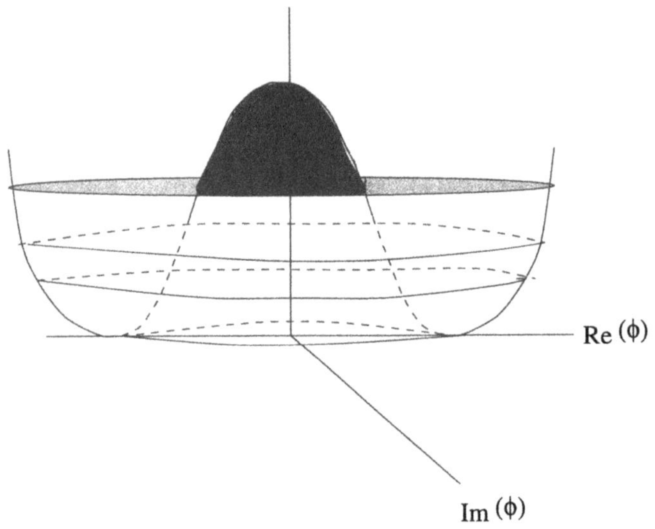

Fig. 4: Mexican Hat Potential: The minimum potential energy configurations lie on the circle $|\phi| = a$.

Note that if the field ϕ is in the "vacuum" state $\phi = -a$ in one region and is in the "vacuum" state $\phi = +a$ in another region we cannot now conclude that there is a domain wall in between as above, the reason being that the field can change continuously from $\phi = -a$ to $\phi = +a$ by rotating through the complex plane, staying always in a state of minimum potential energy. We shall see shortly however, that by arranging the initial conditions correctly it is possible to trap the field away from the minimum of the potential in a localised stringlike region. The basic idea is very simple. If we can draw a large circle in space with the complex phase vectors all pointing outwards, then we know that somewhere in the middle the field must pass through $\phi = 0$, which implies that there will be potential and gradient contributions to the energy density, or in other words a vortex configuration (cosmic string).

Let's see how this works in more detail. We will assume for the moment that the scalar field is living in 2-D. Consider a large circle (at infinity) of radius R. Imagine that the field on the boundary is specified to be

$$\phi = ae^{in\theta} \quad \text{for polar coordinates } (r, \theta) \tag{3.4}$$

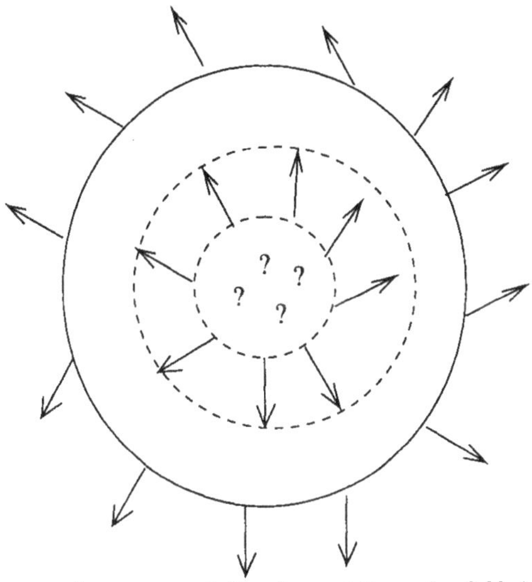

Fig. 5: The arrows here represent the phase of the scalar field ϕ. If the initial conditions are set up so that we can draw a circle in space and the phase of the complex scalar field wraps around it at least once then somewhere inside, the field will be forced away locally from the minimum of the potential energy.

Since ϕ should be single valued and continuous, n has to be an integer. We are interested in solutions to the field equations which will match on to the specified boundary condition at infinity. Hopefully we will find something which looks somewhat like the qualitative picture of figure 5. Note that the field ϕ on the boundary defines a map from the circle at infinity $S_1 \to U(1)$ where $U(1)$ is the vacuum manifold for the scalar field. Since $U(1)$ is the same as S_1 then we have a map from $S_1 \to S_1$. Explicitly the map f say, can be written $f : (R, \theta) \to$ phase of ϕ. On the circle at infinity the field sits at the minimum of the potential but because of the twist there will be somewhere inside where the field moves away from the minimum. If

$$\phi = ae^{in\theta} \quad R \to \infty \tag{3.5}$$

then the gradient of ϕ is in polar coordinates (r, θ)

$$\nabla\phi = \left(0, \frac{(inae^{in\theta})}{R}\right) \quad R \to \infty \tag{3.6}$$

Until now we have been neglecting the gradient energy, since it seems clear that locally at least the gradient energy density will be small as $R \to \infty$. However there is a catch. Namely the *total* gradient energy diverges as $R \to \infty$. To see this consider the energy density for a static configuration $\mathcal{H} = \frac{1}{2}|\nabla\phi|^2 + V(\phi)$. As $R \to \infty$, $V(\phi) = 0$ but the kinetic energy contributes since $\mathcal{H} = \frac{1}{2}|\nabla\phi|^2 = \frac{n^2a^2}{2R^2}$.

Therefore the total energy which is the integral over all of space of the energy density diverges as $E \approx n^2 a^2 \int \frac{1}{r} dr$ i.e. the total energy is infinite! Since the idea was initially that as $R \to \infty$ the field should be in the "vacuum" state things look pretty grim. Whatever such a solution represents it's certainly not a localised string type configuration.

The good news though is that it's possible to add a gauge field A_μ to "cancel" out the divergences. Such a theory has the Lagrangian

$$\mathcal{L} = -\frac{1}{2}(D_\mu \phi)^* D^\mu \phi - V(\phi) - \frac{1}{4} F_{\mu\nu} F^{\mu\nu}$$

where the field strength is defined in the usual way and the covariant derivative is $D_\mu \phi = \nabla_\mu \phi + ieA_\mu \phi$.

The gauge field A_μ can now be chosen to "cancel" the dangerous derivative term. Specifically if

$$\mathbf{A} = -\frac{1}{e}\nabla(n\theta) = (0, \frac{-n}{eR}) \quad \text{as} \quad R \to \infty \tag{3.7}$$

then as $R \to \infty$

$$D_\theta \phi = \frac{1}{R}\partial_\theta \phi + ieA_\theta \phi = 0 \quad \text{and} \quad D_r \phi = 0 \tag{3.8}$$

so the energy functional for large R

$$\mathcal{H} = \frac{1}{2}|\mathbf{D}\phi|^2 + V(\phi) + \frac{1}{4}(\mathbf{B}^2 + \mathbf{E}^2) \to 0 \tag{3.9}$$

(Note that the electric and magnetic field components do not contribute for large R since A_μ is pure gauge as $R \to \infty$.) So it looks like it's at least possible that the specified boundary conditions could match onto a finite energy solution of the field equations. An interesting feature of such a solution is that it would carry magnetic flux. Consider the line integral $\int \mathbf{A} \cdot d\mathbf{l}$ which from Stokes theorem is equal to the magnetic flux $\Phi = \int \mathbf{B} d\mathbf{S}$. It follows immediately from equation (3.7) that the magnetic flux would be $\Phi = -2\pi n/e$. By embedding the two dimensional example in three dimensions in the obvious way it can be seen that the boundary conditions that we have specified match onto solutions that carry tubes of magnetic flux.

To show that such solutions do exist is non-trivial, and was first demonstrated by Nielsen and Olesen[13]. The first thing to do is to write down the field equations,

$$\begin{aligned} D^\mu (D_\mu \phi) &= 2\lambda(|\phi|^2 - \frac{1}{2}a^2)\phi \\ ie(\phi^* D_\mu \phi - \phi^* \overleftarrow{D}_\mu \phi) &= 2\nabla^\nu F_{\mu\nu} \end{aligned} \tag{3.10}$$

It turns out that it is not possible to solve these equations for our specified boundary conditions in closed form, however it is possible to find either an approximate analytical solution or use numerical techniques. The solution is sketched in figure 6 where the z-component of the magnetic field is denoted $B(r)$ and the scalar field is taken to have the form $\phi(r, \theta) = \chi(r)e^{in\theta}$. For typical parameter values it can be characterised as "a tube of magnetic flux inside a tube of false vacuum". In the cosmological context such configurations are examples of cosmic strings.

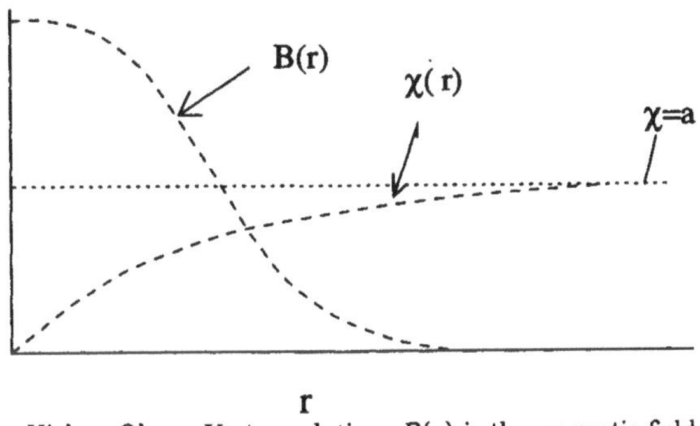

Fig. 6: The Nielsen-Olesen Vortex solution. $B(r)$ is the magnetic field whilst $\chi(r)$ is the modulus of the scalar field.

Well so far there has been no mention of the word topology. Why are solutions, such as the domain wall discussed in the previous section and the Nielsen-Olesen vortex, known as topological defects? We shall see shortly that it is related to questions of stability. A cosmic string configuration is prevented from relaxing to the true vacuum state for topological reasons. Recall that earlier we defined a map f from the circle at infinity to the gauge group $U(1) \equiv S_1$. Although it is not proved here it seems intuitively clear that the number of "twists" n in the phase of the scalar field as we go around the circle at infinity characterises different solutions and that solutions with a different number of twists in the field cannot be deformed into each other continuously. For example to go from a $n = 0$ true vacuum solution to the $n = 1$ solution indicated schematically in figure 5 would involve simultaneously changing the phase by a macroscopic amount over an infinite region (which would of course take an infinite amount of energy). Thus solutions characterised by a particular n cannot decay in time to a solution with a different n. In this sense they are stable. Another way of stating this using the language of Algebraic Topology is to say that the first homotopy group of $U(1)$ is nontrivial ($\Pi_1(U(1)) = Z$). So to check whether or not a particular theory admits cosmic string solutions we need to examine the first homotopy group of the vacuum manifold. (What I've said here is not in fact strictly true! It *is* possible given certain conditions to have a kind of cosmic string called a Semi-Local string[5] even if the first homotopy group is trivial, however this is outside the scope of these notes.)

4. Formation of Cosmic Strings

The Lagrangian defined in equation (3.1) defines a theory which can undergo *spontaneous symmetry breaking*. As the temperature is increased the effective potential acquires temperature dependent corrections as illustrated in the figure. For temperatures greater then T_c such as would have been found in the early universe the "vacuum state" is $<\phi>= 0$.

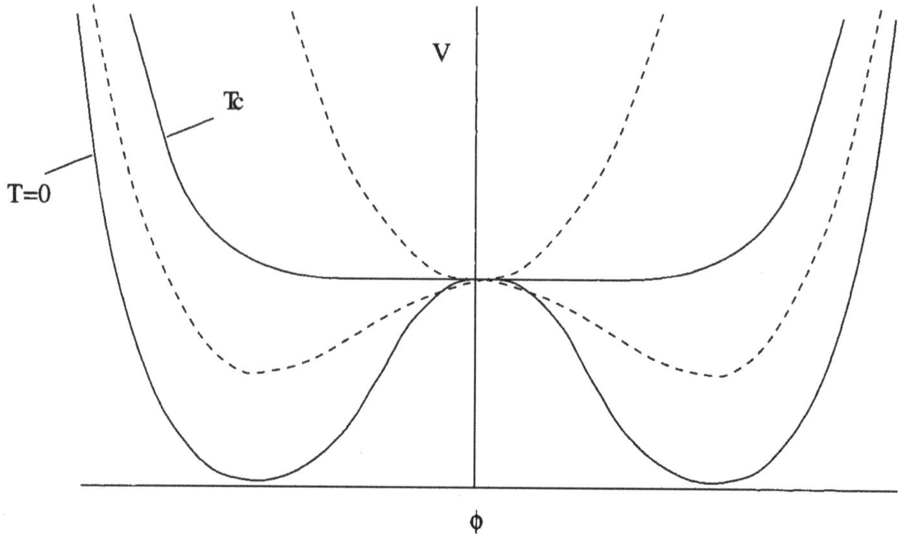

Fig. 7: $T \neq 0$ Potentials: For $T > T_c$ the temperature corrected potential has a minimum at $\phi = 0$.

As the universe cools below the critical temperature T_c the field is forced to choose a vacuum phase $< \phi >= ae^{i\alpha}$. For nearby points the phase will be more or less the same; however for areas of the universe far away from each other the relative phases will be uncorrelated. Thus we can imagine going around a large loop in space and tracing the phase of the field out to see whether or not it is twisted. Eventually, if we make the loop large enough, we will find a loop with non trivial winding number which, from the previous section, we know signals the presence of a cosmic string somewhere inside the loop. The cosmic strings formed after the phase transition cannot have ends, which means that they will either be infinitely long or form closed loops. Monte Carlo[14] simulations indicate that most of the length of the resulting string network consists of long strings (about 80 precent of the total length of the string network) with numerous small cosmic string loops accounting for the remainder of the network.

5. Monopoles and Domain Walls

Imagine now a sphere S^2 at large R. A monopole is a point like defect, so to trap the field in a false vacuum state at a point we need a theory with a vacuum manifold $\mathcal{M}$ such that the field in the "vacuum state" can have a winding number on the sphere S^2 at infinity.

For example we can take our field to be $O(3)$ isovectors $\phi^a \equiv (\phi^1, \phi^2, \phi^3)$ with the Lagrangian

$$\mathcal{L} = -\frac{1}{2}(D_\mu \phi)^a D^\mu \phi^a + \frac{m^2}{2}\phi^a\phi^a - \lambda(\phi^a\phi^a)^2 - \frac{1}{4}F^a_{\mu\nu}F^{\mu\nu a} \tag{5.1}$$

The field strength for the nonabelian field A_μ takes the usual form

$$F^a_{\mu\nu} = \partial_\mu A^a_\nu - \partial_\nu A^a_\mu + e\epsilon^{abc}A^b_\mu A^c_\nu \tag{5.2}$$

while the gauge derivative is given by

$$D_\mu \phi^a = \partial_\mu \phi^a + e\epsilon^{abc}A^b_\mu \phi^c \tag{5.3}$$

The vacuum states are given by

$$(\phi^1)^2 + (\phi^2)^2 + (\phi^3)^2 = \frac{m^2}{2\lambda} \tag{5.4}$$

Thus the vacuum manifold is S^2, i.e. the isovectors that satisfy equation (5.4) can be identified with 3-D vectors of length $m^2/2\lambda$.

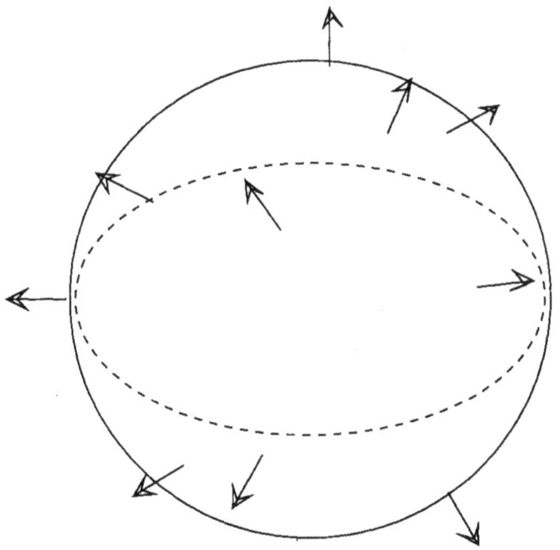

Fig. 8: The 'hedgehog' configuration for the isoscalar field ϕ as $r \to \infty$. It should be remembered that the arrows strictly speaking live in isospace.

By identifying isospace with space itself we can specify a configuration which is "radial", or "hedgehog" like, as illustrated in figure 8. Explicitly we have for the scalar field $\phi^a = \sqrt{(\frac{m^2}{2\lambda})}r^a/r$. The gauge field can be chosen, as in the vortex solution, so that the covariant derivative of the gauge field vanishes at infinity which will ensure that the total energy is not infinite;

$$A_i^a = \epsilon_{iab}\frac{r^b}{er^2} \quad A_0^a = 0 \quad r \to \infty \tag{5.5}$$

It is easily checked that the covariant derivative of the scalar field does vanish as $r \to \infty$. The time derivative vanishes trivially whilst the spatial derivative is

$$D_i\phi^a = \sqrt{\left(\frac{m^2}{2\lambda}\right)}\left[\partial_i\left(\frac{r^a}{r}\right) + e\epsilon^{abc}A_i^b\left(\frac{r^c}{r}\right)\right] = 0 \tag{5.6}$$

It can also be shown that the electromagnetic field does not vanish at infinity and in fact has the form of a radial magnetic field. The interested reader might like to consult the excellent account of the t'Hooft-Polyakov Monopole in Ryder's[10] book for the details. As above it also needs to be checked that there is a solution to the field equations with these boundary conditions, and that such a solution is stable. To check for stability we need to know if the specified boundary conditions are "twisted" on the sphere at large radius. For cosmic strings we examined maps from the circle at large radius into the vacuum manifold $U(1) \equiv S_1$. In a similar way here, we would need to consider maps from S^2 into the vacuum manifold $O(3) \equiv S^2$ to see if they are non trivial. In other words the condition for the existence of monopoles is that the second homotopy group of the vacuum manifold $\mathcal{M}$ is non trivial, i.e. monopoles exist in general if $\Pi_2(\mathcal{M}) \neq 0$.

The condition for the existence of domain walls is that there should be (as seen in the introduction) a *discrete* set of vacuum states. In these lectures we will concentrate on cosmic strings since these have potentially important cosmological consequences. On the other hand domain walls and monopoles are in general disastrous for cosmology!

6. Physical Description of Cosmic Strings

Hopefully by now it should be clear from a mathematical point of view what a cosmic string is. How though would we recognize one if we fell over it?

The following is a brief "users guide" to cosmic strings

• Cosmic strings are linear defects formed during phase transitions in the early universe. The phase transitions can occur at various different scales giving rise to strings with different properties.

• In the standard electroweak theory there are no cosmic strings, however simple generalisations (eg. SUSY) of it do give rise to cosmic strings.

• Strings arising from phase transitions at the GUT scale ($\approx 10^{16} - 10^{15}$ GeV) are extremely interesting from the cosmological point of view. (GUT = Grand Unified

Theories. It has been suggested that above the GUT scale the electroweak theory and QCD are unified. When the universe cools below the GUT scale there is spontaneous symmetry breaking and an associated phase transition and the electroweak theory wanders away from QCD until at a much lower energy another phase transition takes place and the weak and the electromagnetic fields take on different characteristics.)

• For many purposes cosmic strings can be characterised by their mass/unit length μ and the string tension which is also μ. In other words its possible to represent straight strings on a macroscopic scale by an effective stress energy tensor $T_\mu{}^\nu = \text{diag}(\mu, 0, 0, \mu)\delta(x)\delta(y)$, assuming say that the string is aligned along the z-axis. The string tension will cause a loop of cosmic string to try and shrink, and will give rise to oscillations which can release gravitational radiation. For a straight string there will be waves up and down the string. In general parts of the string will be moving at speeds close to the speed of light.

• Cosmic strings in general are *very* thin e.g. a GUT string will have a characteristic radius of $\approx 10^{-31}$ m!

• Although the radius is in general extremely small cosmic strings are *very* massive. A GUT string typically has mass per unit length $\mu \approx 1 \times 10^{21}$ kg m^{-1}. If its length was equal to that of the solar system diameter it would be as massive as the sun. On the other hand an electroweak string would only weigh 10 mg.

• Despite being so massive straight strings exert no gravitational force! Recall that for a static mass distribution if

$$T^\nu{}_\mu = \text{diag}(\rho, -P_1, -P_2, -P_3) \tag{6.1}$$

the Newtonian limit for the gravitational potential ϕ from Einstein's equations is

$$\nabla^2\phi = 4\pi G(\rho + P_1 + P_2 + P_3) \tag{6.2}$$

where P_i is the matter pressure along direction i, ρ is the matter density and G is Newton's constant. For nonrelativistic matter $P_i \ll \rho \Rightarrow \nabla^2\phi \cong 4\pi G\rho$ which is of course Poisson's equation. For strings, though, $T^\mu_\nu = \text{diag}(\mu, 0, 0, \mu)\delta(x)\delta(y)$ and the appropriate limit is $\nabla^2\phi \cong 0$! (However string loops do have an effective mass due to the stress associated with bending.)

• Even though space is flat outside a straight string there are important gravitational effects characterised by $G\mu$ which for GUT strings is $\sim 10^{-6}$. We shall see in detail later how this arises and its implications.

7. Evolution of the String Network

Recall that in the Robertson-Walker universe the metric takes the form

$$ds^2 = -dt^2 + R^2(t)d\Omega^3. \tag{7.1}$$

The matter density scales with "volume" as $\rho_m(t) \sim 1/R^3$ and the radiation density scales as $\rho_r(t) \sim 1/R^4$, the extra factor of $1/R$ coming from the redshift.

An important question to consider is how the density of topological defects scales with the expansion of the universe. For monopoles or point defects the density scales like ordinary matter:

$$\rho_{monopole} \sim \frac{1}{R^3}. \tag{7.2}$$

Since the early universe was radiation dominated it follows that $\rho_{monopole}/\rho_r \sim R$. Thus if a sizable fraction of the universe is initially made up of monopoles (which would be expected for theories which give rise to monopoles) the universe would very shortly be dominated by monopoles. This is known as the "monopole problem" and was the main reason why Alan Guth invented inflation (to get rid of monopoles)! This is also the main reason why we don't consider monopoles much here. However, it is logically possible that whatever mechanism is responsible for solving the monopole problem leaves a very small residual flux of monopoles. For a detailed analysis of the observational consequences of such a flux the interested reader should consult Kolb and Turner's book on the early universe[12].

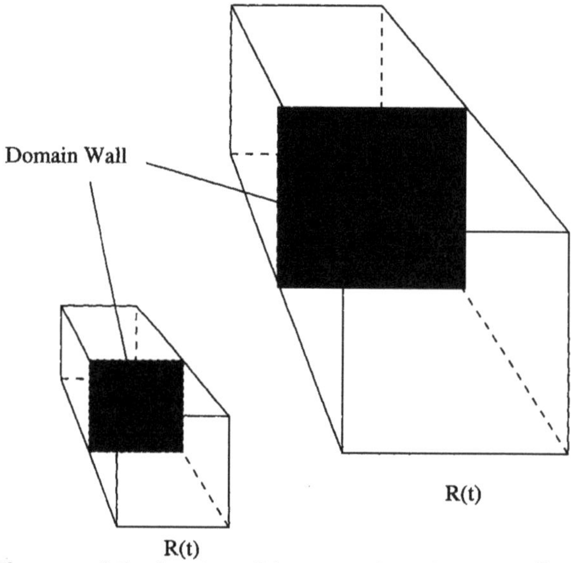

Fig. 9: The mass of the domain wall in a comoving volume is σR^2. The figure illustrates a comoving volume with a domain wall stretched across it at two different times.

For domain walls the disaster seems even more serious. Domain walls are characterised by mass per unit area σ. As indicated in the diagram the mass in a comoving volume will scale as σR^2, hence the ratio of the matter density in domain walls to the radiation density is $\rho_{DW}/\rho_r \sim R^3$ which is quite clearly a disaster!

What about cosmic strings? An initial estimate is that the mass in a cubic volume with sides of length l of a straight cosmic string would be μl, for μ the mass per unit length of the string. Hence the mass in a comoving volume is proportional to μR^2 and the matter density of strings scales as $\rho_{string} \sim 1/R^2$, which is clearly another disaster! However, as we shall see shortly, if we work a bit harder we can find the factor of R^2 which we need to make strings interesting from a cosmological point of view.

Recall that after the phase transition the typical pattern is a random network of strings which trace out a random walk, i.e. they are very wiggly. Numerical simulations indicate that most of the mass of cosmic strings is in long strings (about 80%) with the remainder in smaller string loops. Therefore instead of considering straight strings we really need to examine how a wiggly string scales with the expansion of the universe. Clearly if the wiggles are small compared to the horizon length the amplitude will stay about the same (governed by the local physics), whilst the wavelength will be stretched by the expansion of the universe, which causes the string density to decrease more rapidly than simple scaling arguments indicate.

It is useful to consider the above arguement slightly more quantitatively and introduce the string equations of motion which, as well as describing the evolution of strings in the universe, can be used to analyse string oscillations. The dynamics of a string are described to a good approximation by the Nambu action

$$S = -\mu \int dA = -\mu \int [-g^{(2)}]^{\frac{1}{2}} d^2\zeta \tag{7.3}$$

where dA is the invariant area swept out by the motion of the string. If the string travels along a trajectory $x^\mu(\zeta)$ in the background spacetime with metric $g_{\mu\nu}$, where ζ^i for $i = 0, 1$ are the string coordinates then

$$g^{(2)} = \det\left[\frac{\partial x^\mu}{\partial \zeta^i}\frac{\partial x^\nu}{\partial \zeta^j} g_{\mu\nu}\right] \tag{7.4}$$

The Nambu action has a very simple interpretation – the string tries to minimize the area of its world sheet as it evolves dynamically.

In a Robertson Walker universe the metric can be written in terms of conformal time as

$$ds^2 = R^2(\tau)\left(-d\tau^2 + dx^2 + dy^2 + dz^2\right)$$

It is possible to choose $\zeta_0 = \tau$ and $\zeta_1 = x$. If we assume that the string is restricted to the x-y plane, then the position of the string can be described by one function $y(x, \tau)$, where of course x and y are comoving co-ordinates. The equation of motion derived from the action becomes

$$\left(\frac{\partial}{\partial \tau} + 2\frac{\dot{R}}{R}\right)\left[\dot{y}(1 + y'^2 - \dot{y}^2)^{-1/2}\right] = \frac{\partial}{\partial x}\left[y'(1 + y'^2 - \dot{y}^2)^{-1/2}\right] \tag{7.5}$$

where a dot means partial derivative with respect to conformal time while the prime is the spatial derivative. The first thing to notice is that $y =$ const. is a solution, so a straight string will remain straight. We wish to consider small perturbations away from this solution. If y'^2 and $\dot{y}^2$ are $\ll 1$, then in a radiation dominated universe $(R \sim \tau)$ equation (7.5) becomes

$$\ddot{y} + 2\tau^{-1}\dot{y} - y'' = 0 \tag{7.6}$$

Plane wave like solutions of the form $y(\tau)e^{ikx}$ can be found with $y(\tau) = A\tau^{-1}\sin(k\tau)$. Recall that for a radiation dominated universe the horizon h scales like τ^2. Since x and y are comoving co-ordinates the physical wavelength goes as $\lambda = 2\pi R(\tau)/k$ and the ratio of the horizon size to the physical wavelength is $\sim k\tau$. If the wavelength of the string is much greater than the horizon length $(k\tau \ll 1)$, and the amplitude $A \gg \tau^2$, then the prefactor $y(\tau)$ reduces to a constant (i.e. in the limit that $k\tau \to 0$, $\sin(k\tau) \sim k\tau$ which implies that $y(\tau) \sim$ const.), which means that the string is being conformally stretched with the expansion and its shape remains the same.

On the other hand if the string scale is much smaller than the horizon scale, ($k\tau \gg 1$ and $A \ll \tau^2$), the physical wavelength will still be proportional to the scale factor, however the physical amplitude will remain constant. Thus the string becomes less and less wiggly, supporting the qualitative analysis above. The analysis here indicates that the energy density of a wiggly string grows more slowly compared to that of a string which is being conformally stretched. A much more important effect however is the role of string loops.

Although most of the mass of the initial string network is found in the long "straight" strings there is still a large number of string loops. Loops which are much smaller than the horizon size will want to collapse due to the string tension, however this can only happen if there is an energy loss mechanism. The idea is that the loops will oscillate, which will result in gravitational radiation being emitted. It can be shown (see below) that small string loops rapidly decay, converting their energy into harmless gravitational radiation. If that was the whole story then this energy loss mechanism wouldn't be much use since we'd very quickly be left with long "straight" strings which would rapidly dominate the energy density of the universe. However new loops can form through self intersections or when two strings cross , which will then radiate away and so on. Thus part of the tangled network of long strings and large loops is converted to small loops which disappear (an energy loss mechanism). Is this enough to make strings interesting from a cosmological point of view? The answer here can only be found through numerical simulations[15,16,17], although much can be learned from various analytic approximations (see ref 8 for a summary). The idea is to define the various scales that are important for the string evolution. The most important scale is the characteristic length scale of the network ζ, which is defined as the length such that within any volume ζ^3 there would be on average a length ζ of string, or to put it in another way the string density is μ/ζ^2. Another important scale $\overline{\zeta}$ can be thought of as representing how wiggly the string is, ie. the average distance between large angle kinks on the string. The analytic approach is to try and write equations describing how the various processes indicated (the Hubble expansion, intercommuting, and loop production) affect the different length scales. The analytic work seems to show that the string density *does* scale with the radiation density. The numerical work also indicates that the density of long strings scales as $1/R^4$ in the radiation dominated era. The answer is tentative however, because the simulations can only be run for about 20 Hubble times so any $\log(R)$ dependence would not be seen.

So far we've seen that monopoles and domain walls are a disaster from the cosmological point of view, the reason being that a sizable fraction of the initial mass density of the universe would initially be in the form of these topological defects, which would then end up being almost all the energy density of the universe, which is clearly far from the case at present! Since GUT theories generically give rise to these kinds of topological defects we are left with something of a problem. It is possible to invoke inflation to get rid of dangerous topological defects, however one is then left with the problem that inflation would also get rid of cosmic strings! All is not lost, since it is possible to arrange things so that cosmic strings form at a phase transition which occurs after the inflationary phase. For example[18] one could add a "mass" term to the scalar field ϕ responsible for the cosmic string, of the form $\zeta R|\phi|^2$ where R here stands for the Ricci scalar. As $R \to 0$ at the end of the inflationary era the phase transition occurs. Alternatively[19] if a potential term of the form $\lambda(|\phi|^2 + v^2 - \alpha\sigma^2)^2$ is added to the action, where σ is the inflating field which initially starts at $\sigma = 0$ and gradually increases during inflation, then a phase transition will occur at $\sigma \sim \sqrt{\alpha}v$, again towards the end of the inflationary era.

In conclusion we can see that cosmic strings are potentially very interesting from the cosmological point of view (especially GUT strings), since in the radiation dominated early universe a constant fraction of the energy density of the universe would reside in cosmic strings which could well be important for galaxy and structure formation in the early universe. If cosmic strings did form at the GUT scale in the early universe then it is relatively straightforward to estimate how many are around today. It turns out[8] that within a redshift of ~ 0.3 there would be a few hundred loops of typical size of the order of hundreds of kiloparsecs, which for GUT strings would imply a mass comparable to a large galaxy!

8. Observational Consequences of Cosmic Strings

For the moment we will concentrate on cosmic strings that may have formed at the GUT scale since these are the most interesting "realistic " strings from the cosmological perspective. For a GUT string the energy scale is $\sim 10^{15}$-10^{16} GeV and the dimesionless parameter $G\mu = (T_c/m_{planck})^2$ is somewhere in the range 10^{-6}-10^{-7}. The most important observational consequences of such strings are gravitational and are governed by this parameter.

The observational effects are in principle quite testable. A short list would include

- Gravitational lensing.
- Density perturbations of the microwave background.
- Density perturbations and galaxy formation scenarios.
- Gravitational Waves.
- Cosmic Rays.

8.1. Gravitational Lensing

The stress energy tensor for a straight cosmic string can be approximated by an effective stress energy tensor distribution

$$T_\nu{}^\mu = \mu\delta(x)\delta(y)\text{diag.}(1,0,0,1) \tag{8.1}$$

where the string is taken to be aligned along the z axis. It is straightforward to verify that the effective stress energy tensor does indeed have this form. Clearly for a string aligned along the z-axis the cosmic string solution to the field equations will depend only on x and y and not on z or t, which implies that $T_t{}^t = T_z{}^z$. This follows immediately by inspecting the field theoretic expression for the stress energy tensor of the coupled scalar gauge field system which gives rise to the cosmic string. It is also clear that the only possible other non-zero components are $T_i{}^j$ where i, j can only take the values x or y. Conservation of the stress energy tensor implies that

$$\nabla_\mu T^\mu_\nu = 0$$

$$\Rightarrow \int \partial_j T^j{}_i x^k dx dy = 0 \Leftrightarrow \int T^k_j = 0 \tag{8.2}$$

and hence the result. Previously it was asserted that the Newtonian limit of GR

$$\Rightarrow \nabla^2\phi = 4\pi G(\rho + \sum P_i) = 0 \tag{8.3}$$

for a cosmic string. We can now see directly from the distributional limit of the stress-energy tensor that this is indeed true (at least if the Newtonian approximation is a reasonable one, which will be true for $G\mu \ll 1$). It turns out however that there is a fascinating twist, since the above argument only shows that space is *locally* flat, i.e. that the metric can be written in the form

$$ds^2 = -dt^2 + dr^2 + dz^2 + A^2r^2d\theta^2 \tag{8.4}$$

Clearly such a metric is indeed locally flat since under the co-ordinate transformation $\tilde{\theta} \rightarrow A\theta$ the metric becomes

$$ds^2 = -dt^2 + dr^2 + dz^2 + r^2d\tilde{\theta}^2 \tag{8.5}$$

However such a metric is not globally equivalent to Minkowski space since the angular variable $\tilde{\theta} \in [0, 2\pi A]$ which for $A < 1 \Rightarrow$ a conical structure for spacetime, in other words flat spacetime with a wedge removed. To calculate A we will use the Gauss-Bonnet Theorem[20] in 2-D which says that for a 2-surface Σ with gaussian curvature

$$K = R^{xy}{}_{xy} = R^{x}{}_{x} = R^{y}{}_{y} \tag{8.6}$$

we have that

$$\int_{\Sigma} K.dS = 2\pi - \int_{\partial\Sigma} k_g.dl \tag{8.7}$$

Here k_g is the geodesic curvature obtained by parallel transporting a vector around the boundary $\partial\Sigma$. Since $k_g = 2\pi A$ and Einstein's equations tell us that

$$R_x^{\ x} = 8\pi G(T_x^{\ x} - \frac{1}{2}T) = 4\pi G\delta(x)\delta(y)\mu$$

it follows that $A = (1 - 4G\mu)$. Hence the cosmic string metric is

$$ds^2 = -dt^2 + dr^2 + dz^2 + (1 - 4G\mu)^2 r^2 d\theta^2 \tag{8.8}$$

which represents flat spacetime with an angular wedge $\delta = 8\pi G\mu$ removed from it.

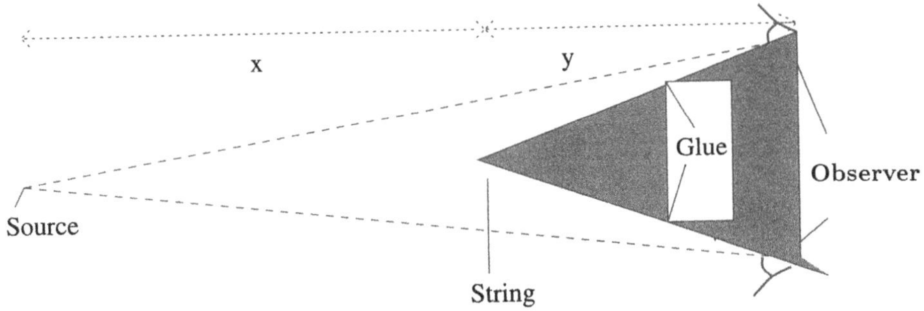

Fig. 10: Gravitational Lensing. The shaded area has been cut out of the spacetime and opposite sides identified. The deficit angle is $8\pi G\mu$.

It should be clear that the conical global structure will cause gravitational lensing of distant objects. If one imagines cutting the deficit region out in figure 10 and gluing the identified points together then it can be seen that light traveling along paths on opposite sides of the string will appear to be coming from different parts of the sky. The lensing angle will be

$$\delta\theta = 8\pi G\mu(1 + \frac{x}{y})^{-1} \tag{8.9}$$

Using equation (8.9) and using the known galaxy distribution with redshift and simulated densities of cosmic strings it is relatively straightforward to calculate the probability of lensing as a function of redshift[8] which is $\sim \pi G\mu(z_g)^2$ where z_g is the redshift of the most distant galaxies which one is likely to be able to observe. Thus for $z_g \sim 0.4$ about 1 in 10^6 galaxies would be lensed for GUT strings. So far there are no definitive candidates for cosmic string lensing. One experimental signal would be several lensed objects in a small region of space. There has been in fact a report[21] of seven apparent lensed objects within a small angular separation, but it is still not clear if these are all real lensing events and not just coincidental superpositions of different galaxies.

8.2. Cosmic Microwave Anisotropies

The energy shift for photons traveling in a background metric $g_{\mu\nu} = \eta_{\mu\nu} + h_{\mu\nu}$ is determined by the geodesic equation

$$\ddot{x}^\mu + \frac{1}{2}\eta^{\mu\nu}\left(h_{\nu\beta,\alpha} + h_{\alpha\beta,\nu} - h_{\nu\alpha,\beta}\right)x^\alpha x^\beta = 0 \tag{8.10}$$

There are several contributions for relic background photons; fortunately the most important is the doppler shift effect which only depends on the current apparent position and velocity of the string and not on the entire history of the cosmic string and the photon. The idea is that comoving frames on different sides of the string move towards each other with characteristic velocity $\sim 8\pi G\mu v_\perp$. Hence an observer should see a step-like discontinuity in the sky, with radiation from the trailing edge blue shifted relative to the radiation coming from just in front of the string. If we put $v_\perp \sim 1$, then

$$\frac{\delta T}{T} \sim 8\pi G\mu. \tag{8.11}$$

Unfortunately the COBE data is not sensitive enough to "see" the step-like discontinuities, however the measured fluctuations are consistent with those expected from GUT strings. More extensive calculations can be done numerically, the most recent of which[22] due to Coulson et. al. finds that the COBE results are consistent with

$$G\mu = (2.0 \pm 0.5) \times 10^{-6}$$

8.3. Density Perturbations

A lot of the initial excitement in cosmic strings was due to the fact that they appeared to be a natural candidate for the primordial density fluctuations responsible for galaxy formation. The initial cosmic string scenario placed a lot of emphasis on the role of loops. The idea was that loops naturally formed with a characteristic size some fraction of the horizon size, which would mean that they would persist for quite a long time before radiating their energy away in the form of gravitational waves. In this case the loops would provide seeds for galaxies to form around. Initially it seemed that this picture could successfully explain the Abell cluster correlation function. Sadly more recent simulations[15,16,17] indicate that loops are much smaller and hence live for a shorter amount of time. They are still important for galaxy formation but the focus has moved back to longer strings.

Longer strings are important in two ways. Firstly a string moving through a region of dust will leave a wake or overdense accretion region behind it. To see this imagine for a moment that the string is stopped and the dust particles are moving past the string. Then it is clear that the conical global structure will cause their trajectories to converge, just as in the gravitational lensing of light rays discussed above. Clearly if the string is moving and the dust is at rest the effect will be the same, as illustrated in the figure.

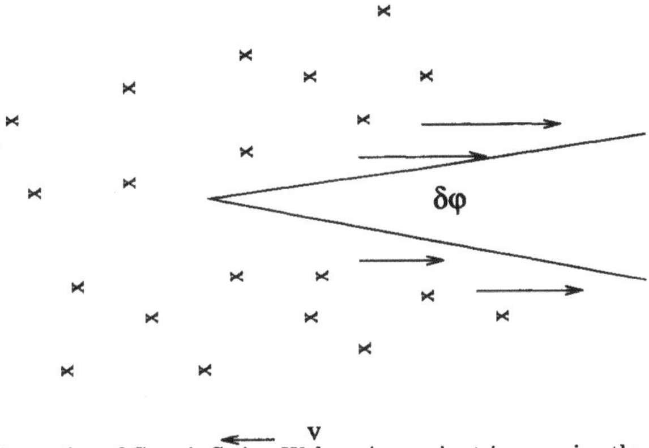

Fig. 11: Formation of Cosmic String Wakes. A cosmic string moving through dust imparts a perpendicular velocity to the trailing dust particles as it "cuts out" a deficit angle from the space it enters.

Particles will be deviated inwards by an angle $\delta\theta = 4\pi G\mu$ or in other words every particle acquires a velocity component $\sim v\delta\theta$ which is independent of the impact parameter. Eventually particles on either side of the wake will collide, resulting in an overdense region or accretion disk in the cosmic string wake, which has the potential to seed galaxy formation.

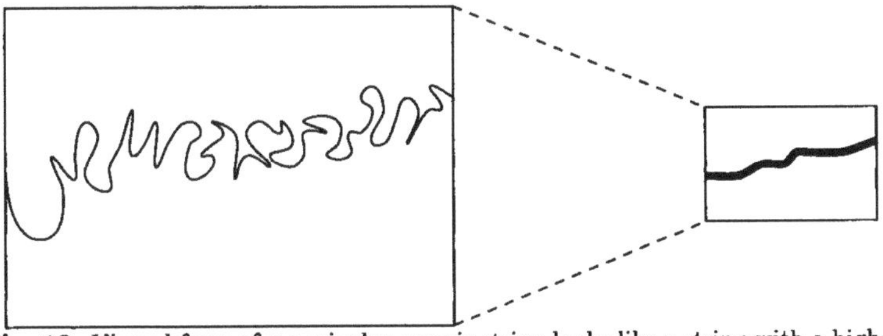

Fig. 12: Viewed from afar a wiggley cosmic string looks like a string with a higher effective mass per unit length and a lower effective tension per unit length

The second way that long strings can be important for galaxy formation is related to the fact that in general it is quite a poor approximation to think of cosmic strings as being straight. As indicated schematically in figure 12 a long cosmic string is more like a random walk then a straight string. Viewed from a large scale as indicated in the figure, the long string network can be approximated for some purposes by an effective string, where the effective energy per length is increased while the effective string tension is reduced. The cancellation between the energy density $T_t{}^t$ and the tension $T_z{}^z$ no longer occurs, and hence a long kinky string would act as an ordinary gravitational attractor generating density perturbations which again could act as seeds for galaxy formation.

Clearly for a fast moving string the second effect is not so important and the first effect will be significant, while for a slow moving string the ordinary gravitational attraction generated by a kinky string will predominate. Hence fast moving strings will tend to generate planar structures or bubbles around which galaxies will nucleate, while slow moving strings will generate filamentary structures. Simulations suggest that the r.m.s. velocity of strings on the small scale is large ($\sim 0.6c$) but that the coherent velocity of long roughly straight strings is much less ($\sim$ 0.1-0.2c) which means that filamentary features should predominate over planar features. Of course extensive numerical simulations with particular choices of dark matter need to be done to quantify such statements. However since the observed universe does seem to have both planar and filamentary structures on large scales it is certainly worth pursuing.

Quantitative predictions are hard to come by in the string scenario. What we would like to calculate is the power spectrum of density perturbations

$$P(k) = |\delta_k|^2 \tag{8.12}$$

where

$$\delta_k = (8\pi V^{-1}) \int d^3x \delta(\mathbf{x}) e^{ik\cdot x} \quad \text{for} \quad \delta(\mathbf{x}) = \delta\rho(\mathbf{x})/\rho(\mathbf{x}) \tag{8.13}$$

Recent calculations[23] indicate that at late times $P(k) \sim k$ for small k, which is the scale invariant Harrison Zeldovich spectrum. Present observations are consistent with an approximately scale invariant primordial perturbation spectrum, which is certainly encouraging for the cosmic string scenario for galaxy formation. The scale invariance of the density perturbation spectrum is a direct consequence of the scaling behaviour of the string network which we considered earlier. (It turns out that scaling occurs eventually during the matter era as well as the radiation era.)

One might also ask if the planar or linear features persist during the evolution of the universe. Here the answer seems sensitive to the kind of dark matter used in the numerical simulations[24]. For Cold Dark Matter (CDM) these features tend to be swamped by the general background of perturbations generated by the strings on smaller scales. However for Hot Dark Matter models (HDM) such features do stand out clearly.

8.4. Gravitational Radiation

As indicated previously a cosmic string loop will tend to oscillate wildly, causing gravitational waves to be radiated which in turn causes the string to decay away completely eventually. The radiation rate for the loop can be estimated using the well known quadrupole formula

$$P \sim \frac{dM}{dt} \sim G\left(\frac{d^3 I_{ij}}{dt^3}\right)^2 \tag{8.14}$$

where R is the radius of the loop and I_{ij} is the quadrupole moment tensor. On dimensional grounds $I_{ij} \sim \mu R^3$ and the time derivatives $\sim 1/R$ (since the loop will oscillate with characteristic frequency $\sim 1/R$). Putting all this together we find that

$$P = \Gamma G \mu^2 \quad \text{where } \Gamma = \text{const.} \tag{8.15}$$

which is (surprisingly) independent of size. It is also possible to derive an analogous expression for gravitational radiation from straight strings[8]. These expressions lead to two distinct observational consequences.

• *Nucleosynthesis*, which depends on a delicate balance between the rate of the various processes involved and the rate of Hubble expansion which in turn depends on the energy density. For example an extra species of neutrino would change the energy density and would lead to an unacceptably high value for the primordial abundance of helium. Energy in the form of gravitational waves would have the same effect, which implies a bound on the density of gravitational waves as a fraction of the critical density at the time of nucleosynthesis

$$\Omega_{gr} \leq 0.01 \tag{8.16}$$

Using equation (8.15) and making an estimate of the time dependence of the number of loops it is possible to estimate the build up of gravitational radiation, which turns out to have a logarithmic dependence on time. The other important factor that needs to be included in calculating the bound is the redshifting of the gravitational radiation with the expansion of the universe. Putting everything together gives a bound on $G\mu$ of

$$G\mu \leq 10^{-6}\text{--}10^{-5} \tag{8.17}$$

• *Pulsar Timing data* gives perhaps the most stringent limit on $G\mu$. Gravitational waves emitted in the early universe by string loops will have been stretched by the expansion of the universe, with wavelengths now possibly of the order of lightyears. The pulsar timing data sets limits on the long wavelength background gravitational radiation. Imagine a pulsar sending out pulses at regular intervals. In general we will see the pulses arriving at the same regular intervals, and if we plotted the number of pulses received versus time in years we would see a straight line. Imagine now that a very long wavelength gravitational wave passes between us and the pulsar. As it does so it will be stretching the spacetime in between us and the pulsar, altering slightly the distance that the photons have to travel to get here, which in turn alters slightly the time of arrival of the photon. If we now plot the number of pulses against time in years we would see a glitch as the gravitational wave passed through. Such measurements, which have been going for a decade or more, give a limit on $G\mu$ of

$$G\mu \leq 7 \times 10^{-7} - 5 \times 10^{-6} \tag{8.18}$$

which is very close to ruling out the cosmic string scenario for galaxy formation altogether!

8.5. Cosmic Rays

An outstanding puzzle in astrophysics is to explain the origin of very energetic cosmic rays with energies $\geq 10^{20}$eV. It has been proposed that evaporation from cosmic string cusps could provide the explanation. As indicated in the diagram, when cusps form the local string velocity will be relativistic. It has been suggested[25] that interactions between the overlapping string segments moving at the speed of light will produce energetic particles. The string would emit quanta of the underlying string gauge or scalar fields. Such quanta would be unstable and would in turn decay into quarks and leptons. The decay quanta would have energies around the GUT scale ($\sim 10^{24}$ eV) and be moving at relativistic speeds so the remnant leptons or quarks could easily have energies in excess of 10^{20} eV, and hence be a significant contributor to the cosmic ray background at these energies. To estimate how significant such a contribution could be, it is important to have some estimate of the fraction (f_{cr}) of the initial energy of the cosmic string which is evaporated away in this manner. Alternatively we could assume that all the cosmic ray background at this energy is due to cosmic string evaporation and see what value of f_{cr} this would correspond to, which yields an upper bound

$$G\mu f_{cr} \leq 10^{-8}$$

For GUT strings this would imply that $f_{cr} \approx 10^{-2}$, which seems implausible.

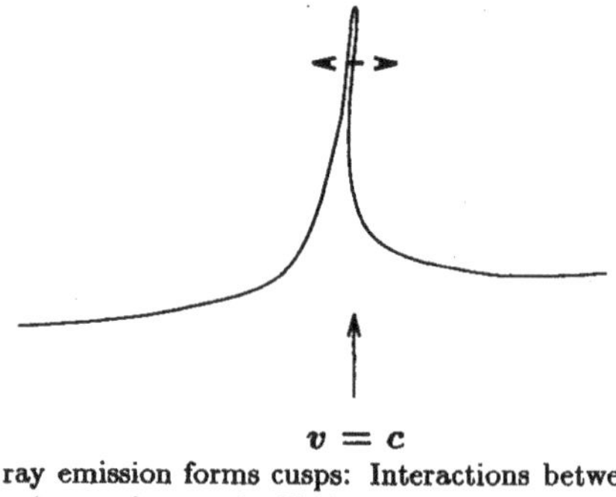

Fig. 13: Cosmic ray emission forms cusps: Interactions between the overlapping string segments moving at the speed of light will produce energetic particles.

9. Global Textures

Finally we consider very briefly another kind of topological defect called global texture[1]. A texture is a configuration for which the field takes vacuum values everywhere but with a non trivial energy density due to the gradient terms in the Lagrangian. In the language of algebraic topology, textures may appear when there are topologically nontrivial mappings from the three-sphere into the vacuum manifold.

To start with we will consider a toy model in 1 spatial dimension[26]. Consider a scalar field with Lagrangian

$$\mathcal{L} = -\frac{1}{2}\partial^\mu \phi^* \partial_\mu \phi - \frac{\lambda}{2}(\phi^*\phi - \eta^2)^2 \tag{9.1}$$

As indicated schematically in the diagram it is possible to imagine a field configuration where the field ϕ always stays in the vacuum manifold and the topology is trivial at spatial infinity but nevertheless winds around the line. The only contribution to the energy density comes from the gradient terms, nevertheless to unwind ϕ would require energy.

A crucial feature of the above construction is that the scalar field is ungauged, otherwise we would have been able to choose the gauge field to cancel out the gradient terms.

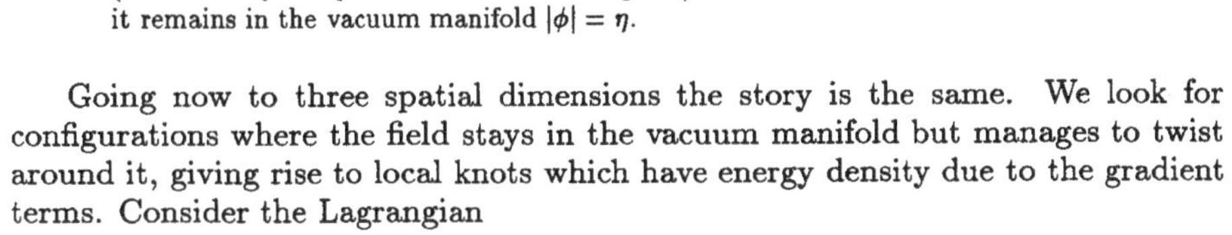

Fig. 14: Texture in one spatial dimension: The scalar field takes complex values (indicated by the phase vector in the diagram) As the field winds around the line it remains in the vacuum manifold $|\phi| = \eta$.

Going now to three spatial dimensions the story is the same. We look for configurations where the field stays in the vacuum manifold but manages to twist around it, giving rise to local knots which have energy density due to the gradient terms. Consider the Lagrangian

$$\mathcal{L} = -\frac{1}{2}\partial^\mu \phi_a^* \partial_\mu \phi_a - \frac{\lambda}{2}(\phi_a^*\phi_a - \eta^2)^2 \tag{9.2}$$

The scalar field ϕ_a is a four dimensional isovector $\phi_a \equiv (\phi_1, \phi_2, \phi_3, \phi_4)$. The vacuum manifold is then given by

$$\phi_1^2 + \phi_2^2 + \phi_3^2 + \phi_4^2 = \eta^2 \tag{9.3}$$

In a Robertson-Walker universe with metric

$$ds^2 = -dt^2 + R^2(t)\left[d\zeta^2 + \sin^2\zeta(d\theta^2 + \sin^2\theta d\varphi^2)\right] \tag{9.4}$$

a texture with winding number 1 is

$$\phi = \eta\,(\cos\varphi\sin\theta\sin\zeta, \sin\theta, \sin\varphi\sin\zeta, \cos\theta\sin\zeta, \cos\zeta)$$

In general textures are unstable since they like to shrink and then tunnel to a trivial vacuum state. To see this note that the energy of a texture goes as

$$E = \frac{1}{2}\int \partial^k\phi\cdot\partial_k\phi d^3x \tag{9.5}$$

If we scale $x \to \alpha x$ then the gradient energy density goes as $1/\alpha^2$ but after the integration the energy scales like αE. Nevertheless there have been suggestions[26] that textures are important for the formation of structure in the early universe. However the COBE measurements seem to have almost ruled this out as a scenario.

References

1. R. Davis, Phys. Rev. **D35** (1987) 3705
2. E. Witten, Nucl. Phys. **B249** (1984) 557
3. E. Copeland, M. Hindmarsh, and N. Turok, Phys. Rev. Lett.**58** (1987) 1910
4. A. Vilenkin, Nucl. Phys. **B196** (1982) 240
5. A. Achucarro and T. Vachaspati, Phys. Rev. **D44** (1991) 3067
6. A. Achucarro, R. Gregory and K. Kuijken, *Abelian Higgs Hair for Black Holes*, gr-qc/9505039
7. A. Vilenkin, Phys. Rep. **D121** (1985) 263.
8. M.B. Hindmarsh and T.W.B. Kibble, to be published in Rep. Prog. Phys., hep-ph/9411342.
9. P. Shellard and A. Vilenkin in *Topological Defects in Cosmology*, (Cambridge University Press), (1994).
10. L. H. Ryder, *Quantum Field Theory*, (Cambridge University Press), (1985).
11. S. Coleman, *Aspects of Symmetry* (Cambridge University Press), (1988).
12. E.W. Kolb and M.S. Turner, *The Early Universe* (Addison- Wesley) (1990).
13. H.B. Nielsen and P. Olesen, Nucl. Phys. **B61** (1977) 45.
14. T. Vachaspati and A. Vilenkin, Phys. Rev. **D30** (1984) 2036
15. A. Albrecht and N. Turok, Phys. Rev. Lett. **54** (1985) 1868; Phys. Rev. **D40** (1989) 973
16. B. Allen and E. Shellard, Phys. Rev. Lett. **64** (1990) 119
17. D. Bennet and F. Bouchet, Phys. Rev. Lett. **60** (1988) 257; Phys. Rev. Lett. **63** (1988) 2776; Phys. Rev. **D41** (1990) 2408.

18. J. Yokoyama, Phys. Lett. **212B** (1988) 273

19. E. Vishniac, K.Olive and D. Seckel, Nucl. Phys. **B289** (1987) 717

20. D. Sokolov and A. Starobinski, Sov. Phys. Dokl. **22** (1978) 312

21. E. M. Hu, Ap. J. **360** (1990) L7.

22. D.Coulson, P. Ferreira, P. Graham and N. Turok, Nature **368** (1994) 27.

23. S. Veeraraghavan and A. Stebbins, Ap. J. **365** (1990) 37

24. A. Albrecht and A. Stebbins, Phys. Rev. Lett. **68** (1992) 2121; Phys. Rev. Lett. **69** (1992) 2615

25. P. Bhattacharjee, Phys. Rev. **D40** (1989) 3968

26. N. Turok, Phys. Rev. Lett. **63** (1989) 2625

COSMOLOGICAL IMPLICATIONS OF NUCLEAR PHYSICS

DAVID N. SCHRAMM
The University of Chicago
5640 S. Ellis Avenue, Chicago, IL 60637
and
NASA/Fermilab Astrophysics Center
Fermi National Accelerator Laboratory, Box 500, Batavia, IL 60510-0500

CRAIG COPI
The University of Chicago
5640 S. Ellis Avenue, Chicago, IL 60637

and

XIANGDONG SHI
Queen's University
Kingston, Ontario K7L 3N6, Canada

ABSTRACT

It is shown that the impact of nuclear physics arguments has been profound. In fact, if one looks at the critical cosmological issues of age and dark matter on a plot of H_0 vs. Ω, it is apparent that most of the constraining bounds come from nuclear arguments. This lecture series reviews the Big Bang Nucleosynthesis constraints on the cosmological baryon density which demonstrate that the bulk of the baryons are dark and also that the bulk of the matter in the universe is non-baryonic. The recent extragalactic deuterium observations as well as the other light element abundances are examined in detail as is the argument on the number of neutrino flavors. Arguments from recent MACHO/EROS observations of halo dark matter seem to imply that our Galaxy's halo is $< 30\%$ baryonic; hence, we seem to need non-baryonic dark matter even on that scale. Comparison of baryonic density arguments with recent x-ray cluster data is also made. The lectures then shift to the issue of the age of the universe where it is shown that nucleochronology and hydrogen burning in old low mass stars establishes that the universe is at least 10 Gyr. This is compared with the Hubble constant determinations for various values of Ω_0 and Λ_0. A lower bound on the H_0 from the nuclear burning of carbon to iron in Type IA supernovae is briefly discussed. The final lecture focuses on the current situation with regard to solar neutrinos. It is shown that the new ^{51}Cr source measurement by GALLEX lends support to the MSW neutrino mixing model. Such a model implies that neutrinos have mass. It is noted that Big Bang Nucleosynthesis constrains the sterile neutrino option. If the Sun's ν_e's mix with ν_μ's, then ν_μ's probably have small masses, $m_{\nu_\mu} \sim 10^{-3} eV$, but a simple "see-saw" scaling then puts μ_{ν_τ} in the perfect range to be the non-baryonic matter of the universe.

1. Introduction

This series of lectures focuses on various cosmological implications of nuclear physics. In particular, four interrelated key topics are discussed in detail:

1. Big Bang Nucleosynthesis (BBN);
2. Dark Matter Constraints;
3. The Age of the Universe;
4. Solar Neutrinos.

The first leads naturally to a constraint on the baryon density which leads naturally to the dark matter arguments. The dark matter arguments require that (a) the bulk of the baryons are not associated with the emission of visible light; and (b) the bulk of the matter in the universe is non-baryonic. These in turn place restrictions on the relationship of density and cosmological age. The age is further constrained by nuclear arguments using radioactive dating and nuclear burning in low mass stars. Furthermore, the Hubble expansion rate is also constrained from below by a nuclear burning argument for Type IA supernovae. The final discussion regarding solar neutrinos picks up another aspect of hydrogen burning in low mass stars (our Sun) and notes that the current situation after the recent ^{51}Cr source measurement by GALLEX leans quite strongly towards a neutrino-mixing explanation and therefore implies that neutrinos have mass. Although the solar neutrino implied mass is very low, it probably concerns the ν_e-ν_μ system. If standard "see-saw" arguments are used, this would argue that the ν_τ mass is in the ideal range to be a significant non-baryonic dark matter candidate.

We will begin with a detailed development of BBN, then proceed to the dark matter implication, then to the age arguments, and finally to the solar neutrino situation.

2. Big Bang Nucleosynthesis (BBN)

2.1 Introduction to BBN

The study of the light element abundances has undergone a recent burst of activity on many fronts. New results on each of the cosmologically significant abundances have sparked renewed interest and new studies. The bottom line remains: primordial nucleosynthesis has joined the Hubble expansion and the microwave background radiation as one of the three pillars of Big Bang cosmology. Of the three, Big Bang Nucleosynthesis probes the universe to far earlier times ($\sim$ 1 sec) than the other two and led to the interplay of cosmology with nuclear and particle physics. Furthermore, since the Hubble expansion is also part of alternative cosmologies such as the steady state, it is BBN and the microwave background that really drive us to the conclusion that the early universe was hot and dense.

Recent heroic observations of ^{6}Li, Be and B, as well as ^{2}D, ^{3}He and new ^{4}He determinations, have all gone in the direction of strengthening the basic picture of cosmological nucleosynthesis. It will be shown that theoretical calculations of cosmic ray production of ^{6}Li, Be and B have fit the observations remarkably well, thus preventing these measurements from disturbing the standard scenario[1]. The recent reports of D/H in quasar absorption systems at redshift $Z \sim 3$ are particularly interesting and will be discussed,[2-4] since BBN requires that fragile deuterium be found in primitive material. However, the possible variation of D/H in different lines of sight at $Z \sim 3$ argues that perhaps hydrogen cloud interlopers may be imitating deuterium, at least in lines of sight having higher apparent D/H. Such an interpretation avoids a potential conflict with ^{3}He arguments. Furthermore, recent theoretical calculations have confirmed that quark-hadron inspired inhomogenous Big Bang Nucleosyntheis does not significantly alter the basic conclusions of standard BBN. We will also briefly discuss the possible impact on BBN of the recent ROSAT and ASCA x-ray satellite cluster results and the recent halo microlensing results. This summary will attempt to put it all together within an historical framework. The bottom line that emerges is how dramatically robust BBN is.

Let us now briefly review the history, with special emphasis on the remarkable agreement of the observed light element abundances with the calculations. This agreement works only if the baryon density is well below the cosmological critical value. This summary draws on the reviews of Walker et al.[5], Schramm,[6] and Copi, Schramm and Turner[7].

2.2 History and Highlights of Big Bang Nucleosynthesis

It should be noted that there is a symbiotic connection between BBN and the 3K background dating back to Gamow and his associates, Alpher and Herman. The initial BBN calculations of Gamow's group[8] assumed pure neutrons as an initial condition and thus were not particularly accurate, but their inaccuracies had little effect on the group's predictions for a background radiation.

Once Hayashi[9] recognized the role of neutron-proton equilibration, the framework for BBN calculations themselves has not varied significantly. The work of Alpher, Follin and Herman[10] and Taylor and Hoyle,[11] preceeding the discovery of the 3K background, and of Peebles[12] and Wagoner, Fowler and Hoyle[13], immediately following the discovery, and the more recent work of our group of collaborators[1,5,7,14-18] all do essentially the same basic calculation, the results of which are shown in Figure 1. As far as the calculation itself goes, solving the reaction network (see Figure 2) is relatively simple by the standards of explosive nucleosynthesis calculations in supernovae, with the changes over the last 25 years being mainly in terms of more recent nuclear reaction rates as input, not as any great calculational insight, although the current Kawano code[18] is somewhat streamlined relative to the earlier Wagoner code.[13] In fact, the earlier Wagoner code is, in some sense, a special adaptation of the larger nuclear network calculation developed by Truran[19] for work on explosive nucleosyntheis in supernovae. With the exception of Li yields and non-yields of Be and B, to which we will return, the reaction rate changes over the past 25 years have not had any major effect (see Yang et al.[17] and Krauss and his collaborators[20,21] or Copi, Schramm and Turner[7] for discussion of uncertainties). The one key improved input is a better neutron lifetime determination.[22] There has been much improvement in the t(α,γ) ^{7}Li reaction rate,[23] but as the width of the curves in Figure 1 shows, the ^{7}Li yields are still the poorest determined, both because of this reaction and even more because of the poorly measured ^{3}He (α,γ) ^{7}Be.

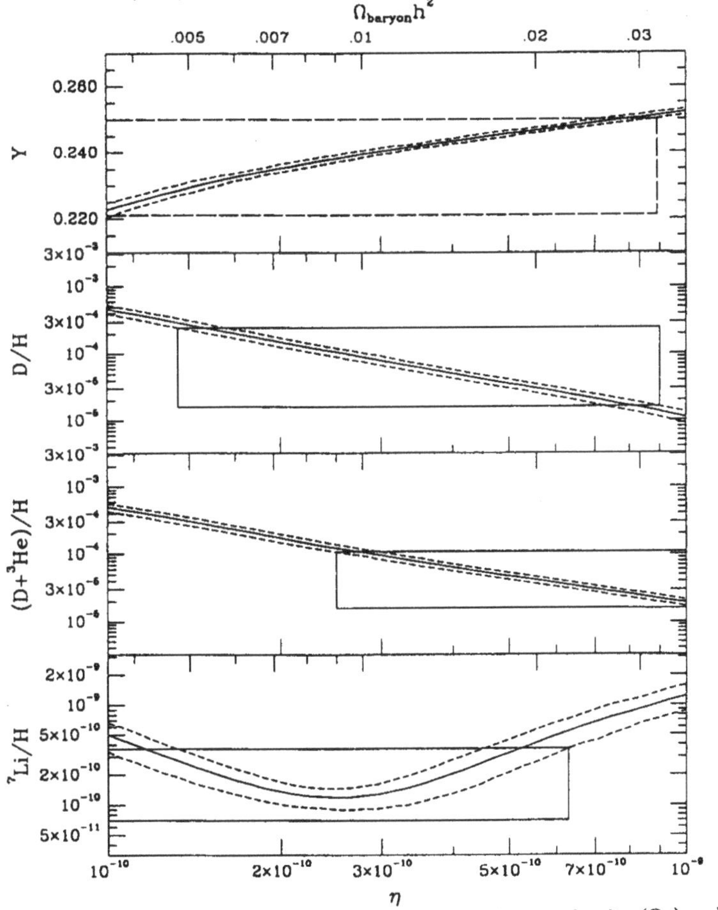

Figure 1. Big Bang Nucleosynthesis abundance yields versus baryon density (Ω_b) and $\eta \equiv \frac{n_b}{n_\gamma}$ for a homogeneous universe. ($h \equiv H_o/100$ km/sec/Mpc; thus, the concordant region of $\Omega_b h^2 \sim 0.015$ corresponds to $\Omega_b \sim 0.06$ for $H_o = 50$ km/sec/Mpc.) Figure is from Copi, Schramm and Turner.[7] Note concordance region is slightly larger than Walker et al.[5] due primarily to inclusion of possible systematic errors on Li/H.

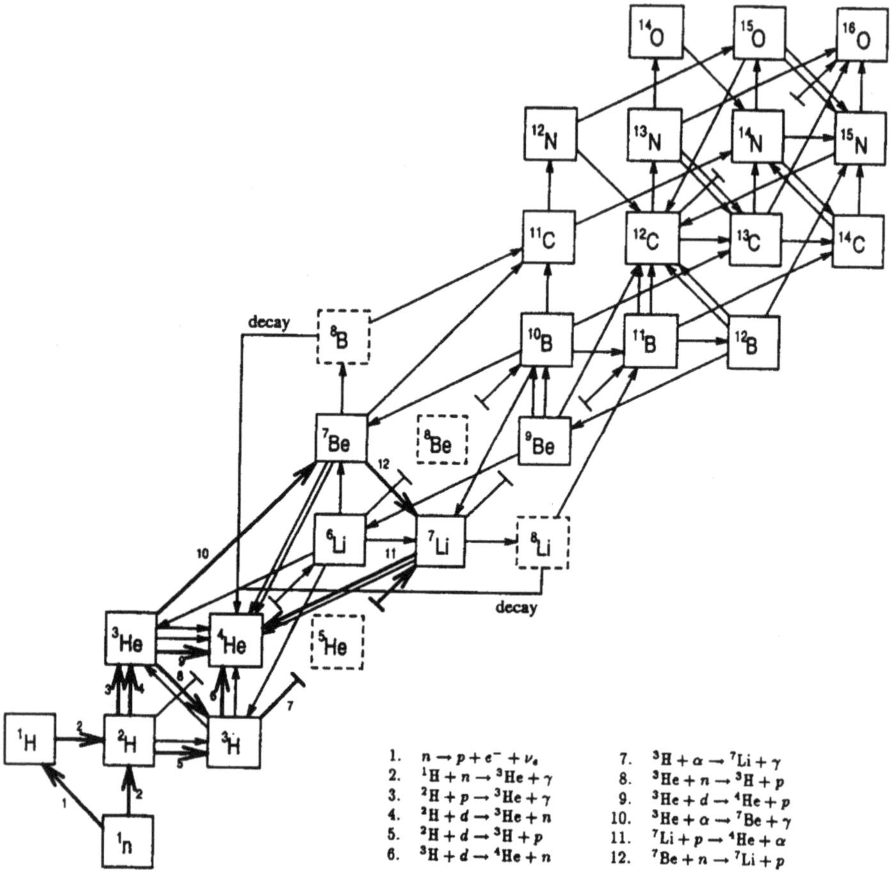

Figure 2. Full reaction network used by Copi, Schramm and Turner.[7] Note that lower half of network is all that matters for standard BBN, but the higher part and the neutron rich side can be relevant in inhomogeneous models and in arguments relating to Be and B and that part is supplemented with additional neutron-rich reactions for those calculations.

With the exception of the effects of elementary particle assumptions, to which we will also return, the real excitement for BBN over the last 25 years has not really been in redoing the basic calculation. Instead, the true action is focused on understanding the evolution of the light element abundances and using that information to make powerful conclusions. In the 1960's, the main focus was on ^{4}He which is very insensitive to the baryon density. The agreement between BBN predictions and observations helped support the basic Big Bang model but gave no significant information, at that time, with regard to density. In fact, in the mid-

1960's, the other light isotopes (which are, in principle, capable of giving density information) were generally assumed to have been made during the T-Tauri phase of stellar evolution,[24] and so, were not then taken to have cosmological significance. It was during the 1970's that BBN fully developed as a tool for probing the universe. This possibility was in part stimulated by Ryter et al.[25] who showed that the T-Tauri mechanism for light element synthesis failed. Furthermore, ^{2}D abundance determinations improved significantly with solar wind measurements[26,27] and the interstellar work from the Copernicus satellite.[28] (Recent Hubble Space Telescope observations reported by Linsky et al.[29] have compressed the ^{2}D error bars considerably.) Reeves, Audouze, Fowler and Schramm[30] argued for cosmological ^{2}D and were able to place a constraint on the baryon density excluding a universe closed with baryons. Subsequently, the ^{2}D arguments were cemented when Epstein, Lattimer and Schramm[31] proved that no realistic astrophysical process other than the Big Bang could produce significant ^{2}D. This baryon density was compared with dynamical determinations of density by Gott, Gunn, Schramm and Tinsley.[32] See Figure 3 for an updated $H_0 - \Omega$ diagram.

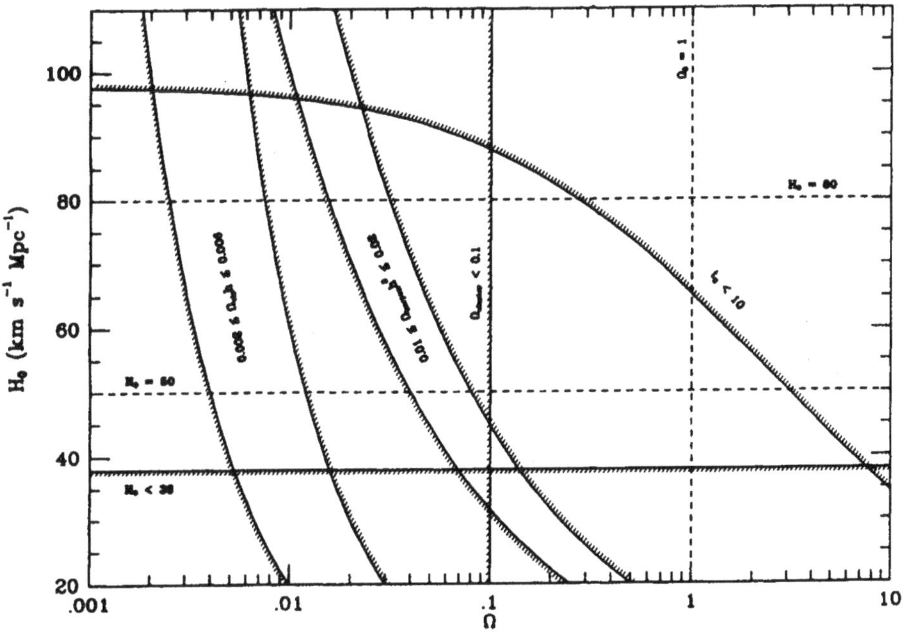

Figure 3. An updated version of $H_0 - \Omega$ diagram of Gott, Gunn, Schramm and Tinsley[32] showing that Ω_b does not intersect $\Omega_{VISIBLE}$ for any value of H_0 and that $\Omega_{TOTAL} > 0.1$, so non-baryonic dark matter is also needed.

By the late 1970's, a complementary argument to ^{2}D had also developed using ^{3}He. In particular, it was argued[33] that, unlike ^{2}D, ^{3}He was made in stars; thus, its abundance would increase with time. Since ^{3}He like ^{2}D monotonically decreased with cosmological baryon density, this argument could be used to place a lower limit on the baryon density[34] using ^{3}He measurements from solar wind[25,26] or interstellar determinations.[35] Since the bulk of the ^{2}D was converted in stars to ^{3}He, the constraint was shown to be quite restrictive.[17] Rood, Bania and Wilson[36] showed that ^{3}He was indeed enriched in planetary nebulae; hence, the argument that ^{3}He increases with time is now strengthened. However, there is nonetheless the worry that interstellar ^{3}He measurements[37] do vary with location more than one might expect for an isotope produced primarily in low mass stars.

It was interesting that the lower boundary from ^{3}He and the upper boundary from ^{2}D yielded the requirement that ^{7}Li be near its minimum of $^7\text{Li/H} \sim 10^{-10}$, which was verified by the Pop II Li measurements of Spite and Spite,[38,39,40] hence yielding the situation emphasized by Yang et al.[17] that the light element abundances are consistent over nine orders of magnitude with BBN, but only if the cosmological baryon density, Ω_b, is constrained to be around 6% of the critical value (for $H_0 \simeq$ 50 km/sec/Mpc). The Li plateau argument was further strengthened with the observation of ^{6}Li in a Pop II star by Smith, Lambert and Nissen.[41] Since ^{6}Li is much more fragile than ^{7}Li, and yet it survived, no significant nuclear depletion of ^{7}Li is possible.[42,43] This observation of ^{6}Li has now been verified by Hobbs and Thorburn,[44] and a detection in a second Pop II star has been reported. Lithium depletion mechanisms are also severely constrained by the recent work of Spite et al.[45] showing that the lithium plateau also is found in Pop II tidally locked binaries. Thus, meridonal mixing is not causing lithium depletion.

The other development of the 70's for BBN was the explicit calculation of Steigman, Schramm and Gunn[46] showing that the number of neutrino generations, N_ν, had to be small to avoid overproduction of ^{4}He. (Earlier work[11,47,48] had commented about a dependence on the energy density of exotic particles but had not done an explicit calculation probing N_ν.) This will subsequently be referred to as the SSG limit. To put this in perspective, one should remember that the mid-1970's also saw the discovery of charm, bottom and tau, so that it almost seemed as if each new detector produced new particle discoveries, and yet, cosmology was arguing against this "conventional" wisdom. Over the years, the SSG limit on N_ν improved with ^{4}He abundance measurements, neutron lifetime measurements, and with limits on the lower bound to the baryon density, hovering at $N_\nu \lesssim 4$ for most of the 1980's and dropping to slightly lower than 4 just before LEP and SLC turned on.[1,5,49,50] This was verified by the LEP results[51] where now the overall average is $N_\nu = 2.99 \pm 0.02$.

An exciting new observation has been reported by Songaila et al.[2] and by Carswell et al.[3] of a possible D/H measurement in a QSO (quasi-stellar object) absorption system at $Z = 3.3$. Such a detection would be an extremely important confirmation of the basic BBN argument that deuterium is primordial. However, as Steigman[52] points out, there is a fair probability that this observation is merely an interloping H cloud with a velocity of 82 km/sec relative to the main cloud and thus mimicking D. Therefore, the implied D/H of $\sim 2.5 \times 10^{-4}$ should be viewed as an upper limit at the present time. It should be noted that such a high value would be hard to fit with relatively low ^{3}He values (few $\times 10^{-5}$) observed in the solar system and in the ISM (interstellar medium) today unless something is seriously wrong with our understanding of ^{3}He evolution (the D primarily burns to ^{3}He in stars) or the ^{3}He observations are in error. While the current ^{3}He ISM values are certainly problematic,[53,54] we do have confidence that the solar system ^{3}He is a good measure (to within a factor of 2) of the galactic ^{3}He abundance 5 Gyr ago. Thus, D/H exceeding 10^{-4} at high Z does seem unlikely. A recent report by Tytler and Fann[4] of another extragalactic D/H determination, this time at $Z = 3.6$ with a metalicity of only 1/500 of solar, appears to put an upper limit on D/H $< 4 \times 10^{-5}$, thus implying the Songalia et al. measurement is indeed a hydrogen interloper. If an extragalactic D/H determination can eventually be confirmed in several directions, it would provide the firmest determination of the baryon density and may enable a collapse of the present range for ρ_b to an even narrower band. The Tytler and Fann[4] value would restrict us to the high baryon density side of our band. It would also have important implications for galactic evolution since the present ISM value is not significantly lower than the high Z value. This would seem to favor infall models or models with variable initial mass functions to explain heavy element production in our Galaxy.

The power of homogeneous BBN comes from the fact that essentially all of the physics input is well determined in the terrestrial laboratory. The appropriate temperature regimes, 0.1 to 1 MeV, are well explored in nuclear physics laboratories. Thus, what nuclei do under such conditions is not a matter of guesswork, but is precisely known. In fact, it is known for these temperatures far better than it is for the centers of stars like our sun. The center of the sun is only a little over 1 keV, thus, below the energy where nuclear reaction rates yield significant results in laboratory experiments, and only the long times and higher densities available in stars enable anything to take place.

To calculate what happens in the Big Bang, all one has to do is follow what a gas of baryons with density ρ_b does as the universe expands and cools. As far as nuclear reactions are concerned, the only relevant region is from a little above 1 MeV ($\sim 10^{10}$K) down to a little below 100 keV ($\sim 10^9$ K). At higher temperatures, no complex nuclei other than free single neutrons and protons can exist, and the ratio of neutrons to protons, n/p, is just determined by $n/p = e^{-Q/T}$, where $Q = (m_n - m_p)c^2 \sim 1.3$ MeV. Equilibrium applies because the weak interaction rates are much faster than the expansion of the universe at temperatures much above 10^{10}K. At temperatures much below 10^9K, the electrostatic repulsion of nuclei prevents nuclear reactions from proceeding as fast as the cosmological expansion separates the particles.

After the weak interaction drops out of equilibrium, a little above 10^{10}K, the ratio of neutrons to protons changes more slowly due to free neutrons decaying to protons, and similar transformations of neutrons to protons via interactions with the ambient leptons. By the time the universe reaches 10^9K (0.1 MeV), the ratio is slightly below 1/7. For temperatures above 10^9K, no significant abundance of complex nuclei can exist due to the continued existence of gammas with energies greater than 1 MeV. Note that the high photon to baryon ratio, η, in the universe ($\sim 10^{10}$) enables significant population of the MeV high energy Boltzman tail until $T \lesssim 0.1$MeV.

Table 1
Current Estimates of Primordial Light Element Abundances
(First $\pm$ is statistical, second is systematic. Limits are extreme of systematic + 2σ statistical.)

Isotope	Primordial Value	Note	Reference
^{4}He	$Y_o = 0.232 \pm 0.03^{+0.01}_{-0.005}$	Ex-H-II region	Olive & Steigman[55] Pagel[50] Skillman[56]
^{2}D + ^{3}H	$\left.\frac{D+^3H}{H}\right\|_0 \leq 1.1 \times 10^{-4}$	Solar wind with Max-correction for possible ^{3}He depletion	Geiss[57]
^{2}D	$D/H\Big\|_0 > 1.4 \times 10^{-5}$	ISM (HST)	Linsky et al.[29]
^{6}Li	$\left.\frac{^7Li}{H}\right\|_0 = 1.4 \pm 0.3^{+1.8}_{-0.4} \times 10^{-10}$	Systematic includes possible depletion in stars and spallogenic addition	Spite & Spite[38] Thorburn[58]

Once the temperature drops to about 10^9K, sufficient abundances of nuclei can exist in statistical equilibrium through reactions such as $n + p \leftrightarrow {}^2D + \gamma$ and $^2D + p \leftrightarrow {}^3He + \gamma$ and $^2D + n \leftrightarrow {}^3H + \gamma$, which in turn react to yield ^{4}He. Since ^{4}He is the most tightly bound nucleus in the region, the flow of reactions converts almost all the neutrons that exist at 10^9K into ^{4}He. (See Figure 2 for full reaction rate network used by Copi, Schramm and Turner.[7]) The flow essentially stops there because there are no stable nuclei at either mass-5 or mass-8. Since the baryon density at Big Bang Nucleosynthesis is relatively low (about 1% the density of terrestrial air) and the time-scale short ($t \lesssim 10^2$ sec), only reactions involving two-particle collisions occur. It can be seen that combining the most abundant nuclei, protons and ^{4}He via two body interactions always leads to unstable mass-5. Even when one combines ^{4}He with rarer nuclei like ^{3}H or ^{3}He, we still get only to mass-7, which, when hit by a proton, the most abundant nucleus around, yields mass-8. (A loophole around the mass-8 gap can be found if $n/p > 1$, so that excess neutrons exist, but for the standard case $n/p < 1$). Eventually, ^{3}H decays to ^{3}He, and any mass-7 made decays to ^{7}Li. Thus, Big Bang Nucleosynthesis makes ^{4}He with traces of ^{2}D, ^{3}He, and ^{7}Li.

(Also, all the protons left over that did not capture neutrons remain as hydrogen.) For standard homogeneous BBN, all other chemical elements are made later in stars and in related processes, such as cosmic ray spallation. (Stars jump the mass-5 and -8 instability by having gravity compress the matter to sufficient densities and have much longer times available so that three-body collisions, $3\,^4\mathrm{He} \rightarrow {}^{12}\mathrm{C}$, can occur.)

Table 1 summarizes our interpretation of the current light element abundances with my estimate of their uncertainties. The numbers are drawn from discussions at Elba and Snowmass and from Copi, Schramm and Turner's[7] summary. The only significant difference with Walker et al.[5] is in the opening of the lithium range due to systematic errors that were not explicitly presented in the observational papers at that time.

The limits shown in Table 1 (or any others for that matter) can be plugged into the following fitting formula to obtain a limit on the number of equivalent neutrino species allowed by BBN:

$$N_\nu \lesssim 3.4 + 0.08\left[\frac{Y_0\big|_{MAX} - 0.245}{0.001}\right] + \left[\frac{\frac{D+{}^3He}{H}\Big|_0 - 1.1\times 10^{-4}}{10^{-4}}\right]$$

or, equivalently

$$N_\nu \lesssim 3.4 + \left[\frac{Y_0\big|_{MAX} - 0.245}{0.012}\right] - \left[\frac{\eta_{10}^{MIN} - 2.5}{5}\right]$$

where η_{10}^{MIN} is the minimum baryon to photon ratio consistent with ^{3}He/H and/or D/H. It should be noted that the N_ν limit obtained here is complimentary to the LEP limit since BBN constrains all relativistic species regardless of their coupling to the Z. This is discussed explicitly by Steigman, Olive and Schramm.[59]

2.3 Inhomogeneous Big Bang Nucleosynthesis

As noted above, BBN yields all agree with observations using only one freely adjustable parameter, $\eta = n_b/n_\gamma$, or equivalently, the baryon density, ρ_b (since n_γ is directly related to the temperature of the microwave background.) Thus, BBN can make strong statements regarding ρ_b if the observed light element abundances cannot be fit with any alternative theory. The most significant alternative that has been discussed involves quark-hadron transition inspired inhomogeneities.[60,61,62] While inhomogeneity models had been looked at previously[14,17] and were found to make little difference, the quark-hadron inspired models had the added ingredient of variations in n/p ratios. Cosmologists are well aware that current trends in lattice gauge calculations imply that the transition may be secondary[63] or not a phase transition at all, although a recent report[64] using Wilson fermions again may imply a strong first order transition. Nevertheless, it is useful to explore the maximal cosmological impact that can occur. This maximal impact requires a first-order phase transition.

The initial claim by Applegate et al.,[61] followed by a similar argument from Alcock et al.,[62] that $\Omega_b \sim 1$ might be possible, created tremendous interest. Their argument was that if the quark-hadron transition was a first-order phase transition, then it was possible that large inhomogeneities could develop at T $\gtrsim$ 100 MeV. The preferential diffusion of neutrons versus protons out of the high density regions could lead to Big Bang Nucleosynthesis occurring under conditions with both density inhomogeneities and variable neutron/proton ratios. In the first round of calculations, it was claimed that such conditions might allow $\Omega_b \sim 1$, while fitting the observed primordial abundances of ^{4}He, ^{2}D, ^{3}He with an overproduction of ^{7}Li. Since ^{7}Li is the most recent of the cosmological abundance constraints and has a different observed abundance in Pop I stars versus the traditionally more primitive Pop II stars,[38,39,40] some argued that perhaps some special depletion process might be going on to reduce the excess ^{7}Li. Today we know from the ^{6}Li observation[41] that such depletion is unlikely but, at the time, this option was more viable.

At first it appeared that if the lithium constraint could be surmounted, then the density constraints of standard Big Bang Nucleosynthesis might disintegrate. To further stimulate the flow through this new loophole, Fowler and Malaney[65] showed that, in addition to looking at the diffusion of neutrons out of high density regions, one must also look at the subsequent effect of excess neutrons diffusing back into the high density regions as the nucleosynthesis goes to completion in the low density regions. (The initial calculations treated the two regions separately.) Fowler and Malaney[65] argued that for certain phase transition parameter values (e.g. nucleation site separations $\sim$ 10m at the time of the transition), this back diffusion could destroy much of the excess lithium.

However, Kurki-Suonio, Matzner, Olive and Schramm,[66] Thomas et al.,[67] the Tokyo group,[68] and the Livermore group[69] have eventually argued that in their detailed diffusion models, the back diffusion not only affects ^{7}Li, but also the other light nuclei as well. They find that for $\Omega_b \sim 1$, ^{4}He is also overproduced (although it does go to a minimum for similar parameter values as does the lithium). Added to this has been the detailed work of Jedamzik et al.[70] on lithium in inhomogeneous models showing that, despite its sensitivity to assumptions, it nonetheless supports the conclusion that inhomogeneous models yield essentially the same constraints on Ω_b as homogenous models do.

One can understand why these models might tend to overproduce ^{4}He and ^{7}Li by remembering that in standard homogeneous Big Bang Nucleosynthesis, high baryon densities lead to excesses in these nuclei. As back diffusion evens out the effects of the initial fluctuation, the averaged result should approach the homogeneous value. Furthermore, it can be argued that any narrow range of parameters, such as those which yield relatively low lithium and helium, are unrealistic since in most realistic phase transitions there are distributions of parameter values (distribution of nucleation sites, separations, density fluctuations, etc.). Therefore, narrow minima are washed out which would bring the ^{7}Li and ^{4}He values back up to their excessive levels for all parameter values with $\Omega \sim 1$. Furthermore, Freese and Adams[71] and Link[72] have argued that the boundary between the two phases may be fractal-like rather than smooth. The large surface area of a fractal-like boundary would allow more interaction between the regions and minimize exotic effects.

Figure 4 shows the updated results of Kurki-Suonio et al.[66] for varying spacing l with the constraints from the different light element abundances. This can be compared to the recent work of Thomas et al.[67] or Jedamzik et al.[70] where essentially the same conclusions are reached. Notice that the Li, ^{2}D, and even the ^{4}He constraint do not allow $\Omega_b \sim 1$. Note also that with the Pop II ^{7}Li constraint, the results for Ω_b are quite similar to the standard model. Only by relaxing the ^{7}Li constraint and by using only the ^{4}He and/or ^{2}D to bound the baryon density can even factors of ~ 2 variations in Ω_b be allowed. Thus, even an optimally tuned first order quark-hadron transition is not able to alter the basic conclusions of homogeneous BBN regarding Ω_b. (It also cannot significantly change the N_ν argument.) In fact, the main role that a quark-hadron option has played for BBN is to show how robust the standard model results are.

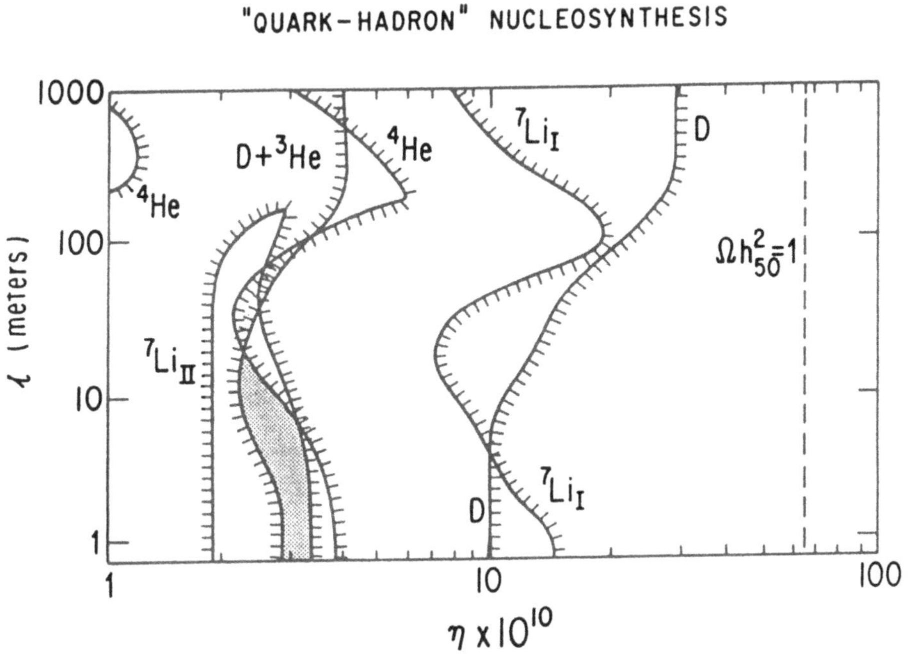

Figure 4. The updated Kurki-Suonio, Matzner, Olive and Schramm[66] results showing that even allowing for a 1st order quark-hadron transition with variable spacing l between nucleation sites, the light element abundances constrain the baryon to photon ratio, η, and thus Ω_b to essentially the same values as those obtained in the homogeneous case with only slightly large Ω_b's possible with $l \sim 100$. Note, even if the Pop II lithium constraint is dropped, the ^{4}He constraint excludes values of η above $\sim 5 \times 10^{-10}$, still far below $\Omega = 1$.

2.4 Boron, Beryllium and the Spallation Process

While quark-hadron inspired variations have not been able to alter the basic conclusions of BBN, an important question remains, namely, is there an observable signature that could differentiate quark-hadron inspired variations from the homogeneous model? On the theoretical side, this point was initially debatable. Several authors[62,73] have argued that because of the high n/p region in the inhomogeneous models, leakage beyond the mass 5 and 8 instability gaps can occur and traces of ^{9}Be, ^{10}B, ^{11}B and maybe even r-process elements can be produced. Thus, detection of nuclei beyond ^{7}Li in primitive objects may be a signature. However, Sato and Tarasawa[68] and Thomas et al.[67] have argued that such leakage is negligible.

Let us now comment on some recent observations of Be and B in primitive Pop II stars. In particular, there has been much recent attention given to reports[74,75,76] of beryllium lines being observed in extreme Pop II stars. For one very metal poor Pop II star, HD 140283, boron was also observed.[77] These observations showed Be/H and B/H ratios that scaled with the metalicity of the observed stars. Thus, contrary to some initial claims, the Be/H and B/H observations[78] do not require cosmological origin, only a scaling with oxygen of the same process that produced Be and B in the Pop I stars. Recent arguments discussed at Elba on non-LTE (local thermodynamic equalibrium) effects in the Be and B determinations do not in any way alter these conclusions, although they may imply a higher B/Be. (See discussion of Fields, Olive and Schramm.[79])

The presumed process that produced Be and B in Pop I stars (as well as the ^{6}Li) is thought to be cosmic ray spallation.[80] For Be and B, such spallation comes from the breakup of heavy nuclei such as CNO and Ne, Mg, Si, S, Ca and Fe by protons and alphas. As noted by Epstein, Arnett and Schramm,[81] for lithium one must also include alpha plus alpha fusion processes as well. This latter point was well noted by Steigman and Walker[82] and by Walker et al.[83] who emphasized that Be and B spallation production on Pop II abundances would imply a significant enhancement of lithium from alpha-alpha relative to the reduced production of Be and B from depleted heavy nuclei. While the ^{6}Li so produced would be destroyed at the base of the convective zones in most stars observed,[84,85] it did manage to survive in at least one extremely high temperature (thin convection zone) star.[41,45]

Perhaps most critical to any spallation origin is the resultant B/Be ratio. It is also known, from actual measurements, that the cosmic rays themselves[86] show B/Be $\sim$ 14 (and B and Be are pure spallation products in the cosmic rays) with a carbon to oxygen ratio exceeding unity (Pop I has C/O $\lesssim 0.5$). Since spallation off carbon favors B relative to Be (mass 11 requires only a single nucleon ejected from mass 12), whereas oxygen being farther from either shows less favoritism, the cosmic ray observations are actually an upper limit on what B/Be ratio one might expect in Pop II cosmic rays. (See also Fields, Olive and Schramm.[79]) However, of more concern here is the lower limit on B/Be achievable by a spallation process.[83] The resultant limit is B/Be $\simeq 7$ which means that cosmic ray spallation can yield $7 \lesssim B/Be \lesssim 14$. In other words, spallation yields do not result in a B/Be ratio below 7. Furthermore, non-spallation mechanisms such as ν-process[87] in supernovae (see also Dearborn et al.[88]) or a low energy cosmic ray spike[89,79] can enhance B/Be, but no known process can produce Be without making B. Thus, the lower bound on B/Be is firm, whereas values above 14 would only be indicative of a B enhancing addition.

Thomas et al.[90] have shown that ^{9}Be/H yields are always less than 10^{-14} regardless of η and that for $\eta \sim 3 \times 10^{-10}$, where agreement with the traditional BBN light element abundances occurs, Be/H and B/H $\sim 10^{-18}$. In other words, homogeneous BBN cannot yield Be/H consistent with the Pop II stellar observations. As Thomas et al.[67] noted, even inhomogeneities do not significantly alter these conclusions. Thus, at first glance an observed plateau in Be/H and B/H versus low metalicity might be a problem for BBN. However, Mathews[23] has argued that galactic evolution can produce such a partial plateau for extremely low metalicities with no cosmological origin for Be and B.

3. Dark Matter

3.1 BBN and Dark Matter

The success and robustness of BBN have given renewed confidence to the limits on the baryon density constraints. Let us convert this density regime into units of the critical cosmological density for the allowed range of Hubble expansion rates. This is shown in Figure 3. In particular, $\Omega_b = (0.06 \pm 0.02)h_{50}^{-2}$ where h_{50} is the Hubble constant in units of 50km/sec/Mpc. Figure 3 also shows the lower bound on the age of the universe of 10 Gyr from both nucleochronology and from globular cluster dating[91] and a lower bound on H_0 of 40 from extreme type IA supernova models with pure 1.4 $M_\odot$ carbon white dwarfs being converted to ^{56}Fe. The constraint on Ω_b means that the universe *cannot be closed with baryonic matter.* (This point was made over twenty years ago[80] and has proven to be remarkably strong.) If the universe is truly at its critical density, then nonbaryonic matter is required. This argument has led to one of the major areas of research at the particle-cosmology interface, namely, the search for non-baryonic dark matter. In fact, from the lower bound on Ω_{TOTAL} from cluster dynamics of $\Omega_{TOTAL} > 0.1$, it is clear that non-baryonic dark matter is required unless $H_0 < 50$. The need for non-baryonic matter is strengthened on even larger scales.[92,93] Figure 3 also shows the range of $\Omega_{VISIBLE}$ and shows that there is no overlap between Ω_b and $\Omega_{VISIBLE}$. Hence, the bulk of the baryons are dark.

The estimate of the Ω_{VIS} in Figures 3 and 5 can be obtained from noting that stellar material tends to have mass-to-light ratios in solar units as shown below

$$\frac{M}{L}|_* \sim 2 - 8 \ \frac{M_\odot}{L_\odot}|_B$$

for blue band light. (The number is greater than one since mass is dominated by low mass stars and light by high mass ones, and initial mass functions favor low mass ones.) The uncertainty comes from the location of the low mass cutoff. Recent observations from HST[94] argue that the cutoff is above the red dwarf limit of several tenths $M_\odot$ since these objects were not found to be ubiquitous. To obtain a density from an M/L one needs to multiply by the average luminosity density, L, in that wave band. Kirshner et al.[95] have determined :

$$L_B \simeq 1.8 \pm 0.2 \times 10^8 h \frac{L_\odot}{Mpc^3}$$

where h is the Hubble constant in units of 100 km/sec/Mpc. One can then obtain Ω_{VIS} by dividing by

$$\rho_{crit} \equiv \frac{3H_0^2}{8\pi G} = 2.8 \times 10^{11} h^2 \frac{M_\odot}{Mpc^3}.$$

Hence,

$$\Omega_{VIS} = \frac{\frac{M}{L}|_B \cdot L_B}{\rho_{crit}} = \frac{0.003 \pm 0.002}{h}.$$

Note that Ω_{VIS} obtained in this way is inversely proportional to H_0.

Another interesting conclusion[32] regarding the allowed range in baryon density is that it is in agreement with the density implied from the dynamics of single galaxies, *including their dark halos*. The recent MACHO and EROS[96,97] reports of halo microlensing may well indicate that at least some of the dark baryons are in the form of brown dwarfs in the halo. However, Gates, Gyuk and Turner,[98] and Alcock et al.[96] show that the observed distribution of MACHOs (4 towards LMC, 45 towards Galactic Bulge) cannot be fit if more than 30% of the halo is MACHOs. Thus, the bulk of the halo must be in some other form such as cold dark matter (see below).

3.2 The Need for Non-Baryonic Dark Matter

The arguments requiring some sort of dark matter fall into separate and quite distinct areas. These arguments are summarized in Figures 3 and 5. First are the arguments using Newtonian mechanics applied to various astronomical systems that show that there is more matter present than the amount that is shining. It should be noted that these arguments reliably demonstrate that galactic halos seem to have a mass $\sim$ 10 times the visible mass.

For dynamical estimates of Ω one estimates the mass from $M \sim \frac{v^2 r}{G}$ where v is the relative velocity of the objects being studied, r is their separation distance, and G is Newton's constant. The proportionality constant out front depends on orientation, relative mass, etc. For large systems such as clusters, one uses averaged quantities. For single galaxies v would represent the rotational velocity and r the radius of the star or gas cloud. Note that since v is measured by relative redshift, it has no H_0 dependence, whereas distances will scale with $1/H_0$. Also note that the luminosity of a system will scale as the surface area and hence is proportional to $1/H_0^2$. Hence, for dynamically measured systems, $\frac{M}{L}|^{DYN} \propto H_0$. Again, $\Omega_{DYN} = \frac{\frac{M}{L}|_B^{DYN} \cdot L_b}{\rho_{crit}}$ which now is independent of H_0. It is this technique which yields the halo and cluster points shown on Figure 5 and the cluster bound on Ω shown on Figure 3. It should be noted that the value of $\Omega_{CLUSTER} \sim 0.2$ is also obtained in those few cases where alignment produces giant gravitational-lens arc. Recent work using weak gravitational lensing by Kayser[99] also supports large Ω. The IRAS/Great Attractor point on Figure 5 comes from the large scale flow arguments. As Davis and Nusser [93] show, if the large scale velocity flows measured from the IRAS survey are due to gravity, then $\Omega_{IRAS} \gtrsim 0.3$. Similar arguments have been obtained using the Great Attactor study or the Potent technique. All imply $\Omega > \Omega_{BARYON}$, hence the need for non-baryonic dark matter. However, there is still considerable uncertainty of the exact value of Ω determined in this way as discussed by Szalay [100] at this summer school. But all groups agree it is greater than 0.3. However, as Figure 3 illustrates, except for $H_0 < 50$, $\Omega_{CLUSTER}$ already required $\Omega_{TOTAL} > \Omega_{BARYON}$ and hence the need for non-baryonic dark matter.

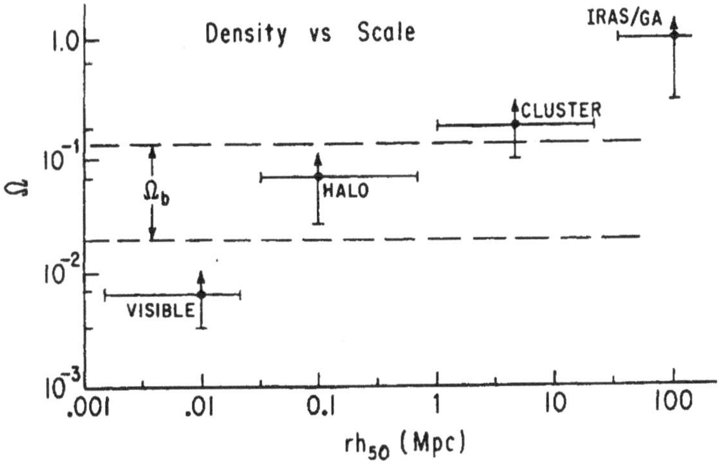

Figure 5. Implied densitites versus the scale of the measurements.

An Ω of unity is, of course, preferred on theoretical grounds since that is the only long-lived natural value for Ω, and inflation[101,102] or something like it provided the early universe with the mechanism to achieve that value and thereby solve the flatness and smoothness problems. Note that our need for exotica is not dependent on the existence of dark galatic halos and that high values of H_0 (see Figure 3) increase the need for non-baryonic dark matter.

Non-baryonic matter can be divided[103] into two major categories for cosmological purposes: hot dark matter (HDM) and cold dark matter (CDM). Hot dark matter is matter that is relativistic until just before the epoch of galaxy formation, the best example being low mass neutrinos with $m_\nu \sim 20$eV. Cold dark matter is matter that is moving slowly at the epoch of galaxy formation. Because it is moving slowly, it can clump on very small scales, whereas HDM tends to have more difficulty in being confined on small scales. Examples of CDM could be massive supersymmetric particles with masses, M_x, greater than several GeV or the lightest super-symmetric particle which is presumed to be stable and might also have masses of several GeV. Following Michael Turner, all such weakly interacting massive particles are called "WIMPS". Axions, while very light, would also be moving very slowly[104] and thus would clump on small scales. Or, one could also go to non-elementary particle candidates, such as planetary mass blackholes or quark nuggets of strange quark matter, possibly produced at the quark-hadron transition.[105,106] Another possibility would be any sort of massive topological remnant left over from some early phase transition. Note that CDM would clump in halos, thus requiring the dark baryonic matter to be out between galaxies, whereas HDM would allow baryonic halos. The MACHO and EROS events now favor at least some CDM. Obviously, mixed models with some HDM and some CDM have even more flexibility and have thus become quite popular as data constraints increase.

Some baryonic dark matter must exist since we know that the lower bound from Big Bang Nucleosynthesis is greater than the upper limits on the amount of visible matter in the universe. However, unless the microlensing efficiencies turn out to be in error, dark baryons cannot all be in halos. If the baryonic dark matter is not in the halo, it could be in hot intergalactic gas, hot enough not to show absorption lines in the Gunn-Peterson test, but not so hot as to be seen in the x-rays. The exciting report by Jakobsen et al.[107] of a Gunn-Peterson effect observed with HST for He-II at high Z showed that at least some hot IGM (intergalactic medium) exists (and verifies that He seems primordial).

Another possible hiding place for the dark baryons would be failed galaxies, large clumps of baryons that condensed gravitationally but did not produce stars. Such clumps are predicted in galaxy formation scenarios that include large amounts of biasing where only some fraction of the clumps shine. Evidence for some hot gas is found in clusters of galaxies from the ROSAT and ASCA satellites. In particular, Mushotzky[108] and White et al.[109] have discussed how certain observed rich clusters have $M_{HOTGAS}/M_{TOTAL} \sim 1/5$, which, with $\Omega_{CLUSTER} \sim 0.2$ and $\Omega_b \sim 0.05$, would imply no conflict with BBN, but that clusters are not fair samples of the baryon to non-baryon ratio in the universe. Apparent variation[103] in M_{HOT}/M_{TOTAL} for small groups relative to clusters seems to support the point of view that megaparsec scales are not always fair samples. If true, this is difficult to reconcile with cold dark matter models but can be fit with topological defects and HDM or with mixed models. Figure 6 shows the hot x-ray gas band on our H-Ω plot. Note that there is even marginal agreement at $H_0 \sim 40$ between Ω_b and the hot gas. Thus, pure CDM does survive for very low H_0. Note also that mixed models with a 30% light HDM component such as two generations of LSND-like* neutrinos[110] can shift the x-ray band to have considerable overlap with Ω_{BARYON}.

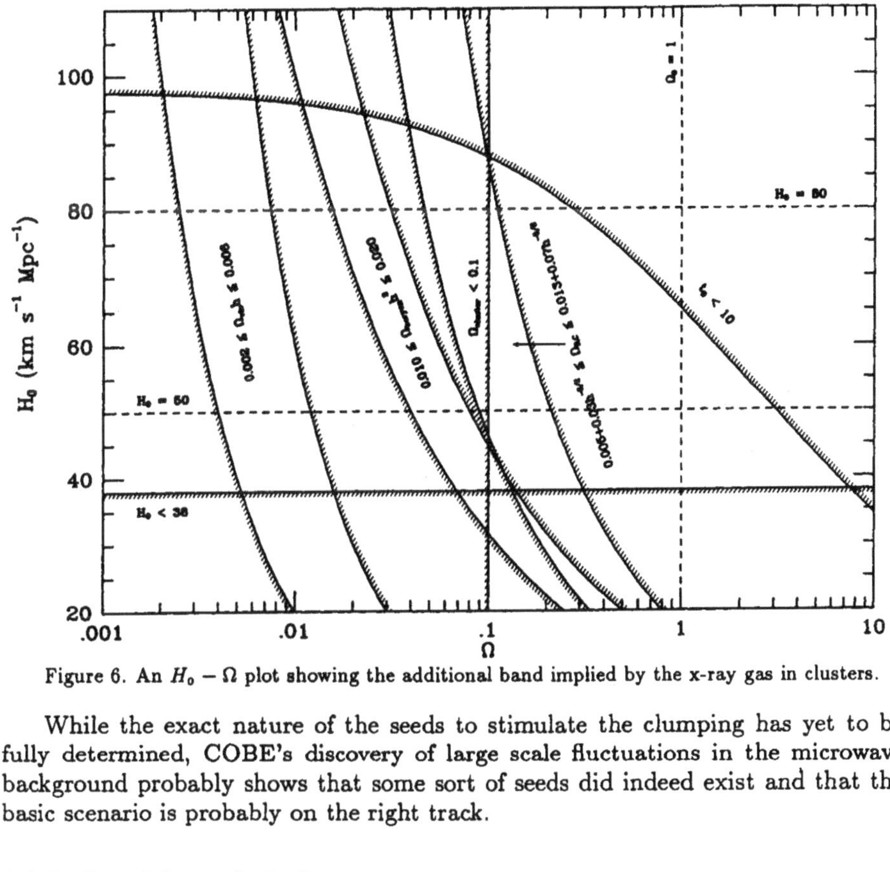

Figure 6. An $H_0 - \Omega$ plot showing the additional band implied by the x-ray gas in clusters.

While the exact nature of the seeds to stimulate the clumping has yet to be fully determined, COBE's discovery of large scale fluctuations in the microwave background probably shows that some sort of seeds did indeed exist and that the basic scenario is probably on the right track.

3.3 Seeds and Large Scale Structure

To form structure in the universe requires some sort of "seed" in addition to the baryonic matter and the non-baryonic dark matter. The types of seeds tend to fall into two broad categories:

1. density fluctuations;
2. topological seeds.

Density fluctuations are presumably produced via quantum fluctuations near the end of the inflation epoch and topological seeds such as cosmic strings or textures are topological defects again left over from some vacuum phase transition. Thus,

---

* LSND is the name of an experiment at Los Alamos National Laboratory looking for neutrino oscillations.

in either case a cosmological phase transition seems to be required. Table 2 shows other names and properties these two seed classifications sometimes go by.

Since COBE only was able to observe anisotropies on angular scales $\gtrsim 7^\circ$, it was not able to directly probe the scales that eventually led to the observed large scale structure.

Table 2
Seed Types

Density Fluctuations	Topological Seeds
Quantum Fluctuations from Inflation	Topological Defects such as strings, textures or domain walls
Gaussian	Non-Gaussian
Adiabatic	Iso-curvature (or almost equivalently isothermal)

The horizon size at the decoupling of the microwave radiation corresponds only to an angular size of less than 2°, so COBE could not probe the causal processes that led to structure. Many new microwave experiments are now probing the angular scales of $\sim 30'$ to $\sim 2^\circ$ to try to differentiate between the models of Table 2 and various CDM/HDM assumptions. These experiments are being done in balloons and at the South Pole and at various other remote terrestrial locations such as Saskatoon, Canada. At present, there is much ambiguity among the different experiments, so no single picture emerges. Non-gaussian models may even claim to predict that different directions should see different results, but the error bars are too large to claim this has been proven yet. Many experiments do find a few times higher signal at $\sim 1^\circ$ than COBE finds for larger scale (COBE $\frac{\Delta T}{T} \sim 10^{-5}$) which is what the fluctuation model predicts. Clearly, the exact nature of the seeds still remains to be resolved. In addition to looking at microwave anisotropies, astronomers are also looking at the distribution of galaxies with large redshift survey, as discussed by Szalay[100]. The intercomparison of statistically significant galaxy distribution data and microwave anisotropy data, plus direct dark matter searches and accelerator searches will hopefully resolve the origin of structure in the universe by the end of this decade.

4. The Age of the Universe

4.1 - Introduction

The problem of estimating the age of the universe is longstanding. For example, in 1650, Bishop James Ussher[111] determined by a technique of summing the Biblical begats and making other corrections and connections based on the then available historical and astronomical records that the universe began in 4004 BC, at the moment that would correspond to sunset in Jerusalem on the evening before October 23. This would correspond to 4 PM U.T. on October 22.

This early determination illustrates a key point which we will also apply to more modern techniques. Namely, while Bishop Ussher was able to obtain a result with reported accuracy of about 8 significant figures, his systematic errors are considerably larger. (Even his intrinsic error is larger than the accuracy of his result indicates, since the Jewish calendar, using essentially the same technique, obtains an age that is over 200 years off from Ussher's.)

Today the age of the universe can be estimated by three independent means:

1. Dynamics (Hubble age and deceleration)
2. Oldest Stars (globular clusters)
3. Radioactive Dating (nucleocosmochronology)

We will see that despite much activity on the dynamical technique, the best age bounds are still those derived from nuclear arguments - namely #2 and #3. Each of these gives a lower bound of $t \gtrsim 10$ Gyr as plotted in Figure 3. Furthermore, the age of the disk also bounds the age of the universe ($t_{disk} \sim 10 Gyr$), as does radioactive dating of the Earth-Meteorite system at $t_{ss} = 4.6 \pm 0.1 Gyr$.

4.2 - The Age from Dynamics

The use of the Hubble constant to determine an age is the most quoted and least accurate of all the age determination methods. The point is that it is not really determining an age but only a dynamic timescale. For perspective let us note that astronomers continue to get values ranging from $H_o \sim 100$ km/sec/Mpc down to values near $H_o \sim 40$ km/sec/Mpc. The higher values tend to come from people using empirical techniques like Tulley-Fisher, whereas the smaller values come primarily from people using supernovae. In principle, supernovae are better understood physically, but some astronomical calibrations inevitably creep in. However, no hidden-variables should creep in since the physics is in reasonable shape, unlike the empirical technology. A critical question tends to be the accuracy of intermediate distance calibrators and the correction for infall into the Virgo cluster. Most of us can't see anything wrong at face value with the Tulley-Fisher techniques other than a possible susceptibility to the so-called Malmquist bias. However, many physicists have a certain fondness for the use of Type-I supernovae as standard candles. Type I's seem to be due to the detonation of a C-O white dwarf star converting its C-O to Fe. Such a model has a physical relationship between its luminosity and basic nuclear quantities that can be measured in the lab. Current best-fit models (c.f. Nomoto) tend to convert about $0.7 M_\odot$ of C-O, which yields $H_o \sim 60$ km/sec/Mpc. However, even in the extreme where the entire $1.4 M_\odot$ Chandrasekhar mass is burned, H_o is never below ~ 40 km/sec/Mpc (see also Nugent et al.[112]). Sandage and Tammann's[113] empirical calibrations, which ignore the nuclear mechanism, yield $H_o \sim 50$ km/sec/Mpc, which fall within the theoretically allowed range and correspond to almost complete burning of a Chandrasekhar core. Recently, Riess, Press and Kirshner[114] have argued that there may be some variation in type IA light curves which shifts H_o up to ~ 66. Kirshner[115] also argues that the expanding photosphere of type II supernova implies $H_o \sim 73$. While selecting between 40 and 75 is still a matter of choice, it does seem that values less than 40 can be reliably excluded. Why these numbers disagree with the best Tulley-Fisher determinations of about 80 remains to be understood.

Most recently there has been much publicity about the Hubble Space Telescope (HST) seeing individual cepheid variable stars in a galaxy (M100) in the Virgo Cluster.[116] This is a truly impressive achievement; however, the final value of H_o they obtain still depends on the Tulley-Fisher technique (which Sandage and Tammann[113] argue is biased). It should also be noted that Sandage and Tammann use HST measured cepheids in M101 to calibrate their type IA distance scale. Over the next few years, HST will find many more cepheids in other galaxies in Virgo so that part of the uncertainty will decrease. However, Freedman et al.[116] were very conscientious in listing both statistical and systematic errors in their quoted value of $H_o = 80 \pm 7 \pm 15$ km/sec/Mpc. Although most astronomers quote this as $H_o = 80 \pm 17$, it is probably more realistic not to add systematic errors in quadrature since they are not gaussian distributed. Hence, a better estimate is 80 ± 22. Even this large error does not include the possibility that cepheids themselves may have a systematic shift in luminosity between the LMC (where the calibration is done) and M100. Historically, Hubble got $H_o \sim 500$ due to using cepheids calibrated from Pop II objects and applied to Pop I in other galaxies. While much of the metalicity effect is now taken into account, the observed trends of cepheids in M31 may hint that there is still some residual effect. Thus, while the systematic error due to the unknown location of M100 within Virgo will be reduced with more HST detections of cepheids in other Virgo galaxies, some systematic errors will remain (including potential differential reddening between the southern LMC direction and the northern Virgo direction).

With all of these systematics, it is clear at the present time that we still do not know H_o to 30 %, probably not to a factor of 2.

Age, t_u, is related to H_o by:

$$t_u = \frac{f(\Omega)}{H_o} \tag{1}$$

where for standard matter-dominated models with cosmological constant $\Lambda = 0$,

$$f(\Omega) = \begin{cases} 1 & \Omega = 0 \\ 2/3 & \Omega = 1 \\ \sim 0.5 & \Omega = 4 \end{cases}$$

and smoothly varies between those values. A range of $40 \lesssim H_o \lesssim 100$ km/sec/Mpc yields 10 Gyr $\lesssim 1/H_o \lesssim$ 25 Gyr.

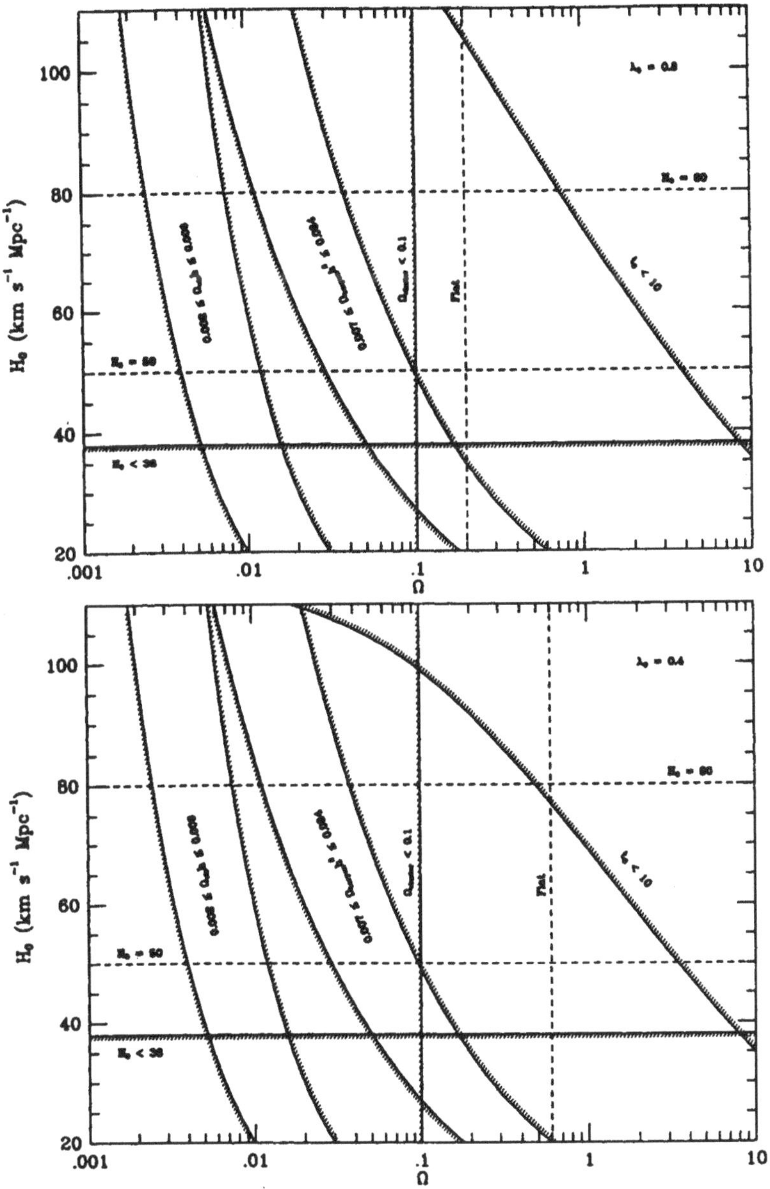

Figures 7a and 7b. Same as Figure 3 but with effective $\lambda_o = \frac{\Lambda_o}{3H_o^2}$ 0.8 (a) and 0.4 (b). Note for $\lambda_o = 0.8$, $\Omega = 0.2$ yields a flat universe and for $\lambda_o = 0.4$, $\Omega = 0.6$ is flat. These figures illustrate that even small λ_o enables high H_o to be consistent with a flat universe.

From dynamics alone we can put an upper limit on Ω by limiting the deceleration parameter q_o. From limits on the deviations of the redshift-magnitude diagrams at high redshift, we know that $q_o \lesssim 2$ (for zero cosmological constant $\Omega = 2q_o$). Thus, we can argue that $\Omega \lesssim 4$ or that $f(\Omega) \gtrsim 0.5$. Therefore, from dynamics alone, with no further input, we can conclude only that

$$5 \lesssim t_u \ (Gyr) \ \lesssim 25 \tag{2}$$

Since the lower bound here could also be obtained from the age of the earth, it is clear that the dynamical technique is not overly restrictive unless one could somehow decide between the supernova approach and Tulley-Fisher.

Even high values of H_o can be consistent with high ages by invoking the cosmological constant. Figures 7a and 7b are the equivalent of Figure 3, but with effective $\lambda_o = \Lambda_o/3H_o^2$ of 0.8 and 0.4, respectively. Note that even $\lambda_o \sim 0.4$ allows high H_o to be consistent with a flat universe. It is interesting to speculate that $\Lambda_o \neq 0$ can be produced by a late-time vaccum phase transition of the type proposed by Hill, Schramm and Fry.[117] However, such models do require a fair degree of tuning.

4.3 - The Age from the Oldest Stars

Globular cluster dating is an ancient and honorable profession. The basic age comes from determining how long it takes for low mass stars to burn their core hydrogen and thereby move off the main sequence. The central temperature of such stars is determined by their composition and the degree of mixing. While there has certainly been some static as to what is the dispersion between the age of the youngest versus the oldest globular cluster in a given calculation, there is a surprising convergence on the age of the oldest clusters. Since the age of the very oldest cluster is the critical cosmological question, it is really somewhat of a red herring as to how much less the youngest cluster may be. The convergence on the age of the oldest does require a consistency of assumptions about primordial helium and metalicity (including O/Fe). Difference between different groups can be explained away once agreement is made on these assumptions. For example, Sandage's[113] oldest ages of $\sim$ 18 Gyr and Iben and Renzin's[118] of $\sim$ 16 Gyr are consistent if the same helium is used. (Lower helium yields lower ages. Iben assumed the Pagel[119] value of y = 0.23.) Another decrease of a billion years occurs if O/Fe is assumed high as current observations show for extreme Pop II.

Another effect is the fact that these old stable stars will have some gravitational settling of their helium which will also shorten the ages about 1 Gyr relative to calculations where core helium enrichment is purely due to nuclear burning. All of these assumptions give a standard model for the oldest globulars of $\sim 15 \pm 1$ Gyr where the ± 1 is only the difference between different groups using the same standard assumptions. However, in addition to the calculational errors, there are also uncertainties in composition/opacity, uncertainties in distance/turnoff luminosity, and uncertainties in reddening/surface temperature at turnoff. Then, there are systematic uncertainties due to model assumptions such as using Big Bang helium abundances (y = 0.24) versus enhanced helium due to helium production accompanying the extreme Pop II metal production and perhaps preferential helium in cluster formation.[120] Also note that the current best fit BBN helium is actually closer to 0.25 than 0.23. Shi et al.[120] showed that assumptions about He could lower the best fit age by as much as 2 Gyr without violating any other constraint (e.g. γ must be ≤ 0.28 to fit RR Lyrae blue edge). Furthermore, there are recent suggestions from the first Keck spectroscopic temperature determinations of globular cluster stars that the true temperatures are as much as 200 K hotter than the photometric determinations. This could also shift the age downward by as much as 2 Gyr. Furthermore, Shi et al.[120] (see also Shi [121]) have shown that mass loss due to the variable strip crossing the main sequence near the cluster turnoff could also shift the age down by 1 to 2 Gyr. However, these combined effects do not add linearly. No matter what, low mass stars can burn their hydrogen only so fast. In fact, one can use the standard solar model to get a quick estimate of an extreme globular age. The main line pp-chain is the main energy generation mechanism for the Sun and the globular clusters. The basic pp part of the solar model is now well confirmed by the GALLEX solar neutrino experiment as we will see in the next section. Since the Sun has a much higher metalicity than the oldest globular clusters, and presumably has higher helium content and is at least as massive, if not more massive, it is paramount that the calculated main sequence lifetime of 10 Gyr for our Sun will always be a lower bound on the oldest globular cluster lifetimes. This 10 Gyr is also consistent with Shi et al.[120] and with an independant study by Chaboyer in 1995,[122] where an age of to 22 Gyr was found. (Chaboyer added his errors in quadrature, which may have minimalized certain systematic effects that would probably push the lower bound down to 10 Gyr.) Thus, it is reasonable to conclude that shifting the fit age for the oldest globulars down to 12 Gyr is quite reasonable considering all the possible age lowering tools available. But an extreme lower bound at 10 Gyr is not able to be broken.

Note that the time delay for cluster formation does not change this limit, since it is certainly possible to hypothesize an isocurvature model where globular clusters are the first objects[123] to form after recombination (their Jeans mass at that time is the globular cluster mass). Their Kelvin-Helmholtz time is only $\sim 10^7$ yr, so in principle, they could be present as early as 10^8 yr rather than the Big Bang. (Of course, standard CDM models extend this to several Gyr.)

4.4 - Nucleocosmochronology

Nucleocosmochronology is the use of abundance and production ratios of radioactive nuclides coupled with information on the chemical evolution of the Galaxy to obtain information about time scales over which the solar system elements were formed. Typical estimates for the Galaxy's (and Universe's) age as determined from cosmochronology are of the order of 9 -18 Gyr (e.g. Meyer and Schramm[124]).

In recent years questions about the role of β-delayed fission in estimating actinide production ratios as well as uncertainties in ^{187}Re decay due to thermal enhancement and the discussion of Th/Nd abundances in stars have obfuscated some of the limits one can obtain. In particular, we note that the formalism of Schramm and Wasserburg[125] as modified by Meyer and Schramm[124] continues to provide firm bounds on the mean age of the heavy elements. In fact, Th/U provides a firm lower limit to the age and Re/Os, a firm upper limit. These limits are based solely on nuclear physics inputs and abundance determinations. To extend these mean age limits to a total age limit requires some galactic evolution input. However, as Reeves and Johns[126] first showed, and as Meyer and Schramm[124] developed further, one can use chronometers to constrain Galactic evolution models and thereby further restrict the age from the simple mean age limits of Schramm and Wasserburg. To try to push further on such ranges and give ages to ±1 Gyr accuracy, as some authors have done, always necessitates making some very explicit assumptions about Galactic evolution beyond the pure chronometric arguments. At the present time such model-dependent ages are not fully justified and should probably not be used as arguments to question (or support) cosmological models, but pure, nuclear derived lower bounds are very useful in this regard.

Let us review what can be said from the nuclear physics without making too many specific model-dependent assumptions.

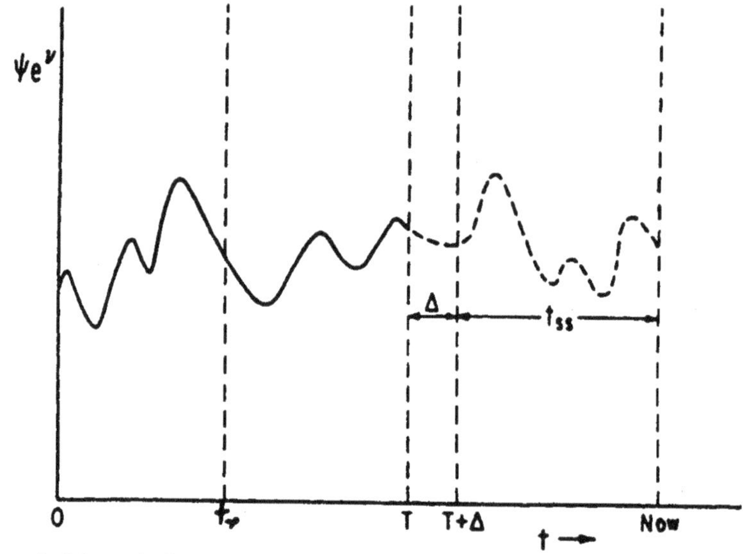

Figure 8. Schematic diagram showing the effective nucleosynthesis rate as a function of time. The quantity T is the total duration of nucleosynthesis, and t_ν is the mean time for the formation of the elements; Δ is the time interval between the end of nucleosynthesis and solidification of solar system bodies; t_{ss} is the age of the solar sysstem solid bodies. The total age of the elements is $T + \Delta + t_{ss}$.

The linearized equation for the time dependence of the abundance N_i of nuclide i in the interstellar medium of the Galaxy[127–129] is

$$\frac{dN_i(t)}{dt} = -\lambda_i N_i(t) - w(t)N_i(t) = P_i\psi(t), \tag{3}$$

where λ is the decay rate of nuclide i, $w(t)$ is a time-dependent parameter representing the rate of movement of metals into and out of the interstellar medium for reasons other than decay, $\psi(t)$ is the amount of mass going into stars per unit time, and P_i is the number of nuclei i produced per unit mass going into stars.

It is now possible to solve for the abundance N_i of nuclide i at a given time by integration of equation (3). This is done in the context of the scenario for evolution of the material making up the solar system shown in Figure 8. In Figure 8, T is the time of the last event contributing to formation of the elements going into the solar system, Δ is the time interval between this last nucleosynthetic event and the solidification of the solar system solid bodies, and $t_{ss}(= 4.6 \pm 0.1$ Gyr) is the age of these solid bodies. In this scenario, Δ is a free decay period for the elements, and, consequently, we might choose to measure meteoritic abundance back to $t = T$. Meteoritic material is a closed system only after time $t + \Delta$. This material thus gives abundances at times as early as $T + \Delta$ with minimal uncertainty due to chemical fractionation, but not before. Integration of the equation for time $t = 0$ to $t = T$ followed by free decay over an interval Δ, yields

$$N_i(T+\Delta) = P_i exp(-\lambda_i\Delta)exp[-\lambda_i T - \nu(T)] \int_o^t \psi(t)exp[\lambda_i t + \nu(t)]dt, \tag{4}$$

where

$$\nu(t) = \int_o^t \omega(\xi)d\xi,$$

and we have assumed P_i to be independent of time.

Age estimates from radionuclides are obtained first by expansion of the normalized effective nucleosynthesis rate $\phi(t)$, defined as

$$\phi(t) \equiv \frac{\psi e^\nu}{\int_o^t \psi e^\nu dt} \tag{5}$$

in moments μ (defined below) about the mean time for formation fo the elements t_ν, given by

$$t_\nu = \int_o^t \phi(t)dt. \tag{6}$$

With this mean time, the moments μ are defined as

$$\mu_n \equiv \int_o^t (t - t_\nu)^n \phi(t) dt. \tag{7}$$

Meyer and Schramm[124] find (analogously to Schramm and Wasserburg[125]) an expression for the mean age of the elements as measured back from $t = T$:

$$T - t_\nu = \Delta_{ij}^{max} - \Delta + \frac{(\lambda_i + \lambda_j)\mu_2}{2} + \frac{(\lambda_i^2 + \lambda_i\lambda_j + \lambda_j^2)\mu_3}{6} \tag{8}$$

$$+\frac{1}{4}(\frac{\mu_4}{6} - m_2^2)(\frac{\lambda_i^4 - \lambda_j^4}{\lambda_i - \lambda_j}) + ..., \tag{9}$$

where

$$\Delta_{ij}^{max} \equiv \frac{ln(\frac{(P_i/P_j)}{N_i(T+\Delta)/N_j(T+\Delta)})}{\lambda_i - \lambda_j} \equiv \frac{lnR(i,j)}{\lambda_i - \lambda_j}$$

and where the subscript j denotes a second radionuclide and $R(i,j)$ is its distance from nuclide i.

Clearly, $T - t_\nu$ depends upon $\phi(t)$ through the moments μ; thus, we require some information about the effective nucleosynthesis rate if we are to continue. We may proceed in one of two fashions. We may choose a specific form or model for $\phi(t)$, in which case our results will be model-dependent. Alternatively, we may attempt to find external constraints for $\phi(t)$ that will allow upper and lower limits to be placed on T essentially independent of any model for $\phi(t)$. This later tack is the one described below.

First, we note that since the moment terms increase $T - t_\nu$ over $\Delta_{ij}^{max} - \Delta$, a lower limit on T is given by

$$T \gtrsim \Delta_{ij}^{max} - \Delta. \tag{10}$$

This is the long-lived limit of Schramm and Wasserburg[125] and gives a model-independent lower limit on the time for nucleosynthesis. With knowledge of t_ν/T, the lower limit is pushed up to

$$T \gtrsim (1 - \frac{t_\nu}{T})^{-1}(\Delta^{max} - \Delta). \tag{11}$$

We derive limits on t_ν/T below. In principle, nucleochronology alone is not able to give a firm upper limit to an age as was demonstrated by Wasserburg, Schramm and Huneke[130] who found consistent ages of $\gtrsim 10^{13}$ yrs. However, by using our constraints on average rates, some statements can be made, assuming that production was relatively smooth with no large gaps. For an upper limit on T, Meyer and Schramm find that

$$T \lesssim (1 - t_\nu/T)^{-1}(\Delta^{max} - \Delta)(1 + \epsilon) \tag{12}$$

where ϵ, which represents the correction to the long-lived limit, is constrained as

$$\epsilon \lesssim \frac{1}{8}(1 - t_\nu/T)^{-2}(\lambda_i + \lambda_j)(\Delta^{max} - \Delta)(1+\epsilon)^2$$

$$+\frac{5}{312}(1 - t_\nu/T)^{-3}(\lambda_i^2 = \lambda_i\lambda_j + \lambda_j^2)(\Delta^{max} - \Delta)^2(1+\epsilon)^3$$

$$+\frac{1}{288}(1 - t_\nu/T)^{-4}\frac{(\lambda_i^4 - \lambda_j^4}{\lambda_i - \lambda_j)}(\Delta^{max} - \Delta)^3(1+\epsilon)^4 + \ldots . \qquad (13)$$

With limits on t_ν/T, equation (13) can be solved and, hence, limits on T will be available from equations (11) and (12). Meyer and Schramm[124] develop such limits on t_ν/T in a method inspired by the work of Reeves and Johns.[126] First, an average nucleosynthesis rate $< \psi >_{r_{i,j}}$, i over the interval $r_i \leq t \leq T$ is defined:

$$< \psi >_{r_{i,j}} \equiv \frac{\int_{r_i}^{T} \psi e^{\nu} e^{-\lambda_i(T-t)} dt}{\int_{r_i}^{T} e^{-\lambda_i(T-t)} dt}. \qquad (14)$$

Then, through use of equation (14), analogous expressions for nuclides j, and variation over all possible intervals $r_i \lesssim t \lesssim T$ and $r_j \lesssim t \lesssim T$, it is found that

$$\frac{e^{(\lambda_i - \lambda_j)\Delta}}{R(i,j)} \frac{(1 - e^{-\lambda_j TT})}{(1 - e^{-\lambda_i T})} \frac{\lambda_i}{\lambda_j} \leq \frac{< \psi >_i}{< \psi >_j} \leq \frac{e^{(\lambda_i - \lambda_j)\Delta}}{R(i,j)} \frac{\lambda_i}{\lambda_j}, \qquad (15)$$

where we choose T to be its smallest possible value, namely, that given by equation (10).

The ratio of average rates $< \psi >_i / < \psi >_j$ constrained in equation (15) is useful because it determines the general trend of ψe^ν over a few lifetimes of nuclide i. Thus, since $\lambda_i > \lambda_j$, if $< \psi e^\nu >_i / < \psi >_j \approx 1$, then ψe^ν is generally falling over a few times r_i, and if $< \psi >_i / < \psi >_j > 1$, then ψe^ν is generally rising over a few times r_i.

To obtain constraints on t_ν/T, we define $r(i,j)$ as the ratio $< \psi >_i / < \psi >_j$. We assume a set of m chronometers. We label the longest by $i = 1$, the next longest by $i = 2$, and so on, to the shortest-lived, labeled $i = m$. We then have as constraints on ψe^ν

$$\psi e^\nu = r(i,l) \text{ for } t_{i-1} \lesssim t \lesssim t_i,$$

where i runs from l to m, t_i is defined as

$$t_i = \frac{T}{r_i}(r_i - r_{i+1}),$$

and $t_m = T$.

From the above, constraints on t_ν/T are available, viz.,

$$\frac{t_\nu}{T} = \frac{1}{2}\frac{\sum_{i=1}^{m} r(i,l)[(r_i - r_i = 1)^2]}{r_i \sum_{j=1}^{m} r(j,l)(r_j - r_j + 1)}.$$

Use of upper limits on $r(i,j)$ gives an upper limit on t_ν/T. For lower limits on t_ν/T, Meyer and Schramm choose to use two chronometers in a slightly different fashion to obtain

$$\frac{t_\nu}{T} = \frac{r(2,1) - \sqrt{r(2,1)}}{r(2,1) - 1}.$$

Use of lower limits on $r(2,1)$ gives lower limits on t_ν/T. Similarly, upper limits on $r(2,1)$ can give upper limits on t_ν/T.

With constraints on t_ν/T, we can, given the requisite input data, derive limits on T and T_{GAL} from the fact that

$$T_{GAL} = T + \Delta + t_{ss}.$$

We turn now to a discussion of the input data.

In Table 3 we present best estimates of decay rates, the ratios $R(i,j)$, and resulting Δ^{max} values for the Re/Os, Th/U, U/U, and Pu/U chronometric pairs. The text that follows gives a brief discussion of the uncertainty in these data.

Table 3

Cosmochronological Input Data and Paramaters		
Pair	$\lambda_i(Gyr^{-1})$	$\lambda_j(Gyr^{-1})$
$^{187}Re/^{187}Os$	0.0159(+0.0005, -0.0004)	
$^{232}Th/^{238}U$	0.0495 (+0.0000, -0.0000)	0.1551 (+0.0002, -0.0002)
$^{235}U/^{238}U$	0.985 (+0.009, -0.009)	0.1551 (+0.00002, -0.00002)
$^{244}Pu/^{238}U$	8.47 (+0.27, -0.27)	0.1551 (+0.0002, -0.0002)
Pair	$R(i,j)$	$\Delta^{max}(Gyr)$
$^{187}Re/^{187}Os$	1.03-1.23	1.8-13.4
$^{232}Th/^{238}U$	0.65(±0.09)	4.1 (+1.4, -1.2)
$^{235}U/^{238}U$	4.7 (+1.3, -0.9)	1.9 (+0.3, -0.3)
$^{244}Pu/^{238}U$	112 (+138, -92)	0.57 (+0.12, -0.21)

A. *Re/Os*

The long-lived chronometric pair $^{187}Re/^{187}Os$ is unique because ^{187}Os is stable. Since $\lambda_j = 0$, and since $\Delta^{max} > \Delta$ (see Symbalisty and Schramm[129] who find $\Delta \lesssim 0.2Gyr$), we may write equation (13), through use of equation (9), as

$$\epsilon \lesssim \frac{1}{8}(1-\frac{t_\nu}{T})^{-2}[lnR(187,187)](1+e)^2 + \frac{5}{312}[(lnR(187,187)]^2(1+e)^3 \tag{16}$$

$$+\frac{1}{288}(1-\frac{t_\nu}{T})^{-4}[lnR(187,187)]^3(1+e)^4....$$

The only necessary data, then, are $R(187,187)$ and λ_{187} (to get $\Delta^{max}_{187,187}$). $R(187,187)$ is given by Schramm[131],

$$R(187,187) = 1 + \frac{(^{187}Os)c}{^{187}Re}, \tag{17}$$

where $(^{187}Os)c$ is the r-process contribution to ^{187}Os.

Unfortunately, both $R(187,187)$ and λ_{187} are uncertain quantities. Bound state β-decay of ^{187}Re occurring due to astration may greatly enhance the decay rate over the lab rate.[132,133] Detailed galactic evolution models are thus required to determine the amount of astration of ^{187}Re and, consequently, the effective decay rate of ^{187}Re. This is difficult and uncertain work. We note instead that the effect of astration is always to increase λ_{187}. Thus, from equation (9) we obtain an upper limit on $\Delta^{max}_{187,187}$. We also emphasize that ϵ in equation (16) is independent of λ_{187}.

The uncertainty in $R(187,187)$ arises from two sources. First, a low-lying, excited nuclear state in ^{187}Os complicates the determination of (^{187}Os)c.[134–136] Second, s-process branchings in the $W - Os$ region[137,138] may contribute to the uncertainty in $R(187,187)$. The range on $R(187,187)$ found by Meyer and Schramm from the analysis of Yokoi et al. and meteoritic data of Luck, Brick and Allegre[139] is $1.06 \lesssim R(187,187) \lesssim 1.14$. The numbers of Arnould et al. lead Meyer and Schramm[124] to the larger range $1.03 \lesssim R(187,187) \lesssim 1.23$. Meyer and Schramm rely mainly on the former range for $R(187,187)$, but also consider the effects of the latter range. Use of cross section data from Winters et al.[140] and a best value for the cross section correction factor f_σ of 0.82[141] gives a best $R(187,187)$ of 1.12. The lab λ_{187} is $1.59^{+0.05}_{-0.04} \times 10^{-11} yr^{-1}$. The bottom line here is that one should not ignore ^{187}Re but use it as an upper bound.

B. *Th/U*

Beta-delayed fission is the cause of the largest amount of uncertainty in the production ratio of $^{232}Th/^{238}U$. The calculations of Thielmann et al.[142–144] give 1.4 as the production ratio. The calculations of Meyer et al.[145] give less delayed fission and, hence, suggest a higher production ratio. Although Meyer et al. do not include barrier penetration in their calculations, a fact which suggests that their production ratios may be too large, their results seem to be favored by Hoff's[146] study of yields from thermonuclear explosions. The implication appears to be that less delayed fission occurs than that given by Thielmann et al.[142–144] We thus conclude that Meyer and Schramm's use of the Thielmann et al. value of 1.4 as a lower limit on the Th/U production ratio is justified. With the probability of some β-delay fission, 1.7 is probably a reasonable upper limit with 1.55 as a good compromise. They also argue from terrestrial isotopic lead ratios and meteoritic ratios that the present solar system value for $^{232}Th/^{238}U$ is 3.9 ± 0.2. The relevant decay rates are $\lambda_{232} = 4.95 \times 10^{-11} yr^{-1}$ (Jaffey et al.[147]).

C. U/U

Meyer and Schramm choose the Schramm and Wasserburg production ratio range $1.5^{+0.4}_{-0.3}$ as the best range for $^{235}U/^{238}U$. The range contains the Thielmann et al. value of 1.24.

The $^{235}U/^{238}U$ abundance ratio is well-known. Meyer and Schramm take it to be $1/(137.88 \pm 0.14)$. $\lambda_{235} = (9.8485 \pm 0.0135) \times 10^{-10} yr^{-1}$ (Jaffey et al.[147]).

D. Pu/U

The $^{244}Pu/^{238}U$ pair is the pair most affected by delayed fission. Meyer and Schramm[124] use the Thielmann et al.[142–144] value of 0.12 as a lower limit and the Symbalisty and Schramm[129] upper limit 1.0 (no delayed fission) as an upper limit with 0.56 as a compromise best value. The abundance ratio is 0.006 ± 0.001 (Hudson et al.[148]), although Meyer and Schramm note that the abundance ratio range may be more uncertain than this. $\lambda_{244} = 8.47 \pm 0.27 \times 10^{-9} yr^{-1}$ (Fields et al.[149]).

Meyer and Schramm derive a range on t_ν/T of $0.43 \lesssim t_\nu/T \lesssim 0.59$. From this range and data from Table 1 for Th/U, we find a lower limit on T_{GAL} of 9.6 Gyr. Also, from the range $1.06 \lesssim R(187, 187) \lesssim 1.14$, Meyer and Schramm derive an upper limit on T_{GAL} of 28.1 Gyr. The range on t_ν/T agrees with the results of Hainebach and Schramm's[128] study of detailed galaxy evolution models. The range 9.6Gyr $\lesssim T_{GAL} \lesssim$ 28.1 Gyr includes the cosmochronological allowed galactic models and shows that the effect of these uncertainties is large. One may conclude from these results that nuclear cosmochronology, so simple in conception, is rendered quite difficult in practice because of input data uncertainties. It should be noted that any author who gives smaller ranges from nucleochronology is making some implicit assumptions about the chemical evolution of the Galaxy, and so, such ages are not pure nuclear dating. (They are also probably underestimating the uncertainty in the production rate determination process.)

4.5 - Age Summary

The age situation at the present time can be summarized by Figures 3 and 6. We see there that an $\Omega = 1$ universe is consistent with $t > 10$ Gyr as long as $H_o \lesssim 66$ km/sec/Mpc. If uncertainties on H_o (including bounds on systematics) ever exclude 66, then one would require $\Lambda_o \neq 0$ to achieve the flat universe favored by inflation models.

Naively, we expect gravitational microphysics on the Planck scale, M_p to determine the scale of Λ_o. An effective $\lambda_o \sim 1$ requires $\rho_\Lambda \sim 10^{-121} M_p^4$. This seems like remarkable tuning. Of course, some late-time transition on the fraction of an eV scale could substitute for M_p if the early $\rho_\Lambda \sim M_p^4$ effects could be surpressed to more than 121 orders of magnitude. Because these problems seem awkward to avoid, most physicists think $\Lambda = 0$.

As an anthropic aside, if it were ever shown that $\Lambda_o \neq 0$, then we may have to appeal to the following anthropic argument (ugh!). While particle physics prefers a large value for $\Lambda_o \sim M_p^4$, the only values consistent with an old universe have to have $\Lambda_o < 10^{-121} M_p^4$. Thus, our existence plus particle theory would make the maximum value consistent with our existence the most likely value. (Hopefully, a better motivated physics explanation for Λ_o will eventually be found.)

5. Solar Neutrinos

5.1 - Introduction to Solar Neutrino Discussion

The observed deficit of the solar neutrino flux with respect to the prediction of the standard solar models has been one of the most interesting problems in both astrophysics and particle physics.[150–154] The "Solar Neutrino Problem" received renewed interest in the fall of 1994 when the GALLEX [155] experiment measured an artificial neutrino source, ^{51}Cr and found precisely the expected number of events (actually 1.04 ± 0.12). This gave a renewed confidence in solar neutrino observations. Other recent activity has included new studies of the crucial ^{7}Be (p, γ) ^{8}B reaction.[156,157] In this section, we attempt to review the current solar neutrino situation in the light of the most recent nuclear experimental results and solar neutrino experimental results, and we examine the minimal rates that successful solar models can yield for the Homestake and the two gallium experiments. We use the results from the Kamiokande experiment without making any connection to solar models to show the reason for the difficulty between the Kamiokande, the Homestake, and the gallium data. We also note that even with the assumption of a very low ^{7}Be(p,γ)^{8}B rate, the Homestake experiment is still in conflict with variations of standard solar models. We show that the gallium experiments will also be in conflict if they are proven to have fluxes significantly below 110 SNU (1 SNU = 10^{-36} capture/target/atom/second) as current data seem to imply.

The observed solar neutrino flux of four currently available solar experiments, the Homestake ^{37}Cl capture experiment,[158] the Kamiokande $\nu - e$ scattering experiment,[159] the GALLEX ^{71}Ga capture experiment,[160] and the SAGE ^{71}Ga capture experiment,[161] have been summarized in Table 4. All of them are lower than the predictions of the most referenced standard solar models of Bahcall and Pinsoneault

(BPSSM),[150] or the standard solar model of Turck-Chieze et al. (TLSSM).[151] It should be noted that alternative solar model calculations give essentially identical results if the same input parameters (nuclear reaction rates, radiative opacities, the heavy element abundance in the sun, the age of the sun, etc.) are used.[150]

Table 4
Solar neutrino fluxes measured by experiments vs. predictions of standard solar models. (Kamiokande rates are normalized to the prediction of BPSSM.)

	Experimental data	BPSSM	TLSSM	A lowflux model
Homestake	2.55 ± 0.25 SNU	8.0 SNU	6.4 SNU	4.7 SNU
Kamiokande	$0.51 \pm 0.04 \pm 0.006$	1.00	0.75	0.54
Gallium exps.	$79 \pm 10 \pm 6$ SNU (GALLEX) $74 \pm 17 \pm 10$ SNU (SAGE)	132 SNU	123 SNU	117 SNU

Among the four experiments, both the Kamiokande experiment and now the GALLEX experiment have been in some real sense, well "callibrated". The two gallium experiments, GALLEX and SAGE, although their initial results were seemingly inconsistent,[162,163] are increasingly consistent with each other after more than three years of observations. The Homestake experiment, which observed the most significant solar neutrino deficit with respect to the standard solar models and is therefore crucial to the solar neutrino problem, naturally draws questions since so much was riding on the experiment. Although no serious questions about the experiment have been proven, some still worry that its event rate seems to be inconsistent with a constant solar neutrino flux; in particular, its event rate after 1985-1986 pump failures (2.8 ± 0.3 SNU[164]) is marginally higher than that before (2.1 ± 0.3 SNU[165]). The latter point was elaborated by some attempts to reconcile the solar neutrino experiments and the standard solar models, arguing that a higher Homestake rate after 1986 might not be far below some modified standard solar models.[166,167]

However, it should be pointed out that the deviation of the Homestake data from a constant rate is not statistically significant[168–170] and there are no grounds besides different rates that discriminate data before 1984 from those after 1986. Different pump configurations and different pumping times have been introduced yielding no obvious change in the capture rate.[171] Various tests by the Davis group also showed no unexpected systematic uncertainties.[171] Furthermore, the rate after 1986 is still statistically consistent with the rate before 1984 within about 2σ. It is therefore completely unjustified to use only the Homestake data after 1986 in discussing the solar neutrino problem, since, if the rate did shift, then there are unexpected systematic uncertainties in the experiment which put the whole experiment into question until it can be resolved.

Attempts to lower the solar neutrino flux predicted by solar models concentrate on two approaches: (1) lowering the core temperature T_c of the Sun; and (2) using a lower ^{7}Be(p, γ)^{8}B rate, because the uncertainties from these two factors are known to be far greater than the uncertainties from the other aspects of solar models.[171] In the context of the standard solar models (i.e., no rotations, no magnetic fields, no exotic particles, and standard nuclear reaction network, etc.[171]), a lower T_c can be achieved by a lower heavy element abundance of the Sun, Z, and/or lower radioactive opacities at the center of the Sun. Most if not all non-standard solar models, such as invoking assumptions like a 10^9 Gauss magnetic field in the solar interior, or a black hole at the center of the Sun, or captures of Weakly Interacting Massive Particles (WIMPs) by the Sun, or even invoking additional core mixing, also end up reducing the neutrino flux by effectively lowering the core temperature T_c of the sun.[171] They therefore can be categorically included in the first approach to lower the predicted solar neutrino flux.

It has been shown by Bahcall *et al.*[171] that in standard solar models, the predicted fluxes from different neutrino sources depend very differently on the core temperature of the models. For example, for pp neutrinos, ^{7}Be neutrinos and ^{8}B neutrinos, which are predicted to constitute most of the neutrinos detected in solar neutrino experiments,

$$\phi(\mathrm{pp}) \propto T_c^{-1.2}, \ \phi(^7\mathrm{Be}) \propto T_c^8, \ \phi(^8\mathrm{B}) \propto T_c^{18}, \tag{18}$$

where ϕ's are neutrino fluxes. Therefore, a $\pm 1.5\%$ variation in T_c alone, which is readily achievable by adjusting standard solar models within a reasonable range (for example, the $\pm 10\%$ 1σ uncertainty in Z alone changes T_c by $\pm 0.8\%$, as our calculations[153,154] show approximately that $Z \propto T_c^{0.08}\%$), will result in a variation in the ^{8}B solar neutrino flux by a factor of 2. On the other hand, the resultant ^{7}Be solar neutrino flux only varies by less than 30%, while the predicted pp neutrino flux is essentially free of uncertainties from a variation in T_c in standard solar models.

The uncertainty in the ^{7}Be(p,γ)^{8}B rate brings significantly more uncertainty to the ^{8}B neutrino flux. This rate linearly affects the ^{8}B neutrino flux (since the ^{8}B neutrinos are the decay product of the resultant ^{8}B excited state). Its uncertainty comes from both theoretical extrapolations and interpretations of experimental data themselves.[172] Theoretical extrapolations for six previous experiments yielded $S_{17}(0)$ for the reaction ranging from 16 to 42 eV·barn, with a weighted average of 22.4 ± 2.1 eV·barn.[172] ($S(E) = \sigma(E)E\exp(-2\pi\eta)$, where $\sigma(E)$ is the cross section, $\eta = e^2 Z_1 Z_2/\hbar v$ where Z_1 and Z_2 are the charges of colliding particles and v is their relative velocity. The subscript 17 denotes reaction ^{7}Be(p,γ)^{8}B.) Only two of the six experiments, namely those of Kavanagh[173] and Filippone *et al.*,[174] went to energies well below the $M1$ resonance at ~0.6 MeV. They yield 25.2±2.4 eV·barn (Kavanagh) and 20.2±2.3 eV·barn (Filippone) respectively, which disagree with each other at 2σ. Several favor the experiment of Filippone *et al.* since it is the latest among the two and was published in a refereed journal.[153,154,167] The preliminary result of a new experiment by Motobayashi *et al.*, which avoids the $M1$ resonance by using the reverse reaction of disassociating ^{8}B by the coulomb field of heavy nuclei, implies a low $S_{17}(0)$. A new extrapolation of the Motobayashi *et al.* data by Langanke and Shoppa,[157] which argues for a large $E2$, if proven correct, could imply $S_{17}(0)$ as low as 12±3eV·barn. Such a low $S_{17}(0)$ would obviously be in conflict with previous measurements. Gai,[175] on the other hand, claims that their data do not favor a large $E2$ contribution. However, given the current situation, one should keep an open mind and possibly allow S_{17} to vary downward by as much as a factor of 2 from the previous average. While such variations can clearly allow easy fits to the Kamiokande data, we will show that they nonetheless do not resolve the apparent conflict between the Kamiokande data and the Homestake data.

5.2 Implications of Solar Neutrino Experiments

Given such uncertainties in the ^{8}B neutrino flux, none of the individual solar neutrino experiments taken by themselves indicates strong evidence for a solar neutrino deficit that cannot be resolved with solar model variations: the Kamiokande experiment observes ^{8}B neutrinos only; hence, its event rate may serve better as a normalization of the absolute ^{8}B neutrino flux than an indicator of a solar neutrino deficit. The Homestake experiment, which should observe mostly ^{8}B neutrinos plus some ^{7}Be neutrinos (and a few pep neutrinos and CNO neutrinos), is also subject to a large uncertainty in its expected flux. The two gallium experiments, although capable of observing the uncertainty-free pp neutrinos, see a solar neutrino flux that is larger than the pp neutrino flux and hence fail to show a deficit in the model independent pp neutrinos.

Nevertheless, a problem for solar model solutions to the solar neutrino problem becomes robust if results from the Kamiokande and Homestake experiments are combined.[153,154,176–179] That is, if one normalizes the ^{8}B neutrino flux predicted by solar models to the Kamiokande result, the ^{8}B neutrinos should still contribute 3.1±0.4 SNU to the Homestake result. Even if one completely ignores neutrinos below the Kamiokande threshold, the part of the solar neutrino spectrum seen by the Kamiokande should contribute 2.85±0.4 SNU to the Homestake result, according to our calculation[153,154] and a calculation of Kwong and Rosen.[180] (However, an argument that the two experiments with different energy thresholds may observe different reductions in the ^{8}B neutrinos with respect to a standard solar model immediately implies a departure in the ^{8}B neutrino spectrum from that predicted by the standard electroweak theory, and thus new particle physics.) The observed Homestake rate of 2.55±0.25 SNU, therefore, indicates that the ^{7}Be neutrinos which should also be seen by the Homestake experiment, suffer more reduction than the ^{8}B neutrinos with respect to a particular standard solar model, in this case, BPSSM.

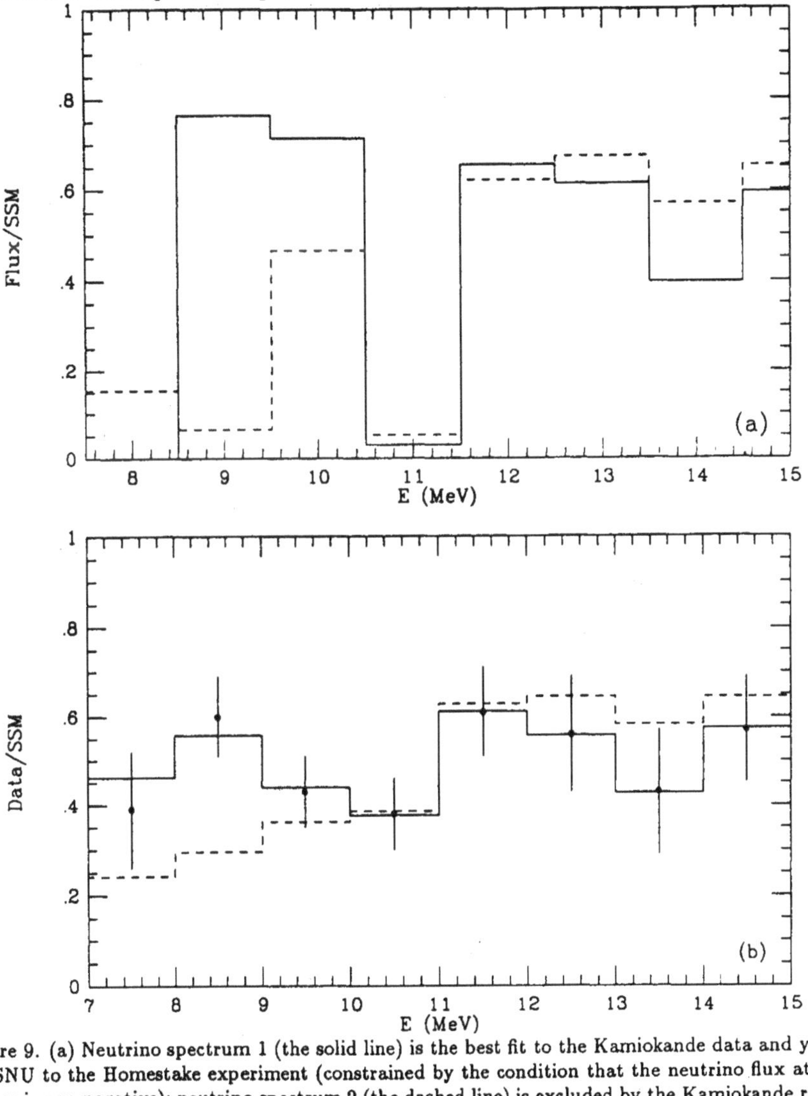

Figure 9. (a) Neutrino spectrum 1 (the solid line) is the best fit to the Kamiokande data and yields 2.6 SNU to the Homestake experiment (constrained by the condition that the neutrino flux at any energy is non-negative); neutrino spectrum 2 (the dashed line) is excluded by the Kamiokande result at 95%C.L. and yields 1.7 SNU to the Homestake experiment. Both spectra are normalized by the prediction of BPSSM.[359] (b) The solid line is the expected spectrum of recoil electrons from neutrino spectrum 1; the dashed line is the expected spectrum of recoil electrons from neutrino spectrum 2. The Kamiokande data are also shown.[181] Although spectra are normalized by the prediction of BPSSM for convenience, the results are completely independent of solar models.

A similar but model-independent argument is also intriguing. With a lower neutrino energy threshold, the Homestake experiment should see all the neutrinos observed by the Kamiokande (namely the ^{8}B neutrinos, but for this argument the source is irrelevant) and some neutrinos that have energies below the Kamiokande threshold (presumably ^{7}Be neutrinos, pep neutrinos and CNO neutrinos). Without any solar model assumption on the solar neutrino spectrum and flux, but only assuming a pure ν_e flux from the sun (i.e., no neutrino mixings), solar neutrinos observed by the Kamiokande should contribute 2.6$\pm$0.45 SNU to the Homestake result. Figure 9(a) shows a neutrino spectrum that yields the best fit to the Kamiokande data and 2.6 SNU to the Homestake experiment (constrained by the condition that the neutrino flux at any energy is non-negative), and a neutrino spectrum that is excluded by the Kamiokande result at 95% C.L. and yields 1.7 SNU to the Homestake experiment. Figure 9(b) shows their resultant recoil-electron spectra compared with the spectrum observed by the Kamiokande experiment. For convenience, spectra in Figure 9 are normalized by the prediction of BPSSM, but the results are completely independent of solar models. The pep neutrinos whose flux is directly tied to the pp neutrino flux, which is essentially uncertainty-free, are insensitive to solar model uncertainties and should contribute another 0.2 SNU to the Homestake result.[171] Therefore, the Homestake result of 2.55$\pm$0.25 SNU relative to the minimum of 2.6$\pm$0.45 SNU from the neutrinos seen by Kamiokande alone suggests that the ^{7}Be neutrinos must be severely suppressed with respect to predictions of standard solar models. As a result, to reconcile the Homestake result and the Kamiokande result requires either an explanation of a larger reduction in the ^{7}Be neutrinos than that in the ^{8}B neutrinos with respect to predictions of standard solar models, and/or an assumption of contributions from other neutrino flavors to the Kamiokande experiment, either of which would imply new neutrino physics or that one of the two experiments is wrong.

There is no known modification in solar models with known physics that can explain a more severe reduction in ^{7}Be neutrinos than in ^{8}B neutrinos with respect to BPSSM.[153,154,176–179] Lowering T_c in standard solar models will only suppress more ^{8}B neutrinos since they are more T_c-sensitive. This conclusion is also valid for non-standard solar models, since the ^{8}B neutrinos are intrinsically more T_c dependent than the ^{7}Be neutrinos due to a higher coulumb barrier in the reaction that produces the ^{8}B neutrinos.[171] In terms of nuclear reactions, the ^{7}Be(e,ν_e)^{7}Li reaction that produces ^{7}Be neutrinos and the ^{7}Be(p,γ)^{8}B reaction that produces ^{8}B neutrinos are the two branches of the ^{7}Be reactions in the sun, with branching ratios of 99.6% and 0.4% respectively.[171] The only artificial way to achieve a greater reduction in the ^{7}Be neutrinos than in the ^{8}B neutrinos is then to suppress the ^{7}Be production rate (by either suppressing ^{3}He(^{4}He,γ)^{7}Be rate at low energy by a least a factor of 2 or increasing the ^{3}He(^{3}He,2p)^{4}He rate at low energy by more than a factor of 4 to account for the ^{8}B neutrinos observed by the Kamiokande[182]) which affects the ^{7}Be neutrinos and ^{8}B neutrinos equally, and at the same time increases the $S_{17}(0)$ or raises T_c. But besides the fact that $S_{34}(0)$ and $S_{33}(0)$ (which are astrophysical factors similarly defined as $S_{17}(0)$ but for the ^{3}He(^{4}He,γ)^{7}Be reaction and the $^3He(^3He,2p)^4He$ reaction) are currently determined to within about 10% and 20% respectively,[150] the possible trend of $S_{17}(0)$ is downward instead of upward, and a higher T_c will increase CNO neutrinos (which have an even higher T_c dependency than the ^8B neutrinos) significantly to escalate the conflict between the Homestake result and the Kamiokande result rather than solve it.

5.3 Variations of Standard Solar Models

Table 5
A comparison between a model in this work and BPSSM w/o diffusion.[151]

	BPSSM w/o diffusion	A model in this work
Predictions for ^{37}Cl exp.		
pep neutrinos	0.2 SNU	0.2 SNU
^7Be neutrinos	1.1 SNU	1.1 SNU
^8B neutrinos	5.5 SNU	5.5 SNU
^{13}N neutrinos	0.1 SNU	0.1 SNU
^{15}O neutrinos	0.3 SNU	0.4 SNU
Total	7.2 SNU	7.3 SNU
Predictions for ^{71}Ga exp.		
pp neutrinos	71 SNU	70.0 SNU
pep neutrinos	3 SNU	3.0 SNU
^7Be neutrinos	34 SNU	33.0 SNU
^8B neutrinos	12 SNU	12.4 SNU
^{13}N neutrinos	3 SNU	3.5 SNU
^{15}O neutrinos	4 SNU	6.6 SNU
Total	127 SNU	129.0 SNU
Z	0.01895	0.0190
$S_{17}(0)$	22.4 eV·b	22.4 eV·b
T_c	1.56×10^7K	1.55×10^7K

To illustrate the above arguments and to show to what extent the solar neutrino prediction of standard solar models can vary, we construct a series of standard solar models with different T_c's and allowed $^7Be(p,\gamma)^8B$ to vary freely. These models are constructed with Dearborn's solar code[183] using the Livermore OPAL opacity table and nuclear reaction rates from Caughlan and Fowler[184] except for a freely varying $^7Be(p,\gamma)^8B$ rate. As a benchmark, we compared one of our models with Bahcall and Pinsonneault's no diffusion model (BPSSM w/o diffusion) in Table 5. They yield quite similar results. To see the uncertainties caused by the rates of $^3He(^4He,\gamma)^7Be$

and $^3He(^3He,2p)^4He$, we also vary $S_{34}(0)$ and $S_{33}(0)$ in our solar model calculations, although the variations in neutrino fluxes from their uncertainties are relatively small.

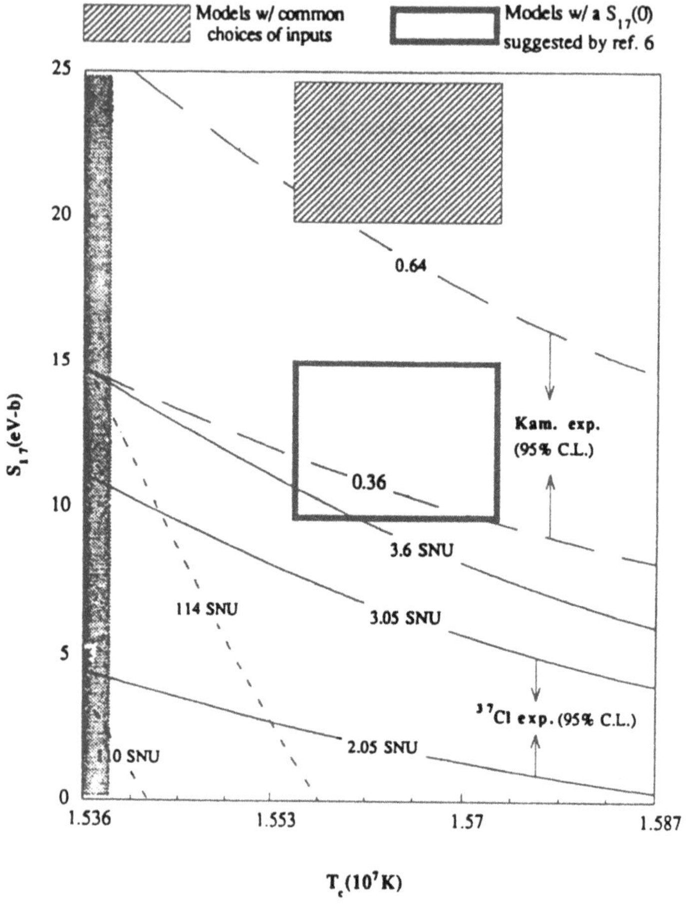

Figure 10. The predictions of standard solar models with different T_c and $S_{17}(0)$. Long-dashed lines: predicted Kamiokande rate, normalized by the solar model of Bahcall and Pinsonneault[150]; Solid lines: predicted Homestake rate in unit of SNU; Short-dashed lines: predicted gallium rate in unit of SNU. Regions allowed by the Kamiokande experiment and the Homestake experiment at 95% C. L. are shown by arrows. The shaded region on the left side is ruled out by the constraint from helioseismology. The hatched region on the upper center is expected by standard solar models with common choices of inputs (including their uncertainties); the rectangular region at the center is expected by standard solar models with common choices of inputs except a small $S_{17}(0)$ suggested by ref. 157.

Models with different T_c are constructed by varying Z between 0.014 and 0.021.

The measured value of Z is 0.0245 times the solar hydrogen abundance, or roughly 0.0177 ± 0.0017.[185] Our lowest T_c model has a Z of 0.014, that is, more than 2σ below the measured value. Models with such a low Z show distinctive structure differences from models with $Z \sim 0.018$. For example, the model with $Z = 0.014$ has a convective zone with a depth between $0.720R_\odot$ and $0.730R_\odot$ (where $R_\odot$ is the radius of the Sun), much shallower than the measured $0.713 \pm 0.003R_\odot$ from helioseismology,[186] whereas models with Z in the range of 0.015 to 0.021 have convective zones as deep as between $0.718R_\odot$ and $0.705R_\odot$. Models with $Z > 0.021$ may also be compatible with helioseismic results but they yield higher neutrino fluxes contrary to the direction of solving the solar neutrino problem.

Figure 10 shows predictions of these solar models with different T_c and $S_{17}(0)$ for the four solar neutrino experiments. Clearly, no overlap region exists between any two experiments at the 2σ level for each experiment, except for a small overlap between the Homestake and SAGE at low T_c and low $S_{17}(0)$. The gap between the Homestake and Kamiokande experiments is significantly larger, not surprisingly as argued before, due to the additional contributions from ^{7}Be neutrinos, CNO neutrinos and pep neutrinos in the Homestake experiment. In fact, as long as the Homestake experiment result is significantly below 3.6 SNU, a conflict between the Homestake and Kamiokande experiments cannot be solved by simply lowering T_c and the $^7\mathrm{Be}(\mathrm{p},\gamma)^8\mathrm{B}$ rate in the standard solar models.

The gaps between the gallium experiments and the other two experiments are not yet as severe as the gap between the Homestake experiment and the Kamiokande experiment. But they are still problematic for the variations to the standard solar models as we have discussed here, if the gallium experiment rate is significantly less than 110 SNU. Current gallium experiment results, therefore, may still be marginally compatible with variations of standard solar models. The ^{51}Cr source experiment tells us we can have a high degree of confidence in the GALLEX result. To move the GALLEX result up will require a real shift in statistics. As statistics improve, it may be possible that gallium experiments like the Homestake chlorine experiment will not be compatible with a standard solar model normalized by the Kamiokande result and hence would suggest new neutrino physics.

Figure 11 shows predictions of standard solar models with different T_c and artificially varied $S_{34}(0)$ and $S_{33}(0)$. (Solar neutrino fluxes are functions of $S_{34}(0)/\sqrt{S_{33}(0)}$.[171,182]) $S_{17}(0)$ is set to be 20 eV·barn. Obviously, 114 SNU is also the minimal gallium capture rate that can be reached by the standard solar models when allowing S_{34} and S_{33} to vary within their uncertainty ranges and allowing T_c to vary within the constraint from helioseismology. The gap between the Homestake and Kamiokande experiments cannot be narrowed even with a wild variation of $S_{34}(0)$ by a factor of 2 or $S_{33}(0)$ by a factor of 4.

Besides lowering Z as we did in our solar models, lower T_c may also be equivalently achieved by lowering the overall opacities, increasing the pp reaction rate, or shortening the age of the sun[187]. It is interesting to note that models with a $Z \gtrsim 0.015$ and opacities further artificially lowered only at the center (which might be possible under certain hypotheses[188]) may achieve a T_c lower than the lowest T_c

discussed above and still satisfy the current helioseismic constraint. But our calculations show that such models cannot suppress ^{7}Be neutrinos and CNO neutrinos as efficiently as the simple low Z models when a similar T_c is achieved. In addition, the opacities at the solar core can only be artificially lowered to the extent that the resultant helium abundance in the sun is $\gtrsim 0.26$.[189] As a result, after normalization by the 2σ lower limit of the Kamiokande experiment, the minimal rates predicted by these models for the Homestake experiment and the gallium experiments remain roughly the same as we discussed above.

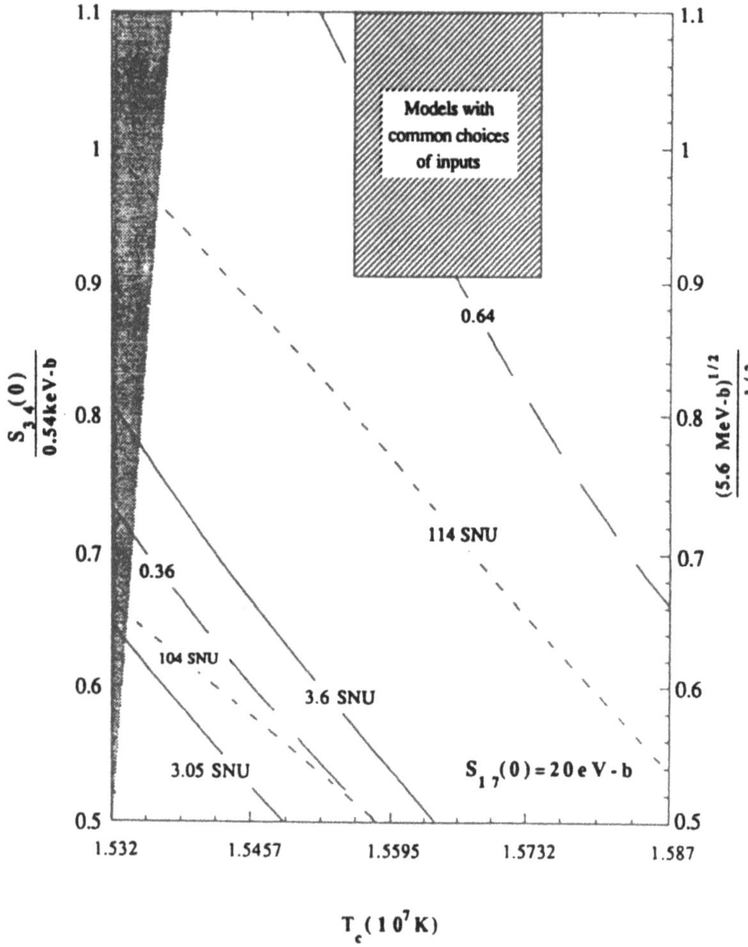

Figure 11. A similar plot as Figure 10 but with different T_c, $S_{34}(0)$ and $S_{33}(0)$ for standard solar models. Legends are similar to those in Figure 10. The shaded region on the left side is ruled out by helioseismology. The hatched region is expected by standard solar models with common choices of inputs (including their uncertainties).

The result we obtained also holds for models that include helium diffusion. A comparison between the Bahcall-Pinsonneault helium diffusion model and the no-helium diffusion model[150] shows that the T_c dependency of various solar neutrino sources, i.e., eq. (18) remains roughly intact after consideration of helium diffusion in solar models and the two classes of models yield similar neutrino fluxes when they have the same T_c. This is confirmed by our preliminary helium diffusion calculations. Having similar input parameters, a model with helium diffusion yields a higher T_c (and hence higher ^{8}B, ^{7}Be and CNO neutrino fluxes) than a model without helium diffusion, due to a higher helium concentration in the solar core in the diffusion model that increases the mean molecular weight.[150] Therefore, to achieve the same T_c of a no diffusion model, a helium diffusion model has to have a lower Z input than the no diffusion model. For example, our calculations show that a helium diffusion model that has the same low T_c as the $Z = 0.014$ no diffusion model should have $Z \approx 0.0134$. On the other hand, since the surface helium abundance of a helium diffusion model is roughly 10% less than the initial helium abundance due to gravitational settling,[150] the constraint on Z deduced from the measured Z/X for helium diffusion models becomes stricter than that for no diffusion models, namely $Z \approx 0.0184 \pm 0.0018$ instead of $Z \approx 0.0177 \pm 0.0017$. As a result, helium diffusion models with $Z \approx 0.0134$ are harder to reconcile with the constraint on Z.

As we discussed previously, all modifications to solar models that do not contradict known physics can only suppress the ^{7}Be neutrinos by at most the same factor as they suppress the ^{8}B neutrinos with respect to BPSSM. Therefore in the most extreme cases, if the Kamiokande experiment sees only 36% of the BPSSM prediction for ^{8}B neutrinos, the Homestake experiment should then expect 2.7 SNU from the ^{8}B neutrinos and the ^{7}Be neutrinos. If we add the 0.2 SNU contribution from pep neutrinos, which has a very small uncertainty,[171] and the contribution from CNO neutrinos, which may be neglected since they are very sensitive to T_c and some nuclear rates, we expect an absolute minimal rate of about 3.0 SNU for the Homestake experiment to accommodate any solar model solutions. Similarly, gallium experiments expect 74 SNU from pp neutrinos and pep neutrinos, 18 SNU from ^{7}Be and ^{8}B neutrinos, and contributions from CNO neutrinos. Therefore, the minimal gallium rate to accommodate solar models after ^{8}B neutrinos have been normalized by the Kamiokande experiment is about 92 SNU, if such solar models can successfully suppress most of the uncertain CNO neutrinos. It should be noted, however, we have not been able to construct any realistic solar model that achieves such extreme reductions.

5.4 Neutrino Physics Solutions

With an overall average Homestake rate of 2.55±0.25 SNU and a Kamiokande rate of 0.50±0.04 ± 0.06 times the BPSSM prediction, there is little space for a convincing solar model solution to the solar neutrino problem whether or not we assign a solar model spectrum to the neutrinos seen by the Kamiokande experiment, unless one of these experiments is in error. Variations of standard solar models yield a minimal rate of 3.6 SNU for the Homestake experiment and 110 SNU for the gallium experiments, when the ^{8}B neutrino flux is normalized to the Kamiokande result. For gallium experiments, their current yields are not yet significantly lower than the minimal 110 SNU but they are already consistently down by 2σ and their accuracy will improve with time.

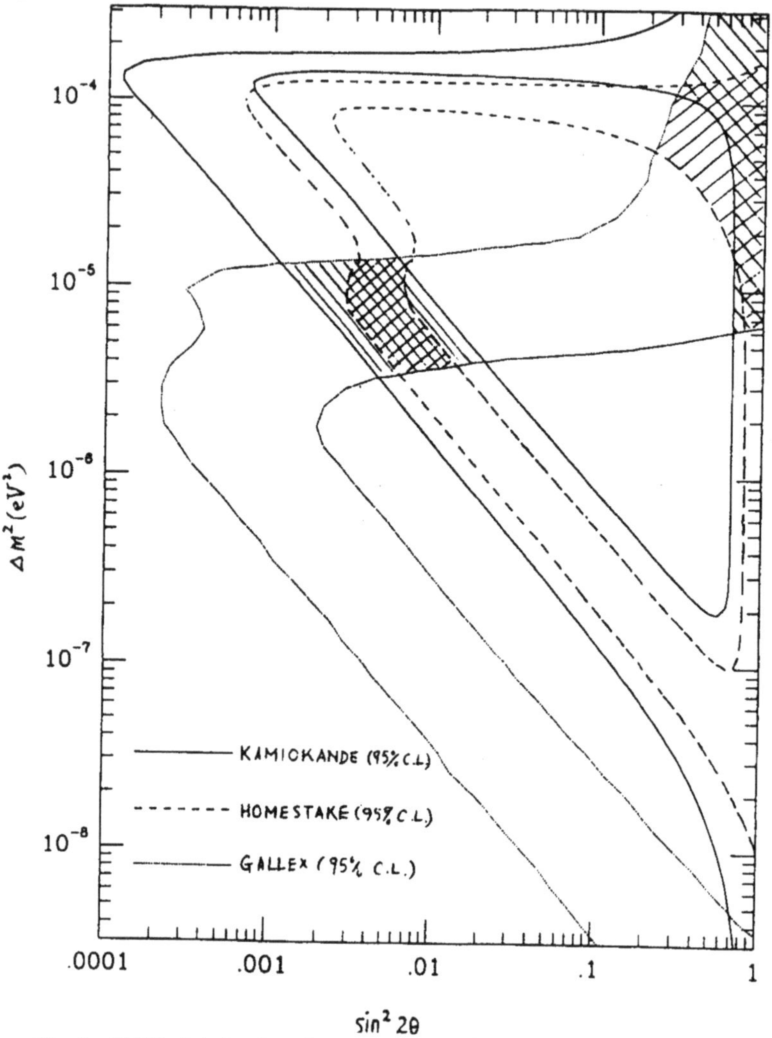

Figure 12. An MSW plot showing the concordant regions for the various experiments. Notice that the triple concordant regions and the pair wise concordant regions for Kamiokande plus GALLEX and for Chlorine plus GALLEX are quite similar. Figure is from Shi and Schramm.[154]

It is nice to know the next generation solar neutrino experiments (SNO, Super-Kamiokande, Borexino, ICARUS, etc.[171]) will be able to distinguish different solutions to the solar neutrino problem within the next five years.[190] Also, the gallium experiments may even be able to exclude 114 SNU in the not too distant future. Helioseismic observations by the on-going Global Oscillation Network Group (GONG)[191] and the future Solar and Heliospheric Observation (SOHO) mission will also provide much more information on the solar interior, thus further constraining solar models.

Since a solar model solution is unable to fit the current data and with the ^{51}Cr test one cannot easily discount experiments, it is beginning to look like we need a neutrino physics solution. The classic solution is the Mikheyev-Smirnov-Wolfenstein (MSW)[192] neutrino matter-induced mixing solution. The critical parameters in such a solution are the mass-difference squared of the mixing neutrinos $\Delta m^2 \equiv m_{\nu_x^2} - m_{\nu_e^2}$ and $\sin^2 2\Theta$ of the $\nu_e - \nu_x$ mixing angle. Figure 12 shows a plot from Shi and Schramm[153,154] of allowed $\Delta m^2 - \sin^2 2\Theta$ parameter space for the three experiments, GALLEX, Homestake and Kamiokande. The "v" is the overlap of the Homestake and Kamiokande experiments. The upper tips of the "v" are also consistent with GALLEX. The figure also shows that those same two regions are approximately obtained by taking GALLEX with either Kamiokande or Homestake. The two MSW concordant regions (there is also a small pure vacuum mixing region at very low ^{2}) are at $\Delta m^2 \sim 10^{-5}$ eV^2, one with large mixing angle, the other with $\sin^2 2\Theta \sim$ few 10^{-3}. As Shi, Schramm and Bahcall[193] pointed out, these basic conclusions remain even if three flavor mixing is taken into account and if solar model uncertainties are included. It has also been noted by Shi, Schramm and Fields[190] that BBN excludes the large angle solution branch if ν_x is a sterile neutrino rather than ν_μ or ν_τ. Besides the MSW solution, there is also a vacuum mixing solution at a narrow range around $\Delta m^2 \sim 10^{-10}$eV^2 and $\sin^2 2\Theta \sim 1$.

5.5 A Possible Cosmological Connection

If Δm^2 really is $\sim 10^{-5}$eV^2, and if the solar MSW mixing is between ν_e and ν_μ, then the question arises as to the mass of ν_τ.

The simplest way to generate neutrino masses is via the so-called see-saw model of Gell-Mann, Ramond and Slansky[194]. In such a model, $m_{\nu_\tau} \sim m_{\nu_\mu}(m_t/m_c)^2$ with $m_{\nu_\mu} \sim (\Delta m^2)^{1/2} \sim 10^{-2.5}$eV and $m_t \sim 170$ GeV and $m_c \sim 1.5$ Gev, we find that $m_{\nu_\tau} \sim$ tens of eV. In other words, it is perfect HDM.

Thus, there is the exciting possibility that the solar neutrino problem is related to fundamental physics and even to the dark non-baryonic matter of the universe.

Acknowledgements

We would like to thank our recent collaborators, David Dearborn, Brian Fields,

Dave Thomas, Gary Steigman, Brad Meyer, Keith Olive, Angela Olinto, Bob Rosner, Michael Turner, George Fuller, Karsten Jedamzik, Rocky Kolb, Grant Mathews, Bob Rood, Jim Truran and Terry Walker for many useful discussions. We would further like to thank Poul Nissen, Jeff Linsky, Julie Thorburn, Doug Duncan, Lew Hobbs, Evan Skillman, Bernard Pagel and Don York for valuable discussion regarding the astronomical observations.

I would like to thank Jim Truran for many discussions on this topic. I would also like to thank my past nucleochronometric collaborators, Jerry Wasserburg, Ken Hainbach, Eugene Symbalisty and Brad Meyer, as well as the founder of the business, Willy Fowler.

We thank John Bahcall, Sid Bludman, Ray Davis, S. Del'Innocenti, Moshe Gai, Ken Lande, Naoya Hata, Paul Langacker, Karlheinz Langanke, Douglas Morrison, Per Rosen and Lincoln Wolfenstein for valuable discussions. We especially thank the Institute for Nuclear Theory at the University of Washington for its hospitality and support during our stay. This work is supported by the NASA and the DoE(nuclear) at the University of Chicago, and by the DoE and NASA grant NAG5-2788 at Fermilab. This manuscript was prepared in part at CERN where one of us, DNS, was a visitor.

References

1. K. Olive, D.N. Schramm, G. Steigman, and T. Walker, Phys. Lett B 236 (1990) 454.
2. A. Songaila, L.L. Cowie, C.J. Hogan, and M. Rugers, Nature 368 (1994) 599.
3. R. F. Carswell et al. MMRAS, in press (1994).
4. D. Tytler and J. Fann, private communication (1995).
5. T. Walker, G. Steigman, D.N. Schramm, K. Olive, and H.S. Kang, Astrophys. J. 376 (1991) 51.
6. D.N. Schramm, in Proc. of Yamada Conf. XXXVII - Evolution of the Universe and Its Observational Quest, Tokyo, June 1993 (Universal Academic Press, Tokyo), in press (1994). See also D.N. Schramm, in Proc. of ESO/EIPC Workshop on the Light Element Abundances, Elba, May 1994 (Springer-Verlag, Heidelberg) in press (1994).
7. C. Copi, D. Schramm, and M. Turner, Science 267 (1994) 192.
8. R.A. Alpher, H. Bethe, and G. Gamow, Phys. Rev. 73 (1948) 803.
9. C. Hayashi, Prog. Theor. Phys. 5 (1950) 224.
10. R.A. Alpher, J.W. Follin, and R.C. Herman, Phys. Rev. 92 (1953) 1347.
11. R. Taylor and F. Hoyle, Nature 203 (1964) 1108.
12. P.J.E. Peebles, Phys. Rev. Lett. 16 (1966) 410.
13. R. Wagoner, W.A. Fowler, and F. Hoyle, Astrophys. J. 148 (1967) 3.
14. D.N. Schramm and R.V. Wagoner, Ann. Rev. of Nuc. Sci. 27 (1977) 37.

15. K. Olive, D.N. Schramm, G. Steigman, M. Turner, and J. Yang, Astrophys. J. 246 (1981) 557.
16. A. Boesgaard and G. Steigman, Ann. Rev. of Astron. and Astrophys. 23 (1985) 319.
17. J. Yang, M. Turner, G. Steigman, D.N. Schramm, and K. Olive, Astrophys. J. 281 (1984) 493.
18. L. Kawano, D.N. Schramm, and G. Steigman, Astrophys. J. 327 (1988) 750.
19. J. Truran, Doctoral Thesis, Yale University (1965);
J.W. Truran, A.G.W. Cameron, and A. Gilbert, Can. Jour. of Phys. 44 (1966) 563.
20. L.M. Krauss and P. Romanelli, Astrophys. J. 358 (1990) 47.
21. P. Kernan and L. Krauss, Phys. Rev. Lett. 72 (1994) 3309.
22. W. Mampe, P. Ageron, C. Bates, J.M. Pendlebury, and A. Steyerl, Phys. Rev. Lett. 63A (1989) 593; W. Mampe et al, JETP Lett. 57 (1993) 82.
23. G. Mathews, in Proc. Snowmass Workshop on Particle Astrophysics in the Next Millenium, in press (1994).
24. W.A. Fowler, J. Greenstein, and F. Hoyle, Geophys. J.R.A.S. 6 (1962) 6.
25. C. Ryter, H. Reeves, E. Gradstajn, and J. Audouze, Astron. and Astrophys. 8 (1970) 389.
26. J. Geiss and H. Reeves, Astron. and Astrophys. 18 (1971) 126.
27. D. Black, Nature 234 (1971) 148.
28. J. Rogerson and D. York, Astrophys. J. 186 (1973) L95.
29. J. Linsky et al., Astrophys. J. 402 (1993) 694.
30. H. Reeves, J. Audouze, W.A. Fowler, and D.N. Schramm, Astrophys. J. 179 (1973) 909.
31. R. Epstein, J. Lattimer, and D.N. Schramm, Nature 263 (1976) 198.
32. J.R. Gott, III, J. Gunn, D.N. Schramm, and B.M. Tinsley, Astrophys. J. 194 (1974) 543.
33. R.T. Rood, G. Steigman, and B.M. Tinsley, Astrophys. J. 207 (1976) L57.
34. J. Yang, D.N. Schramm, G. Steigman, and R.T. Rood, Astrophys. J. 227 (1979) 697.
35. R. Wilson, R.T. Rood, and T. Bania, in Proc. of the ESO Workshop on Primordial Healing, eds. P. Shaver and D. Knuth (European Southern Observatory, Garching, 1983).
36. R.T. Rood, T. Bania and J. Wilson, Nature 355 (1992) 618.
37. D.S. Balser et al., Astrophys. J., submitted (1994).
38. J. Spite and M. Spite, Astron. and Astrophys. 115 (1982) 357.
39. R. Rebolo, P. Molaro, and J.Beckman, Astron. and Astrophys. 192 (1988) 192.
40. L. Hobbs and C. Pilachowski, Astrophys. J. 326 (1988) L23.
41. V.V. Smith, D.L. Lambert, and P.E. Nissen, Astrophys. J., 408 (1982)

262.
42. K. Olive and D.N. Schramm, Nature 360 (1992) 434.
43. G. Steigman, B. Fields, K. Olive, D.N. Schramm, and T. Walker, Astrophys. J. 415 (1993) L35.
44. L. Hobbs and J. Thorburn, Astrophys. J. Lett. 428 (1994) L25.
45. M. Spite, F. Pasquini, and F. Spite, Astron. and Astrophys., in press (1995).
46. G. Steigman, D.N. Schramm, and J. Gunn, Phys. Lett. B 66 (1977) 202.
47. V.F. Schvartzman, JETP Letters 9 (1969) 184.
48. P.J.E. Peebles, Physical Cosmology (Princeton University Press, 1971).
49. D.N. Schramm and L. Kawano, Nuc. Inst. and Methods A 284 (1989) 84.
50. B. Pagel, in Proc. of 1989 Rencontres de Moriond (1990).
51. ALEPH, L3, OPAL, DELPHI results, 1993 Lepton-Photon meeting at Ithaca, NY (1993).
52. G. Steigman, MNRAS 269 (1994) L53.
53. E. Vangioni-Flam, K. Olive, and N. Prantzos, Astrophys. J. 427 (1994) 618-627.
54. K. Olive, D. Schramm, J. Truran, R. Rood and E. V-Flam, U. Minnesota preprint (1994).
55. K. Olive and G. Steigman, U. Minnesota preprint UMN-TH-1230/94 (1994).
56. E. Skillman, Talk at particle and Nuclear Astrophysics and Cosmology in the Next Millennium, Snowmass, Colorado, July 1994.
57. J. Geiss in Origin and Evolution of the Elements, eds. N. Prantzos et al. (Cambridge Univ. Press, Cambridge, 1993), p. 89.
58. J. Thorburn, Astrophys. J. 421 (1994) 318.
59. G. Steigman, K. Olive and D. Schramm, Phys. Rev. Lett. 43 (1979) 239
60. R. Scherrer, J. Applegate, and C. Hogan, Phys. Rev. D 35 (1987) 1151.
61. J. Applegate, C. Hogan, and R. Scherrer, Astrophys. J. 329 (1988) 572.
62. C. Alcock, G. Fuller, and G. Mathews, Astrophys. J. 320 (1987) 439.
63. K. Karsch and E. Laerman, Phys. Rev. D 50 (194) 6954.
64. Y. Iwasaki, Tsukuka preprint (1994) UTHEP-219, December 1994.
65. W.A. Fowler and R. Malaney, Astrophys. J. 333 (1988) 14.
66. H. Kurki-Suonio, R. Matzner, K. Olive, and D.N. Schramm, Astrophys. J. 353 (1990) 406.
67. D. Thomas, D.N. Schramm, K. Olive, G. Mathews, B. Meyer, and B. Fields, Astrophys. J. 430 (1994) 291.
68. K. Sato and N. Tarasawa, in The Birth and Evolution of Our Universe: Proceedings of Nobel Symposium 79, eds. J.S. Nilsson et al. (World Scientific, Singapore, 1991) 60.
69. B. Meyer, C. Alcock, G. Mathews, and G. Fuller, Phys. Rev. D 43

(1991) 1079.
70. K. Jedamzik, G. Fuller, G. Mathews, and T. Kajino, Astrophpys. J. 422 (1988) 423.
71. K. Freese and F. Adams, Phys. Rev. D 41 (1990) 2449.
72. B. Link, Phys. Rev. Lett. 68 (1992) 2425.
73. R. Boyd and T. Kajino, Astrophys. J. 359 (1990) 267.
74. G. Gilmore, B. Edvardsson, and P. Nissen, Astrophys. J. 378 (1991) 17.
75. S. Ryan, M. Bessel, R. Sutherland, and J. Norris, Astrophys. J. Lett. 348 (1990) L57.
76. S. Ryan, J. Norris, M. Bessell, and C. Deliyannis, Astrophys. J. 388 (1992) 184.
77. D. Duncan, D. Lambert, and D. Lemke, Astrophys. J. 401 (1992) 584.
78. D. Kiselman, Astron. and Astrophys., in press (1994).
79. B. Fields, K. Olive, and D.N. Schramm, Astrophys. J. 439 (1995) 854.
80. H. Reeves, W.A. Fowler, and F. Hoyle, Nature 226 (1970) 727.
81. R. Epstein, W.D. Arnett, and D.N. Schramm, Astrophys. J. Lett.190 (1974) L13; Astrophys. J. Supp. 31 (1976) 111.
82. G. Steigman and T. Walker, Astrophys. J. Lett. 385 (1992) L13.
83. T. Walker, G. Steigman, D.N. Schramm, K. Olive, and B. Fields, Astrophys. J. 413 (1993) 562.
84. L. Brown and D.N. Schramm, Astrophys. J. Lett. 329 (1988) L103.
85. M. Pinsonneault, C. Deliyannis and P. Demarque, Astrophys. J. Supp. 78 (1992) 179.
86. J. Simpson, Ann. Rev. Nuc. and Part. Sci. 33 (1983) 323.
87. S.E. Woosley, D.H. Hartmann, R.D. Hoffman, and W. Haxton, Astrophys. J. 356 (1990) 272.
88. D. Dearborn, D.N. Schramm, G. Steigman, and J. W. Truran, Astrophys. J. 347 (1989) 455.
89. M. Meneguzzi, J. Audouze, and H. Reeves, Astron. and Astrophys. 15 (1971) 337.
90. D. Thomas, D.N. Schramm, K. Olive, and B. Fields, Astrophys. J. 406 (1993) 569.
91. X. Shi, D.N. Schramm, D. Dearborn, and J.W. Truran, Comments on Astrophysics, in press (1995).
92. C. Fisher, Ph.D. Thesis, University of California at Berkeley, and references therein (1992).
93. M. Davis and A. Nusser, in Proc. Maryland Symposium on Dark Matter, in press (1995).
94. J. Bahcall et al., IAS preprint, October 1994, and Astrophys. J., in press (1995).
95. R. Kirshner, A. Oemler and P.L. Schechter, Astronom. J. 84 (1979) 951.
96. C. Alcock et al., Nature 365 (1993) 621.
97. E. Aubourg et al., Nature 365 (1993) 623.
98. E. Gates, G. Gyuk, and M. Turner, Phys. Rev. Lett., 74 (1995) 3724;

D. Bennett et al., LLNL preprint (1995).
99. N. Kayser in Proc. of the Texas Symposium on Relativistic Astrophysics, Munich, December 1994, in press (1995).
100. A. Szalay in Proc. of the Eighth Physics Summer School on Cosmology: The Physics of the Universe, Canberra, Australia, January 1995 (World Scientific, Singapore, 1995).
101. A. Guth, Phys. Rev. D 23 (1981) 347.
102. A. Linde, Particle Physics and Inflationary Cosmology (Harwood Academic Publishers, N.Y., 1990).
103. R. Bond and A. Szalay, in Proc. Texas Relativistic Astrophysical Symposium, Austin, Texas (1992).
104. M. Turner, F. Wilczek, and A. Zee, Phys. Lett. B 125 (1993) 35; 125 (1993) 519.
105. M. Crawford and D.N. Schramm, Nature 298 (1982) 538.
106. C. Alcock and A. Olinto, Ann. Rev. Nuc. Part. Phys. 38 (1988) 161.
107. P. Jakobsen, A. Boksenberg, J.M. Deharveng, P. Greenfield, R. Jedrzewski, and F. Paresce, Nature, in press (1994).
108. R. Mushotsky, in Relativistic Astrophysics and Particle Cosmology: Texas PASCOS 92, eds. C. W. Akerlof and M. A. Srednicki, Annals of the N.Y. Academy of Sciences 688 (1993) 184.
109. S.D.M. White, J.F. Navarro, A.E. Evrard, and C.S. Frenck, Nature 366 (1993) 261.
110. LSND collaboration result reported in New York Times (1995) January.
111. J. Ussher, The Annals of the World (E. Tyler for J. Crook, London, 1658).
112. P. Nugent, D. Branch, E. Boran, A. Fisher, T. Vaughn, and D. Hauschildt, Phys. Rev. Lett., submitted (1995).
113. A. Sandage and G. Tamman, Carnegie Institute preprint (1995).
114. A. G. Riess, W. H. Press, and R.P. Kirshner, Astrophys. J. Lett. 438 (1995) L17.
115. B. Schmidt, R. Kirshner and R. Eastman, Astropys. J. 395 (1992) 366.
116. W.L. Freedman et al., Nature 371 (1994) 757.
117. C. Hill, D.N. Schramm, and J. Fry, Comm. Nuc. Part. Phys. 19 (1989) 25.
118. I. Iben and A. Renzini, Physics Reports 105 (1984) 329.
119. B. Pagel, NORDITA preprint (1989).
120. X. Shi, D.N. Schramm and D. Dearborn, Phys. Rev. D 50 (1995) 2414
121. X. Shi, Astrophys. J., in press (1995).
122. B. Chaboyer, CITA preprint CITA-94-52 (1994) and Astrophys. J., submitted (1995).
123. S. Lee, G. Mathews, and D.N. Schramm, Astrophys. J, in press (1995).
124. B.S. Meyer and D.N. Schramm, Astrophys. 311 (1986) 406.
125. D.N. Schramm and G.J. Wasserburg, Astrophys. J. 162 (1970) 57.
126. H. Reeves and O. Johns, Astyrophys. J. 206 (1976) 958.

127. B.M. Tinsley, Astrophys. J. 198 (1975) 145.
128. K. Hainebach and D.N. Schramm, Astrophys. J. 212 (1977) 347.
129. E.M.D. Symbalisty and D.N. Schramm, Rep. Prog. Phys. 44 (1981) 293.
130. G.J. Wasserburg, D.N. Schramm, and J. Huneke, Astrophys. J. 157 (1969) 191.
131. D.N.Schramm, Ann. Rev. Astr. Ap. 12 (1974) 303.
132. K. Takahashi and K. Yokoi, Nucl. Phys. A404 (1981) 578.
133. K. Yokoi, K. Takahashi, and M. Arnould, Astr. Ap. 117 (1983) 65.
134. W.A. Fowler, in Explosive Nucleosynthesis, ed. D.N. Schramm and W.D. Arnett (University of Texas Press, Austin, 1973).
135. J.A. Holmes, S.E. Woosley, W.A. Fowler, and B.A. Zimmermann, B.A., Atomic Data Nucl. Data Tables 18 (1976) 306.
136. S.E. Woosley and W.A. Fowler, Astrophys. J. 233 (1979) 411.
137. M. Arnould, Astr. Ap. 31 (1974) 371.
138. M. Arnoud, K. Takahashi, and K. Yokoi, Astr. Ap. 137 (1984) 51.
139. J.M. Luck, J.L. Birch, and C.J. Allegre, Nature 283 (1980) 256.
140. R.R. Winters, R.L. Macklin, and J. Halperin, Phys. Rev. C21 (1980) 563.
141. R.R. Winters, in Neutron-Nucleus Collisions: A Probe of Nuclear Structure, ed. J. Rapaport, R.W. Finlay, S.M. Grimes, and F.S. Dietrich, (1984) p. 495.
142. F.-K.Thielmann, J. Metzinger, and H.V. Klapdor, Astr. Ap. 123 (1983) 162.
143. F.-K.Thielmann, J. Metzinger, and H.V. Klapdor, Z.S. Phys. A. 309 (1983) 301.
144. F.-K.Thielmann, and J.W. Truran, in Nucleosynthesis and Its Implications on Nuclear and Particle Physics, ed. J. Audouze and N. Mathieu (Reidel, Dordrecht, 1985).
145. B.S.Meyer, W.M. Howard, G.J. Mathews, P. Moller, and K. Takahashi, in Proc. Meeting of the American Chemical Society (1985).
146. R.W. Hoff, in Proc. of the International Symposium on Weak and Electromagnetic Interactions in Nuclei (Heidelberg, 1986).
147. A .H. Jaffey, D. F. Flynn, L. E. Glendenin, W. C. Bentley, and A. M. Essling, Phys. Rev. C4 (1971) 1989.
148. B. Hudson, C.M. Hohenburg, B.M. Kennedy, and F.A. Podosek, F.A., Lunar Panetary Sci. XIII (1982) 346.
149. P. Fields, A.M. Friedman, J. Milsted, J. Lerner, C.M. Stevens, and D. Metta, Nature 212 (1966) 131.
150. J.N. Bahcall and M.H. Pinsonneault, Rev. Mod. Phys., 64, (1992) 885.
151. S. Turck-Chieze and I. Lopes, Astrophys. J., 408 (1993) 347.
152. P. Langacker, in Proc. of the 4th International Symposium on Neutrino Telescopes, Venice, Italy, 1992, ed. M. Baid-Celin (University of Padua, Padua, 1992).

153. D.N. Schramm and X. Shi, in Proc. of the Inst. for Nuclear Theory Program INT94-1 on Solar Neutrinos and Neutrino Astrophysics, Seattle, WA, March 1994, in press (1994).
154. X. Shi and D.N. Schramm, Particle World, 3, (1994) 151; Nucl. Phys. B, Proc. Suppl., 35 (1994) 321.
155. T. Kirsten et al., GALLEX preprint (1994).
156. T. Motobayashi, N. Iwasa, Y. Ando et al., Phys. Rev. Lett. **73** (1994) 2680.
157. K. Langanke, S.E. Koonin, T.D. Shoppa, and R. Seki, Phys. Rev. C 48 (1994) 837.
158. The Homestake group report on Neutrino '94, Eilat, Israel, May 1994.
159. The Kamiokande report on Intl. Workshop on Neutrino Telescopes, Venice, February, 1994.
160. The GALLEX report, ibid.
161. The SAGE report, ibid
162. A.I. Abazov et al., Phys. Rev. Lett. 67 (1991) 3332.
163. P. Anselmann et al., Phys. Lett. B285 (1992) 376.
164. K. Lande, private communication.
165. J.K. Rowley, B.T. Cleveland and R. Davis, Jr., in Solar Neutrino and Neutrino Astronomy, eds. M. L. Cherry, K. Lande and W.A. Fowler, AIP Conf. Proc. No. 126, AIP, New York, 1984.
166. A. Dar and G. Shaviv, Technion-Ph-94-95; astro-ph 9401043.
167. D.R.O. Morrison, Particle World, Vol.3 (1992) 20.
168. J.N. Bahcall, G.B. Field and W.H. Press, Astrophys. J., 320 (1987) L69.
169. B.W. Filipone, P. Vagel, Phys. Lett. B 246 (1990) 546.
170. X. Shi, D.N. Schramm, R. Rosner and D.S. Dearborn, Comments on Nuclear and Particle Physics, Vol. 21 (1993) 151.
171. J.N. Bahcall, Neutrino Astrophysics, (Cambridge Univ. Press, Cambridge, 1989).
172. C.W. Johnson et al., Astrophys. J., 392 (1992) 320.
173. R.W. Kavanagh et al., Bull. Am. Phys. Soc., 14 (1969) 1209.
174. B.W. Filippone et al., Phys. Rev. C 28 (1983) 2222.
175. M. Gai, private communication.
176. S. Pakvasa, in Proc. 25th Intl. Conf. High Energy Phys., eds, K.K. Phua and Y. Yamaguchi (World Scientific, Singapore, 1991) 698.
177. J. N. Bahcall and H.A. Bethe, Phys. Rev. D 47 (1993) 1298.
178. S.A. Bludman, N. Hata, D.C. Kennedy and P.G. Lanacker, UPR-0572T and Phys. Rev. D. **47** (1993), 2220.
179. V. Berezinski, Comm. Nucl. Part. Phys., 21 (1994) 249.
180. W. Kwong and S.P. Rosen, University of Texas at Austin preprint, UTAPHY-HEP-9, 1994.
181. K. Nakamura, ICRR-Report-312-94-7, 1994.
182. V. Castellani, S. Del'Innocenti and G. Fiorentini, Astron. Astrophys. 271 (1993) 601.

183. D. Dearborn, K. Griest and G. Raffelt, Astrophys. J., 368 (1991) 626.
184. G. Caughlan and W. Fowler, Atomic Data and Nuclear Data Tables, 40 (1988) 283.
185. N. Grevesse and A. Noels, in Origin and Evolution of the Elements, eds. N. Prantzos, E. Vangioni Flam and M. Casse (Cambridge University Press, 1993).
186. J. Christensen-Dalsgaard, D.O. Gough and M.J. Thompson, Astrophys. J., 378 (1991) 413.
187. V. Castellani, S. Degl'Innocenti, G. Fiorentini and Ricci, in preparation.
188. G. Marx and D.S. Dearborn, Acta Physica Hungarica 65 (1989) 315.
189. E. Anders and N. Grevesse, Geochim. Cosmochim. Acta, 53 (1989) 197.
190. X. Shi, D. N. Schramm and B.D. Fields, Phys. Rev. D 48 (1993) 2563.
191. Seismic Investigation of the Sun and Stars (GONG Workshop), eds. T. M. Brown et al., Boulder, Colorado, 1992.
192 S.P. Mikheyev and A. Yu. Smirnov, Yad. Fiz. 42 (1985) 1441 [Sov. J. Nucl. Phys. 42 (1985) 913]; L. Wolfenstein, Phys. Rev. D 17 (1978) 2369.
193. X. Shi, D.N. Schramm and J. Bahcall, Phys. Rev. Lett. 69 (1992) 717.
194. M. Gell-Mann, P. Ramond and R. Slansky in Sanibel Talk, CALT-68-709, Feb. 1979, and in *Supergravity* (North Holldan, Amsterdam, 1979).

THE EPOCH OF GALAXY FORMATION

Joseph Silk
Departments of Astronomy and Physics,
and Center for Particle Astrophysics,
University of California, Berkeley, CA 94720

1 Introduction

These lectures are devoted to exploring the epoch of galaxy formation. I begin by describing the *ab initio* approach, commencing with the growth of infinitesimal density fluctuations in the early universe. Gravitational instability eventually leads to collapse of massive clouds of matter and galaxy formation. We see fluctuations in the last scattering surface of the background radiation, at $z \sim 1000$. The period between then and the epoch of structure formation at $z \lesssim 5$ may be thought of as the medieval period of cosmology, where little is known, and even less is observed.

A complementary approach commences with the galaxies in our vicinity, and, treating them as fossils, attempts to extrapolate back in time to the formation epoch. This phenomenological approach also enables us to examine galaxy formation, and approach the dark ages by going further into the past. Galaxies in our vicinity and at look-back times of up to about half a Hubble time provide the best understood connection to the past. More primitive objects such as Lyman alpha clouds, and rare objects, such as radio galaxies and quasars, can be studied back to $z \sim 4$ or greater. These objects are as yet poorly understood with regard to their role in the framework of galaxy and structure formation, but nevertheless are indispensable as our unique tools for probing the dark ages of fluctuation growth and structure evolution.

The first lecture (Section 2) reviews the theory of the origin and growth of density fluctuations in the expanding universe. The evolution of fluctuations in a baryon-dominated universe is described. In the second lecture (Section 3), I discuss the effect of a dominant density of non-baryonic dark matter on linear fluctuation growth. Simple nonlinear models are described, and scaling relations are derived for dark halos. Section 4 describes cold dark matter in its various incarnations, with regard to the spectrum of primordial density fluctuations. Evidence on the distribution of dark matter and on its fluctuating component is reviewed over scales from halos of galaxies, to clusters and up to the horizon scale. Armed with an initial power spectrum of density fluctuations, I now describe the theory of the origin of the mass function of nonlinear structures. This leads to the prediction of the redshifts at which halos and clusters form. The smallest objects are the first to form, and the role of these dwarf galaxies in reionizing the universe is reviewed.

The phenomenological approach to galaxy formation is developed in Sections 5 and 6, respectively devoted to formation of disk galaxies and of ellipticals. In the first of these sections, I describe the angular momentum and cooling problems for disks, and how star formation can resolve these difficulties. A self-regulation model can account for the characteristics of star-forming disk galaxies. Chemical abundance signatures help discriminate the star-forming processes of these disks, namely in understanding the stellar abundances found in thick disks and halos. The Tully-Fisher correlation provides a challenge to models of disk formation. Elliptical galaxy formation is discussed in Lecture 5 in the context of galaxy mergers. The successes and failures of simulations are reviewed. Winds play a role in accounting for the properties of ellipticals. While cooling arguments can account for the characteristic galaxy luminosity L_*, the fundamental plane presents more of a puzzle. The relevance of starbursts to bulge and elliptical formation is assessed.

2 Gravitational Instability in the Expanding Universe

Fluctuation origin

The theory of inflationary cosmology was developed in 1980, and ushered in a new era in the study of the evolution of density fluctuations in the early universe. Inflation meant that one had a theory which connected quantum fluctuations to macroscopic scales. Inflationary cosmology provides the backdrop for the origin of the fluctuations that gave rise to galaxies.

There are many variations on the original inflationary models, which failed because of the generation of an excessively inhomogeneous universe at the end of the inflationary epoch. Additional levels of complexity have led to such variants as new inflation with a gently rolling potential, chaotic inflation, power-law inflation and other spin-offs. All have the built-in aim of producing primordial density fluctuations at a level of a few parts in 10^4 on galactic scales. The spectrum of density fluctuations must be nearly scale-invariant, as is expected for the amplitude of fluctuations as they leave the time-translation invariant de Sitter horizon. Any small deviations from scale-invariance are of great interest, since there is of course immense leverage over the range from the inflationary horizon to the horizon at the present epoch.

Gaussian density fluctuations driven by quantum fields are inevitably present at the threshold of classical cosmology, the Planck epoch at $\sim 10^{-43}$ sec. Inflation is triggered by a phase transition that sets in at the epoch of grand unification symmetry breaking, $\sim 10^{-35}$ sec, and is completed by $\sim 10^{-33}$ sec. During this phase, the universe undergoes a strongly accelerated phase of expansion, which effectively enlarges the particle horizon to a comoving scale far larger than its present value.

Inflation predicts that the universe should be isotropic, nearly homogeneous, devoid of magnetic monopoles, and should be dark-matter dominated with $\Omega = 1$; Gaussian scale-invariant curvature fluctuations are its primary legacy for large-scale structure.

This prediction arises as a consequence of the de Sitter phase, characterised by an equation of state $p = -\rho$, that describes the inflationary phase of the universe. For a massless and minimally coupled scalar field ϕ which drives inflation, the quantum fluctuations are necessarily Gaussian, and of amplitude

$$|\Delta\phi| = \frac{H}{2\pi}.$$

The potential energy $V(\phi)$ is approximately equal to ρ_ϕ, the energy density of the ϕ-field, since the potential is very flat, in order to result in sufficient inflation, and the kinetic energy of the ϕ- field is negligible:

$$\dot{\phi}^2 = \rho + p << V(\phi).$$

The resulting density fluctuations on the horizon scale are

$$\frac{\delta\rho}{\rho + p} = \frac{\delta V}{\dot{\phi}^2} = \frac{\delta\rho V'}{\dot{\phi}^2},$$

whence application of the ϕ-field equation of motion $(V' \approx -3H\dot{\phi})$ leads to $\delta\rho/\rho \approx H^2/\dot{\phi}$. Since both H and $\dot{\phi}$ vary very slowly during inflation, a constant amplitude of density fluctuations is generated on the horizon scale. It is a consequence of the de Sitter model, there being no preferred time-scale for horizon crossing, and all scales have similar amplitudes imprinted on them.

The precise amplitude of the resulting $\delta\rho/\rho$ depends on the flatness of the scalar field potential, which requires very weak field coupling in order for the entire observable universe to be contained within one giant bubble of true vacuum (the new phase). The price one pays for the weak field coupling is that reheating and baryon number generation become problematic. The dynamics of inflation are also important, and variations on canonical inflation appeal to modified gravity that changes the rate of inflation from exponential to power-law in time. In this case, the relic density fluctuations arise, not from boosted quantum fluctuations but from the kinetic energy of interacting bubbles of true vacuum that arise during the phase transition. The minimal fluctuations produced by inflation are associated with graviton production. These are produced thermally and result in associated metric fluctuations during the de Sitter phase amounting to

$$\Delta h = |\Delta\phi|\sqrt{16\pi G}.$$

Recall that the Friedmann equation is $H^{-1} \approx m_{pl} M_{GUT}^{-2}$, and $G = m_{pl}^{-2}$ in fundamental units ($h = c = 1$) whence

$$\Delta h \approx \frac{H}{m_{pl}} \approx (\frac{M_{GUT}}{m_{pl}})^2.$$

With $T_{GUT} \approx 10^{16}$ GeV, one expects a quadrupole CMB anisotropy of approximately 10^{-6} to result from horizon–scale gravity waves. They are an inevitable consequence of the de Sitter inflationary phase. There are also tensor metric fluctuations associated with the scalar density fluctuations $\delta\rho/\rho$ if scale-invariance is violated. In one of the simplest inflationary models (with a ϕ^4 potential), the ratio of tensor to scalar amplitudes is

$$T/S = 7(1-n),$$

but there is no generic prediction of the amplitude of tensor fluctuations, or even of their inevitability.

Inflation generically results in $\Omega = 1$, and indeed resolution of the "flatness problem" was a prime motivation in its initial justification. However recent inflationary models have succeeded in generating open universes by combining successive epochs of inflation or by fine-tuning of the inflaton potential. An interesting and unavoidable feature of such models is an excess in the power spectrum of density fluctuations that corresponds, in one example, to constant gravitational potential fluctuations on supercurvature scales. On scales less than the comoving curvature scale $cH_0^{-1}(1-\Omega_0)^{-1/2}$, the scale-invariant spectrum $n \approx 1$ applies, but on larger scales, the effective value of the power spectrum index, defined for fluctuations in energy density, is close to [1] -1 or even [2] -2. The corresponding angular scale at which this effect sets in is $\frac{1}{2}\Omega_0(1-\Omega_0)^{-1/2}$ radian. This leads to potential difficulties for open models with the large angular scale CMB anisotropies as measured by COBE.

Fluctuation modes

The inflationary phase is time–translation–invariant, and the residual quantum fluctuations that emerge from the horizon, as the standard Friedmann-Robertson-Walker expansion resumes, retain a scale–invariant spectrum. This is equivalent to a fluctuation distribution with constant metric (or gravitational potential) fluctuations, the amplitude of which can, in principle, be specified in terms of the inflationary field, but, in practice, is treated phenomenologically as a free parameter pending development of a compelling theory of inflation.

If the universe is dominated by baryonic matter, the preferred mode of fluctuations is via entropy perturbations. The rationale is simple: adiabatic fluctuations would be incompatible with observations of CMB anisotropy. If moreover $\Omega_B \lesssim 0.1$, as required by canonical primordial nucleosynthesis, inflation could not have occurred in the absence of a cosmological constant. Hence the primordial fluctuation spectrum is weakly constrained by theory, in that rather special inflationary models must be devised. Primeval entropy perturbations are associated with the so-called isocurvature mode, which leaves a negligible imprint on the universe at very early epochs. Because these fluctuations are dynamically unimportant at early epochs, inflationary models provide little guidance to the late-time spectrum. In the absence of a plausible model for primordial entropy fluctuations, the slope of the isocurvature power spectrum is specified phenomenologically by the requirements of large-scale structure.

The density fluctuation spectrum is conventionally parameterised as

$$\delta\rho/\rho \propto M^{-\frac{1}{2}-\frac{n}{6}},$$

where n is the primordial power-law spectral index of the power spectrum of density fluctuations, written in terms of a Fourier decomposition as

$$\frac{\delta\rho}{\rho} = \frac{V}{(2\pi)^3}\int \delta_k\, e^{i\mathbf{k}.\mathbf{x}}\, d^3k; \quad |\delta_k|^2 \propto k^n.$$

An underlying, smoothly varying density field $\rho(x)$ is assumed, containing Gaussian fluctuations $\delta\rho(\mathbf{x})$.

The equivalent power spectrum of density fluctuations is described after decoupling by

$$|\delta_k|^2 \propto k^n.$$

One infers a mass distribution

$$\delta\rho/\rho \equiv \left(\int e^{i\mathbf{k}\cdot\mathbf{r}} d^3\mathbf{k}\right)^{1/2} \propto M^{-\frac{1}{2}-\frac{n}{6}}.$$

Comparison with the observed galaxy correlations limits the power law index to lie in the range $0 \gtrsim n \gtrsim -2$, with $n = 0$ being white noise: convergence requires $n > -3$, and a causal origin for the fluctuations imposes $n < 4$. The lack of distortions in the CMB spectrum further constrains the power spectrum of isocurvature fluctuations to be bounded by [3] $n \lesssim 1.5$. Inflation generically predicts a power spectrum fluctuation spectral index n close to unity. Because of the logarithmic roll-off in the almost flat potential, a more precise fit to scale-invariant initial conditions is given by setting $n \approx 0.95$.

The principal alternative to isocurvature fluctuations is that of primordial curvature fluctuations in an inflationary universe with $\Omega = 1$. Inflation completely defines the spectrum of primordial curvature fluctuations with $n = 1$: note however that specifying a primordial spectral index n is equivalent to primordial curvature fluctuations with spectrum $\delta\kappa \propto M^{(n-1)/6}$. The curvature perturbations are equivalent to

$$\delta\kappa \sim GM/LC^2 \sim (\delta\rho/\rho)(L/ct)^2 \sim (\delta\rho/\rho)(M/M_h)^{2/3},$$

and hence to sub-horizon density fluctuations during the matter-dominated era with an effective index equal to $n - 4$.

Fluctuations in an Einstein-de Sitter universe

Qualitative analysis.

While $n = 0$ yields a Poisson distribution, $n = 1$ is the requisite scale-invariant index. This follows from the definition of metric, or equivalently over small scales, gravitational potential fluctuations associated with density fluctuations δM :

$$|h_{ij}| \equiv |\delta\phi| = |\frac{G\delta M}{rc^2}|$$

over a scale containing mass M. One has $\delta\rho/\rho = constant$, if $\delta M/M \propto M^{-\frac{2}{3}}$, that is, if $n = 1$.

After setting $M = \frac{4}{3}\pi\rho r^3$, it is instructive to rewrite $|\delta\phi|$ as

$$|\delta\phi| \approx \frac{\delta M}{M}\left(\frac{r}{a}\right)^2\left(\frac{a}{ct}\right)^2,$$

where I have set $\rho \sim 1/Gt^2$. Here r is the physical scale of a fluctuation, and $\lambda \equiv r/a(t)$ is the comoving scale.

One can now write, if $|\delta\phi|$ is constant,

$$\frac{\delta M}{M} \propto \lambda^{-2} t \propto M^{-\frac{2}{3}} t$$

in the radiation-dominated regime ($a \propto t^{1/2}$), and

$$\frac{\delta M}{M} \propto \lambda^{-2} t^{\frac{2}{3}} \propto M^{-\frac{2}{3}} t^{\frac{2}{3}}$$

in the ensuing matter-dominated epoch ($a \propto t^{\frac{2}{3}}$).

More generally,

$$|\delta\phi| \propto M_h^{\frac{1-n}{6}}$$

where M_h is the mass contained within a horizon-scale density fluctuation. The scale-invariant spectrum is now seen to explicitly specify the same amplitude for density fluctuations at horizon crossing.

Hence constant metric fluctuations, $|h_{ij}|$ = constant, are equivalent to gravitational potential fluctuations, $\delta\Phi$, evaluated at horizon crossing. I deduce that for a comoving sphere of physical diameter $\lambda(1+z)^{-1}$ containing the invariant mass

$$M(\lambda) = 1.54 \times 10^{11} \Omega h^2 (\lambda/1Mpc)^3 \, \mathrm{M}_\odot,$$

one has

$$|\delta\Phi| = \left|\frac{\mathrm{G}\delta\mathrm{M}}{\lambda\mathrm{c}^2}\right|_{\mathrm{h}} = \mathrm{constant}$$

The resulting fluctuations δM over the sphere M are given by

$$\frac{\delta M}{M} \propto \left(\frac{ct}{\lambda}\right)^2 \propto M^{-2/3}\left(\frac{t}{1+z}\right)^2$$

I conclude that a scale invariant spectrum has mass fluctuations $\propto M^{-2/3}$, and moreover that the fluctuations grow as either $t^{2/3}$ (matter-dominated) or t (radiation-dominated). These remarks apply provided pressure effects are neglected (a concern on sub-horizon scales). Their validity on super-horizon scales is evident once one realizes that arbitrary linear perturbations of a de Sitter model (K = 0) can be decomposed into a superposition of plane wave curvature perturbations, which can in turn be regarded as a comparison of two Friedmann models with slightly different curvature.

Formal analysis ($p = 0, K = 0$).

The K = 0 background with superimposed fluctuations may be examined by expanding the parametric solutions for the Friedmann model with arbitrary curvature:

$$a(t) = A(1 - cos\theta) \approx \frac{A}{2}\theta^2\left(1 - \frac{\theta^2}{12}\right),$$

$$t = B(\theta - sin\theta) \approx B\frac{\theta^3}{6}$$

This immediately yields the growing mode

$$\frac{\delta\rho}{\rho} = -3\frac{\delta a}{a} \propto t^{2/3},$$

in the matter–dominated regime. In general, there is also a decaying mode; one can write more generally

$$t - \tau = B(\theta - sin\theta) \approx B\frac{\theta^3}{6},$$

whence

$$\frac{\delta\rho}{\rho} \propto t^{2/3}\left(1 - \frac{2}{3}\frac{\tau}{t}\right).$$

The decaying mode corresponds physically to non-simultaneous (on some initial hypersurface at $t \to 0$) initiation of fluctuations, where the growing mode describes curvature or gravitational potential fluctuations.

Newtonian analysis ($K = 0$).

Inclusion of pressure gradients is expected to stabilize gravitational instability, and is most simply described in the Newtonian approximation, valid over small scales ($\ell \ll ct$) and velocities ($v \ll c$). The following general equations are linearized about the Friedmann equation, including expansion in lowest order:

the equation for mass conservation,

$$\frac{\partial \rho}{\partial t} + \nabla \cdot (\rho v) = 0;$$

the momentum equation,

$$\frac{dv}{dt} = \nabla\phi - \nabla p/\rho;$$

and the generalized Poisson equation,

$$\nabla^2\phi = 4\pi G(\rho + 3p/c^2),$$

together with an appropriate equation of state, $p = p(\rho)$.

After linearizing these equations about the Friedmann background, one can make the following statements about the first order perturbed quantities of interest. Decomposition of the perturbed velocity field into rotational v_{rot} and irrotational parts leads to a decaying mode,

$$\nabla \cdot \mathbf{v}_{rot} = 0,$$

and a compressible mode,

$$(\nabla \cdot \mathbf{v}_{irrot})/a = -\dot{\delta},$$

where a comoving coordinate $x = r/a$ has been introduced. Rotational velocities therefore decay as $d(a\omega)/dt = 0$ or $\omega \propto a^{-1}$, where $a^{-1}\nabla_x \times \mathbf{v}_{rot} = \omega$. The irrotational component of velocity drives compressions and rarefactions of the density fluctuations, which are governed by

$$\frac{\partial^2}{\partial t^2}\delta + \frac{2\dot{a}}{a}\frac{\partial \delta}{\partial t} = 4\pi G\rho\delta + \frac{1}{a^2}\frac{dp}{d\rho}\nabla^2\delta.$$

This equation explicitly reveals the damping effect of the expansion, the destabilizing role of self-gravity, and the stabilizing effect of pressure gradients. Definition of the physical wavelength $\lambda = 2\pi a/k$, with arbitrary perturbations described by

$$\delta(\mathbf{x}, t) = \sum_k \delta_k e^{i\mathbf{k}\cdot\mathbf{x}},$$

leads to the generalized Jeans criterion for instability:

$$\lambda > \lambda_J \equiv \left(\frac{\pi v_s^2}{G\rho}\right)^{1/2}$$

The "Jeans mass" is $M_J = \frac{\pi}{6}\lambda_J^3\rho$. If pressure is negligible ($v_s^2 = dp/d\rho = 0$), we recover the zero-pressure solutions: $\delta \propto t^{2/3}, t^{-1}$. Mass scales above M_J are gravitationally unstable.

In the radiation era, $v_s = c/\sqrt{3}$, and $M_J \sim M_{horizon} \propto t$. After matter-radiation equality at $z_{eq} = 4 \times 10^4 \Omega h^2$, the Jeans mass is approximately constant, since

$$v_s = \left(\frac{d(p_m + p_r)}{d(\rho_m + \rho_r)}\right)^{1/2} \approx \frac{c}{3^{1/2}}\left(1 + \frac{3}{4}\frac{\rho_m}{\rho_r}\right)^{-1/2} \propto (1+z).$$

The maximum value of the Jeans mass is

$$M_J^{max} = 10^{16}(\Omega h^2)^{-2}\,\mathrm{M}_\odot,$$

and drops abruptly after decoupling to $\sim 10^6(\Omega h^2)^{-1}\,\mathrm{M}_\odot$ (assuming $T = T_r$).

Fluctuation growth: $K \neq 0$

At redshift $\lesssim 1/\Omega$, the expansion rate differs significantly from that in the $K = 0$ model, and the fluctuation growth rate is correspondingly affected. One can study this simply in the zero pressure limit. In terms of the parametric solution for $a(t)$, the master equation for $\delta \equiv \frac{\delta\rho}{\rho}$, which was derived for arbitrary K, is transformed to

$$\frac{d^2\delta}{d\eta^2} = 3\left(\begin{matrix}\cos^2\eta \\ \cosh^2\eta\end{matrix}\right)\delta \qquad \text{for } K = \begin{cases}+1 \\ -1\end{cases}$$

This equation is easily solved: for example, if $K = +1$,

$$\delta_+ = \frac{-3\eta\sin\eta}{(1-\cos\eta)^2} + \frac{5+\cos\eta}{1-\cos\eta}, \qquad \delta_- = \frac{\sin\eta}{(1-\cos\eta)^2}$$

For small η, one recovers $(\delta_+, \delta_-) \sim (t^{2/3}, t^{-1})$. The asymptotic behaviour of the solution is, for $K = +1$, that $\delta_+ \gtrsim 1$ only when $a(t) \to 0$, and for $K = -1$ that $\delta_+ \to$constant when $\eta \to \infty$. The physical meaning of these long-time limits is that if $K = +1$, the density fluctuations only diverge when the Universe recollapses, and if $K = -1$, gravity becomes unimportant when the expansion is curvature–dominated at late times. In this latter case, perturbations, along with the unperturbed background, coast along in undecelerated expansion with self–gravity being unimportant.

The transition regime for late growth in a low Ω ($\gtrsim 0.1$) universe is of interest for large–scale peculiar velocities. In this case the perturbed continuity equation for peculiar velocities driven by the growing mode can be written

$$\nabla.\vec{v}_p = -H_0 f(\Omega)\delta,$$

where $f(\Omega) \approx \frac{3}{2}H_0 t \approx \Omega^{0.6}$, whence $v_p/v_{Hubble} \approx \frac{1}{3}\delta\Omega^{0.6}$.

Adiabatic fluctuations in a baryon-dominated universe

In a purely baryonic universe, acoustic modes are important on scales below the Jeans length, the characteristic scale of gravitational instability, $R_J \approx v_s(G\rho)^{-1/2}$.

At epochs when $\rho_r > \rho_m$, $R_J \approx ct$. As fluctuations enter the horizon, they enter the acoustic regime $\frac{\delta\rho}{\rho} \sim A\,exp(ikv_s t)$. The adiabatic invariant ΔE associated with sound waves of wavelength λ is proportional to frequency. The adiabatic approximation yields

$$A^2 v_s^2 \rho \lambda^3 \propto \frac{v_s}{\lambda}.$$

With $\lambda \propto a(t)$, one infers that in the radiation-dominated epoch,

$$A \approx constant,$$

while in the matter-dominated epoch, we have

$$A \propto (v_s a)^{-1/2}.$$

The matter-dominated acoustic modes are of decaying amplitude. Prior to decoupling, between z_{eq} and $z_{dec} \sim 1000$, one has

$$v_s \approx \frac{2}{3} c \left(\frac{\rho_r}{\rho_m}\right)^{1/2} \propto a^{-1/2}$$

and

$$A \propto a^{-1/4}.$$

In general, acoustic modes decay in amplitude for $\frac{5}{3} > \gamma > \frac{4}{3}$. After decoupling, if the matter temperature decays adiabatically as $p \propto \rho^{5/3}$ or $T \propto a^{-2}$, one has $v_s \propto T^{1/2} \propto a^{-1}$ and $A \approx constant$ for the acoustic mode. The demarcation between acoustic and unstable wavelengths occurs at the Jeans scale.

The physics of the "early" universe plays an important role in modifying the primordial spectrum of density fluctuations. For the case of baryonic adiabatic fluctuations, diffusive damping destroys fluctuations up to a maximum smoothing scale of

$$R_D \approx c\left(t \frac{1}{n_e \sigma_T c}\right)^{1/2}.$$

This results in a maximum damping mass at redshift z, which must be evaluated at the latest epoch that matter and radiation are coupled.

Damping is partially compensated on this scale by the generation of peculiar velocities that arise from the acoustic modes of density fluctuations when decoupling occurs. The typical amplitude is $v_p \sim v_s \frac{\delta\rho}{\rho}$, where v_s is the sound speed immediately prior to decoupling. One has

$$v_s = \frac{c}{3^{1/2}} \left(\frac{\rho_r}{\rho_m + \rho_r}\right)^{1/2}.$$

Thomson scattering-induced drag counteracts these peculiar velocities in the still partially ionized universe over a time scale

$$t_{drag} \sim \frac{\rho_m}{\rho_r} \frac{1}{n_e \sigma_T c}.$$

Recombination and matter-radiation decoupling occur fairly abruptly, over a timescale $t_{dec} \sim \frac{t_{exp}}{10}$, so that comoving fluctuations require damping to be suppressed by drag, or $t_{drag} < t_{rec} \sim \frac{t_{exp}}{10}$. Hence the maximum damping scale is

$$R_D^{max} \sim c \left(t_{dec} \frac{1}{n_e \sigma_T c} \right)^{1/2} \sim c \frac{t_{exp}^{LS}}{10} \left(\frac{\rho_r}{\rho_m} \right)^{1/2}.$$

Compton drag is actually effective to redshift

$$z < z_d \equiv 160(\Omega_0 h^2)^{\frac{1}{5}} x^{-\frac{2}{5}},$$

where x is the fractional ionization. Note that the Compton drag redshift has no explicit dependence on baryon density in a universe dominated by cold dark matter.

It is instructive to compare this scale with the Jeans length that describes the maximum scale of gravitational instability:

$$R_J \sim \frac{v_s}{(G\rho)^{1/2}} \sim v_s t_{exp}^{LS} \sim c t_{exp}^{LS} \left(\frac{\rho_r}{\rho_m + \rho_r} \right)^{1/2}.$$

Evidently, $R_D^{max} \sim \frac{R_J}{10}$, so that the relevant comoving mass scales for damping is

$$M_D \sim 10^{-3} M_J^{max} \approx 10^{13} (\Omega h^2)^{-2} \, M_\odot,$$

where I have assumed that the universe is matter-dominated at decoupling, a valid approximation provided that $\Omega \gtrsim 0.1$. A more sophisticated analysis given below takes account of cold dark matter.

Primordial isocurvature fluctuations

Isocurvature fluctuations are not subject to diffusive damping, but their growth is inhibited by radiation drag until after matter-radiation decoupling. The minimum scale of unstable isocurvature fluctuations is that of the post-decoupling Jeans mass, $M_J^{min} \sim 10^6 (\Omega h^2)^{-1} \, M_\odot$. However, the constancy of $M_J^{max} \sim 10^{16} (\Omega h^2)^{-2} \, M_\odot$ between the epoch of matter-radiation equality and decoupling, when the horizon contains $M_{dec} \sim 4 \times 10^{17} (\Omega h^2)^{-1/2} \, M_\odot$, results in development of a large-scale feature in the density fluctuation spectrum. During this phase of the universe, fluctuations are unstable and grow in amplitude on scales between M_J^{max} and M_{dec}, while all smaller scales remain at constant amplitude. Scales larger than M_{dec} are fully compensated isocurvature perturbations, with the radiation density deficit balancing the baryonic overdensity. Only within the horizon does the developing pressure gradient create a curvature fluctuation that grows at a rate proportional to $t^{2/3}$. After decoupling of the radiation, the large-scale isocurvature perturbations resume their growth as the compensating radiation density redshifts away relative to the matter overdensity.

The net effect is the following. A primordial power law, described by $\frac{\delta\rho}{\rho} \propto M^{-\frac{n+3}{6}}$ with $0 \gtrsim n \gtrsim -1$, describes the fluctuations prior to epoch z_{eq}, while one might imagine the Poisson case $n = 0$ would be most appealing from the theoretical point of view. Phenomenology requires $n \approx -1$ in order to match the large-scale distribution of galaxies, specifically the galaxy two-point correlation function. After z_{eq}, a feature develops in the form of excess power between z_{eq} and z_{dec}. On small scales, the fluctuation spectrum diverges towards lower masses. Consequently, immediately after decoupling, mass-scales of $M_J^{min} \approx 10^6(\Omega h^2)^{-1}\,\mathrm{M}_\odot$ are unstable. Collapse and star formation in these relatively small clouds most likely results in prolific dark matter formation in the form of compact stellar remnants and copiously produces enough energetic photons to reionize the universe. Compton drag will then be effective at inhibiting fluctuation growth and collapse on larger scales until $z \lesssim 100$.

3 Non-baryonic Dark Matter and Fluctuation Growth

Dark matter

Despite some six decades of intensive speculation about the nature and form of the dark matter, it remains almost as elusive as ever. Candidates span the range from supermassive black holes to weakly interacting elementary particles.

If supersymmetry provides a valid description of matter, and hence of the contents of the universe, at sufficiently high energies, then the lightest SUSY particle provides a compelling candidate for cold dark matter. Moreover the CDM should be ubiquitous: it is inevitably present in all gravitational potential wells. Whether it dominates or not depends on what scale one is surveying. Dense objects such as galactic disks and galaxy spheroids may contain only a minor component of CDM. However there is evidence in these systems for dark matter, amounting up to about 50 percent of the contained mass.

WIMPs, with masses on the order of the SUSY breaking scale ($\sim 100\,\mathrm{GeV}$), should eventually be found in direct detection experiments. The WIMP abundance can be predicted, once the annihilation cross-section, sensitive to WIMP mass, is specified. One can have $\Omega = 1$ in WIMPs over a wide range in mass, the interaction strength necessarily being strong if the mass exceeds $\sim 100\,\mathrm{TeV}$. However, once the WIMP mass exceeds $\sim 1\,\mathrm{TeV}$, the predicted fluxes make these particles sufficiently elusive that direct detection is probably unfeasible. Alternative strategies include the study of high energy gamma-rays produced by WIMP annihilations in the halo and high energy neutrinos from annihilations in the solar core as the sun captures WIMPs that are intersected along its orbit around the Galaxy.

It is convenient to divide non-baryonic dark matter candidates into hot and cold categories, defined as follows: if the relic particle velocity dispersion $v_s << 100$ km s^{-1} (typical of the potential well of a protogalaxy) at z_{eq} (when fluctuation growth can first commence), the dark matter is said to be cold. Examples of cold dark matter (CDM) are massive relic particles ($m_x > 1$ keV) or primordial black holes. The hot dark matter (HDM) prototype is the light neutrino, with the mass range $m_\nu = 1-100$ eV being of interest for dark matter. The predicted neutrino number density is 110 cm^{-3} per neutrino species, and

$$\Omega_\nu h^2 = m_\nu/100\,\mathrm{eV}.$$

Hot dark matter

Primeval density fluctuations in hot dark matter are erased by free streaming of neutrinos when their velocities are near-relativistic. Fluctuations survive only on scales above the maximum value of the neutrino Jeans length. The neutrino momentum distribution is described by an occupation number

$$n(p,t) = \frac{1}{exp\frac{pc}{T_\nu(t)} + 1},$$

so that the neutrino "temperature", initially fixed at weak interaction freeze-out, decays as $T_\nu \propto a^{-1}$ while the neutrinos are relativistic, and as $T_\nu \propto a^{-2}$ when they stream non-relativistically. The transition occurs at $T_\nu < m_\nu c^2/k$, at a redshift $z_{ns} = 1.8 \times 10^5 (m_\nu/30\,\mathrm{eV})$.

At present, the neutrino velocity dispersion is

$$\langle v^2 \rangle^{\frac{1}{2}} = 6(1+z)(30\,\mathrm{eV}/m_\nu)\,\mathrm{km\ s}^{-1}.$$

This enables one to define the comoving neutrino Jeans length

$$L_{J\nu} = \left(\frac{\pi \langle v^2 \rangle}{G\rho}\right)^{\frac{1}{2}} a^{-1},$$

which increases as a in the relativistic limit, and declines as $a^{-\frac{1}{2}}$ in the non-relativistic limit. The maximum value of $L_{J\nu}$ is the distance neutrinos free stream:

$$L_{J\nu}^{max} = \frac{\int_0^t v(t)}{a(t)} = 40\,\mathrm{Mpc}\,(30\,\mathrm{eV}/m_\nu),$$

and defines the maximum neutrino Jeans mass,

$$M_{J\nu}^{max} \approx \frac{m_{pl}^3}{m_\nu^2} = 3 \times 10^{15}\,\mathrm{M}_\odot \left(\frac{30\,\mathrm{eV}}{m_\nu}\right)^2,$$

above which HDM fluctuations survive to eventually form large-scale structure.

Fluctuation evolution in an HDM-dominated universe with $\rho_\nu >> \rho_b$ is described by

$$\frac{d^2\delta_b}{dt^2} + \frac{2\dot{a}}{a}\frac{d\delta_b}{dt} = 4\pi G\rho_\nu\left(\delta_b\frac{\rho_b}{\rho_\nu} + \delta_\nu\right)$$

where $\delta_b = \delta\rho_b/\rho_b, \delta_\nu = \delta\rho_\nu/\rho_\nu$.
One recovers the following solutions: for wavelengths

$$\lambda > L_{J\nu}^{max}, \delta_b \propto \delta_\nu \propto t^{\frac{2}{3}};$$

for $L_{J\nu} < \lambda < L_{J\nu}^{max}$, neutrino fluctuations are damped but baryonic fluctuations grow as long as $\lambda > R_D^{max}$ for baryons, according to

$$\delta_b \propto t^{(1+24\rho_b/\rho_\nu)^{\frac{1}{2}}-\frac{1}{6}}.$$

Baryon self-gravity allows inhibited growth on scales above the instantaneous neutrino Jeans length.

On smaller scales ($\lambda < L_{J\nu}$), growth is completely suppressed. This leads to a top-down scenario for structure formation, larger scales being the first to collapse.

Cold dark matter

Typical cold dark matter candidates are weakly interacting particles in the mass range $m_x = 10 - 100$ GeV that are produced thermally in the very early universe, and whose abundance is frozen when thermal equilibrium eventually breaks down at $kT \lesssim 0.1 m_x$. The relic abundance is determined primarily by the particle interaction rate: one infers

$$\Omega_x h^2 \approx (10^{-36}\mathrm{cm}^2/\sigma_f)(0.1 m_x/kT),$$

where σ_f is the annihilation cross-section at freeze-out (defined when the annihilation rate equals the Hubble expansion rate and there are too few particles in the Boltzmann tail to maintain equilibrium).

One may subsequently neglect any coupling to the photons (if the CDM particles are stable), and the growth of CDM fluctuations is described on sub-horizon scales by

$$\frac{d^2\delta}{dt^2} + \frac{2\dot{a}}{a}\frac{d\delta}{dt} = 4\pi G\rho\delta,$$

where ρ is the density of CDM plus any baryons. This equation is valid both in radiation and matter-dominated eras, and a solution is

$$\delta = A(1 + \frac{3}{2}a/a_{eq}).$$

I conclude that the subhorizon growth factor at $a \sim a_{eq}$ is ~ 2.5; at $a >> a_{eq}$, the usual $\delta \propto a$ growth is recovered. In other words, unlike in the baryon-dominated situation where the pressure gradient opposes any sub-horizon growth in the radiation era, CDM fluctuations grow weakly, CDM being subdominant during the early expansion of the universe. The transition of growth rate from radiation to matter-dominated limits is a gradual one, and imposes a characteristic scale on the relic CDM fluctuation spectrum at $L_{eq} \approx 12(\Omega_0 h^2)^{-1}$Mpc, the comoving horizon scale at the epoch of equal matter and radiation densities, when $1 + z_{eq} = 3 \times 10^4 \Omega_0 h^2$.

The horizon scale at z_{eq} is consequently a critical scale for the fluctuation spectrum: for example, in a cold dark matter–dominated universe, it characterizes the (gradual) transition between larger and larger scales ($L \gtrsim 2ct_{eq}$) which first enter the horizon at later and later times, subsequently undergoing uninterrupted sub-horizon growth, and those scales ($L \lesssim 2ct_{eq}$) which have their sub-horizon growth frozen until t_{eq}, subsequently all sharing a common growth factor. One therefore expects that the comoving length scale $L_{eq} = 2ct_{eq}(1 + z_{eq}) = 12(\Omega_0 h^2)^{-1}$Mpc will be the principal surviving imprint of the early universe on the residual linear density fluctuation spectrum that seeds the large-scale structure of the universe via the action of gravitational instability. On larger scales, the effective spectral index is $n = 1$ or $\delta\rho\ /\rho \propto M^{-2/3}$, reflecting the primordial scale-invariant spectrum. On smaller scales, the spectral index flattens to $n = -3$ or $\delta\rho/\rho = constant$, a consequence of the diminished subhorizon growth in the radiation era.

The matter fluctuations continue to grow by gravitational instability, eventually achieving non-linearity at $z \lesssim 100$. The sequence of structure formation is bottom-up, the scale-invariant spectrum generating $\delta\rho/\rho \propto M^{-2/3}$ density fluctuations that result in formation of galaxies, galaxy clusters, and even larger structures. Normalization of the fluctuation spectrum to the observed large-scale structure was traditionally performed via the galaxy correlation function or via large-scale peculiar velocities, either of which directly probe the linear fluctuation regime. However the modern normalization procedure utilizes the COBE DMR detection of cosmic microwave background temperature anisotropies as a measure of linearized density fluctuations at $z \sim 1000$. This procedure, as described below, results in excessive power in CDM density fluctuations on the scales of galaxies and galaxy clusters.

Mixed dark matter

A compromise model asserts that one can improve on CDM by combining the best attributes of two models that individually are unsatisfactory for reasons explored below. Mixed dark matter is an admixture of nonbaryonic dark matter that is 20 percent hot and 80 percent cold. This combination succeeds in providing an acceptable description of large-scale structure.

Λ-dominated cold dark matter

Another means of rescuing CDM is to lower Ω, thereby reducing small-scale power if the density fluctuations are COBE-normalized, but at the same time maintaining a spatially flat universe by introducing the cosmological constant, Λ. One may interpret Λ as the contribution of the vacuum to the dark matter density, amounting to an energy density $\lambda \equiv \Lambda/3H_0^2$ such that $\lambda + \Omega_0 = 1$, where Ω_0 is mostly in the form of CDM. The preferred combination requires $\Omega_0 \approx 0.3$ in order to account for large-scale structure.

Generalized damping scale

The detailed physics of recombination modifies the baryonic fluctuation damping scale, R_D^{max}. For an open CDM-dominated universe, one finds

$$R_D^{max} \approx \frac{2}{H_0\Omega_0}\ell_D^{-1},$$

where the angular scale subtended by R_D^{max} is $\sim \pi/\ell$ radians and the spherical harmonic wavenumber ℓ appropriate to the damping scale is given by

$$\ell_D = 10^3 \left(\frac{\Omega_B}{0.05}\right)^{1/4} \Omega_0^{-3/4}.$$

This angular scale is relevant for cosmic microwave background anisotropy measurements. In a Λ-dominated flat CDM universe, the expression for ℓ_D must be multiplied by a factor

$$f_\Lambda \equiv \Omega_0^{1/2}(1 + ln\Omega_0)^{0.085},$$

whereas R_D^{max} is multiplied by a compensating factor f_Λ. The corresponding comoving scale in either case is

$$R_D^{max} \approx 5h^{-1} \left(\frac{0.05}{\Omega_B}\right)^{1/4} \Omega_0^{-1/4}\,\mathrm{Mpc}.$$

Baryonic dark matter

The microlensing experiments identify the dark matter component in the inner galactic disk to be low mass stars, including a possibly large stellar contribution by degenerate remnants. Of course, the low mass stars are not strictly dark objects, but they do not contribute significantly to the integrated light.

The MACHO content of our halo is estimated to be 20 percent, in the simplest model, but with very large uncertainty [4]. The time-scales of the few confirmed halo microlensing events point to a contribution by M dwarfs or substellar objects, such as brown dwarfs. There is no conflict with recent attempts to image the faintest visible halo stars, provided either that there is an upturn in the initial Population II stellar mass function below the hydrogen-burning limit, or else that the calculations of mass-to-luminosity ratios for M dwarfs are far more uncertain than have hitherto been considered.

From the theoretical perspective, there is little objection to a dark halo consisting of brown dwarfs. One would certainly expect BDM to be produced in the vicinity of luminous regions, and in particular, in galaxy halos, as opposed to, say, intergalactic space. Primordial gas clouds present in dark halos soon after the halos first formed provide a physically plausible environment for the formation of brown dwarfs. Indeed there is a natural coincidence between the nucleosynthesis prediction of baryonic dark matter and the amount of dark matter in galaxy halos. Status: RO

How the theory of star formation can result in a region containing exclusively low mass stars is an unresolved question, simply because we lack any fundamental theory of how stars form that can be applied to environments as diverse as galaxy halos. A hint that there is a possibility for forming clusters of exclusively very low mass stars and brown dwarfs is suggested by considerations of accretion rates and protostellar feedback effects in dense, cold clouds. Premain sequence stars generate considerable amounts of energy, and especially momentum in the form of outflows, as a consequence of Kelvin-Helmholtz contraction. If indeed feedback, such as might be associated with the interaction of bipolar outflows from premain sequence stars with their environment, balances accretion, then a comparison of the accretion energy release onto a protostellar core ($\propto \Delta V^5 G^{-1}$) with the Kelvin-Helmholtz luminosity ($\propto M^3$) suggests that low mass cores would predominate in very quiescent (low ΔV) cloud cores. Here ΔV represents the generalized sound speed, including turbulent and magnetohydrodynamic contributions, that controls the spherical accretion rate onto a protostellar core. Dense clouds in the halo would be colder than clouds in the galactic disk, where the interstellar radiation field is an important heat source.

If dark clusters were to form in this manner, one would have to increase the estimates of the halo brown dwarf fraction that is detectable in the microlensing experiments. Whether low mass star formation could be a robust process in a cold cloud is less clear: a chance perturbation producing a massive star could energize the cloud sufficiently that the enhanced accretion would occasionally induce more massive star formation. The ensuing mix of mild and more extreme protostar-driven turbulence could subsequently generate a broad range of stellar masses, as is indeed the case in the galactic disk.

Residual fluctuation spectrum

Large-scale structure, from galaxies to clusters and superclusters of galaxies, arises from the growth of small density fluctuations in the early universe. Inflation, at $T \gtrsim 10^{15}$GeV, provides a setting for the production of curvature fluctuations. Fluctuation growth effectively begins once the universe is matter-dominated. At the transition epoch of equal densities of radiation and matter, the comoving horizon size is $L_{eq} = 12(\Omega h^2)^{-1}(g_*/3.36)^{1/2}$Mpc. Here Ω is the density in units of the critical density for closure, h is the Hubble constant H_0 in units of $100\,\mathrm{km\,s^{-1}\,Mpc^{-1}}$, and g_* is the effective number of relativistic species of neutrinos present at the epoch of matter-radiation equality, equivalent to a redshift of $z_{eq} = 3\times 10^4 \Omega h^2 (3.36/g_*)^{1/2}$.

The scale L_{eq} is imprinted on the emergent fluctuation spectrum. Since L_{eq} is comparable to the scale at which density fluctuations are presently just becoming nonlinear, one can use large-scale galaxy distribution observations to study the shape of the present epoch primordial density fluctuation spectrum, over the range 10 – $100h^{-1}$Mpc. On larger scales, cosmic microwave background anisotropies measure the fluctuations when the radiation was last scattered, at redshift $z_{LS} \sim 1000$. The redshift of last scattering $z_{LS} \approx 10^3$ is insensitive to Ω_0 and depends only weakly on the baryon density: for example, if $\Omega_0 \sim 1$,

$$z_{LS} \approx 10^3 \Omega_B^{-0.027/(1+0.11 ln \Omega_B)}.$$

While the primary peak in the emergent density fluctuation power spectrum is determined by L_{eq}, there are secondary peaks on larger scales which correspond to oscillations on the sound horizon at the epoch of last scattering. The radiative damping of large-scale fluctuations generates several oscillations of the phase-coherent baryonic fluctuations that act as sound waves on sub-Jeans wavelengths. The phase coherence occurs, according to one interpretation, because only the growing fluctuation mode is present: any inflation-generated decaying modes have long since decayed. A more precise interpretation is that it arises from special timing in the Boltzmann source terms: for example in the case of isocurvature fluctuations, once the universe is matter–dominated, a curvature mode is spawned from infall at horizon crossing. The amplitude of such peaks is only significant if the baryons represent a substantial contribution to Ω_0. There are corresponding ripples in the matter fluctuation power spectrum, that may eventually generate observable signatures over large comoving scales in deep, well-sampled redshift surveys.

However, the resulting peaks in the radiation power spectrum are observable in pending CMB anisotropy experiments. The sound horizon generates a first peak, for adiabatic fluctuations, at a comoving scale of approximately [5]

$$1.2\pi^{-1} z_{LS}^{-1/2} \Omega_0^{-1/2} H_0^{-1} \approx 120 \Omega_0^{-1/2} h^{-1}\mathrm{Mpc},$$

or an angular scale corresponding to

$$\ell \approx 200 \Omega_0^{-1/2}$$

in an open universe. Any isocurvature mode fluctuations are 90 degrees out of phase with the adiabatic mode. Thus measurement of the "adiabatic/doppler" peaks in the

CMB anisotropies on subdegree scales could potentially distinguish between isocurvature and adiabatic modes, and provide a direct measure of Ω_0.

Large-scale structure in effect measures a combination of Ω_0, H_0 and g_*, while the combination of last scattering and present epoch probes requires specification of the growth rate between $z \sim 1000$ and the present, and hence the background curvature. The fluctuation growth rate is cosmological model-dependent: for most applications, one may simply constrain the additional degree of freedom by introducing the cosmological constant Λ. This enables one to retain the generic inflationary prediction of a spatially flat universe and of a scale-invariant primordial density fluctuation spectrum even if $\Omega_0 < 1$.

More exotic models can be constructed by introducing further parameters. For example, in addition to CDM and/or HDM, one may have topological defects, with cosmic strings or textures having been studied in most detail, which seed early structure formation. Topological defects can be produced at the end of inflation, but also would arise generically in the absence of inflation. Hence such models are considered in the context of both open and flat cosmologies. Perhaps the most exotic alternative, in the sense of requiring most parameters, commences with the most astronomically compelling assumptions, that the universe be open and that the dark matter be exclusively baryonic, in which case one has to resort to primordial isocurvature fluctuations in a Λ-dominated universe. In order to avoid observational conflict between an isocurvature low Ω_0 baryonic model and the CMB constraints on temperature fluctuations and spectral distortions, introduction of a cosmological constant is unavoidable. Topological defects are one manifestation of such fluctuations, and primordial entropy fluctuations constitute another. Models for large-scale primordial entropy fluctuations are somewhat contrived, and the allowable parameter space of such models is far more tightly constrained by microwave background and large-scale structure observational limits than is the case for rival models.

Non–linear evolution: spherical collapse

For a spherical geometry, one can write down a fully non–linear solution to perturbations in an expanding universe with $p = 0$. For simplicity, take the background to be $K = 0$. A spherical shell of radius $r = r_i a_s(t)$ satisfies

$$\ddot{a}_s = \frac{-4\pi G}{3}\rho_0 a_s(1+\delta),$$

where δ is the overdensity interior to the shell. The parametric solution of this equation is

$$a_s = \frac{(1-\cos\theta)}{2\delta}; \qquad \frac{t}{t_0} = \frac{3(\theta - \sin\theta)}{4\delta^{3/2}},$$

and the ratio of $a_s(t)$ for the shell to $a(t)$ ($\propto t^{2/3}$) for the background is

$$\frac{a_s(t)}{a(t)} = \left(\frac{2}{9}\right)^{1/3} \frac{1-\cos\theta}{(\theta-\sin\theta)^{2/3}}.$$

Maximum shell expansion occurs when $\theta = \pi$, at epoch $t_i/t_0 = 3\pi/4\delta^{3/2}$ when the overdensity is

$$\frac{\rho_s}{\rho} = \left(\frac{3\pi}{4}\right)^2 .$$

Smaller initial amplitudes δ collapse later. Collapse occurs at $\theta = 2\pi$, or at epoch $t_i/t_0 = 3\pi/2\delta^{3/2}$, when the linear theory density contrast is $\frac{3}{5}\left(\frac{3\pi}{2}\right)^{2/3} = 1.69$.

If there is no dissipation, the final radius r_f is related to the maximum radius r_i by the virial theorem ($2T_f + V_f = 0$) and energy conservation ($T_f + V_f = V_i$). This results in $V_f = +2V_i$, and since $V = -GM/r$, it follows that $r_f = r_i/2$. The density contrast with the background universe at the epoch of virialization, $t_f \approx 2t_i$, is

$$\frac{\rho_f}{\rho(t_f)} = 18\pi^2,$$

and t_f is usually identified with the epoch of "galaxy formation". For a density fluctuation spectrum

$$\frac{\delta\rho}{\rho} \propto M^{\frac{-(n+3)}{6}} t^{\frac{2}{3}}$$

(corresponding to a primordial power spectrum of index n), one infers virialized scaling relations

$$M \propto \rho_f^{-\frac{2}{(n+3)}} \propto v^{\frac{12}{(1-n)}} \propto R^{\frac{6}{(n+5)}} .$$

For CDM, $n \approx -2$ on galactic scales, resulting in a superficially attractive scaling law of mass with respect to velocity dispersion or circular velocity in the dark halo ($M \propto v^4$) that resembles the Tully–Fisher scaling of (infrared) luminosity with maximum rotational velocity, but however has far too large a dispersion (if the primordial fluctuations are Gaussian) to yield a physically relevant model.

Non–linear evolution: pancake collapse

A non-linear solution for one–dimensional collapse of a perturbation in the expanding universe was first given by Zeldovich. It describes an approximate ballistic solution (if $p = 0$) for particle trajectories, given by

$$Z(t, Z_i) = a(t)\left(Z_i - \frac{a(t)}{a_p}\frac{\pi}{L_d}\sin\left(\frac{\pi}{L_d}Z_i\right)\right),$$

where Z_i is the Lagrangian coordinate for particles on a plane–wave fluctuation of wavelength L_d. A pancake, or caustic surface, develops where the trajectories become focussed, the density profile being given by

$$\rho = \rho_0\left(1 - \frac{a(t)}{a_p}\cos\left(\pi\frac{Z_i}{L_d}\right)\right)^{-1} .$$

This follows from mass conservation: $dM = \rho_0 a dZ_i = \rho(Z,t)dZ$, and shows formally that the density becomes infinite in the midplane at time $a(t) = a_p$. Numerical solutions show that for dissipationless matter, the midplane crossing is followed by sequential formation of additional caustics, as matter traverses the plane and repancakes, eventually heating up in all three dimensions.

More realistically, one expects dissipation to occur via gravitational shocking. In the case where gas is included, the dissipational physics is straightforward, and results in the formation of a dense, cooled, shocked layer in the midplane, provided L_d is not too large. The condition for efficient cooling is that the surface density of shocked gas ($\sim \rho L_d$) satisfies the criterion for a radiative shock:

$$\rho L_d > \frac{v_s^3}{\Lambda(T)},$$

where $\Lambda(T)$ is the cooling function and v_s is the sound speed. The gas cools to $\sim 10^4$K and is unstable to fragmentation. The cooled gas fraction amounts to

$$\sim 0.25\left(\frac{5\,\mathrm{Mpc}}{L_d}\right)^{1/3},$$

and the characteristic fragment scale is $M_{frag} \approx 1.2v_s^4/(G^{3/2}p^{1/2}) \sim 10^7 M_\odot$ for typical parameters corresponding to $L_d \sim 5$ Mpc, at redshift $z \sim 1$. This leads to a top-down sequence of structure formation, since pancake scales in HDM correspond to cluster mass scales because of the imprint of the neutrino damping scale.

Interestingly, transient caustics also develop in the CDM model during the bottom-up sequence of hierarchical structure formation, near the transition regime where the effective spectral index of the fluctuation power spectrum is near to -1. For effective values of n larger than this, large–scale power dominates over small-scale power in terms of the contributions to peculiar velocity perturbations, so that there is considerable large-scale coherence on supercluster mass scales.

Biasing

It is customary to fix the normalization empirically by specifying the bias parameter b that gives the ratio of the amplitudes of fluctuations in the luminous to dark matter densities, defined to be $b = \sigma_{gal}/\sigma_8 = \sigma_8^{-1}$ at 8 h^{-1}Mpc, where σ_8 denotes the variance in $\delta\rho/\rho$ evaluated at $8h^{-1}$Mpc, the scale on which the rms galaxy count fluctuations σ_{gal} have unit variance. According to linear theory, the amplitude of the large-scale peculiar velocity field is approximately proportional to $\Omega^{0.6}/b$ at a specified overdensity. Large-scale velocity studies suggest a normalization with [6],[7] $\Omega^{0.6}/b \approx 0.4 - 0.9$. If $\Omega_0 \approx 1$, a larger value of $2 \lesssim b \lesssim 2.5$ is favored by galaxy-scale constraints, such as pairwise velocities and the number density of bright galaxies above a specified potential-well depth [8],[9],[10]. With this normalization, one can specify the epoch of galaxy formation, provided one can extrapolate the fluctuation spectrum to the relevant comoving scale. Collapse of massive halos, mass $\sim 10^{12}\,M_\odot$, now occurs at $1+z=8/b$.

If Ω_0 is low, on the other hand, the attractive possibility emerges that the mass in the Universe might actually trace the light, so that $b \approx 1$. This removes one of the most *ad hoc* parameters of phenomenological cosmology. In low Ω_0 models, galaxy formation occurs earlier than in high Ω_0 models by a factor $\sim \Omega_0^{-1}$ earlier in redshift.

Inflationary models generally result in Gaussian fluctuations. If galaxy formation is *biased* so that galaxies form only from rare high peaks of a Gaussian distribution, the correlations of luminosity (galaxies) can exceed the mass correlations [11]. Consider the Gaussian distribution

$$\delta(\mathbf{x}) = \frac{(\rho(\mathbf{x}) - \langle\rho\rangle)}{\langle\rho\rangle}. \tag{2.5.1}$$

with dispersion $\sigma^2 = \langle|\delta|^2\rangle$. The auto-correlation function of the density fluctuations is

$$\xi_{\rho\rho}(r) = \langle\delta_1\,\delta_2\rangle, \tag{2.5.2}$$

where $|\mathbf{x_1} - \mathbf{x_2}| = r$. If we identify galaxies with rare peaks, $\delta > \nu\sigma$ where $\nu > 1$, then the two-point correlations are given approximately by

$$1 + \xi_{gg} = e^{\nu^2 \xi_{\rho\rho}}, \tag{2.5.3}$$

so that $\xi_{gg} \simeq \nu^2 \xi_{\rho\rho}$ gives the enhancement factor. Thus rarer peaks are also more clustered, and a prescription [12] that assigns elliptical galaxies to, say, 3σ fluctuations, while spiral galaxies are 2σ peaks, could account for the observed morphological segregation of galaxies, although it lacks rigorous physical justification.

Suppose, for example, that galaxies form from $\gtrsim 2\sigma$ peaks. The luminosity correlations are enhanced by a factor ~ 4 relative to the mass correlations. We see that ν can be identified with b for scales that are just going nonlinear. A physical model to explain the preferential formation of luminous galaxies in high-σ peaks was developed by Dekel and Silk [13]. Supernova-driven winds systematically suppressed star formation in the low mass, $\sim 1\sigma$, clouds, which constitute the bulk of the mass at the epoch of galaxy formation in a CDM density fluctuation model. One consequence is the prediction of a widespread population of "failed" dwarf galaxies.

An alternative scheme relies on "natural" biasing. Suppose that galaxies of higher luminosity form in deeper potential wells, in such a way as to satisfy the Tully-Fisher relation for spiral galaxies, $L \propto v_c^\epsilon$, [14] where v_c is the maximum disk rotation velocity, proportional to the halo velocity dispersion [10]. One finds $\epsilon \approx 2$ in the B band, increasing to about 3 in the I band. Numerical simulations indeed confirm that high luminosity "galaxies" formed preferentially from peaks in denser regions, thereby generating a substantial bias against bright galaxy formation in regions of average density. However, this model suffers from a lack of theoretical insight into the physical origin of the adopted Tully-Fisher relation, which observationally is a remarkably good correlation. A more serious difficulty, perhaps, is that more massive systems form later, in disagreement with the observation that the most massive galaxies, as judged by potential well depth, are ellipticals and are the oldest systems, already in place and unevolving by $z \sim 1$.

Yet another scheme appeals to gas heating via ultraviolet photons or fast particles, attributed to active galactic nuclei or even just the first generation of galaxies, to augment the gas pressure over scales of up to 10-20 Mpc and thereby inhibit galaxy formation – a phenomenon that may be labelled 'anti-bias' [15], [16],[17]. Such a biasing scheme involving an early generation of dwarf galaxies or AGNs would heat the intergalactic medium and raise the Jeans mass sufficiently to favor massive galaxy formation especially in galaxy clusters where mergers occur, with the lower mass systems mostly forming later as the IGM cooled adiabatically. This scheme does provide a more reasonable ordering scheme for early formation of ellipticals by mergers and late formation of disks by infall.

In COBE–normalized $\Omega_0 = 1$ CDM, the mass is more strongly clustered than the galaxies: such a model must be strongly antibiased to succeed. No physical prescription has been provided for the requisite degree of antibiasing. We have seen that CDM may however be revived by modest adjustment of the large-scale power parameters: *e.g.*, by reducing n from the "best" inflationary value of approximately 0.95 to 0.9 and thereby introducing the implied tensor mode to renormalize COBE to a lower value, and simultaneously setting the Hubble constant to a suitably low value of, say, 45 $\mathrm{km\,s^{-1}Mpc^{-1}}$. Indeed, as will be discussed more extensively below, the resulting model gives an excellent fit to all of the large-scale power constraints, including a value of $\sigma_8 \approx 0.7$.

Scaling relations for virialized objects

Mass scales that are presently nonlinear ($\sigma_\rho > 1$, where $\sigma_\rho \equiv \langle |\delta\rho/\rho|^2 \rangle$ as defined by linear theory) went nonlinear at redshift $1 + z = \sigma_\rho$, while the epoch of galaxy formation, or non-linearity of $\nu\sigma$ peaks, occurred at $1 + z_{gf} = \nu\sigma_\rho$. Here, the subscript ρ denotes dark matter, as opposed to the luminous component observable via galaxy counts. The comoving number density of objects of mass in the range M to $M + dM$, that are newly nonlinear, is approximately given by [18]

$$\frac{dn(M,t)}{dM} \approx \frac{\rho_0}{M^2} \frac{1}{\sigma_\rho} (1+z) \exp[-\delta_c^2 (1+z)^2 / 2\sigma_\rho^2], \qquad (2.5.4)$$

where ρ_0 is the background density and $\delta_c = 1.69$ is the linear theory overdensity corresponding to when a spherical fluctuation would have collapsed in an $\Omega = 1$ universe. For dissipationless matter, this corresponds to the virialization epoch of the dark halo. This agrees well with the estimates for the instantaneous number of galaxy halos or clusters derived from both numerical simulations [19] and more sophisticated overdensity-threshold excursion set analyses, albeit depending on the technique used to identify a mass [20].

The Press-Schechter expression for the mass function is derived from an initial assumption that the probability distribution governing whether a given volume contains a mass of specified overdensity is a Gaussian. This approach assumes successive mass scales that go nonlinear subsume earlier stages of the evolving hierarchy of density fluctuations, producing self-similarity. An alternative approach, the 'peaks formalism', retains substructure, identifying peaks in the primordial density field, smoothed over a suitable scale to remove smaller-scale structure, as the sites of future galaxy formation. Within the latter formalism, the statistics of Gaussian maxima determine wide-ranging properties of galaxies. These have been explored analytically by Bardeen *et al.* [9] and numerically by Peacock and Heavens [21]. Neither approach properly accounts for the evolution of 'clouds within clouds'.

The density profile within newly collapsed, non-dissipative spherical objects in an Einstein-de Sitter universe, for which

$$\frac{\delta\rho}{\rho} \propto M^{-(n+3)/6}(1+z)^{-1} \tag{2.5.5}$$

and which "turned around" when $\delta\rho/\rho = (3\pi/4)^2$, is

$$\rho \propto r^{-3(n+3)/(n+4)}, \tag{2.5.6}$$

with corresponding 'rotation curve'

$$V_c \propto r^{-(1+n)/(4+n)}. \tag{2.5.7}$$

Thus the observed flat rotation curves suggest that $n \sim -1$, consistent with CDM over galaxy scales (*c.f.* [24]. However it should be emphasized that observational constraints on the dark matter density profile from rotation curve analyses are rather weak, as noted by Bahcall, Schmidt and Soneira [25] and by Lake and Feinswog [26].

4 Where is the Dark Matter?

Dark matter in disks

Oort first used the acceleration of disk stars to measure the amount of matter in the solar neighborhood. The recent determinations of the Oort limit show that the total amount of matter in a cylinder perpendicular to the galactic plane near the sun amounts to 75 ± 25 solar masses per square parsec. The observed mass in stars and interstellar matter amounts to about 50 solar masses per square parsec. Hence there may be as much as 50 percent of local matter in unknown form. Recent experiments have shed further light on the nature of dark matter in the galactic disk and halo. These involve gravitational microlensing of stars in the bulge of our galaxy and in the Large Magellanic Cloud. Amplification events have been seen which are produced by intervening compact dark objects of mass about 0.1 solar mass, with, however, a large uncertainty. The LMC events probe MACHOs in the galaxy halo and have detected a frequency of these objects which appears to fall short by about a factor of 3 of the amount of dark matter needed to account for the amplitude of the galactic rotation curve. Towards the bulge, however, approximately three times more microlensing events were found than anticipated by standard models of the stellar distribution in the galactic disk. The additional events may require the galactic bar to make a significant contribution to the frequency of microlensing events, by bar-bar lensing due to a nearly aligned bar (relative to the line of sight). Indeed the COBE map of the inner galaxy supports such a bar. In addition, an additional, smaller contribution may come from a "maximum" disk model in which the surface density of low mass stars saturates the Oort limit, the local disk surface density inferred from measuring the velocity dispersion and scale height of old Population I stars.

Dark matter in halos

Rotation curves of spiral galaxies are generally flat out to several disk scale lengths. Interpretation of the rotation curves requires a substantial halo dark mass component. In terms of the ratio of mass to light in the blue spectral band, the dark matter content inferred is $M/L \sim 50h^{-1} - 100h^{-1}$ in solar units, where $h = 100\,\mathrm{km\,s^{-1}Mpc^{-1}}$. For comparison, in the luminous regions of spiral galaxies, the mass to light ratio is about 3. Older stellar populations have lower luminosity per unit of stellar mass and $M/L \approx 5 - 10$. Recent observations of diffuse x-ray emission from the halos of elliptical galaxies confirm that ellipticals also have substantial dark halos.

Some uncertainties remain. Rotation curves occasionally decline. This may be due to a dark halo of somewhat lower central density. Dwarf spirals appear to have a finite core as inferred from their rotation curves. Cold dark matter simulations do not reveal such a core, indicating that the dark matter in the halo may have undergone some dissipation [29],[30].

Halos of luminous galaxies extend to $\gtrsim 100\,\mathrm{kpc}$, where rotation curves become unmeasurable. Binary galaxies and satellite galaxies are used to study the dark matter content of halos on scales up to 100 kpc. The results agree with direct measurements from rotation curves, but are more uncertain because of uncertainties in inclinations and orbital parameters.

Groups and clusters of galaxies

On scales of about 1 Mpc, the pairwise velocities of galaxies probe the dark matter distribution. One finds that the mass to light ratio is approximately $200h$ at a scale of 1 Mpc. On scales of $1 - 10$ Mpc, aggregations of galaxies can be used to study the dark matter distribution. X-ray observations of the hot gas in groups and clusters of galaxies help determine the gravitational potential wells, provided that the temperature of the x-ray emitting gas can be measured. The scale from which a galaxy cluster of mass $10^{15}M_{15}M_\odot$ has condensed is $10M_{15}^{1/3}(\Omega h^2)^{1/3}$ Mpc. The dark matter distribution on this comoving scale is being probed by observations of galaxy clustering.

The rich cluster abundance effectively measures the amplitude of density fluctuations on the comoving scale from which clusters condensed. Even if galaxies do not form in cluster mass potential wells, accretion of gas results in x-ray emission. The number density of rich clusters, either optically or x-ray-selected, has been used to constrain the amplitude of the rms density fluctuations on a scale of about 10 Mpc. The velocity or temperature distribution of clusters is also sensitive to the shape of the underlying power spectrum near this scale. One finds that $\sigma_8 \approx 0.5$, where, as before, σ_8 is the rms level of density fluctuations on scale $8h^{-1}$Mpc. For spheres of this radius, the galaxy count fluctuations have unit amplitude. The resulting mass-to-light ratio is approximately 400h for a well-studied cluster such as Coma.

These systems are usually be assumed to be in virial equilibrium, as inferred from the overdensity, central concentration, and relaxed appearance of the rich clusters. In this case, the masses of dark matter and of gas are inferred from measurements of the x–ray surface brightness and temperature profile. Hydrostatic support requires $M_{total}(r) = \frac{kT(r)}{\mu m_H G}\left(\frac{\partial ln\rho_g}{\partial lnr} + \frac{\partial lnT}{\partial lnr}\right)$ and the gas density ρ_g is measured from the x–ray profile. Typical gas temperatures $T \sim 10^8$K are measured directly from the x-ray spectrum, although the radial resolution is currently inadequate to construct precise models. Dark matter is found to dominate the stellar mass by 20 to 1, but exceeds the gas mass by only about 5 to 1 within a radius of $\sim 2h^{-1}$ Mpc. One uncertainty is that x-ray maps indicate considerable deviations from isothermality and temperature asymmetries, possibly reflecting deviations from hydrostatic equilibrium of the gas distribution. However the mass modelling cannot be too far from the truth, as recent gravitational lensing maps of the gravitational potential, determined by measuring distortions of the images of background galaxies, have generally, but not invariably, resulted in similar mass-to-light ratios to those derived by other techniques.

A surprise coming from studies of x-ray clusters has been the high baryon content, which is approximately $0.1h^{-3/2}$. Since the primordial nucleosynthesis of helium, deuterium and lithium requires a baryon density of $\Omega_B h^2 \approx 0.015$, one infers either that $\Omega_0 \approx 0.2$, or that there has been substantial separation and concentration of the baryons on cluster scales if $\Omega_0 = 1$. X-ray selection of clusters may be biased towards unusually x-ray luminous clusters, so that this latter inference may not have universal implications for Ω_B. Clusters show strong negative (*i.e.* dimmer in the past) evolution in x-ray luminosity and considerable substructure in x-ray morphology and temperature. This is consistent with expectation if $\Omega_0 = 1$ as a consequence of recent and ongoing infall [22], but is more difficult to reconcile with a low Ω_0 universe, where infall ceased at $z \lesssim \Omega^{-1} - 1$.

<u>Large-scale redshift surveys</u>

Two techniques have been applied to probe dark matter on scales between 10 and 100 Mpc. The correlation function of galaxy clusters measures the mass power spectrum to a scale of about $40h^{-1}$ Mpc. Observations are performed in redshift space, and the conversion to physical space introduces some uncertainty because of galaxy peculiar motions within and near clusters. This creates less of a problem as the fluctuations become more linear on larger scales, where the peculiar velocities of galaxies, and especially of galaxy clusters, measure the dark matter distribution out to $\sim 200h^{-1}$Mpc. Infall to our local supercluster requires $\Omega \sim 0.3$, otherwise an excessive velocity field would be measured. On these smaller scales, determination of the effective Ω is coupled to the density fluctuation amplitudes, the peculiar velocities measuring the clumped component of dark matter, and effectively constraining $\Omega^{0.6}\sigma_8$ to be about unity.

The local infall in the Virgo Supercluster generates a pattern velocity

$$v_p \approx \frac{1}{3}\Omega_0^{0.66} v_H \frac{\delta\rho}{\rho}(1 + \frac{\delta\rho}{\rho})^{-1/4},$$

where $\delta\rho/\rho = 1.5 - 2$ is the spherically-averaged local oversdensity out to $20h^{-1}$Mpc around Virgo. From the data on galaxy redshifts and distances determined by various direct measurement techniques (Tully-Fisher relation for spirals, surface brightness fluctuations for early-type galaxies, etc.), an infall speed of ~ 250km s^{-1} is inferred. The resulting limit on the density parameter is $\Omega_0 \gtrsim 0.3$, this being an actual determination of Ω_0 if mass traces light. Cluster and supercluster observations do not constrain a component of the dark matter that is uniform over $\sim 10h^{-1}$ Mpc, thereby allowing the global value of Ω to possibly be larger. Status: RO

Several galaxy and cluster redshift surveys out to a depth of about $200h^{-1}$Mpc have been combined to measure the shape of the density fluctuation power spectrum [31] $P(k) \equiv \langle|\delta_k|\rangle^2$, where $\delta_k = V^{-1}\int \frac{\delta\rho}{\rho}(\mathbf{x})e^{i\mathbf{k}.\mathbf{x}}d^3x$. Comparing these data to the CDM prediction, one effectively constrains the ratio of L_{eq} to a typical survey scale, thereby measuring $\Omega h g_*^{-1/2}$. If $g_* = 3.36$ for the usual number of relativistic degrees of freedom, including 3 neutrino species, the best fit value is $\Omega h = 0.25(\pm 0.1)$. There is unambiguous evidence for a change in shape of the power spectrum, which flattens from [23] $n \approx -2$ on small scales ($\lambda \equiv 2\pi/k \lesssim 30h^{-1}$Mpc) to $n \approx -1.1$ on intermediate scales, $\lambda \lesssim 120h^{-1}$Mpc. There is very tentative evidence for a maximum in the deepest samples studied, at $\lambda \sim 300h^{-1}$Mpc.

A second technique utilizes maps of the peculiar velocities of galaxies, that is, deviations from the Hubble flow, enabling one to study dark matter that is clumped over these scales. The results depend on either application of the Tully-Fisher relation to spiral galaxies or of the fundamental plane ($D_n - \sigma$) analysis to elliptical galaxies in order to obtain true distances. If there are no systematic biases in the distance determinations, the typical bulk flows of ~ 400 km s^{-1} over $\sim 50h^{-1}$ Mpc require that $\sigma_8 \cdot \Omega^{0.6} \approx 0.4 - 0.9$. However there are uncertain systematic errors that have arisen in attempts to compare optically-selected galaxy samples and the associated velocity fields, derived from distance indicators and redshifts, with the velocity field predicted from the all-sky IRAS-selected galaxies via the inferred gravitational acceleration on the assumption that mass traces light on large scales.

Cosmic microwave background fluctuations

The COBE detection of fluctuations in the microwave background in 1992 on angular scales of between 10 and 90 degrees heralded a new era in dark matter studies. This is because, for the first time, one had measured the amplitude of the primordial fluctuation spectrum from which large-scale structure arose. Temperature fluctuations in the cosmic microwave background effectively measure the density fluctuation amplitude at the last scattering epoch [32], once a model is adopted that relates $\delta T/T$ to $\delta\rho/\rho$. The physics is straightforward: the Boltzmann equation couples matter to radiation, and must be evolved through the epoch of last scattering of the radiation up till the present.

COBE has measured $\delta T/T$ over scales of 10–90 degrees at a level of about 1 part in 10^5. In the post-COBE era, cosmic microwave background fluctuations have been detected by a variety of experiments, ground-based and on balloons, at both millimetric and centimetric wavelengths, over scales ranging from 0.2 to 10 degrees. Typical values of $\delta T/T$ are about 2×10^{-5}. The equivalent comoving length scales sampled extend down to about 50 Mpc, a linear scale of 100 Mpc subtending approximately $\Omega^{1/2}$ degree at redshift $z \gg 1$. The relation between $\delta T/T$ and large-scale structure parameters is model-dependent, but some general features may be inferred.

On scales larger than several degrees, the fluctuations directly represent primordial gravitational potential fluctuations, although these are modified by the curvature of the universe if $\Omega < 1$. Even in the absence of primordial curvature fluctuations, the presence of topological defects or strings imprints entropy fluctuations on the microwave background, with an approximately Poisson signature on large angular scales. On degree scales, baryon-photon oscillations on the maximum sound horizon boost the temperature fluctuations by a factor of typically 2 or 3. The amplitude of the resulting adiabatic peaks is a measure primarily of the baryon density, but is also sensitive to the unknown ionization history of the universe. On small scales, the temperature fluctuations are suppressed by adiabatic damping due to photon diffusion.

Dark matter plays an indirect but important role in accounting for the observed amplitude of $\delta T/T$. Without the presence of cold dark matter, there would be no growth of density fluctuations between matter-radiation equality and last scattering. Hence primordial adiabatic density fluctuations would result in excessively large degree-scale temperature fluctuations. One can only avoid this requirement of a dominant component of non-baryonic dark matter if the initial fluctuations are not adiabatic, but are primordial entropy fluctuations in a low Ω universe.

Armed with CMB anisotropies and a theory that connects $\delta\rho/\rho$ to $\delta T/T$, one can now attempt to reconstruct the primordial density fluctuation spectrum on large scales. It is remarkable that, at least to within a factor of 3 in power, the CMB reconstruction of $P(k)$ overlaps with that obtained directly at the present epoch from the redshift surveys. The microwave background fluctuations on degree scales effectively complement the fluctuations measured in deep redshift surveys. In the early universe, the radiation and matter are tightly coupled, and the radiation fluctuations on the last scattering surface have a redshift of approximately 1000. In effect, primordial density fluctuations are being directly measured by mapping cosmic microwave background temperature fluctuations on large angular scales.

Implications for large-scale structure

Comparison of fluctuations from both the microwave background studies and redshift surveys enables one to reconstruct the underlying fluctuation power spectrum over scales from 10 to 10^4 Mpc. Comparison with the simplest cosmological models is at first sight relatively reassuring. Consider, for example, an inflationary cosmology in which the universe contains the critical density of cold dark matter. The initial fluctuation spectrum is expected to be scale invariant. There is, consequently, a peak at late times in the fluctuation spectrum that emerges from the early universe, at a scale corresponding to the horizon size when the universe first became matter-dominated, namely,

$$L_{eq} \approx \frac{12Mpc}{\Omega h^2}\left(\frac{g_*}{3.36}\right)^{1/2} .$$

Here g_* is the effective number of relativistic species that are present. The reconstructed power spectrum shows a peak at a somewhat larger scale, although the qualitative similarity in shape is quite encouraging. However, it is in the detailed fit that interesting constraints emerge on the nature of the dark matter.

Close inspection of the observed matter power spectrum with that predicted by $\Omega = 1$ CDM reveals serious quantitative discrepancies. COBE normalization requires $\sigma_8 = 1.4(\pm 0.1)$, and results in generation of excessive power on $1 - 10$ Mpc scales, not just from the reconstructed power spectrum but also from the various non-linear indicators that include pairwise velocities, cluster abundances, massive halo rotation curves and halo frequency. All of these are manifestations of the excessive power driven by mass tracing light when $\Omega_0 = 1$ and the dark matter is cold, or equivalently the statements that $\sigma_8 \approx 1$ and $\Omega_0 = 1$. The problem is that we simply do not measure a critical density in dark matter on scales below 10 Mpc.

On $10-100$ Mpc scales, the situation is still ambiguous. Some analyses of bulk flows using galaxy clusters as probes favor the unbiased CDM model, [27], with $\sigma_8\Omega^{0.6} \approx 1$, whereas the shape of the power spectrum requires [31] $\Omega h \approx 0.25$. The zero crossing in the correlation function for CDM occurs at $\sim 17\ (\Omega h^2)^{-1}$ Mpc, and, if $\Omega h = 0.5$, is in possible conflict with the galaxy-galaxy and cluster-cluster correlation functions. The latter remains positive to $\gtrsim 40\, h^{-1}$ Mpc with a correlation length of $\sim 15h^{-1}$ Mpc. The galaxies have a similar correlation function, rescaled to a correlation length of

$\sim 5h^{-1}$ Mpc. Galaxy clusters can be interpreted as $\sim 3\sigma$ fluctuations of the underlying Gaussian galaxy distribution. The correlation functions are well fit by the low density CDM model.

Evolutionary effects also provide a probe of CDM, in which large-scale structure is undergoing significant evolution at low redshift. Of especial interest is the evolution of galaxy clusters and of galaxies. X-ray selected clusters display strong evolution of the x-ray luminosity function, possibly consistent with CDM, whereas examples of rich optically-selected clusters are found at redshifts of ~ 1 or larger. Some of the distant clusters appear to contain large populations of spectroscopically old galaxies, and there are individual examples of radio-selected galaxies containing apparently old stellar populations at $z \sim 2$, 3 and even 4. Such early evolution of stars and galaxies favors a low density universe.

The recent determinations that favor a high value of H_0, if sustained, can only be reconciled with a low Ω universe. However it should be noted that systematic errors in the various measures of H_0 have undoubtedly been underestimated. For example, use of such empirical calibrators as the globular cluster luminosity function, the planetary nebula luminosity function or the $D_n - \sigma$ relation in old stellar populations are subject to bias if the galaxies studied have an underlying younger stellar component that differs from the stellar mix in the woefully few local calibrators. Even the low dispersion in surface brightness fluctuations in early-type galaxies that yields apparently accurate distances may be subject to bias if the stellar populations are inhomogeneous as a consequence of, for example, a recent or even not-so-recent merger. Some methods are subject to environmental bias concerns: for example, galaxy clusters are dense regions where the star formation history, mean galaxy age and mean luminosity may differ from that in the field. The situation will only improve when more Cepheid distances are obtained for galaxies in and beyond the Virgo cluster, or when geometrical techniques for measuring the distance scale, such as the Sunyaev-Zel'dovich effect in x-ray clusters or the expanding photosphere method applied to supernovae, are fully developed.

Is Ω equal to unity?

Cold dark matter normalized to COBE overproduces the power and density fluctuations on the smaller (galaxy and cluster) scales, although it provides an excellent fit on larger scales. One suggested modification to improve the situation involves addition of a component of hot dark matter amounting to about 20 percent of the critical density. On large scales, the resulting mixed dark matter model is indistinguishable from cold dark matter, but on smaller scales, the power is reduced because of the free streaming of the hot component. A critical test for the mixed model lies in its predictions of structure formation at high redshift, where the reduced power means that structure formation occurs later than in the cold dark matter model.

The principal argument that favors $\Omega = 1$ stems from theory. An Einstein-de

Sitter universe is a natural consequence of inflation. Cold dark matter is a particularly favored dark matter candidate, that needs minimal fine-tuning, but one might, as already mentioned, invoke a 20 percent admixture of hot dark matter to result in better observational agreement with the inflationary $\Omega = 1$ prediction. Another rescue scheme involves tinkering with the fluctuation spectral index defined by $\delta\rho/\rho \propto M^{-\frac{1}{2}-\frac{n}{6}}$: reducing n to 0.9 helps reduce σ_8 from the disastrous COBE-normalized value of 1.4 in standard CDM to the more acceptable range of $0.7 - 0.9$, if $H_0 \lesssim 50\,\mathrm{km\,s^{-1}Mpc^{-1}}$. Recall that while $n = 1$ is the formal scale-invariant choice, whereby primordial potential fluctuations are of order $(|\delta\rho/\rho|_h)^{\frac{1-n}{6}}$, where $|\delta\rho/\rho|_h$ is the density fluctuation amplitude at horizon crossing, standard inflationary models are only approximately scale-invariant and result in $n \approx 0.95$.

More radical alternatives involve tinkering with the various parameters that enter into L_{eq}, namely, Ω, h, or g_*. For example, a satisfactory model might set $h = 0.3$. This is in contradiction with local observations of the Hubble constant, but cannot be excluded as a possible global solution. An even more radical approach adds extra neutrino species that are unstable and result in $g_* \approx 10$. Lowering Ω provides a more reasonable way to obtain extra large-scale power. In order to maintain a flat universe as required by the simplest inflationary models, one would have to add a cosmological constant to contribute about 70 or 80 percent of the critical density in terms of the vacuum contribution to the energy density of the universe.

None of these modifications are entirely successful. The baryonic mass fraction in rich clusters argues against all $\Omega_0 = 1$ models, as does the failure to find a critical density of dark matter in the vicinity of galaxy clusters and in the Local Supercluster. The CDM model only just succeeds in forming galaxies sufficiently early, and some of its variants fall short in this regard. For example, MDM (mixed dark matter) suffers from late galaxy formation, as it has far less small-scale power than does CDM: in fact, MDM on small scales ($1 - 10\,\mathrm{Mpc}$) resembles biased ($\sigma_8 \approx 0.5$) CDM, although power decreases more rapidly on subgalactic scales, while on large scales ($> 100\,\mathrm{Mpc}$) it is indistinguishable from unbiased CDM. Whether the late epoch of massive galaxy formation in the MDM model poses a problem for observations such as the mass fraction in damped Lyman alpha absorption clouds at high redshift remains to be clarified as higher resolution observational data is acquired. A potentially more serious problem is that MDM does not appear to give a satisfactory shape for the measured power spectrum in deep redshift surveys, that is so well-fitted phenomenologically by $\Omega h = 0.25(\pm 0.1)$ CDM. However the uncertainties are sufficiently large that this may not yet be a serious issue. The N-body simulations certainly indicate agreement with the data for the more quantitative measures of large-scale structure, such as the cluster mass function and large-scale peculiar motions.

ΛCDM (the flat, cosmological constant-dominated model) does better than MDM with regard to galaxy formation, but does relatively poorly with regard to accounting for large-scale peculiar velocities. For example, while the power spectrum measured in the redshift surveys and the excess angular correlations found in the APM survey at $\sim 10-20$ Mpc scales are well fit by the phenomenological $\Omega h = 0.25$ parametrization, such a model cannot account for the bulk flows, as characterized for example by [28] the 6 clusters out to $z \sim 0.1$ with peculiar velocities that allegedly exceed $\sim 1000\,\mathrm{km\,s^{-1}}$.

The tilted cold dark matter (TCDM) variant, as originally formulated, is incompatible with the COBE-DMR-Tenerife determination of n ($\gtrsim 0.85$). However TCDM may also contain a substantial amplitude of tensor (gravitational wave) fluctuations that generate large-scale $\delta T/T$. This would alleviate the need to tilt n. As previously noted, a value of n of about 0.9 in combination with the approximate inflationary ϕ^4 potential prediction of the ratio of amplitudes of tensor to scalar modes, $T/S = 7(1-n)$, suffices to rescue CDM, yielding a value for $\sigma_8 \approx 0.8$ that is reconcilable with most large-scale structure data, provided that $h \approx 0.45$.

There is no compelling version of standard ($\Omega = 1$) CDM that accounts for *all* of the observational constraints. Either some of the data is wrong, or, more precisely, is subject to unknown systematic error, or else a more radical model is needed. Such a model may contain several ingredients. For example, a cosmological constant plus a modest tilt plus a tensor component certainly improves on the ΛCDM model. Fine-tuning of MDM could be accomplished by tilting n : this would, for example, allow H_0 to be increased [34]. While more extreme options, such as topological defects, fail to rescue CDM, producing even more small-scale power because of the presence of early seeds for structure growth, there is little doubt that a CDM model with several parameters could fit all of the data on large-scale structure.

The most radical approach is to argue that the dark matter is entirely baryonic, and amounts to no more than what is observed in galaxies, galaxy clusters and galaxy superclusters. This option leads us to the low Ω universe. Such a universe is beloved by astronomers, since it incorporates the matter that is well observed by virtue of it being either luminous, or else in close proximity to luminous matter. Whether or not there is an additional widespread diffuse component of dark matter that is uniform over larger scales ($\gtrsim$ 10Mpc) remains to be observationally clarified. Only large-scale flows have been found to favour high Ω, and even their interpretation is disputed. A baryonic universe does marginally satisfy the most recent nucleosynthesis constraints, $\Omega_B h^2 = 0.02 \pm 0.005$. Indeed, if the primordial deuterium abundance D/H is as low as a recent high redshift measurement claims, one would need to raise Ω_B by an additional factor of ~ 2. In this case, one could readily have $\Omega = 0.1 - 0.2$ entirely in baryons. Of course, such a low value of Ω is precisely what is measured by surveys of luminous matter and dynamical studies, out to $\sim 10h^{-1}$Mpc scales.

The scepticism that arises when such a model is proposed stems from its theoretical underpinnings. Primordial curvature fluctuations are unacceptable in a baryon-dominated universe, on account of the excessive cosmic microwave anisotropies that are predicted. This leaves one with primordial entropy fluctuations, for which there is no model that predicts the spectral distribution or amplitude of the associated density fluctuations. One has to resort to the ugly procedure of phenomenologically fitting the data from deep redshift surveys and microwave background fluctuations in order to extract the primordial fluctuation spectrum. If a single value of n is selected, observationally constrained to be near -1, the open BDM model is unacceptable. However by introducing the cosmological constant Λ, one can rescue the model. The extra time available at a given redshift in a Λ-dominated universe allows more growth between last scattering and today, and thereby reduces the amplitudes of intermediate scale $\delta T/T$ once normalization to $\sigma_8 \approx 1$ is specified. This latter condition is simply equivalent to requiring mass to trace light on ~ 10 Mpc scales, a principal motivation of the low Ω model. Alternatively, one could tinker with a scale-dependent n: a satisfactory model would have $n \gtrsim 1$ on scales $\gtrsim 100$ Mpc but $n \approx -1$ on smaller scales.

An interesting prediction of the BDM models is that there is considerable large-scale power, with a peak in the matter power spectrum on the scale of the horizon at last scattering rather than matter-radiation equality as in the CDM models. This is a consequence of the baryon–coupling to radiation, the Jeans mass being constant between equality and last scattering, and thereby allowing additional growth as fluctuations first enter the horizon. Moreover, there are undamped acoustic oscillations just below the Jeans scale, that result in residual wiggles in the relic matter power spectrum. Future well-sampled and deep galaxy redshift surveys may have the capability of testing this prediction.

5 Formation of Disk Galaxies

The topic of galaxy formation has traditionally focused on cosmological issues, perhaps primarily because non–dissipative dark matter is relatively simple to incorporate into large–scale structure simulations. The absence of any definitive, or even plausible, detailed theory of galactic star formation has however meant that no realistic galaxy formation models have hitherto been constructed. As a result there has been relatively little contact between theory and the new deep surveys and discoveries of high redshift galaxies. What theory there has been may be said to be largely empirical: e.g. using population synthesis models. A recent development has combined analytic theory, based on the Press-Schechter formulation of mass functions, with population synthesis models [35][36][37]. Even this ignores the guts of the problem, namely the fundamental issues of star formation. The acknowledgement of the star formation issue typically involves invoking a cooling time argument, and is sometimes combined with triggering by tidal interactions or mergers and feedback from supernovae. One can surely improve on this approach by utilizing models for star formation based more closely on what is actually observed. This approach can be developed for the formation and evolution of disk galaxies.

Rotation

An obvious distinction between spiral disk galaxies and ellipticals is that spirals are rotationally supported, whereas ellipticals are supported primarily by the anisotropic velocity dispersion of the stars. Studies of the origin of angular momentum have a thorough pedigree, beginning with a pioneering study by Hoyle nearly a half century ago. Angular momentum arises from tidal torques between neighboring, asymmetrically shaped protogalaxies. The dimensionless angular momentum characterizes the degree of rotational suport for a protogalaxy of mass M and potential energy E, and is defined by

$$\lambda \equiv \frac{|E|^{1/2} J}{G M^{1/2}},$$

where J is the angular momentum. For a de Vaucouleurs profile, this expression for λ reduces to

$$\lambda \approx 0.4 v_{rot}/\sigma,$$

where v_{rot} is the rotational velocity and σ is the velocity dispersion; for a rotationally supported disk, $\lambda \approx 0.4$. Simulations of hierarchical clustering show that the median value of λ for a protogalaxy is initially

$$\lambda \approx 0.05 \Omega^{0.1},$$

with a broad dispersion, $0.01 \lesssim \lambda \lesssim 0.1$.

Self-gravitating collapse leads to a disastrous outcome for disks. Typical radii are $R_{disk} \approx J/Mv_{rot}$ for centrifugal balance. Since in a self-gravitating collapse, $v_{rot} \propto R^{-1}$, to conserve angular momentum, whereas $\sigma \propto R^{-1/2}$, one would infer $\lambda \propto R^{-1/2}$, and require collapse from an initial radius of 500 kpc if $\lambda \approx 0.05$ initially.

The commonly accepted resolution is via torquing of baryonic matter against dark, non-dissipative halos. Baryonic collapse within a dark halo transfers global angular momentum to the dark matter and alows more eficient collapse. In the dark matter potential, the gas velocity dispersion σ is maintained but specific angular momentum conservation of a volume element of gas means that again $v_{rot} \propto R^{-1}$, so that one now has $\lambda \propto R^{-1}$. In this case, an initial halo extent of 50 kpc is required in order to form the observed disks. The heavy element quasar absorption line systems studied at $z \sim 1$ are inferred to have extents of about this magnitude and to be associated with the halos of luminous galaxies.

Overcooling

While the origin of the angular momentum of disk galaxies is understood in terms of protogalactic and pregalactic tidal torques, simulations of galaxy formation in dark halos have encountered another problem. The gas cools too efficiently, and the cooled gas mass is excessive. One can demonstrate this result analytically. Consider an isothermal halo of velocity dispersion σ, with density

$$\rho = \sigma^2/2\pi G r^2.$$

Suppose that the gas temperature is given by $\sigma^2 = kT/\mu$, and that the gas density is proportional to ρ. The gas cooling timescale is

$$t_{cool} = \frac{3}{2}\frac{nkT}{\mu}[n^2\Lambda(T)]^{-1},$$

where $\Lambda(T)$ is the gas cooling function, approximately proportional to $T^{-0.5}$ over the temperature range $10^6 - 10^{7.5}$ appropriate for virialised gas of density n (cm^{-3}) in the dark halos of normal galaxies. I assume that the gas cools for up to a Hubble time. All gas within a radius r_{cool} cools, where r_{cool} is determined by setting

$$t_{cool} = t_{Hubble}, \quad \text{where } t_{cool} \propto (n\Lambda(T))^{-1} \propto r_{cool}^2/\Lambda(T).$$

I now infer that since $\Lambda(T) \propto \sigma^{-1}$,

$$r_{cool} \propto t_{Hubble}^{1/2}\sigma^{-1/2}.$$

The mass of cooled gas is given by $r_{cool}\sigma^2 G^{-1}$, and is therefore proportional to $\sigma^{3/2}$, and hence to $v_{rot}^{3/2}$. This demonstrates the problem: observations of disks show that $M_{disk} \propto v_{rot}^3$, so that far more gas cools than can be accounted for by the stellar mass of the galaxy disk.

Star formation in disk galaxies

The resolution of the gas overcooling problem appeals to star formation and feedback of energy into the gas via ejecta from supernova explosions. Unfortunately, the parametrization of energy feedback in numerical simulations of galaxy formation is an *ad hoc* procedure, although it generally results in pretty images.

One can attempt to study star formation analytically in order to derive a more intuitive understanding of the feedback process. An important byproduct of supernovae is chemical enrichment of their surroundings. One can study the role of supernovae in chemical evolution of disk galaxies by using the observed metallicity distribution of stars and gas to probe the past history of star formation in disks.

Recent observations of Hα emission from galaxies indicate that the gravitational instability of disks plays an important role in the rate of star formation [38]. In particular, it is found that Toomre's instability criterion correlates well with the onset of star formation. The significance of these observations is twofold. Firstly, they provide a physical foundation for theories of global star formation based on gravitational instability. Previous attempts to model disk star formation have been mostly based on empirical formulations. For example, the Schmidt law, which has the star formation rate proportional to some power of gas density, has been commonly used. Secondly, since Hα emission can be related to massive star formation, one can compare directly with model calculations for the star formation rate.

The global star formation rate based on the theory of gravitational instability has been recently formulated for a cold self-gravitating disk [39]. In general, the amount of interstellar matter per unit area per unit time converted into stars can be written as $\phi = \epsilon\mu_{gas}/t_{inst}$, where ϵ is the efficiency of star formation, μ_{gas} is the gas surface density, and t_{inst} is the timescale of the instability which eventually leads to star formation. In the case of gravitational instability of a gaseous disk, it can be shown that the fastest growth mode has a timescale $t_{\rm inst}^{-1} = Q^{-1}(1 - Q^2)^{1/2}\kappa$, where $Q = \alpha\kappa\sigma_{gas}/\pi G\mu_{gas}$ is the Toomre Q parameter, κ is the epicyclic frequency, σ_{gas} is the gas velocity dispersion, G is the gravitational constant, and $\alpha \simeq 0.3$ corrects for the stellar contribution to the self-gravity of the disk. If self-regulation is maintained, $Q \sim 1$, and the star formation rate is inferred to be proportional to the product of κ and gas surface density. If the gas surface density decreases below some critical value, $\mu_{cr} = \alpha\kappa\sigma_{gas}/\pi G$, differential rotation of the disk and random motions of clouds will prevent the gas from further collapsing, halting formation of stars. As the disk becomes less gas-rich, the effective value of Q increases and helps prolong the gas supply. The chemical evolution of disks is an important constraint on this model. Note that infall is not an essential ingredient in such a prescription for disk star formation, although it can readily be incorporated.

One can generalize the preceding discussion by self-consistently including the stabilizing effects of the stellar component of the disk. Self-regulation of star formation arises naturally because the star formation rate is an explicit function of gas velocity dispersion, which is itself generated both by disk gravitational instabilities on large scales and on smaller scales by gas acceleration produced by energy input from H II regions and supernovae. This results in feed-back into star formation: active star formation is both associated with, and induces, a high velocity dispersion in the gas, which in turn hampers the further increase of the gravitational instability in the gas disk.

The star formation efficiency may be estimated as follows. Assume that heating from supernova remnants (and HII regions) provides momentum input to the interstellar medium, effectively accelerating the gas. The interstellar medium is very dissipative, and one may therefore estimate the efficiency of momentum input, identified with the star formation efficiency, as

$$\epsilon = \sigma_{gas}/p_{SN},$$

where

$$p_{SN} = 2E_{SN}v_{cool}^{-1}m_{SN}^{-1}$$

is the specific momentum injected, that for simplicity I have identified with supernova injection. E_{SN} is the initial supernova remnant energy, m_{SN} is the mass found in stars per supernovae, and v_{cool} is the supernova remnant velocity at the onset of the approximately momentum-conserving snowplow phase. Typical numbers are $E_{SN} = 10^{51}$ ergs, $m_{SN} = 100\,M_\odot, v_{cool} \approx 300\mathrm{km\,s^{-1}}$, and $\sigma_{gas} \approx 10\,\mathrm{km\,s^{-1}}$, so that $p_{SN} \approx 1000\,\mathrm{km\,s^{-1}}$ and $\epsilon \approx 0.01$. Of course, this efficiency is more or less what is observed in giant molecular clouds, and also, not coincidentally, what is required to infer the observed galactic star formation rate from the galactic molecular cloud mass that would otherwise collapse over a cloud free-fall time.

The Tully-Fisher relation

One can in fact do a little better than extract the star formation rate. I shall argue that the Tully-Fisher relation is a natural consequence of the adopted star formation law. I approximate the star formation rate by

$$SFR = \epsilon \mu_{gas} \Omega,$$

and identify blue luminosity with star formation rate in spirals. I assume self-regulation, $Q \approx 1$, so that $\mu_{gas} \approx \Omega\sigma_{gas}/\pi G$ for a nearly flat rotation curve and rotation rate $\Omega(r)$.

I now can write the luminosity of a disk galaxy as

$$L = \beta\epsilon v_{rot}^2 \frac{\sigma_{gas}}{G},$$

where $\beta \equiv L/SFR$. The derived value for ϵ gives the correct normalization for the Tully-Fisher relation:

$$L = 10^{10}\left(\frac{\epsilon}{0.01}\right)\left(\frac{v_{rot}}{220\mathrm{km\,s^{-1}}}\right)^2\left(\frac{\beta}{5\times10^9 \mathrm{L_\odot yr M_\odot^{-1}}}\right)\left(\frac{\sigma_{gas}}{10\,\mathrm{km\,s^{-1}}}\right) L_\odot.$$

Note that $L \propto v_{rot}^2$ fits the observed Tully-Fisher relation (in the B band) where the light is dominated by ongoing and recent star formation, whereas the naive virial theorem plus constant surface density and mass-to-light ratio assumptions would yield the incorrect result that $L \propto v_{rot}^4$. To be fair in this comparison, one should consider the I-band Tully–Fisher relation, which is more sensitive to the old disk population that dominates the disk mass. In this case, one oberves a relation that is close to $L_I \propto v_{rot}^3$.

The dependence on efficiency is expected to be constant within a Hubble type, if ϵ is given by σ_{gas}/p_{SN}. However σ_{gas} is sensitive to bulge/disk ratio, and is observed to increase towards earlier Hubble type. This could be due to the enhanced shear associated with a massive bar that may have constituted a transient early phase in the formation of a typically triaxial bulge.

There is a large dispersion within a Hubble type in current star formation rate as measured by the ratio $SFR/\langle SFR \rangle$, and presumably therefore of star formation efficiency. Hence despite the fact that the self-regulation arguments suggests that the Tully-Fisher relation should be relatively robust and internally determined, one has to be concerned that environment could influence the normalization, although perhaps not the slope, by modulating the star formation history, and hence efficiency. This could occur, for example, via the influence of ram pressure of intracluster gas or via gas cloud orbital stirring induced by tidal interactions with neighboring cluster galaxies. Either process could result in a period of enhanced star formation that would reduce the current epoch gas supply, and hence the present star formation rate at a fixed Hubble type or rotational velocity. Such an effect could be relatively coherent in space over cluster or even supercluster scales.

The Tully-Fisher relation applies not just in the B and R bands, but also at longer wavelengths. It appears to steepen slightly in the I and in the H bands, to a slope of about $\epsilon \approx 3$ [40]. The fact that a universal relation is seen in the different bands means that the Tully-Fisher correlation is satisfied for both the young and the old stellar populations in spiral galaxies. One can understand the coincidence in a phenomenological way as follows. Dynamically hot stellar populations in spheroids and elliptical galaxies are found to lie in the fundamental plane of parameter space, defined by [41]

$$r_e \propto v^{1.4} \mu_L^{-0.85},$$

where r_e is effective radius, σ is velocity dispersion and μ_L is surface brightness of the old stellar population. Applying the virial theorem, $\sigma^2 = GM/r_e$, one can rewrite this as

$$L \propto \sigma^{10/3} \mu^{-5/6},$$

where μ is surface mass density averaged within r_e. Comparison of the latter with the predicted I-band Tully-Fisher relation shows concordance at fixed μ. Indeed, μ only varies weakly for luminous spheroids, although the dispersion is much larger than for the surface brightness (and surface density) of disks. These L, μ relations are consistent with the idea that ellipticals form via disk mergers without any accompanying global formation of new stars, and also that bulges and disks share a commom origin.

Future applications should include a study of the dispersion in chemical abundances of old disk stars. It should be feasible to compute the star formation rate in disk galaxies at early times, obtaining the star-forming history as a function of galaxy age and angular size. This is particularly interesting because a history of star formation is essential for interpreting observations of distant galaxies. For example, the disk star formation model that is generic to reproducing the Milky Way properties is proportional to disk rotation rate and hence to roughly the inverse of galactocentric distance [42]. This means that disks form inside out. Thus at a lookback time of $\gtrsim 5$ Gyr, disks should appear to be substantially smaller than their nearby counterparts. Another implication is for chemical evolution at high redshift: the model can be applied to predict the abundances in protodisks, which are often identified with damped Lyman alpha absorption line systems, with gas column density comparable to the surface density in nearby stellar disks, and at lookback times that correspond to $\gtrsim 10$ Gyr. A comparison of the integrated stellar radiation from galaxies with the observed background radiation at various wavelengths will constrain the history of galaxy formation.

6 Formation of elliptical galaxies

An admittedly more speculative step is to develop a model for spheroid formation. Here one ventures into unknown territory, to a far greater extent than was encountered previously in discussing disk formation. There is no theory for star formation in dynamically hot systems. Did these systems form by mergers of cold disks? Or did stars form within dense molecular clouds orbiting within the spheroidal potential well? One can only try to address the issue of protospheroids phenomenologically. There are several clues. Star formation was far more efficient than in disks, and there are recent hints about chemical abundance signatures, including correlations with escape velocity, metallicity gradients, and abundances, both stellar and in the diffuse intracluster and halo gas, that bear fossil witness to the star formation history. There are in addition the clues locked into the fundamental plane for ellipticals and bulges, interpretable as a slow increase with luminosity of the mass-to-light ratio, that may well be attributable to variations in the IMF or to systematic mass loss, or to some combination of the two effects.

Spheroids are denser than disks. The central surface brightness is higher ($\sim 1000\,\mathrm{L_\odot\,pc^{-2}}$ as opposed to $\sim 100\,\mathrm{L_\odot\,pc^{-2}}$), and the distribution of circular velocities is systematically higher for ellipticals than for spirals, although there is broad overlap. Yet despite being denser (both in luminous and in dark matter) than spirals, ellipticals are generally not rotationally suported within their half-light radii. The conventional explanation is that ellipticals form by mergers of disks. The dense cores form as a consequence of dynamical friction of massive and dense clumps of stars. The merging substructures include the nuclei of the premerger galaxies as well as of star clumps whose formation is triggered in the merger as gas clouds collide, dissipate orbital energy and coalesce. The dense clumps efficiently transfer angular momentum, and result in formation of a dense spheroid with low v_{rot}/σ. The dissipation and associated star formation produce a high central density as well as an enriched nucleus in the final system. However because of our ignorance of the detailed mechanism of star formation, this scenario is necessarily only schematic.

Characteristic luminosity

However there is at least one property of ellipticals that can be simply understood. The characteristic luminosity $L_* \approx 10^{10}h^{-2}\,\mathrm{L_\odot}$ is explained by a cooling argument similar to the argument that resulted in disk overcooling. There is a crucial distinction in the case of ellipticals and spheroids. Disks form slowly, over a Hubble time, whereas ellipticals formed stars efficiently and rapidly over a dynamical timescale appropriate to the observed system. Population synthesis of ellipticals, predominantly consisting of an old and nearly coeval stellar population, quantifies the inference from the broadband colours that these systems formed at least 10 Gyr ago. The small dispersion in colours for ellipticals and S0s in the Virgo and Coma clusters demonstrates their near-coevality [43]. Star formation must therefore have proceeded with at least an order of magnitude higher efficiency than in disks. Theoretical arguments suggest that most of the extant gas will be induced to form stars within a galaxy merger timescale, and indeed the analogy between ultraluminous starbursts and spheroid formation, discussed below, provides observational support for this thesis.

Thus it seems reasonable to require that gas cooling occurs within a dynamical timescale during formation of the spheroidal component. We now require that

$$t_{cool} \equiv \frac{3}{2}\frac{nkT}{\mu}[n^2\Lambda(T)]^{-1} \lesssim \frac{r}{\sigma}.$$

Setting the density $\rho = f_b\sigma^2/2\pi G r^2$, where f_b is the baryonic gas fraction, one infers that the cooling condition defines a radius within which cooling occurs that is given by

$$r_{cool} = \frac{\Lambda(T) f_b \sigma}{6\pi k G m_p T}.$$

The cooled gas mass is then given by

$$M_{cool} \approx f_b r_{cool} \sigma^2 G^{-1} \approx \frac{f_b^2 \sigma^3 \Lambda(T)}{6\pi G k m_p T}.$$

Hence for $\Lambda(T) \propto T^{-1/2}$, M_{cool} is constant.

This provides the characteristic mass for an elliptical galaxy: one finds that

$$M_{cool} \approx (f_b/0.1)^2 10^{11}\,\mathrm{M}_\odot.$$

For a typical mass-to-light ratio characteristic of an old stellar population, one therefore recovers a universal value for L_*, simply by assuming that there is a universal baryon fraction. The cooling radius is

$$r_{cool} \approx 50 \left(\frac{f_b}{0.1}\right)\left(\frac{300\,\mathrm{km\,s^{-1}}}{\sigma}\right)^2 \mathrm{kpc}.$$

This is sufficiently large for dwarf ellipticals ($\sigma \lesssim 100\,\mathrm{km\,s^{-1}}$) that their gas supply will be drastically curtailed by tidal interactions, implying that dwarf mass-to-light ratios should be systematically larger than those for luminous ellipticals, as indeed is observed.

Starbursts and spheroid formation

My approach to star formation in spheroids will begin by developing a phenomenological model for starbursts, where one may well, in rare examples, be witnessing the nearby counterpart of a protospheroid. There are several reasons for pursuing this analogy: the tidal interaction trigger, observed for starbursts and presumed to be a feature common to bottom-up models for galaxy formation; the star formation rate and efficiency, directly inferred for starbursts and indirectly deduced in order to account for the population content and stellar dynamical support of ellipticals, and even the surface brightness profiles. Tidal interactions are responsible for the elevated rates of star formation that are seen, for example, in many IRAS galaxies, and that are inferred to have occurred during the early gas–rich phases of galaxy formation.

Disks formed inside-out according to the simple self-regulation model developed above. However in a merger, the prexisting disk stars form the spheroid, whereas the gas is rapidly concentrated to within a few hundred parsecs of the merged center by the viscous drag induced by cloud collisions, as demonstrated in simulations, where it forms the galactic nucleus and inner bulge. Hence ellipticals form outside-in, the nucleus providing the environment for the dust-enshrouded starburst to develop and, in many cases, for gas dissipation to apparently result in the formation of an active galactic nucleus around a central massive black hole.

A prominent feature of the starburst model is undoubtedly the momentum input from supernovae in pressurizing the massive, dense molecular clouds that must have at least initially been present in order to have star formation proceed at the requisite efficiency. Indeed, infrared and millimeter wavelength observations of starbursts provide crude estimates of gas mass, molecular fraction, and pressure. Starbursts tend to be centrally concentrated, and simple estimates of the pressure, based on the observed luminosity converted to star formation and hence to supernova rate, immediately yield pressures that are far in excess of the local interstellar medium pressure within the central kiloparsec. This already is sufficient to overcome the Jeans mass barrier, one of the major obstacles to efficient star formation in our own galactic disk. One can scale the star formation rate to metallicity, via a simple chemical evolution model, and this will then suffice to scale the various parameters associated with the starburst. One can then model spheroid formation as a starburst. Since supernovae are sources of cosmic rays, one can also deduce the radio properties of forming galaxies, as well as study other aspects of cosmic ray interactions with protogalactic gas. One finds that the cosmic ray pressure is comparable to the thermal pressure, as directly measured via far infrared recombination lines and estimated from the inferred supernova rate in well-studied nearby starbursts.

Fundamental plane

The fundamental plane relation is equivalent to

$$M/L \propto L^{1/5}.$$

There is no simple explanation for this relation. It may be due to baryonic dissipation being more efficient at lower galaxy mass, to star formation efficiency decreasing with increasing galaxy mass, or even to a systematic variation in the IMF. Hints in support of the latter, possibly most radical, alternative come from several directions. There are indications of gradients in the ratio of α-nuclei to iron in the central regions of ellipticals [44], and the low dispersion in the ratio of α-nuclei to iron abundances in old disk stars [45] may be suggestive of a common origin for these elements. The high mass of iron in intracluster gas was presumably ejected from cluster galaxies but is not compatible with the standard IMF yields that are predominantly of a Type I supernova origin, these supernovae being the major present-day contributors of iron [46]. Most recently, there are hints of a predominantly Type II supernova yield characterizing the heavy element abundance ratios in intracluster gas [47]. All of these observations could plausibly be more easily understood if the initial mass function in starbursts favoured somewhat more massive stars at earlier epochs, and thereby had both a higher nucleosynthetic yield and one more characteristic of Type II supernovae.

Dwarfs

For extreme dwarf galaxies, the relation changes dramatically. The discussion of the cooling radius suggests that dwarfs should have a curtailed gas supply. This leads to formation of fewer stars for a given dark matter halo. Dwarf spheroids have high M/L values, and one finds that M/L rises as $L^{-1/3}$ or $L^{-1/4}$ towards low L for these low mass systems. Indeed for dwarfs, where $10^5 \lesssim T \lesssim 10^6$ K, one can set $\Lambda(T) \approx constant$, whence $r_{cool} \propto \sigma^{-1}$. The baryonic fraction remaining after tidal stripping is therefore of order $r_{cool}\sigma^2/G$ and so is proportional to *sigma*, for an isothermal halo. A very different explanation of dwarf mass-to-light ratios via wind-driven mass loss, something most easily accomplished from shallower potential wells [13], where the retained baryonic fraction should be proportional to σ/p_{SN}.

In either case, one arrives at a similar conclusion: $M_b/M \propto \sigma$, The M/L trend then arises naturally from the $M \propto \sigma^3$ prediction in the CDM context, since $\sigma \propto M^{\frac{1-n}{12}}$, and $n \approx -3$ for dwarfs. Hence by identifying M_b with L, we infer that $M/L \propto \sigma^{-1} \propto M^{-1/3} \propto L^{-1/4}$. Confirmation that this trend of baryon fraction increasing with velocity dispersion is likely to be correct comes from abundance studies. For dynamically hot galaxies, one finds that $Z \propto v$, in a relation that has surprisingly low dispersion [48]. This is precisely what one expects if the mass loss driven by winds interrupts chemical enrichment before it saturates at the nucleosynthesis yield of about $2Z_\odot$, appropriate to a closed box model, as found for luminous elliptical galaxy cores and inner disks.

The epoch of galaxy formation

What finally can be said about the epoch of galaxy formation? Observations drive the response. It is apparent from the recent deep redshift surveys which are probing the universe to $z \sim 1$ that there are two distinct populations of galaxies, one of which is evolving at low redshift, whereas for the second, no evolution is detectable. The evolving population is that of the blue galaxies (as defined by rest frame colours). Hubble Space Telescope observations suggest that predominantly irregular galaxies contribute to the deep counts fainter that $I \sim 22$, about the same limit as that of the deep redshift surveys. MDS observations indicate that many of these galaxies show evidence of recent or ongoing tidal interactions. It is logical to assume that it is the early-type galaxies that contribute to the non-evolving population. Cluster observations at $z \sim 0.4$ with HST confirm that distant ellipticals are relatively undisturbed in appearance, whereas the late-type population appears to constitute a much higher fraction of the cluster population than is the case for nearby clusters and, at least anecdotally, reveals a high frequency of morphological irregularities that are suggeative of tidal interactions.

However while it seems clear that spirals are evolving at low redshift ($z \sim 0.5$), the lack of broad-band colour or morphological evolution for ellipticals at low redshift does not unambiguously indicate a high redshift of formation. The problem is that high spectral resolution studies of nearby ellipticals reveal traces of an intermediate age population. Hence one cannot assign a meaningful redshift of formation. Although there is little doubt that the bulk of the stars are at least 10 Gyr old, ages of $\lesssim 4$ Gyr can be assigned to a not necessarily insignificant part of the stellar population. A precise deconvolution of age and metallicity remains to be performed before more quantitative statements can be made.

Finally, I note that searches for the luminous phases of protogalaxies via Lyman alpha emission have attempted to constrain the epoch of formation of elliptical galaxies. The model predictions assume that Lyman alpha photons escape. For elliptical and spheroid formation, the starburst interpretation suggests that this may be a very dubious assumption. If spheroid formation is shrouded by a dusty starburst, the emission will only be detectable at far infrared wavelengths.

Successes and failures

Armed with a theory of starbursts and spheroid formation, as well as a theory of disk star formation, one should be able to calculate the star formation rate in galaxies at high redshift and derive the galaxy luminosity function as a function of epoch, to make predictions of the expected characteristics of both nearby and high redshift galaxies, of the epoch of galaxy formation, and of the deep galaxy counts.

The scorecard is mixed. Clusters of galaxies are well-simulated, to the extent of being able to account for their x-ray morphology. Successes of the simulations of galaxy formation include the dependence of galaxy morphology on environment, with ellipticals and S0s dominating clusters, and spirals dominating groups and the field. One can understand the Butcher-Oemler effect, whereby the fraction of star-forming galaxies rises at $z \gtrsim 0.3$, both in clusters as originally proposed, but also in the field, as recently realized, as well as in the outlying regions of nearby clusters where evidence for infall is seen. Disk surface brightnesses and bulge-to-disk ratios can be modelled, provided one includes feedback from star formation. This in turn can lead to plausible descriptions of the gas fraction as a function of morphological type and redshift, and also of chemical evolution.

Of course, until an improved formulation is available for star formation, the sceptic may assert that one has simply assumed the desired answer without explaining anything. More specifically, there are notable failures. Massive and luminous galaxies generically form late in a hierarchical model, and there is no accepted prescription for resolving this contradiction with what is observed. There is an excess of low luminosity galaxies. The generic mass function is fit by a Schechter function function with parameter $\alpha \approx -2$, whereas observations indicate that $\alpha \approx -1$, at least to a luminosity brighter than $M_B \approx -17$. The resolution is presumably in the conversion of mass to luminosity, but there is no unique prescription for achieving a good match to the local galaxy luminosity function. Conversely, at high redshift ($z \sim 1$), such an excess in low luminosity galaxies may be precisely what is needed in order to explain the number counts of faint galaxies. The low luminosity galaxies form early in a hierarchical model, and hence are predicted to be red, in contrast to the observed predominance of blue star-forming dwarfs (at least in the vicinity of the Local Group and in the field).

7 Acknowledgements

These lectures were presented at the ANU Summer School in January 1995, and completed during a visit to Mount Stromlo Observatory and the ANU in July 1995, hosted by the ANUATC. Some of the material is based on a series of lectures that I originally gave at the invitation of the former Director, Alex Rodgers, at Mount Stromlo Observatory in late 1990. I thank my students and collaborators for innumerable discussions of many of the problems reviewed in these notes. In particular, I am indebted to E. Bunn, L. Cayon, S. Charlot, W. Hu, D. Scott, N. Sugiyama, R. Wyse, M. White, S. Zaroubi and S. Zepf. My research has been supported in part at Berkeley by grants from NASA, NSF and DOE.

References

[1] B. Ratra and P. J. E. Peebles, *Astrophys. J.* **432** (1995) L5.

[2] M. Bucher *et al.*, *Phys. Rev. D* (1995) in press.

[3] W. Hu, D. Scott and J. Silk, *Astrophys. J.* **430** (1994) L5.

[4] C. Alcock *et al. Phys. Rev. Lett.* **74** (1995) 2867.

[5] W. Hu and N. Sugiyama, *Astrophys. J.* **444** (1995) 489.

[6] C.S. Frenk, *Physica Scripta* **T36** (1991) 70.

[7] M.A. Strauss, A. Yahil, M. Davis, J. P. Huchra, and K. Fisher, *Astrophys. J.* **397** (1992) 395.

[8] M. Davis, G. P. Efstathiou, C. S. Frenk and S. D. M. White, *Astrophys. J.* **292** (1985) 371.

[9] J. M. Bardeen, J.R. Bond, N. Kaiser and A. S. Szalay, *Astrophys. J.* **304** (1986) 15.

[10] S.D.M. White, C.S. Frenk, M. Davis and G. P. Efstathiou, *Astrophys. J.* **313** (1987) 505.

[11] N. Kaiser, *Astrophys. J.* **284** (1984) L9.

[12] G.R. Blumenthal, S.M. Faber, J.R. Primack, and M.J.Rees, *Nature* **311** (1984) 517.

[13] A. Dekel and J. Silk, *Astrophys. J.* **303** (1986) 39-55.

[14] R. B. Tully and J. R. Fisher, *Astron. Astrophys.*, **54** (1977) 661.

[15] M.J. Rees, *Mon. Not. R. astr. Soc.* **213** (1985) 75P.

[16] A. Dekel and M.J. Rees *Nature* **326** (1987) 455.

[17] A. Blanchard, D. Valls-Gebaud, and G. Mamon, *Astron. Astrophys.* **264** (1992) 365.

[18] W. Press and P. Schechter *Astrophys. J.* **187** (1974) 425.

[19] G. Efstathiou, C. S. Frenk, S. D. M. White. and M. Davis, *Mon. Not. R. astr. Soc.* **235** (1988) 715.

[20] J.R. Bond, S. Cole, G. P. Efstathiou and N. Kaiser, *Astrophys. J.* **379** (1991) 440.

[21] J. A. Peacock and A. F. Heavens, *Mon. Not. R. astr. Soc.* **217** (1985) 805.

[22] J. Mohr, A. Evrard, D. Fabricant and M. Geller, *Astrophys. J.* **447** (1995) 8.

[23] M. S. Vogeley, in *Wide-Field Spectroscopy and the Distant Universe*, Proc. 35th Herstmonceaux Conference, in press (1995).

[24] Y. Hoffman *Astrophys. J.* **328** (1988) 489.

[25] J.N. Bahcall, M. Schmidt and R. Soniera, R. *Astrophys. J.* **258** (1982) L23.

[26] G. Lake and L. Feinswog *Astron. J.* **98** (1989) 166

[27] A. Dekel, *Ann. Rev. Astr. Astrophys.* **32** (1994) 371.

[28] G. Efstathiou and R. A. C. Croft, *Mon. Not. R. astr. Soc.* **268** (1994) L23.

[29] R. Flores and J. Primack *Astrophys. J.* **412** (1995) L1.

[30] B. Moore, *Nature* **370** (1994) 629.

[31] J. A. Peacock, and D. D. Dodds, *Mon. Not. R. astr. Soc.* **267** (1994) 1020.

[32] D. Scott, J. Silk and M. White *Science*, in press (1995).

[33] J. Primack *et al.*, *Phys. Rev. Lett.*, in press (1995).

[34] D. Pogosyan and A. Starobinsky *Astrophys. J.*, in press (1995).

[35] C. Lacey and J. Silk, *Astrophys. J.* **381** (1991) 14.

[36] C. Lacey, B. Guiderdoni, J. Silk and B. Rocca-Volmerange, *Astrophys. J.* **402** (1993) 15.

[37] G. Kauffmann, S. D. M. White, and B. Guiderdoni, *Mon. Not. R. astr. Soc.* **264** (1993) 201.

[38] R. Kennicutt, *Astrophys. J.* **344**, (1989) 685-703.

[39] B. Wang and J. Silk *Astrophys. J.* **427** (1994) 759-769.

[40] D. Burstein, in *The Opacity of Spiral Disks*, ed. J. I. Davies (Kluwer Academic, Dordrecht) (1995), in press.

[41] R. Bender, D. Burstein, and S. M. Faber, *Astrophys. J.* (1992) **399** 462-477.

[42] N. Prantzos, N. and O. Aubert, *Astron. Astrophys.*, in press (1995).

[43] R. Bower, J. Lucey and R. Ellis, *Mon. Not. R. astr. Soc.* **254** (1992) 601

[44] G. Worthey, S. M. Faber and J. Gonzales, *Astrophys. J.* **398**, (1992), 69-73.

[45] B. Edvardsson, *et al.*, *Astron. Astrophys.*, **275**, (1993) 101-152.

[46] M. Arnaud *et al.* *Astron. Astrophys.*, **254**, (1992) 49-64.

[47] Y. Tanaka, in *Proc. 17th Texas Symposium, N.Y. Acad. Sci.* (1995) in press.

[48] R. Bender, in *The Stellar Populations of Galaxies*, eds. B. Barbuy and R. Renzini (Kluwer: Dordrecht) (1992) 267-270.

STATISTICAL ASPECTS OF LARGE SCALE STRUCTURE

Alexander S. Szalay
Department of Physics and Astronomy
The Johns Hopkins University

INTRODUCTION

The study of large scale structure is one of the most dynamically evolving areas of astrophysics today. Cosmology and large scale structure is growing into an accurate science and requires correspondingly more sophisticated methods of analysis. Twenty years ago the estimates of the fluctuation amplitude were about 10^{-3}, almost a factor of 100 off of today's measurements. Ten years ago we could only hope for high precision measurements of large scale structure, there were less than 5000 redshifts measured, and only a handful of normal galaxies with $z > 1$ were known. Computer models of structure formation had just begun to consider non-power-law spectra based on physical models like hot/cold dark matter. As a consequence there was considerable freedom in adjusting parameters in the various galaxy formation scenarios. In contrast, many of today's debates are about factors of 2 and soon we will be arguing about 10% differences. The shape of the primordial fluctuation spectrum, first derived from philosophical arguments (Harrison 1970, Zeldovich 1972), can now be quantified from detections of fluctuations in the CBR made by COBE (Smoot *et al.* 1991). The number of available redshifts is beyond 50,000, and soon we will have redshift surveys surpassing 1 million galaxies. N-body simulations are becoming more sophisticated, of higher resolution, and incorporating complex gas dynamics. The unprecedented number of new observations currently under way give us hope that over the next decade we will gain a clear understanding of the shape and evolution of the primordial fluctuation spectrum, understand from first principles how galaxies were formed, and make quantitative comparisons and tests to differentiate among the various galaxy formation scenarios. The different forms of evolution (luminosity, clustering and cosmological) are hard to separate, we need more accurate measurements, and larger samples. Here we would like to primarily focus on a few main questions, all directed towards these goals:

Measure galaxy density fluctuations on COBE scales

Determine the evolution of luminosity up to $z = 1$
Determine the evolution of structure up to $z = 1$
Quantify the non-Gaussian nature of the large scale structure

QUANTIFYING LARGE SCALE STRUCTURE

Key questions

Structure in the universe evolves from the initially small primordial fluctuations. These fluctuations can arise during an inflationary expansion or come from topological defects later. They grow in amplitude, due to gravitational instability, and the shape of the fluctuation spectrum is altered by different physical processes. The nature of the dark matter, whether hot or cold, believed to dominate the mass density of the universe, determines the shape of the power spectrum on small (< 100 Mpc) scales. On the other hand, the shape of the large scale part of the fluctuations (> 200 Mpc) remains remarkably unchanged, because no scale in the evolutionary process becomes this large.

The COBE measurements constrain both the amplitude and the initial spectrum of the fluctuations in this regime, and demonstrate extremely good agreement ($n = 1.1 \pm 0.4$, Gorski *et al.* 1994) with the Harrison-Zeldovich predictions of $P(k) = k^n$, with $n = 1$. These fluctuations are due to differences in the gravitational potential at the surface of last scattering (Sachs and Wolfe 1967), reflecting the state of the universe at a redshift of ≈ 1000. Galaxy surveys (at $z < 0.3$) are rapidly increasing in size, thus providing increasingly better measurements of the fluctuations on small scales (CfA slices - Geller & Huchra 1989, IRAS - Saunders *et al.* 1992, APM - Maddox *et al.* 1989 , APM redshift surveys - Loveday *et al.* 1992, LasCampanas - Schectman *et al.* 1992). One can use theoretical scenarios to evolve and extrapolate the large scale CBR measurements into the structure of the local universe, but the two regimes do not yet overlap directly.

The currently most popular scenario is the Cold Dark Matter dominated universe, where most of the mass is dark, interacting only via gravity, consisting of particles of such a large mass that their thermal motion is negligible. To match the observed clustering of galaxies without producing too large a velocity dispersion, the concept of 'biasing' has been invoked (Kaiser 1984, Bardeen *et al.* 1986): mass is converted into light only at the densest regions in the universe, creating a luminous component more clustered than the mass. This scenario, modulo a properly chosen initial normalization, has been remarkably successful over the last fifteen years.

The COBE measurements create a conflict with the minimal biased CDM model: if a Harrison-Zeldovich spectrum is assumed and the normalization is locked to COBE, then the biasing parameter must be unity to match the small scale part of the fluctuation spectrum, leading to very large small scale velocities. Several alternative models have been rapidly suggested. Gravity waves, which decay with time, may contribute to the largest scale modes observed by COBE and produce a 'tilt' of the spectrum (Davis *et al.* 1992). Alternative scenarios invoke a large cosmological constant (Kofman *et al.* 1993) or a Hubble constant in the 30's (Bartlett *et al.* 1994). A mixture of cold and hot dark matter would also help, because the growth of fluctuations on small scales would be retarded due to the presence of a hot component with $\Omega_\nu \approx 0.2$ (Klypin *et al.* 1994).

The fluctuation spectrum today: power on COBE scales

What are the most important measurements we can make in order to differentiate between proposed models? Overlap between scales probed by CBR experiments and redshift surveys in the 'local' universe would place strong constraints on the power spectrum. Measurements of galaxy clustering on scales of 200-500 Mpc from redshift surveys would tell us whether the gravity wave/tilted model is correct, measure the bias factor, and determine the shape of the spectrum on scales where most of today's models differ but which are too small for COBE and beyond the scale of current galaxy measurements. For the same reason, many CBR experiments are probing 1-2 degree scales, corresponding to a comoving scale of about ≈ 120 Mpc. These experiments are quite difficult, and even if they yield a precise number, the interpretation of these Doppler-peaks will depend on details of the reionization, whereas the fluctuations in the galaxies have a robust interpretation: their amplitude will measure the total fluctuation growth since recombination. In the next section we outline how novel statistical techniques under development will bring this goal within reach, using galaxy redshift surveys.

Measuring the evolution of structure

The observations of the CFRS group (e.g. Lilly *et al.* 1995) increased the available deep redshift sample by a huge factor, and provided the first convincing measurement that the blue half of the galaxy population evolves quite rapidly, while the red half is almost static. Deep K-band selected samples are becoming available, which are both less sensitive to bursts of star-formation and provide the red rest frame flux for redshifts above $z = 1$. With the imminent release of the Hubble Deep Field data, a survey reaching to almost 30th magnitude, we will have access to the deepest optical observations being made for a very long time. Other observations on HST, like the Groth field (Groth etal 1996) provide deep, large area samples, with exquisite morphological information. Spectra for objects fainter than 24 mag are now routinely being done at the Keck telescope. A handful of normal galaxies with redshifts at $z > 3$ have been recently discovered (Steidel *et al.* 1996). With the planned DEIMOS spectrograph the efficiency for deep redshift surveys will be increased by huge factors. Still, most evolutionary studies are very much a numbers game: beyond the obvious difficulties with systematic errors, the statistical noise remains very high, due to the extreme difficulties in obtaining spectra for high redshift, faint galaxies.

Another way to distinguish between the different cosmological scenarios is to use clustering evolution to constrain the models. The growth of the fluctuations at low to moderate redshifts ($z < 1$) is a sensitive function of the global properties of the universe (Ω_0, Λ_0). There are several possible ways to proceed: one method is to do extremely deep redshift surveys and analyse clustering with a high level of shot noise; another method is to extract statistical redshift indicators using broadband colors for a very large number of galaxies, where the small shot noise more than makes up for the lost redshift accuracy.

How non-Gaussian is the Universe today?

Given the rapid decrease with scale of galaxy clustering shown in earlier measurements of the correlation function, (Peebles 1980), the universe was expected to be quite homogeneous on large scales. The discovery of various large scale features like the Bootes void (Kirshner *et al.* 1980), the 'Great Wall' – a supercluster with an extent over 100 Mpc (deLapparent *et al.* 1986) – and the 'spikes' in the pencilbeam surveys (BEKS) indicate not only that there is more large scale power, but also that the structure of the universe on these scales is more complex than expected.

The customary second-order statistical measures, like the correlation function or the power spectrum, are very powerful to quantify Gaussian fluctuations, and measure the overall clustering power in a sample, but are quite inadequate to describe sharp, 'wall-like', coherent structures. The non-Gaussianity can be characterised by higher order moments, like skewness, but no satisfactory statistic has been found to quantify coherence. We discuss various ways to attack this problem.

MEASURING STRUCTURE ON $100 - 1000h^{-1}$ Mpc SCALES

Problems in Power Spectrum Estimation

To estimate the power spectrum from a galaxy redshift survey, we must take into account the sampling density (determined by the magnitude limit) and geometry of the survey (determined by the angular coverage and depth). The sampling process introduces shot noise into the power spectrum (the noise per mode is constant and thus easily subtracted, but contributes to the uncertainty). The observed power spectrum is a convolution of the true power with the Fourier transform of the spatial window function of the survey ($W(\mathbf{x}) = 1$ inside the survey and 0 outside), $P_{obs}(\mathbf{k}) = \int P_{true}(\mathbf{k}')|W(\mathbf{k} - \mathbf{k}')|^2 d^3k'$. We can attempt to deconvolve the true power spectrum or compare to convolved theoretical spectra, but in either case the survey geometry limits both the resolution and the largest wavelength for which we obtain an accurate measurement.

The standard methods for power spectrum estimation (e.g., Park et al. 1994; Feldman et al. 1994; Fisher et al. 1993) work reasonably well for data in a large, contiguous, three-dimensional volume, with homogeneous sampling of the galaxy distribution. Using these techniques, nearby wide-angle redshift surveys (CfA, SSRS, IRAS 1.2, QDOT) yield strong constraints on the power spectrum on scales

$\sim 100h^{-1}$. Because the uncertainty in the power spectrum depends on the number of independent cells of a given wavelength which we sample, constraints on larger scales require deeper surveys. Due to the difficulty of obtaining redshifts for fainter galaxies and limited telescope time, deep redshift surveys typically have complex geometry, e.g., deep pencil beams or slices.

However, the standard methods are not efficient when applied to data in oddly-shaped and/or disjoint volumes, or when the sampling density of galaxies varies greatly over these regions. Convolution of the true power with the complex window function causes power in different modes to be highly coupled. In other words, plane waves do not form an optimal eigenbasis for expansion of the galaxy density field sampled by redshift surveys. We desire methods for power spectrum estimation that optimally weight the data in each region of the survey, taking into account our prior knowledge of the nature of the noise and clustering in the galaxy distribution.

Cross-Correlation Power Spectrum

One problem that arises when one combines disjoint survey regions, e.g., the set of narrow beams in a survey comprised of multiple pencil beams, is that the window function $W_i(\mathbf{k})$ of each subregion has very wide sidelobes, thus the power spectrum of the full data set is plagued by aliasing. However, nearly all of the sensitivity to large wavelength fluctuations arises from comparing the density in different regions, not from the fluctuations within each region (Szalay et al. 1991). The power spectrum for the full survey is

$$P(\mathbf{k}) = \delta(\mathbf{k})\delta^*(\mathbf{k}) = \sum_i \delta_i(\mathbf{k})\delta_i^*(\mathbf{k}) + \sum_{i>j} \left[\delta_i(\mathbf{k})\delta_j^*(\mathbf{k}) + \delta_i^*(\mathbf{k})\delta_j(\mathbf{k})\right] .$$

The first term sums the "auto-correlations" of the individual survey regions, which contribute most of the aliased power due to their narrow geometry. Thus we obtain a much cleaner estimate of the power on large wavelength scales by using only the second term, the "cross-correlation power spectrum," which uses only the contrast in density between different subregions to estimate the power on large scales.

Eigenmode Analysis of Redshift Surveys

The cross-correlation power spectrum method effectively downweights the fluctuations within each region in favor of the fluctuations between disjoint regions. We can generalize this procedure to find an optimal set of spatial filters to probe the density fluctuations (see Vogeley and Szalay 1996). Rather than directly compute the Fourier transform of the distribution of objects, we expand the observed density field in the natural orthonormal basis determined for each survey from our prior knowledge of the survey geometry, selection function, and clustering of galaxies, and find the most likely power spectrum model in a Bayesian fashion. Expansion of the observed density field in this basis is known as the Karhunen-Loève transform (e.g., Therrien 1992). Dividing the survey volume into cells V_i, we compute the correlation matrix of expected counts

$$R_{ij} = \langle N_i N_j \rangle = \langle N_i \rangle \langle N_j \rangle \left(1 + \langle \xi_{ij} \rangle\right) + \delta_{ij} \langle N_i \rangle,$$

where $\delta_{ij} = 0$ for $i \neq j$, N_i is the galaxy count in the ith cell, and

$$\langle \xi_{ij} \rangle = \frac{1}{V_i V_j} \int \xi(\mathbf{x_i} - \mathbf{x_j}) dV_i dV_j .$$

We assume a model for ξ that is consistent with previous observations. The column vectors $\mathbf{\Psi}_j$ of the unitary tranformation that diagonalizes the correlation matrix are the eigenfunctions of the density field of the survey (solving the equation $\mathbf{R} \cdot \mathbf{\Psi}_j = \lambda_j \mathbf{\Psi}_j$).

Figure 1 shows the first 6 density eigenmodes for the geometry, selection function, and correlation function of the first CfA slice (de Lapparent *et al.* 1986). Each of these eigenmodes samples a narrow range of Fourier wavemodes. As we increase the volume of the survey, the Fourier windows of the eigenmodes sharpen; in the limit of an infinite or periodic volume, the eigenmodes are plane waves and we recover the Fourier expansion. Thus, we expand the observed counts in this orthonormal basis: $N_i = \beta^j \Psi_{ij}$ (Einstein summation convention), which defines the transform $\beta^j = \Psi^{ij} N_i$. Sorting these functions by decreasing eigenvalue λ yields the set of eigenfunctions in order of decreasing signal to noise. Figure 2 shows the expected power per mode of the Karhunen-Loève transform analogous to the power spectrum of the Fourier expansion. The total power per mode is the sum of the true clustering power, shot noise, and the mean density (for an infinite or periodic volume the $n = 1$ K-L mode corresponds to the $k = 0$ Fourier mode).

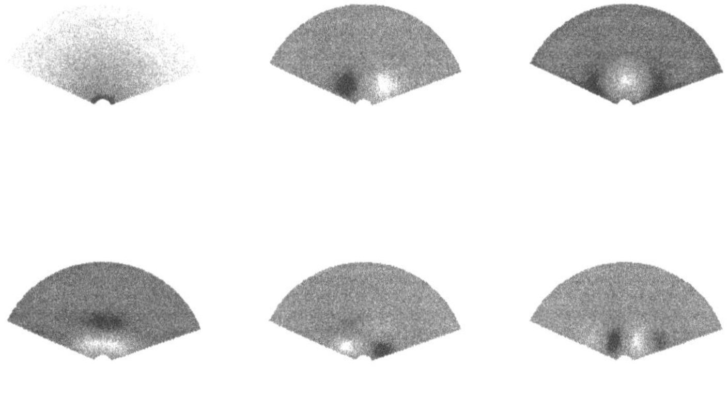

Fig. 1. Eigenmodes of the CfA slice. The Karhunen-Loève transform uses the geometry, selection function, and known clustering of galaxies to find the unique statistically orthogonal basis set for the density fluctuations in a particular survey. Here we plot the six most significant eigenfunctions.

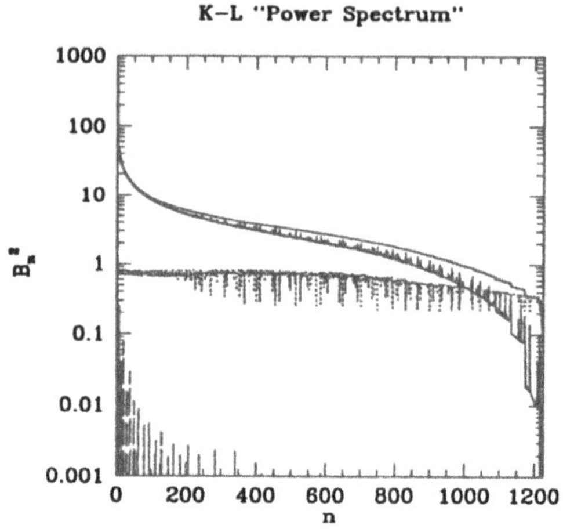

Fig. 2. Power per mode of the K-L expansion. The upper solid line plots the eigenvalue for each mode, which is the expectation value of the total power per mode. Successively lower lines show the contributions to the power from galaxy clustering, shot noise, and the mean density.

Because the β^j are statistically orthogonal and because we can easily compute the expectation value and variance of the power per eigenmode for any power spectrum model

$$\langle \beta_j^2 \rangle = \Psi_j^{-1} \cdot \mathbf{R}^{model} \cdot \Psi_j,$$

hypothesis testing is a straightforward process. Note that this method requires an initial guess at the power spectrum, but the form of the eigenfunctions does not depend sensitively on this assumption, and we can easily iterate the process. Summarizing the main features:

The K-L transform automatically determines the 'correlation eigenmodes' of a complex survey geometry, *each optimally weighted* to measure power on a certain scale

The expansion of the density field in terms of these eigenmodes still contains phase information, that can be used to describe non-linearities

The modes are orthogonal, independent, thus statistical hypothesis testing is quite easy, using either Bayesian or maximum likelihood techniques

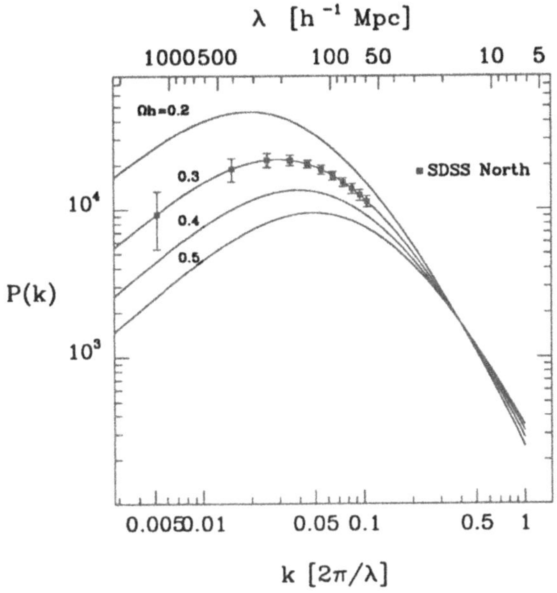

Fig. 3. Uncertainty of the SDSS power spectrum. The 1σ uncertainty expected for a volume limited (to M^*) sample of the SDSS northern redshift survey, assuming Gaussian fluctuations and a $\Omega h = 0.3$ CDM model, compared to power spectra for CDM with different Ωh. Error bars on smaller scales are of similar or smaller size than the symbols.

PHOTOMETRIC REDSHIFTS

Background

To fully understand the evolution of field galaxies we need to consider their evolution as a function of magnitude, redshift, spectral type and morphology. Furthermore, we need to do this in a coherent fashion. Currently the only way to address these questions has been to undertake time intensive spectroscopic observations (the deepest published spectroscopic surveys are to a limit of $B = 24$, Glazebrook *et al.* 1995). Because of the amount of time required surveys tend to cover small areas and have only a limited number of photometric bandpasses. If, however, we can use broadband photometry in terms of a crude, low resolution spectrum we may be able to derive statistical estimates for the physical properties of galaxies (spectral type, luminosity and redshift) to the limit of photometric surveys (i.e. a further 5 magnitudes beyond spectroscopic limits).

From a study of the distribution of galaxies in the U, B_J, R_F and I_N photographic catalogs of Koo and Kron (Kron 1980, Koo 1985, 1986, Munn *et al.* 1995) we have shown that galaxy colors and fluxes are highly correlated for $B_J < 22.5$ (Connolly *et al.* 1995). We find that they form an almost planar distribution within this photometric 4 space and that the physical properties of these galaxies (i.e. their spectral type, luminosity and redshift) can be estimated directly from their position within this plane. In Figure $4a, b$ and d we demonstrate this by plotting the distribution of U, B_J and R_F magnitudes for galaxies with spectroscopic redshifts (from the Koo-Kron catalogs). The redshift of each galaxy is encoded in the color of its data point, with blue corresponding to $z = 0$ and red to $z = 0.5$. The color table is set such that each color represents an interval of 0.1 in redshift.

Planes of constant redshift are seen to project through the multicolor distribution. If we consider only those galaxies at a given redshift (i.e. those galaxies with points of the same color in Figure 4) we find that they form an iso-redshift plane (or 'slab') in the UB_JR_F space. The natural coordinates within this iso-redshift slab are the luminosity and intrinsic color, or spectral type. As galaxies are redshifted, their fluxes decrease (simply the $1/r^2$ effect) and their colors change (i.e. K corrections). These two components shift the planar distributions both in the flux and color directions. The result is a 'shifted stack of slabs' of constant redshift passing through the multicolor distribution.

A consequence of this is that galaxies at different redshifts separate out in multicolor space. We can, therefore, assign a statistical measure (with an associated error) of the redshift, spectral type and luminosity. To show this we derive an estimate of the redshift of a galaxy by fitting a second order function to the broadband photometry. The correlation between spectroscopic and photometric redshifts is shown in Figure 5.

The dispersion about the one-to-one correlation is $\sigma_z = 0.046$ (Connolly *et al.* 1995). The dominant source of error is in the plate-based photometry. This

---

Fig. 4. The distribution of galaxies within the three color space U B_J R_F derived from the Koo and Kron data is shown on panels (a, b, d). The redshift of each galaxy is encoded by the color of its data point (blue at z=0 and red at z=0.5) The color table maps onto an interval of 0.1 in redshift. Panel (c) shows a schematic of the distribution. The position of a galaxy within the three color space is determined by its redshift, luminosity and spectral type. For a given redshift the data form thick slabs in the UB_JR_F space. Redshifting the galaxies moves these slabs through the color space (due to dimming and K corrections). The simulated distribution of galaxy colors, as a function of redshift, for the deep HDF field filters is shown on the panels (e) and (f). Simulations are derived using the spectral energy distributions of Kinney *et al.* (1995). The colors of the tracks refer to the different spectral templates (*violet* = blue starburst, *green*= Sc, *red*= elliptical) Each point covers a redshift interval of 0.05, the white points represent $z = 0$. The orange lines indicate redshifts of 0.3, 0.8 and 1.4 respectively. Panel (e) shows that using the F300W, F450W and F606W filters we can use broadband colors to differentiate galaxies out to $z = 0.6$. Panel (f) shows we can extend this to $z = 1.5$ using the F450W, F606W and F814W filters.

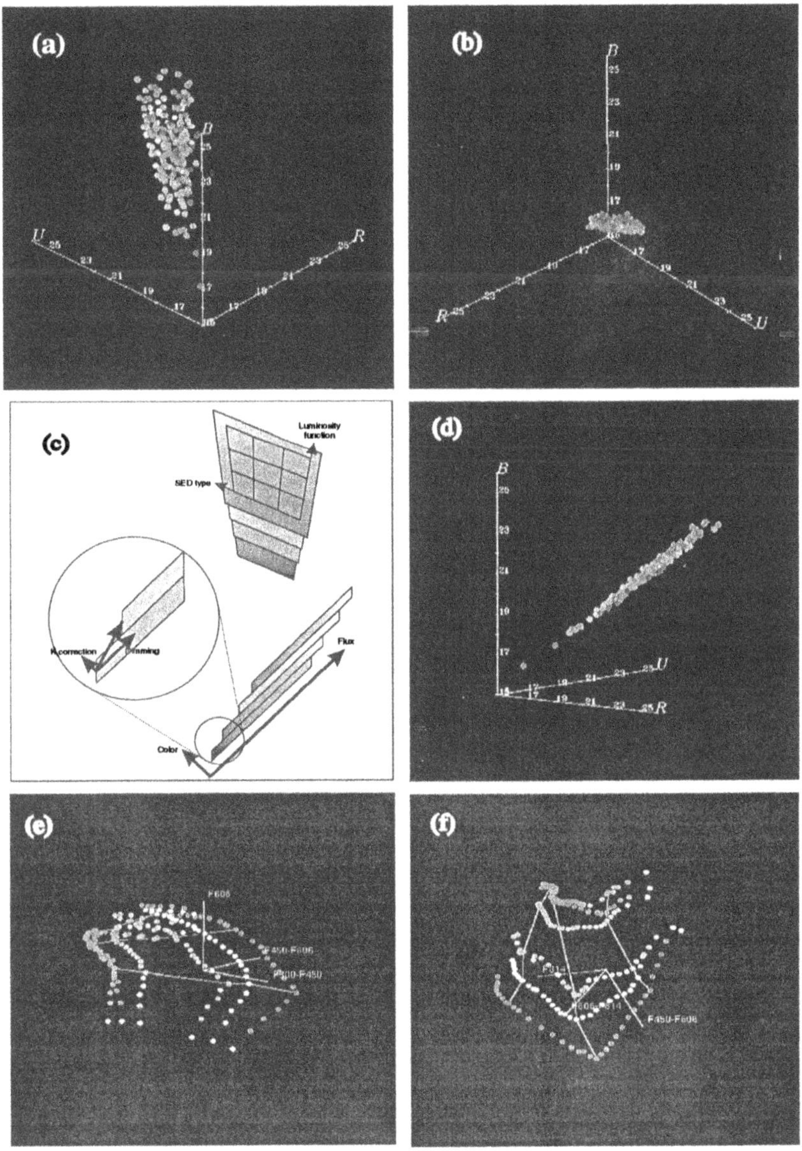
(a)
B
U
R
(b)
(c)
Luminosity function
SED type
K-correction
Flux
Color
(d)
(e)
F606
F450-F606
(f)
F450-F606

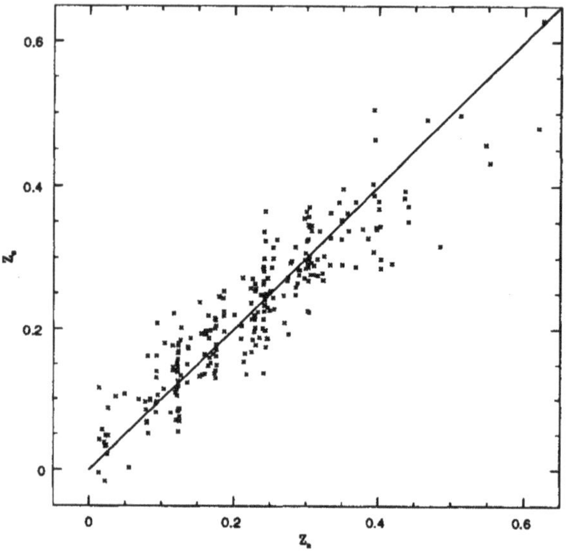

Fig. 5. Fitting a quadratic relation to the four magnitudes $UB_JR_FI_N$ we derive a photometric-redshift for the spectroscopic sample of galaxies. The correlation between the measured (Z_s) and estimated (Z_e) redshifts is shown. The dispersion about this relation is $\sigma_z = 0.046$ for galaxies to B_J=22.5.

situation should improve in the near future quite dramatically: over the last year we have obtained deep CCD-based multicolor observations of several fields with a lot of deep spectroscopic redshifts (see later). While we have demonstrated our ability to estimate the physical properties of galaxies from their broadband colors for a sample of galaxies to $B_J < 22.5$, use of multicolor HST data (supplemented by ultra deep ground based observations) provides an opportunity to extend this technique by 6 magnitudes, to the limits of current photometric surveys ($B < 29$).

The physical reason for the correlations

All these derived dimensions are a natural consequence of the intrinsic properties of galaxies. The fact that the galaxies are located in a subspace with fewer dimensions than the number of photometric bands implies that there is redundant information present. This is consistent with results found by Connolly *et al.* (1995b) who show that galaxy spectral energy distributions can be reconstructed from a one-parameter linear combination of three orthogonal spectra, of which the first two are dominant. They used 10 galaxies which had spectral energy distributions from 1200 Åto about 10,000 Å, from a combination of IUE and optical data, using matched apertures.

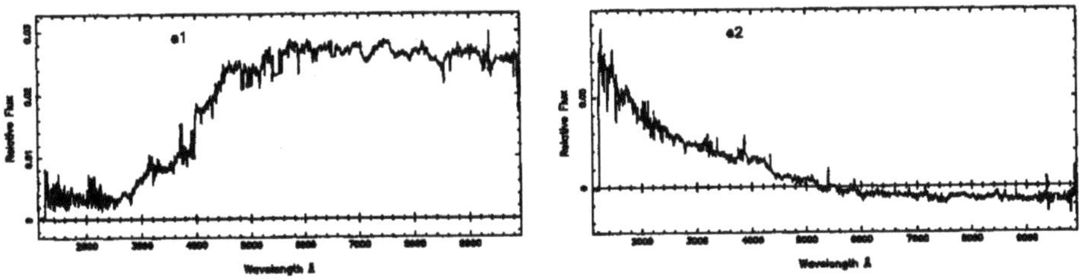

Fig. 6. The first two eigenspectra in the KL expansion of 10 galaxies of various spectral types. These two components contain 99% of the statistical weight. The first eigenspectrum corresponds to a mixture of an E-type spectrum and a young stellar population. The second component adds blue light, and subtracting red, the inverse of the aging of the stellar populations.

A Karhunen-Loeve expansion was applied to the 10 spectra, in similar fashion to the fluctuation eigenmodes, discussed in the previous sections. The first eigenspectrum corresponds to a mixture of an E-type spectrum and a young stellar population. The second component is basically adding blue light, and subtracting red, the inverse of the aging of the stellar populations. These two are shown on Fig. 6. The third component represents the spectral energy distribution characteristic for a post-starburst population. When the spectra are expanded in terms of the first three components, the rms deviations are less than 0.5%, and they form still a one-parameter sequence, that we can naturally associate with age. The galaxies expanded in terms of these KL coefficients are shown on Fig. 7.

This sequence provides the first intrinsic dimension for the multicolor data. Redshift, via the K-correction adds another nontrivial dimension, and thus we can understand why the dimensionality with colors and the tightly correlated redshift are so close to 2. Luminosity is an additional intrinsic parameter, which is known to correlate with the SED type. The measured fluxes are further correlated with redshift. This partially correlated nature of the fluxes with both colors and the redshift accounts for the fractional increase in the dimensionality, when flux is mixed with colors and/or redshift.

Complications at higher redshifts

One can easily see, that the curvature within the multicolor distribution is inherent. The bands $UB_JR_FI_N$ form a roughly logarithmic sequence of central wavelengths. If the galaxy distribution at low redshifts ($z < 0.4$) forms a roughly planar distribution in UB_JR_F, then at higher redshifts ($z > 0.4$), the galaxy SED is shifted about one band up. What used to be in the U band is shifted to B_J, what was in B_J is now in R_F, etc. For no evolution, the geometric distribution of galaxies should remain the same, therefore the galaxies have to form a planar

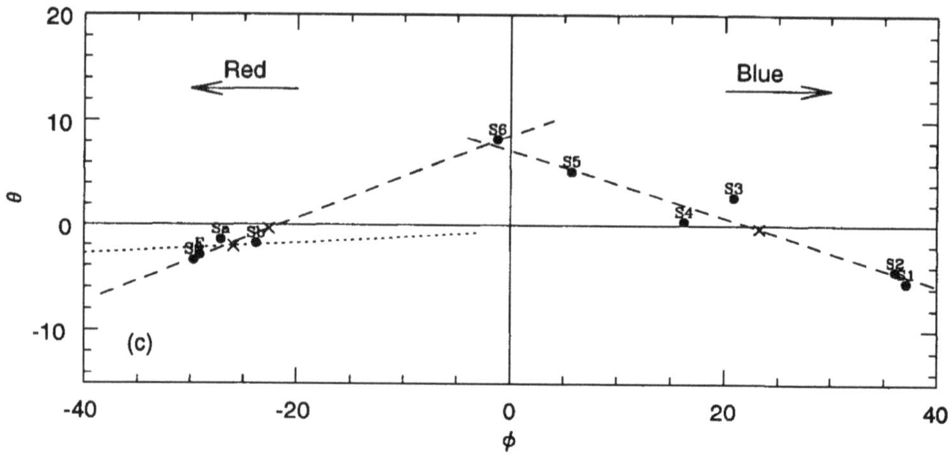

Fig. 7. The representation of the 10 galaxies in Connolly *et al.* (1995), in terms of the first three KL eigenspectra. The mixing angles θ and Φ represent how much the second and third KL components are mixed in with the first. All the galaxies are denoted by a dot on the surface of the unit sphere. Note, that all galaxies form a curved, one-parameter sequence, corresponding to age.

distribution, but now in $B_J R_F I_N$ rather than $U B_J R_F$. This is equivalent to a 90 degree rotation of the axes, which of course takes place continuously. Thus, curvature must be present in the multicolor distribution. This becomes even more pronounced at higher redshifts, this can be well seen on the last two panels of Fig.4. When the H,K break moves to a new band, the trajectories describing a galaxy in multicolor space take a sharp turn.

MEASURING EVOLUTION

Luminosity evolution of faint galaxies

The determination of the galaxy luminosity function is of crucial importance to cosmology. Knowledge of the local luminosity function is necessary to interpret the result of faint galaxy counts and their implications for cosmography and galaxy evolution. A more explicit measure of galaxy evolution is the variation of the luminosity function with redshift, and such an analysis is just beginning to become possible. Considering that the luminosity function may vary with redshift, color, and local density, and that we are interested in as large a range in absolute magnitude as possible, many redshifts to faint limits are required. The difficulty in determining the luminosity function from a magnitude limited sample of galaxies is that intrinsically bright galaxies can be seen out to great distances while intrinsically faint galaxies can be seen only relatively nearby. One way to account for this observational bias is to weight each galaxy's contribution by the inverse of the volume over which it could have been observed. This is the $1/V_{max}$ technique (Schmidt 1968). The disadvantage of this technique is that it does not properly account for galaxy clustering. Despite this fact it is still widely used today (e.g. Lilly *et al.* 1995). There are several more recent non–parametric techniques which can account for galaxy clustering, like the stepwise maximum–likelihood method of Choloniewski (1986), and the C-method (Lynden-Bell 1971). All three techniques are closely related, the stepwise maximum-likelihood being the one most commonly used today.

The developments in the experimental determinations of the luminosity function have been tied to large redshift surveys. These include the Stromlo-APM redshift survey (Loveday *et al.* 1992) which was limited at $b_J = 18.66$ and found a luminosity function well fit by a Schecter function (Schecter 1976) with parameters

$$\phi^* = 1.75 \times 10^{-3} h_{50}^{-3}, \qquad M_{B_J}^* = -21.0, \qquad \alpha = -1.0$$

The luminosity function from the CfA redshift survey (deLapparent *et al.* 1989) later extended by Marzke *et al.* (1994) consists of 9063 galaxies at a limit of $m_z = 17.0$. It is also fit by a flat Schecter function but with a much higher normalization $\phi^* = 5.0 \times 10^{-3} h_{50}^{-3}$ and three times as many faint galaxies as predicted by the extrapolation of the luminosity function. In contrast to these local samples the Canada-France redshift survey (Lilly *et al.* 1995) is much deeper, containing 591 galaxies with a median redshift of $z_{med} = 0.56$. This revolutionary sample of galaxies allows them to explicitly measure the evolution of the luminosity function with redshift and even to divide the sample by color.

Here we propose the use of a technique which will allow the determination of the luminosity function from high redshift samples with much more galaxies than in the CFRS, but with far less observational cost. The use of the photometric redshift technique provides 300 times as efficient use of telescope time over spectroscopic

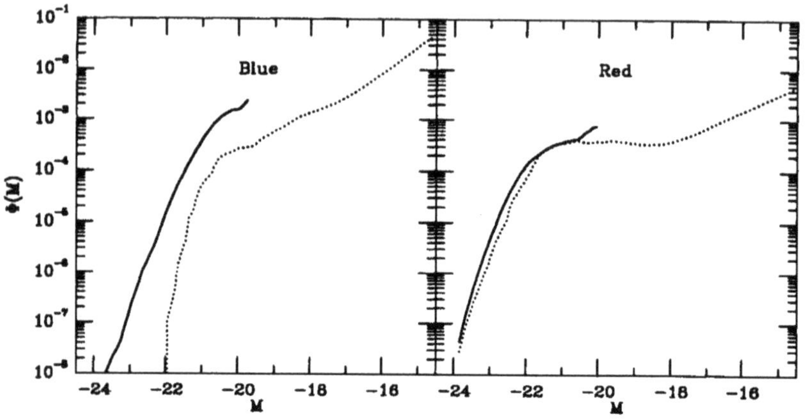

Fig. 8. In order to demonstrate the power of our approach, we have done a small pilot project to study the evolution of the luminosity function with redshift and color, using photometric redshifts, and our modified C-method. The sample of 800 galaxies in the Koo-Kron sample at the Selected Area SA68, with $B_J < 22.5$ has been split into a blue and red half, roughly independent of redshift, resembling the choice of Lilly *et al.* (1995) for comparison. The dotted and solid lines correspond to the redshift ranges of [0.1,0.3] and [0.3,0.7], respectively. The red half of the galaxies hardly evolve, while there is a strong evolution in the blue half. The agreement with Lilly *et al.* (1995) is striking.

techniques. The tradeoff for the increased efficiency is an error in the individual redshift determinations. However, this error, as long as it is well determined, can be taken into account, while due to the large sample size the more critical shot-noise errors are negligible. Our technique accounts for this stochastic error by considering each galaxy as a distribution in redshift. This distribution in redshift — folded in with the K–corrections — leads to a distribution in absolute magnitude. The derived luminosity function, using an appropriate generalization of the C-method, is the ensemble average over the distributions represented by all the galaxies in the sample (SubbaRao *et al.* 1996). Since the relative photometric redshift errors are becoming smaller at higher redshifts, our method is becoming more and more accurate where it matters most.

Measuring evolution and clustering

Current studies of clustering evolution are limited by the uncertainties in the m(z) relation (i.e. whether a magnitude limited sample includes different populations of galaxies at distinct redshifts). Even though spectroscopic redshifts can be determined for galaxies down to a limiting magnitude of $B = 24$ (Lilly *et al.* 1991, Colless *et al.* 1991), it is not yet possible to collect large enough samples to obtain meaningful limits on clustering evolution (Cole *et al.* 1994). The long integration times required to reach these depths limit the volume of the universe that can be covered with reasonable allocations of telescope time.

On the other hand, with the advent of large format CCD detectors, accurate broadband magnitudes for galaxies are relatively easy to obtain, and can be used as a low resolution spectrograph to estimate redshifts and spectral types of galaxies. We have shown in a pilot project using data with deep redshifts and photographic broadband magnitudes that we can derive galaxy redshifts photometrically to an accuracy of $\Delta z < 0.05$. This technique represents a 300-fold increase in observing efficiency over multiobject spectrographs. Because evolutionary tests (e.g., clustering and luminosity evolution, density of high redshift clusters), are typically more sensitive to the numbers of galaxies than errors of a few percent in distance, these tests can significantly benefit from our technique.

To illustrate the power of deriving redshifts this way when applied to future surveys, one could use Keck deep redshift surveys to calibrate such a relation beyond B=25. Photometric redshifts could then be determined for the Sloan Digital Sky Survey (a photometric survey to B$\sim$ 23.5, and a spectroscopic survey to B$\sim$ 19): the number of galaxies with known distance would increase from $< 10^6$ to $> 10^8$!

Galaxy colors and angular correlation functions

The scaling of the angular correlation function of galaxies as a function of the limiting magnitude is well understood: as the depth of the survey increases there are more random, projected pairs, weakening the correlations. Thus, the clustering amplitude is related to an integral over the second moment of the redshift distribution of the galaxies dn/dz as

$$w(\theta) \propto \int dz f(z)^{-0.8} g(z)^{-1} (1+z)^{-3-\epsilon} (dn/dz)^2$$

where $f(z)$ is the angular diameter distance, $g(z)$ is the derivative of the proper distance with z, and ϵ describes clustering evolution. At the same time, the number counts scale with dn/dz. If we choose a subset of the galaxies based upon color, that subset will have a different dn/dz from the rest, resulting in a different amplitude of the angular correlation function.

In order to separate the clustering and luminosity evolution, we would like to use photometric techniques to split galaxies into different, broad redshift bands and spectral classes. One can then evaluate the angular auto- and cross-correlations between these classes. The expected amplification of the individual autocorrelations can be either due to reduced number of random projections or to the specific intrinsic correlations of the particular galaxy type or to clustering evolution. Normally these are hard to disentangle. On the other hand, the cross correlations can be used as well: galaxies in different physical volumes should be uncorrelated, a check on how good the redshift segregation is. The correlation matrix with the respective number counts provides a very strong set of constraints, which we would use to separate the different evolutionary effects, and constrain both clustering and luminosity evolution. For example, in the SDSS survey we expect to have over 10^6 galaxies in the photometric survey with $z > 1$!

The trend toward significantly bluer galaxies beyond $B = 22$ discovered in the early 1980's (Kron 1980) has been confirmed in recent surveys (Cowie *et al.* 1991, Efstathiou *et al.* 1991). The hypothesis that these are the precursors normal galaxies undergoing enhanced star formation at an earlier epoch, $z > 1$, is suspect since their number counts exceed those expected from the density of ordinary luminous galaxies (Koo 1991). This suggests that these galaxies may represent a different population which lie at more moderate redshifts (Broadhurst *et al.* 1988).

Given the difficulties in obtaining redshifts of these faint galaxies, we have studied (Landy *et al.* 1994) the two-point angular correlation properties of several thousand faint galaxies, $20 < J < 23.5$, in two fields as a function of color using data from 4 meter plates in three bandpasses (U, J and F). We find (Figure 9) that the bluest galaxies show surprisingly strong angular clustering among themselves, and appear to be only weakly correlated with other more 'normal' galaxies. The strength of these correlations together with estimates of number count-redshift distributions from standard models suggest they are from a nearby, redshift $z < 0.3$ population.

Fig. 9. (a&b) Results for A_w, the amplitude of the angular correlation function measured at 1 degree, assuming a power law slope of -0.8 are shown, integrated between 9 and 144 arcseconds, in units of 10^{-4}. Error bars are Poisson in the bin counts. The dashed line shows the result for the combined data sets. The signals are calculated using all galaxies less than or greater than a given U-F color for red and blue respectively. Notice enhancements over a factor of 10 for both blue and red extreme subsets. The fraction of the data used at each limit is shown in (d). (c) illustrates the signal with the data sets divided into bands of color of one magnitude width in U-F. These results show that the enhancements in both blue and red are robust over 1.5 magnitudes in U-F. (e) and (f) show $w(\theta)$ for U-F>1.5 (Red) and U-F<-0.25 (Blue) for both data sets broken up into five bins of equal log width between 18 and 289 arcseconds along with a fiducial line for a no-evolution model with q_o=0.01 and H_o=50 km sec^{-1}Mpc^{-1}.

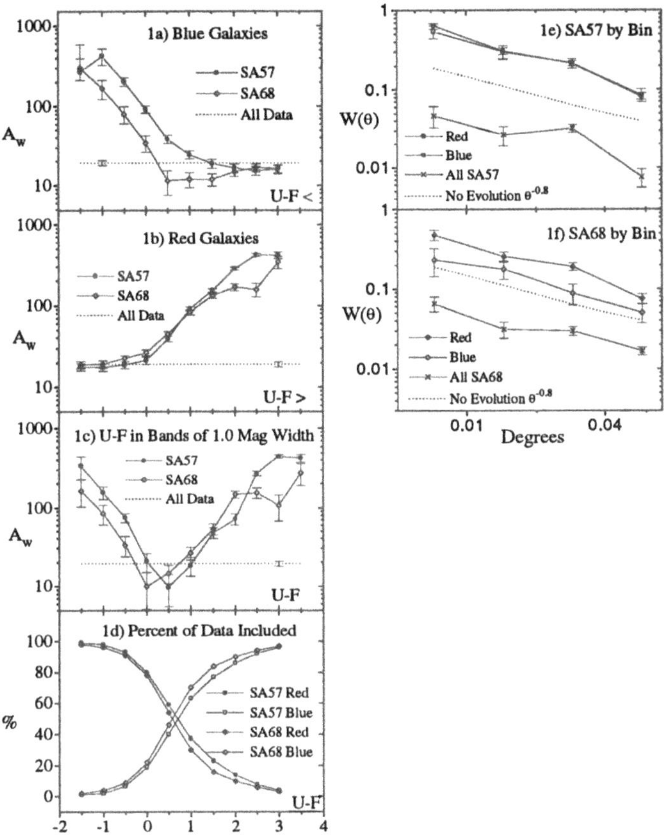

PRACTICAL APPLICATIONS OF HIGHER ORDER CORRELATIONS

Effects of sparse sampling

If fluctuations in the universe are strictly Gaussian, their full statistical description is contained in the two-point correlation function or in its Fourier transform, the power spectrum. The phases of the individual Fourier components are random for such a process, and all high-order measures of clustering vanish. In this case, Kaiser (1986) shows that measuring the redshifts of a small fraction of the galaxies is the most efficient way to measure the power spectrum on large scales. However, if the universe contains sharp large scale features like the "Great Wall;" a sparsely sampled survey may fail to identify them because it is less sensitive to the higher-order correlations (equivalently, the phase correlations) that characterize such structures. A distribution with high-order clustering can be very different from a homogeneous, isotropic, Gaussian random field with the same second-order statistical properties. A sparse survey optimized to measure the two-point correlation function would miss the very real differences between these distributions.

Sparse sampling can severely compromise one's ability to recover the underlying galaxy density field (de Lapparent *et al.* 1991; Ramella *et al.* 1990). Suppose that we are interested in the observed distribution of galaxy counts in cells that contain N galaxies above a given absolute magnitude limit. If we observe the galaxy distribution with a sampling rate $p < 1$ then the number of galaxies M detected in a cell that actually contains N galaxies will have a binomial distribution

$$P(M|N,p) = \binom{N}{M} p^M (1-p)^{N-M}.$$

The frequentist estimator for the true number of galaxies in the cell is simply $K = M/p$. These inferred counts K have the probability

$$P(K|p) = \sum_N P(Kp|N,p)P(N),$$

where $P(Kp|N,p)$ is the binomial distribution given above and $P(N)$ is the original distribution. At $p \approx 1$ the distribution of inferred counts is quite close to the true distribution, but as the sampling rate decreases the distribution of inferred counts becomes very broad. For large values of K, $P(K|p)$ for low sampling rates exceed the true value by orders of magnitude (Szapudi & Szalay 1994). Furthermore, it can become virtually impossible to infer the actual galaxy density field — the true number of galaxies in each spatial cell — because sparse sampling leaves only a tenuous connection between the number of galaxies present and the number observed.

Bayesian density reconstruction

Cluster finding algorithms typically use a filtering/thresholding scheme (Abell 1956), with some corrections for sampling effects (e.g. $[m_3, m_3 + 2]$). To date, most catalogs of superclusters have been created by applying some type of percolation algorithm to a cluster or a group catalog (Huchra & Geller 1982; Bahcall & Burgett 1986).

Prior knowledge of the count probabilities $P(N)$ can improve the density reconstruction: we can apply Bayesian reasoning to the problem of identifying cells that lie above overdensity thresholds in redshift space. Given our prior knowledge of the count probabilities, and the measured count M, we can compute the inverse Bayesian probability

$$P(N|M,p) = P(M|N,p)P(N)/P(M)$$

Voronoi foam, smoothed original

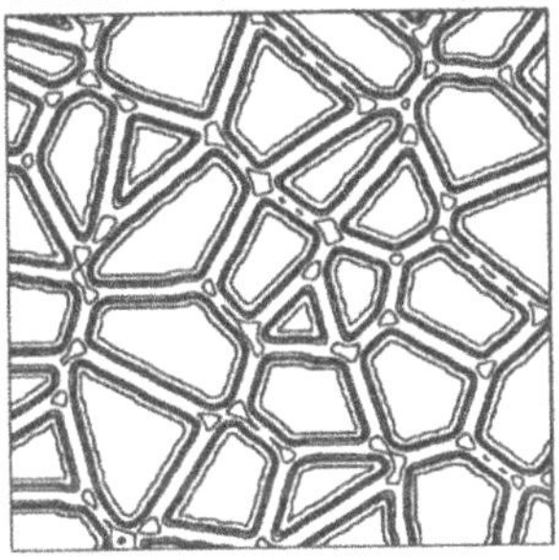

Voronoi foam, random phases

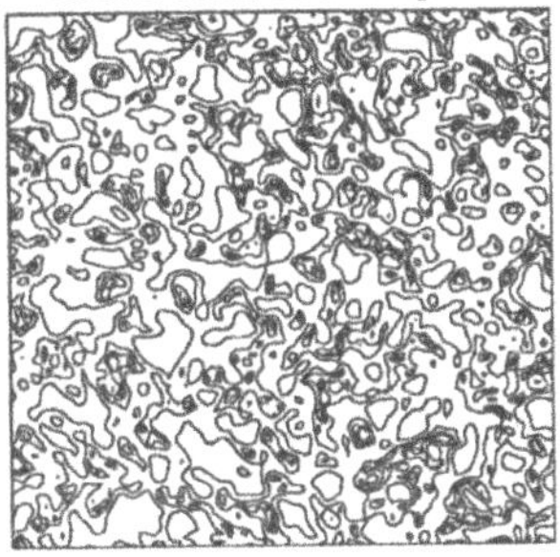

Fig. 10. The left hand figure is a two dimensional Voronoi foam, generated by the median surfaces between Poisson 'seeds' at the mean separation of 100 Mpc. In this simple toy model galaxies reside only on the walls of the foam, smoothed, so the walls have a finite thickness. The structure has a well defined second order statistic, but also has well correlated phases. This picture has been Fourier transformed, all the phases randomized, then transformed back again. The result is another two-dimensional density plot, shown on the right hand side, with the same second order properties, but with a Gaussian distribution.

where $P(M)$ is the probability of detecting M galaxies, for all N. This is an *unbiased estimator* for the inferred count probabilities.

The probability distribution for the inferred counts for each observed cell is independent of its neighbours. However, we know galaxies to be correlated and can include this information via an iterative procedure. Initially, we assign to cells their most probable Bayesian count. We can then correct for correlations between nearby cells using an iterative algorithm in which neighboring cells 'interact' with a strength proportional to their correlation, pulling anomalous low-density cells in a high-density environment up, and vice versa, by minimizing a 'cost function.' Small fluctuations are suppressed, and when the procedure reaches a steady state one is left with coherent groups of cells that represent superclusters, take into account the correlated nature of the neigbouring cells, but also reflect the observed counts. By applying various density thresholds (which can also be chosen from Bayesian principles: optimal Bayesian classifiers) we obtain an optimal, objective reconstruction of the largest scale coherent structures in the universe. Such techniques have been widely applied in non-linear image reconstruction, like the Iterative Conditional Modes (ICM), or the Maximum a Posteriori (MAP) algorithm (see Ripley 1988, Marroquin 1993).

NEW SURVEYS, NEW METHODS

Over the next few years several new large scale surveys, in the design phase today, will start producing data. These data will have much higher quality than available today, and thus improved analysis techniques, which go beyond just compensating for the sparseness of today's data sets may be appropriate.

The SDSS project will cover about 10,000 square degrees on the Northern sky, centered on the North Galactic Pole, measure 10^8 galaxies with 5 color CCD photometry, and obtain spectra of 10^6 of those. The redshift catalog will have an extremely high sampling rate, making it particularly suitable for the analysis techniques described above. Figure 3 shows, how well SDSS will measure the amplitude of the power spectrum at the COBE scales, with our analysis technique. The 5-color photometry catalog will be an ideal sample for photometric redshifts. We can subdivide 10^8 galaxies on a fine grid, without being strongly affected by shot noise.

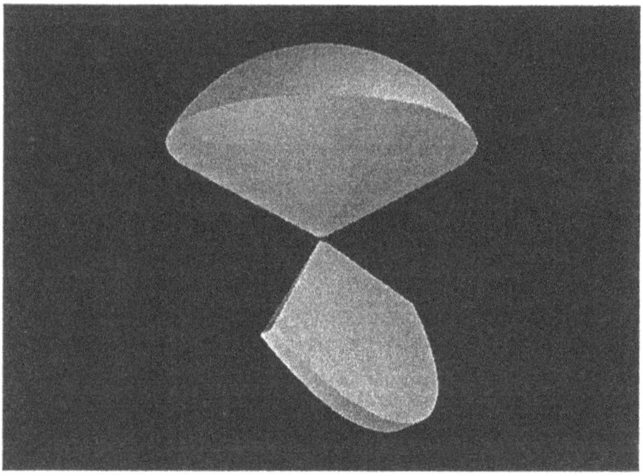

Fig. 11. A schematic plot of the volume of the SDSS redshift survey. The wide angle cone is centered on the North Galactic Pole, and the effective depth is about 450 Mpc. The opening angle of the cone is about 120 degrees, slightly elongated. A deep slice at 0° declination will also be surveyed.

The DEEP project is an extension of the deep pencilbeam surveys, out to much higher redshifts, using a high throughput spectrograph (DEIMOS), under construction for the Keck Telescope (S. Faber, G. Illingworth, D. Koo). The goal is to collect about 20,000 redshifts in several small areas, down to a limiting magnitude of $B_J = 25.5$, over the next six years. The design of the survey is such that the sample can be used for several projects: to follow galaxy populations at equal masses as

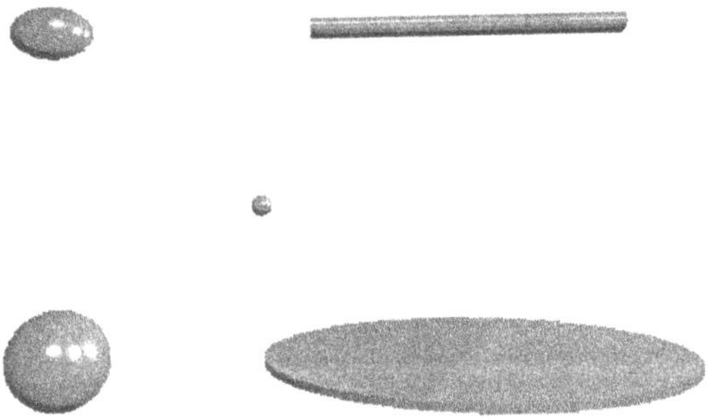

Fig. 12. A schematic three dimensional comparison of window functions for several redshift surveys. The window function is the Fourier transform of the survey volume, it can be thought of as the point spread function for measurements of the power spectrum. The narrower the window function, the better the sensitivity to large scale clustering. Surveys with highly anisotropic geometry (slices, pencil beams) have also highly anisotropic window functions; as a result estimates of the large scale clustering can be polluted by aliased power from small scales. The window functions of five surveys are shown here (the smaller, the better...) lower left is QDOT, upper left is CfA2, upper right is Las Campanas, upper right is the BEKS survey, and the small dot int he center is SDSS.

a function of z (using high resolution, thus dispersion can be measured accurately), to constrain the geometry of the universe (Ω_0, and Λ_0), and measure spatial clustering at $z = 1$. The deep redshifts, combined with multicolor photometry can also calibrate the color-redshift relation to $B_J = 25.5$.

The analysis techniques discussed here are important to understand the nature of the fluctuations on scales over 100 Mpc. They can measure the shape of the fluctuation spectrum in an overlap region with COBE, and quantify the level of coherence (non-Gaussianity) present in the large scale structure of the universe. Many of the ideas presented here have considerable practical applications, and provide new ways to constrain the various evolutionary processes acting in the universe between now and $z = 1$.

ACKNOWLEDGEMENTS

The author would like to thank the organizers for all their efforts, especially Brian Robson, without whom this manuscript would never have been finished. A lot of the material presented here is part of collaborative effort with Michael Vogeley, Andrew Connolly, Stephen Landy, David Koo, Mark SubbaRao and István Szapudi. This research has been supported by the NSF, OTKA and the Seaver Foundation.

REFERENCES

Abell, G.O. 1956, *Ap. J. Supp.*, **3**, 1956.
Bahcall, N.A., & Burgett, W.S. 1986, *Ap. J. Lett.*, **300**, 1986.
Bardeen, J.M., Bond, J.R., Kaiser, N., & Szalay, A.S. 1986, *Ap. J.*, **304**, 1986.
Bartlett, J.G., Blanchard, A., Silk, J., & Turner, M.S. 1994, submitted to Science
Broadhurst, T.J., Ellis, R.S. and Shanks, T. 1988, *M.N.R.A.S.*, **235**, 1988.
Broadhurst, T.J., Ellis, R.S., Koo, D.C., & Szalay, A.S. 1990, *Nature*, **343**, 1990.
Cole, S., Ellis, R.S., & Broadhurst, T.J. 1994, *M.N.R.A.S.*, **267**, 1994.
Colless, M, Ellis, R.S., Taylor, K., & Shaw, G. 1991, *M.N.R.A.S.*, **253**, 1991.
Connolly,A.J. Szalay,A.S., Bershady,M.A., Kinney,A.L. and Calzetti,D. 1995, *Astron. J.*, **110**, 1995.
Connolly,A.J., Csabai,I., Szalay,A.S., Koo,D.C., Kron,R.G. & Munn,J.A. 1995, *Astron. J.*, **110**, 1995.
Cowie, L.L. and Lilly, S.J. 1991, in *The Evolution of the Universe of Galaxies: The Edwin Hubble Centenary Symposium* (ed. R. G. Kron) (Astron. Soc. Pac., San Francisco), 212
Davis, R.L., Hodges, H.M., Smoot, G.F., Steinhardt, P.J., & Turner, M.S. 1992, *Phys. Rev. Lett.*, **69**, 1992.
de Lapparent, V., Geller, M.J., & Huchra, J.P. 1986, *Ap. J. Lett.*, **301**, 1986.
de Lapparent, V., Geller, M.J., & Huchra, J.P. 1991, *Ap. J.*, **409**, 1991.
Efstathiou, G., Bernstein, G., Katz, N., Tyson, J.A. & Guhathakurta, P. 1991, *Ap. J. Lett.*, **380**, 1991.
Feldman, H.A., Kaiser, N., & Peacock, J.A. 1994, *Ap. J.*, **426**, 1994.
Fisher, K.B, Davis, M., Strauss, M.A., Yahil, A., & Huchra, J.P. 1993, *Ap. J.*, **402**, 1993.
Geller, M.J. & Huchra, J.P. 1989, *Science*, **246**, 1989.
Gorski, K.M., Hinshaw, G., Banday, A.J., Bennett, C.L., Wright, E.L., Kogut, A., Smooth, G.F. & Lubin, P. 1994, *Ap. J. Lett.*, **430**, 1994.
Groth, E. *et al.* 1996, *Ap. J.*, in preparation
Harrison, E.R. 1970, *Phys.Rev.D*, **1**, 1970.
Huchra, J.P., & Geller, M.J. 1982, *Ap. J.*, **257**, 1982.
Kaiser, N. 1984, *Ap. J. Lett.*, **284**, 1984.
Kaiser, N. 1986, *M.N.R.A.S.*, **219**, 1986.
Kirshner, R.P., Oemler, A.J., Schechter, P.L. & Shectman, S.A. 1983, *Astron. J.*, **88**, 1983.

Klypin, A., Holtzman, J., Primack, J., & Regös, E. 1993, *Ap. J.*, **416**, 1993.
Kofman, L., Gnedin, N., & Bahcall, N.A. 1993, *Ap. J.*, **413**, 1993.
Koo, D.C. 1985, *Astron. J.*, **90**, 1985.
Koo, D.C. 1991, *The Evolution of the Universe of Galaxies: The Edwin Hubble Centenary Symposium* (ed. R. G. Kron) Astron. Soc. Pac., San Francisco 268
Kron, R.G. 1980, *Ap. J. Supp.*, **43**, 1980.
Lilly, S.J., Cowie, L.L., & Gardner, J.P. 1991, *Ap. J.*, **367**, 1991.
Lilly, S.J., Cowie, L.L. & Gardner, J.P. 1991, *Ap. J.*, **367**, 1991.
Lilly, S.J., Tresse, L., Hammer, F., Crampton, D. & Le Fevre, O. 1996, *Astron. J.*, in press
Loveday, J., Efstathiou, G, Peterson B.A. & Maddox, S.J. 1992, *Ap. J.*, **400**, 1992.
Landy, S.A., Szalay, A.S. & Koo, D.C. 1996, *Ap. J.*, in press
Maddox, S.J., Efstathiou, G., Sutherland, W.J. & Loveday, J. 1990, *M.N.R.A.S.*, **242**, 1990.
Marroquin, J. L. 1993, *Graphics and Image Processing*, **55**, 1993.
Park, C., Vogeley, M.S., Geller, M.J., & Huchra, J.P. 1994, *Ap. J.*, **431**, 1994.
Peebles, P.J.E. 1980, *The Large Scale Structure of the Universe*, (Princeton: Princeton University Press)
Ramella, M., Geller, M.J., & Huchra, J.P. 1990, *Ap. J.*, **353**, 1990.
Ripley, B. D. 1988, *Statistical Inference for Spatial Processes*, (Cambridge: Cambridge University Press)
Sachs, R.K., & Wolfe, A.M. 1967, *Ap. J.*, **147**, 1967.
Saunders, W., Frenk, C., Rowan-Robinson, M., Lawrence, A. & Efstathiou, G. 1991, *Nature*, **349**, 1991.
Shectman, S.A., Schecter, P.L., Oemler, A., Tucker, D., Kirshner, R.P., and Lin, H. 1992, in Clusters and Superclusters of Galaxies, A.C. Fabian, ed. (Dordrecht: Kluwer)
Smoot, G.F. et al. 1992, *Ap. J. Lett.*, **396**, 1992.
Steidel *et al.* 1996, *Ap. J.*, submitted
SubbaRao, M.A., Connolly, A.J., Szalay, A.S. & Koo, D.C. 1996, *Ap. J.*, submitted
Szalay, A.S., Broadhurst, T.J., Ellman, N., Koo, D.C., & Ellis, R.S. 1993, *Proc. Natl. Acad. Sci.*, **90**, 1993.
Szapudi,I. and Szalay, A.S. 1993, *Ap. J.*, **408**, 1993.
Therrien, C.W. 1992, *Discrete Random Signals and Statistical Signal Processing*, (New Jersey: Prentice-Hall)
Vogeley, M.S. & Szalay, A.S. 1996, *Ap. J.*, in press
Zel'dovich, Ya.B. 1972, *M.N.R.A.S.*, **160**, 1972.

Gravitational Lensing

Rachel Webster & Daniel Mortlock†
School of Physics,
University of Melbourne

Abstract

These lectures describe the basic ideas and formalisms of gravitational lensing. The physical ideas behind gravitational lensing will be discussed, and the specific formulæ describing two standard mass distributions, the point mass and the singular isothermal sphere will be given.

1 Introduction

A substantial textbook has been written on gravitational lensing[1], as well as a number of excellent reviews, the most recent of which is by Blandford and Narayan[2]. While the material described in these lectures is by no means disjoint from that already published, an effort has been made to provide a tractable and complete introduction to the subject area.

The bending of light is predicted by Newtonian gravity since the finite mass-energy of a photon must be accelerated by a gravitational field; thus in principle Newton might have predicted the multiple-imaging of background objects by foreground masses. In Newtonian theory the total angular deflection is $2GM/c^2\xi$, where M is the lensing mass and ξ is the impact parameter of the light beam at that mass. Einstein, in his theory of General Relativity, realised that the gravitational field affects not only the rate at which clocks run, as with Newtonian gravity, but also the shape of space. Thus the total deflection in General Relativity[3] is given by $4GM/c^2\xi$. In 1919 the British Solar Expedition obtained values of 1.98 ± 0.12 arcsec and 1.61 ± 0.30 arcsec for two different sets of observations of the bending of light at the limb of the Sun, essentially verifying the General Relativistic prediction of 1.745 arcsec[4].

A detailed account of the historical development of ideas in gravitational lensing is given by Schneider *et al.* [1], however the seminal insights of Zwicky are worth noting[5]. In 1937, he realised that " ...[gravitational lensing] would enable us to see nebulae at distances greater than those ordinarily reached by even the greatest telescopes... Observations on the deflection of light around nebulae may provide the most direct determination of nebular masses." In essence, Zwicky had forseen the power of gravitational lensing observations, which we are realising 60 years later.

Gravitational lensing is likely to make a substantial impact to the solution of three different classes of problems. It may be possible to determine, or at least

†These notes form the basis of lectures presented by Rachel Webster

place constraints on the cosmological parameters in Friedmann-Robertson-Walker (FRW) models. The time delay between the observation of intrinsic variations in two different images of a multiply-imaged quasar depends on the Hubble constant, H_0. Gravitational lensing may provide input into the debate on the value of Ω_0 (the ratio of the mean mass density of the universe to the critical density) by determining the total mass in objects with compact matter distributions; several experiments have been proposed which may lead to a global value of Ω_0. There is a promising possibility that gravitational lensing will place limits on the allowed values of the cosmological constant, Λ, since non-zero values of Λ give observable perturbations to background source counts[6].

A second class of problems uses lensing characteristics to determine properties of high redshift sources which are otherwise beyond the limits of observation. For example, it may be possible to determine the structure of quasars on micro-arcsecond scales using microlensing observations. In addition, gravitational lensing studies provide the most unambiguous measure of the dark matter distribution, whether it is in our own galaxy or in clusters of galaxies half-way across the universe. Such measurements may ultimately direct the search for dark matter candidates.

Unlike the direct measurements of the bending angle, such as those performed in the 1919 Solar Expedition, observations of multiple imaging and image distortion do not provide good tests of General Relativity. This is because the masses of the lensing objects are not well-known, and most observed instances of cosmological lensing are well-described by the weak field limit of General Relativity. Thus, in gravitational lensing models, General Relativity and a metric are assumed, and the observed effects are used to measure the mass of the lensing objects.

2 Vector Formalism

There are a number of different formalisms used to describe instances of gravitational lensing. Two of these are described below, while a detailed exposition of the third, the optical scalar formalism[7], is beyond the scope of these lectures. In all cases of astrophysical interest to date it has been sufficient to use the weak field approximation to Einstein's equations in describing the trajectory of a photon near a point mass.

Here we give an outline of the derivation of that result. Consider a point mass (of mass M) lying at the centre of a spherical co-ordinate system (*i.e.*, r, θ and ϕ). Due to spherical symmetry, any photon in this region will move in a plane that includes the point mass and, as the co-ordinates are arbitrary, we can define this to be the $\theta = \pi/2$ plane (as shown in Figure 1). Hence the standard Schwarzschild metric reduces to

$$dS^2 = -\left(1 - \frac{2r_M}{r}\right) dt^2 + (1 - \frac{2r_M}{r})^{-1} dr^2 + r^2\, d\phi^2, \quad (1)$$

where dS is the proper interval (zero for a photon), dt is the time interval and $r_M = 2GM/c^2$ is the gravitational radius of the mass. From this, we can obtain the components of the four-momentum of the photon. Firstly, p^θ and p_θ are zero. From the time independence of the metric p_0 is a constant, and, from the radial symmetry of the metric, p_ϕ is also a constant of the motion. Now $p^0 = g^{00}p_0 = -p_0/(1-2r_M/r)$ and, similarly, $p^\phi = g^{\phi\phi}p_\phi = p_\phi/r^2$. It is useful at this stage to introduce the affine parameter of the motion, λ, which is defined by $p^r = dr/d\lambda$, (and $p_r = g_{rr}p^r$) or $p^\phi = d\phi/d\lambda$.

For a photon, $p^\alpha p_\alpha = 0$. Expanding this using the expressions for the p's given above we get

$$p^0p_0 + p^rp_r + p^\phi p_\phi = (-p_0)^2(1-\frac{2r_M}{r})^{-1} + \left(\frac{dr}{d\lambda}\right)^2(1-\frac{2r_M}{r})^{-1} + \frac{p_\phi^2}{r^2} = 0, \quad (2)$$

which can be rearranged to give

$$\frac{dr}{d\lambda} = \sqrt{p_0^2 - \left(1-\frac{2r_M}{r}\right)\frac{p_\phi^2}{r^2}}. \quad (3)$$

The ratio of $d\phi/d\lambda$ to this gives

$$\frac{d\phi}{dr} = (r^2\sqrt{\frac{p_0^2}{p_\phi^2} - \frac{1}{r^2}\left(1-\frac{2r_M}{r}\right)})^{-1}, \quad (4)$$

the differential equation that defines the photon's path. This can be separated and integrated, using the substitution $y = 1/r(1-r_M/r)$ and the fact that terms of order $(r_M/r)^2$ can be ignored. This last follows from the fact that we are obtaining the weak field limit of the photon's behaviour, and the gravitational field is weak for $r \gg r_M$. This is also termed the linear approximation, as only first order terms are used. The integration is not a trivial process, and is left to the reader as an exercise.

The result of this integration is

$$\phi(r) = \frac{2r_Mp_0}{p_\phi} + \arcsin\left(\frac{p_\phi}{p_0}\frac{1}{r}\left(1-\frac{r_M}{r}\right)\right) - 2r_M\sqrt{\left(\frac{p_0}{p_\phi}\right)^2 - \left(1-\frac{r_M}{r}\right)^2}, \quad (5)$$

where it is assumed that the photon's initial position is given by $r \to \infty$ and $\phi \to 0$. In order for the last term to be real, as $r_M/r \ll 1$, r must be at least p_ϕ/p_0, and so this can be seen to be the minimum value of r along the photon's path. Calling this r_{min}, we can obtain $\phi(r_{min}) \approx \pi/2 + 2r_M/r_{min}$. So the photon moves through this co-ordinate angle from its starting point to its closest approach to the deflector. From the mirror symmetry of the photon path, it will move through the same angle again on its "outward" path, so $\phi_{final} - \phi_{initial} \approx \pi + 4r_M/r$. But we would consider light to be undeflected if it moved in a straight line, and hence moved through an angle of π. So we measure the photon to have been bent through an angle of $\alpha = \phi_{final} - \phi_{initial} - \pi \approx 4r_M/r$. Note that r_{min} is indistinguishable from the photon's

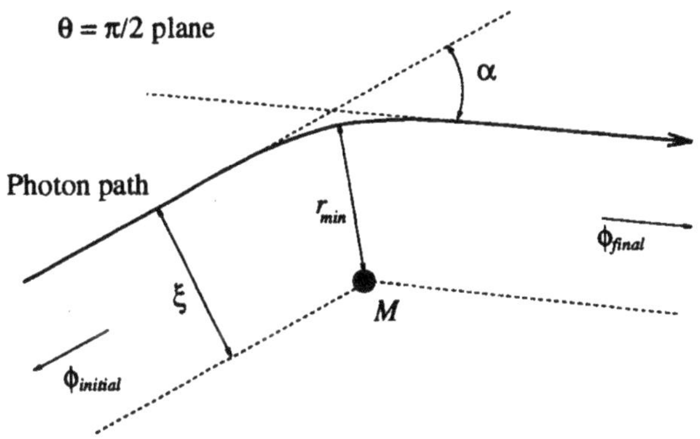

Figure 1: Photon trajectory near a point mass.

classical impact parameter, ξ, in this weak field limit. Thus we get the standard relativistic result, and the starting point for gravitational lensing: the bending angle is given by

$$\alpha = \frac{4GM}{c^2\xi}. \tag{6}$$

Using this bending angle, a thin lens may be represented by a 2-D mass distribution:

$$\vec{\alpha}(\vec{\xi}) = \frac{4G}{c^2}\int_{\mathcal{R}^2}\Sigma(\vec{\xi}')\frac{\vec{\xi}-\vec{\xi}'}{|\vec{\xi}-\vec{\xi}'|^2}d^2\xi', \tag{7}$$

where Σ is the surface density. For the special case of a spherically symmetric mass distribution, this reduces to

$$\vec{\alpha} = \frac{4GM(<\xi)\vec{\xi}}{c^2|\vec{\xi}|^2}. \tag{8}$$

This follows from Birkhoff's theorem, which states that for a spherically symmetric mass distribution only the mass inside the impact radius will effect the light path. Eq. 7 shows that the bending angle is the sum of the angles due to each element of mass.

In order to determine the images which would be observed for a particular distribution in the deflecting mass, solutions are determined for the 'Lens Equation',

$$\vec{\theta}_s D_{os} + \vec{\alpha} D_{ds} = \vec{\theta}_i D_{os} \tag{9}$$

as first described by Young *et al.*[8]. This equation can be derived from Figure 2 by equating the projection in the source plane of $\vec{\theta}_i$ with the sum of the projections of $\vec{\alpha}$ and $\vec{\theta}_s$. Subscripts (used throughout) refer to *o*bserver, *s*ource, *d*eflector (or lens) and

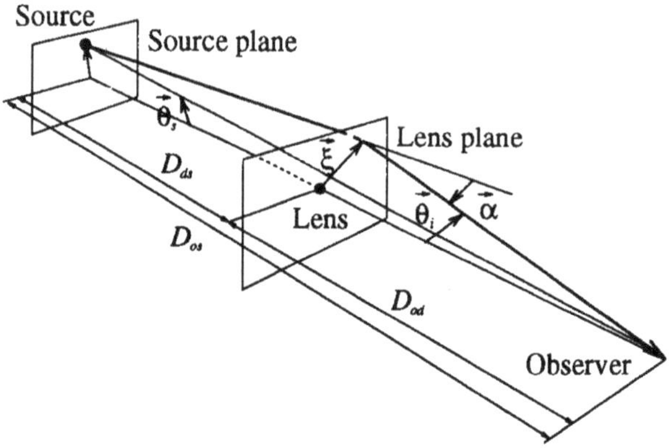

Figure 2: Standard bending picture, labelled with the parameters defined in the text.

image. The distances are affine distances, which are given by the angular diameter distances. A description of distances used in gravitational lensing problems is given in Section 9. For a circularly symmetric mass distribution, the solution to this equation can be determined analytically, and can be represented graphically. Since surface brightness is conserved (see Section 4), the magnification of a source is determined by the relative change in the image area. The magnification matrix is given by the Jacobian of the mapping of $\vec{\theta}_s$ to $\vec{\theta}_i$, vis

$$\mu = \left| \frac{\partial \vec{\theta}_i}{\partial \vec{\theta}_s} \right| = \frac{\theta_i}{\theta_s} \frac{d\theta_i}{d\theta_s}, \tag{10}$$

for a symmetric lens. The probability of observing a particular event is given by the region of the source plane which satisfies the event criteria, *e.g.* ,

$$\sigma = \int d^2\theta_s = \int \mu^{-1} d^2\theta_i. \tag{11}$$

If the observer, lens and source are directly aligned, then the source is observed as an annulus of light, called the 'Einstein ring'. We derive specific formulæ for different mass distributions in Sections 7 and 8.

3 Fermat's Principle

The structure of a wavefront which has been gravitationally lensed by a single, thin, non-spherical lens can be described by a scalar formalism based on Fermat's principle. In this formalism, the wavefront can be considered as a continuous deformable sheet,

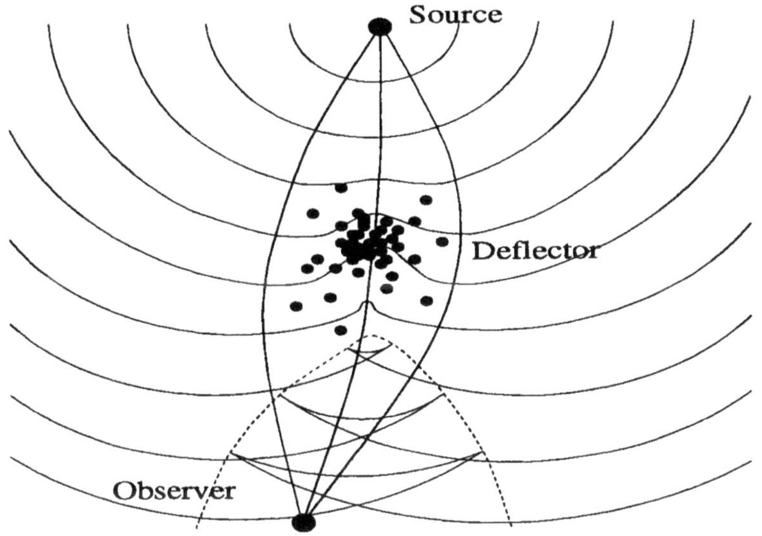

Figure 3: Schematic showing the changes in the shape of the wavefront as it passes by a deflector. The light rays from the source to the observer are also shown, as well as the caustic in the region of the observer.

where the deformations are caused by regions of stronger gravitational potential where the light travels more slowly. As the wavefront passes over a region in space of increased gravitational potential a fold will ensue, corresponding to a caustic in the light pattern (see Figure 3). Fermat's principle states that images will be observed at locations corresponding to the positions where the light path are extrema. A full description of the application of the scalar formalism to gravitational lensing is given by Blandford and Narayan[9].

If the weak field approximation is used to describe the gravitational field, the source is at redshift z_s, the lens at redshift z_d and the source position, in the absence of a lens, is θ_s, then it is possible to describe the time delay surface in some particular direction with respect to an unlensed ray as

$$\begin{aligned} t(\theta_i) &= t_{geom}(\theta_i) + t_{grav}(\theta_i) \\ &= (1+z_d)\left(\frac{D_{os}D_{od}}{2cD_{ds}}(\theta_i - \theta_s)^2 - \frac{2}{c^3}\int \phi(\theta_i D_{od}, s)ds\right). \end{aligned} \quad (12)$$

This expression, which is the sum of the geometric time delay due to the difference in path length and the gravitational time delay due to the difference in gravitational potential, gives the time delay in the observer frame and ϕ refers to the Newtonian potential of the lens. Fermat's principle, which will determine the location of images seen by the observer, can now be simply stated by $\partial t/\partial\theta_i = 0$. The time differences between different images is given by the difference in the time surfaces.

The above equation can be rescaled using

$$\tau = \frac{D_{ds}c}{(1+z_d)D_{os}D_{od}}t \tag{13}$$

and

$$\Phi(\theta_I) = \frac{2D_{ds}c^3}{D_{os}D_{od}}\int \phi(\theta_i)ds, \tag{14}$$

to give

$$\tau(\theta_i, \theta_s) = \frac{1}{2}(\theta_i - \theta_s)^2 - \Phi(\theta_i), \tag{15}$$

and the condition for observing images translated into

$$\theta_i = \theta_s + \nabla\Phi(\theta_i) = \theta_s + \frac{D_{ds}}{D_{os}}\alpha(\theta_s), \tag{16}$$

which is the standard Lens Equation. Thus the scalar formalism is equivalent to the vector formalism.

Locally the time-surface can be described by the curvature tensor $\partial^2 t/\partial\theta_{i1}\theta_{i2}$. If λ_1 and λ_2 are the principal radii of curvature then the magnification of the image is given by $\mu = \lambda_1\lambda_2$. The eigenvalues of the mapping $\theta_s \rightarrow \theta_i$, $\lambda_1^{-1}, \lambda_2^{-1}$ give the partial parities of the image. If they are both positive, the image looks similar to the source, and corresponds to a minimum in the time delay surface. If both are negative, the image is reflected about both axes, and the image corresponds to a maximum in the time delay surface. One positive and one negative eigenvalue corresponds to a reflection about one or other axis, and to a saddlepoint in the time delay surface.

The topology of the time delay surface is well understood[9]. If there is no lens, τ is represented by a paraboloid surface, centred on the source. If three images are formed, then only two generic shapes are possible for the surface. One, comprising a maximum, a minimum and a saddlepoint has the shape of a limaçon. Such a configuration is expected if the lens is an isothermal galaxy. The second type of surface, termed a lemniscate, comprises two minima and one saddle point, and would be formed if the lens were a cosmic string. If five images form, there are six possible surfaces which may be constructed by replacing each of the maxima or minima in the above two configurations by either a lemniscate or a limaçon.

Using this scalar formalism, it is possible to show a number of useful results (more fully described in Blandford and Narayan[9]):

(i) There are always $2n+1$ images, n corresponding to saddle points, and $n+1$ corresponding to either maxima or minima, with at least one minima.

(ii) The first image to vary (i.e. with the shortest path) is always a minimum, and has positive parity. This part of the wavefront does not pass through a caustic.

(iii) If a contour is constructed at the lens which defines the region in which the surface density Σ is greater than critical (see Section 6), then images corresponding to maxima pass through regions with $\Sigma > \Sigma_{crit}$; images corresponding to minima pass through regions with $\Sigma < \Sigma_{crit}$, and images corresponding to saddle points may pass through either region.
(iv) It is possible to have $\Sigma < \Sigma_{crit}$ everywhere, and still create multiple images. An example is a cosmic string or a very elongated image[10].
(v) Images corresponding to both maxima and to saddlepoints can be 'infinitely' deamplified. For example, the central image formed when the lens is a black hole corresponds to a maximum in the time delay surface and for a cosmic string the central image corresponds to a saddlepoint.
(vi) The more magnified a pair of images, the smaller the differential time delay. Physically this corresponds to a pair of images formed near a caustic either adjacent to one another or on opposite sides of the lens.

4 General Ideas

There are a number of important ideas which determine the physics of gravitational lensing. Firstly, the gravitational lensing of radiation from a background source is achromatic. This result follows from the application of the Principle of Equivalence. For the purposes of gravitational lensing, the Principle of Equivalence can be stated as: a gravitational field will affect the trajectory of all photons in the same way. This is the relativistic analogue of Galileo's observation that all objects fall with the same acceleration in a gravitational field.

Secondly, gravitational lensing by a static deflector does not change the surface brightness of the source. This result follows from the inverse square law of geometric optics, which states that the total flux passing through a surface normal to the direction of propagation in a tube of light rays is conserved. This is a consequence of Gauss' theorem, which states that if no radiator or absorber of energy is contained within a closed curface, then the total energy flux is zero. Since surface brightness is conserved, the observed area of a source changes when it is lensed. Indeed in the case of multiple images, the background source is seen from two different directions, though the difference in direction may only be of the order of arcseconds. If an infinite background source has a uniform surface brightness, it is not possible to induce brightness changes with a static gravitational lens. However it is possible to change the surface brightness temperature, which is equivalent to a change in wavelength, if the lens is moving with respect to the observer-source line[11]. The brightness temperature of the microwave background would change by

$$\Delta T_{max} \sim 10\left(\frac{v_t}{1000\,\mathrm{km\,s^{-1}}}\right)\mu\mathrm{K}, \tag{17}$$

where the lens is an isothermal cluster mass with a transverse velocity of v_t. More generally, transverse motion of the deflector will induce a wavelength change of

$$\frac{\Delta\lambda}{\lambda} \sim \frac{v_t}{c}\left(\frac{2GM}{c^2\theta_i D_{od}}\right). \tag{18}$$

For quasar emission lines, for example, this shift is unlikely to be measureable.

Polarisation is uneffected by gravitational lensing, except in the strong field case where the position angle of the polarisation is changed[12]. Interference of the light rays at the observer position will occur when $\lambda \sim GM/c^2$. In such instances, physical optics is required to describe the resulting patterns. Diffraction at the lens[13] will occur when the image separation is of order one Fresnel zone, i.e.,

$$\Delta\theta \sim \left(\frac{\lambda}{D_{od}}\right)^{\frac{1}{2}}, \tag{19}$$

where $\Delta\theta$ is the image separation. This reduces to the same condition as above, and, when the wavelength of the radiation is ~ 1 cm, corresponds to a mass of $\sim 10^{-5}\,M_{\odot}$. The diffraction limit also gives a maximal amplification of

$$\mu_{max} \sim \frac{r_s}{\lambda}, \tag{20}$$

where r_s is the source size.

5 Three Theorems

Three theorems will be discussed in detail. The first theorem[14] states that for a transparent, thin, non-singular mass distribution between a source and an observer an odd number of null geodesics will connect the source and the observer. That is, the observer sees an odd number of images. Intuitively, images are observed when the wavefront passes over the observer. If the wavefront is continuously deformed then additional images must be formed in pairs. The proof of this theorem can be shown by considering two vector fields in the deflector plane: the bending angle due to the mass, and the bending angle required for the ray to reach the observer. The zeros of the difference between these two fields will give the locations of images. The zeros of that vector field are constrained by the Poincaré-Hopf index theorem which assigns an index to a region by counting the net rotation of the vector field around the boundary of the region. For the vector field in question, at large angles from the centre of the deflecting mass, the difference is dominated by the latter angle, as the former will approach zero. Thus the vector on the boundary will complete a single rotation to give an index of +1. So if all images have a simple index of ± 1, there must be an odd number of images.

Observationally, however, an even number of images are almost always seen. There are several possible explanations: the central image may be highly deamplified, and therefore below detection levels, two (bright) images may be very close together and unresolved, extinction through the central part of a galaxy might deamplify an image or a massive black hole or a dense star cluster at the centre of the mass distribution might preclude the existence of the central image. In the case of a star, which is opaque, the central image is obscured; for a black hole, the central image is 'swallowed' by the black hole.

The second theorem states that for a transparent, thin, non-singular mass distribution between a source and an observer, there is at least one image brighter than the source would appear to be in the absence of this matter distribution[15]. The proof of this relies on the observation that images with positive parities each contain more flux than the original image, and that there must be at least one image with positive parities. The time delay surface tends to a large positive number for large angles with respect to the observer-lens axis. Thus the image with the smallest time delay relative to the unlensed ray, must be a global minimum; using the results described above this image must have positive parities.

The final theorem, due to Weinberg[16], states that in an inhomogeneous universe the average apparent magnitude is the same as would be given in an homogeneous universe with the equivalent mean matter density. Essentially this theorem says that the total flux of energy in a given volume cannot depend on whether the matter inside is distributed evenly or in lumps. Assume the universe is transparent, and observe luminous bodies with RW coordinates of r_1. Consider a spherical wavefront as it reaches us. The average fraction of all photons reaching us is $f = A/4\pi r_1^2 R^2(t_0)$, where A is the collecting area, and R is the cosmic scale factor. Then the average power is $f\,L/(1+z_s)^2$ where L is the object's luminosity. This is simply $LA/4\pi d_L^2$ where d_L is the luminosity distance to the object, thereby proving the theorem.

These theorems have a number of important implications. If some sources are amplified relative to the mean, then some must also be deamplified. Assuming the universe is transparent, then the maximum deamplification is given by the 'empty cone' value. In an inhomogeneous universe, small amounts of shear from nearby inhomogeneities have the same effect as a homogeneous mass distribution[17].

6 Basic Observational Details

In order for the observational effects of gravitational lensing to be measureable, the distribution of mass in the deflector must satisfy several conditions. Firstly, the surface density must be close to the *critical surface density,* Σ_{crit}. The definition of Σ_{crit} is that there must be sufficient matter in the lensing object to focus light rays from the source at the observer. Using Eq. 8 in the Lens Equation (Eq. 9), we have

$$\alpha = \frac{4G}{c^2}\frac{M(<\xi)}{\xi},$$

$$D_{ds}\alpha = \frac{D_s}{D_d}\xi,$$
$$\Sigma_{crit} = \frac{M(<\xi)}{\pi\xi^2} \qquad (21)$$

Hence

$$\Sigma_{crit} = \frac{c^2}{4\pi G}\frac{D_{os}}{D_{od}}D_{ds}. \qquad (22)$$

If we assume that the source and the lens are at cosmological distances, this corresponds to $\Sigma_{crit} \sim D_d^{-1} \sim 1\,\mathrm{g\,cm^{-2}} \sim 10^{3.7}\,\mathrm{M_\odot\,kpc^{-2}} \sim 10^{23.5}\,\mathrm{HI\,cm^{-2}}$ (the neutral hydrogen column density). If the source redshift is fixed, then there is a minimum in the critical surface density, which occurs approximately halfway between source and observer. The minimum value is approximately constant with the source redshift, except for low redshift sources, where it increases exponentially as $z_s \to 0$.

One of the great strengths of gravitational lensing observations is the possibility of using observations of either multiply-imaged background sources or distortions of extended sources to determine the foreground matter distribution. In the case of multiple images, we can use the separation of the images as a first-order estimate of the mass. If we assume that the two most widely separated images define the diameter of the Einstein ring, then we can estimate the mass enclosed within a cylinder of that diameter, M_{encl}, by:

$$M_{encl} \sim \Sigma_{crit}\pi\theta^2 D_{ol}^2. \qquad (23)$$

This estimate only sums mass inside the Einstein radius, and neglects the results of shear due to masses away from the line-of-sight to the image.

One powerful early lensing result, due to Press and Gunn[18], determined that the probability of lensing by a population of compact objects is $\sim \Omega_c$. We can derive that estimate in the following way: assume that there is a population of compact masses, m, distributed at cosmological distances, with a normalised density Ω_c. Then the number density of these lenses is given by $n = \Omega_c\rho_{crit}/m$. If the critical density is $\rho_{crit} = 3H_0^2/8\pi G$, the typical cosmological distance is $R = c/H_0$ and the gravitational mass is $M = Gm/c^2$, then $n \sim \Omega_c/MR^2$. Then the bending angle is $\alpha = 4M/r \sim M/\theta R \sim \theta$, where θ is the angular separation of the images. Thus $\theta^2 \sim M/R$.

We can then estimate the probability as $P \sim n\theta^2 R^3 \sim \Omega_c$. A more careful calculation of the probability must take account of the luminosity function of the source objects: for example if the sources are quasars, which show a steep slope at the bright end, then a significant bias will be observed in a magnitude-limited sample. For a population of objects with a flatter luminosity function such as Gamma-ray Bursts (if they are at cosmological distances[19]), then no amplification bias will be observed, and the observed number of multiply-imaged sources will directly measure Ω_c.

For gravitational lensing by compact masses the geometric size of the source is important. Essentially, signficant lensing effects will only be observed if the source size

r_s is smaller than the Einstein ring. If source and lens are at cosmological distances, this condition reduces to

$$r_s \lesssim 10^{-2}\left(\frac{\mathrm{M}}{\mathrm{M}_\odot}\right)^{\frac{1}{2}} \mathrm{pc}. \tag{24}$$

Since we estimate the size of a quasar continuum region to be $\sim 10^{-4}$ pc, this region can be significantly lensed by a sun-like star.

In order to see a complete Einstein ring, at least part of the source must cover the line-of-sight from the observer through the lens to the source plane. Quasar continuum regions are small relative to distant galaxies; thus it is unlikely that a lensed ring of quasar light would be observed, though we have seen many instances of rings of galaxy light.

7 The Point Mass Lens

If the lens can be described as a point mass, then light that passes within a minimum distance, ξ, of a point mass, M, will be deflected by

$$\alpha = \frac{4GM}{c^2\xi} = 2\frac{r_M}{\xi}. \tag{25}$$

If the angular position of the source is given by β, then the point mass lens produces images at angular positions given by

$$\theta_\pm \equiv \frac{\xi}{D_{od}} = \frac{\beta}{2} \pm \sqrt{\beta^2 + 4\alpha_0^2}, \tag{26}$$

where

$$\alpha_0 \equiv \sqrt{2R_S\frac{D_{ds}}{D_{od}D_{os}}} \propto \sqrt{M}, \tag{27}$$

where R_S is the Schwarzschild radius. The angular separation of the images, $\Delta\theta$, is

$$\Delta\theta = |\theta_+ - \theta_-| = \sqrt{\beta^2 + 4\alpha_0^2}. \tag{28}$$

Hence the Einstein ring has an angular diameter of

$$\Delta\theta_E = \sqrt{\frac{16GM}{c^2}\frac{D_{ds}}{D_{od}D_{os}}}. \tag{29}$$

This gives the approximate result that $\Delta\theta$ increases as $\alpha_0 \propto \sqrt{M}$.

Define $y = \theta_s/\alpha_0$ and $x = \theta_i/\alpha_0$ to be scaled source and image angles, respectively, measured with respect to the line of sight to the lens. Then the magnifications of the two images are

$$\mu_{1,2} = \frac{1}{4}\left[\frac{y}{(y^2+4)^{1/2}} + \frac{(y^2+4)^{1/2}}{y} \pm 2\right], \tag{30}$$

where the subscripts 1 and 2 refer to the brighter and fainter of the two images respectively.

The delay between arrival times along the two image directions is

$$\Delta t = \frac{2r_M}{c}(1+z_d)\tau(y), \tag{31}$$

where

$$\tau(y) \equiv \frac{1}{2}y(y^2+4)^{1/2} + \ln\left[\frac{(y^2+4)^{1/2}+y}{(y^2+4)^{1/2}-y}\right]. \tag{32}$$

8 The Singular Isothermal Sphere (SIS)

One of the most commonly used lens models is the Singular Isothermal Sphere (SIS), which gives a good approximation of the structure of galaxies or clusters of galaxies.

Consider a spherically symmetric mass distribution which is in equilibrium due to its self-gravity balancing its thermal pressure, $P(r)$. The temperature of the mass distribution is characterised by the line-of-sight velocity dispersion, σ, and given by $T = \sqrt{\sigma^2 m/k_b}$, where m is the mass of the constituent componenets and k_b is Boltzmann's constant. $P(r) = \rho(r)\sigma^2$, and the force due to the (gas) pressure, $\vec{F}_p(r)$, is given by $-\nabla P$. Radial symmetry means this becomes

$$\vec{F}_p(r) = -\frac{dP(r)}{dr}\hat{r} = -\sigma^2\frac{d\rho(r)}{dr}\hat{r}. \tag{33}$$

For a mass distribution in equilibrium, the gravitational force at r must balance the pressure at the same point. Hence

$$-\sigma^2\frac{d\rho(r)}{dr} = |\vec{F}_p(r)| = -|\vec{F}_G(r)| = \frac{G\rho(r)}{r^2}\int_0^r \rho(r')4\pi r'^2 dr'. \tag{34}$$

Equating powers of r on either side of the expression and implicitly assuming that ρ has a power law dependence on r gives the characteristic mass distribution of an isothermal sphere,

$$\rho(r) = \frac{\sigma^2}{2\pi G}\frac{1}{r^2}. \tag{35}$$

This mass distribution is unbounded and the corresponding surface density is

$$\Sigma(r) = \frac{\sigma^2}{2G}\frac{1}{r}. \tag{36}$$

A scale length in the lens plane is

$$\xi_0 \equiv \frac{4\pi G}{c^2}\frac{\sigma^2}{G}\frac{D_d D_{ds}}{D_s} = \frac{1}{\Sigma_{crit}}\frac{\sigma^2}{G}. \tag{37}$$

From this, we get the dimensionless mass density,

$$\kappa(x) \equiv \frac{\Sigma(\frac{\xi}{\xi_0})}{\Sigma_{crit}} = \frac{1}{2|x|}, \tag{38}$$

where x and y are as previously defined. Finally, we get the similarly scaled bending angle,

$$a(x) \equiv \frac{\alpha(x\xi_0)}{\xi_0} = \frac{m(< x)}{x}. \tag{39}$$

Having made these definitions, we can find an explicit expression for $a(x)$

$$a(x) = \frac{1}{\pi x} \int_0^x 2\pi x' \frac{1}{2|x'|} dx' = \frac{x}{|x|}. \tag{40}$$

Now the lens equation, $y = x - a(x)$, becomes

$$y = x - \frac{x}{|x|}. \tag{41}$$

We can assume, without loss of generality, that $y > 0$. That is, the source is on the "positive side" of the lens. The image positions will be given by x, which can take any real value, allowing images on either side of the lens.

For an axially symmetric mass distribution, the magnification, μ, is given by

$$\mu(x) = \left(\frac{y}{x}\frac{dy}{dx}\right)^{-1} = \left(1 - \frac{a(x)}{x}\right)\left(1 - \frac{d}{dx}a(x)\right). \tag{42}$$

With $a(x)$ defined in Eq. 40, this becomes

$$\mu(x) = \frac{|x|}{|x| - 1}. \tag{43}$$

Eq. 41 has two distinct regimes: for $y > 1$ there is one image solution, and for $y < 1$ there are two image solutions. Each case is dealt with separately.

$y > 1$. In this case, there will only be one magnified image. As it must be on the same side of the lens as the source, $|x| = x$, and the image is located at $x = y + 1$. The magnification is

$$\mu = \frac{x}{x-1} = \frac{y+1}{y} = 1 + \frac{1}{y}, \tag{44}$$

implying all single images will be magnified, and more so if they are closer to being lensed.

$(0 <)y < 1$. Two images are observed of the source, at $x_1 = y + 1$, and $x_2 = y - 1$ on either side of the lens. For the $x > 0$ image,

$$\mu_1 = \frac{x}{x-1} = \frac{y+1}{y} = 1 + \frac{1}{y}, \tag{45}$$

and for the $x < 0$ image,

$$\mu_2 = \frac{-x}{-x-1} = \frac{y-1}{y} = 1 - \frac{1}{y}. \tag{46}$$

Note that the image separation will be a constant

$$|x_1 - x_2| = |(y+1) - (y-1)| = 2. \tag{47}$$

So,

$$\Delta\theta = |x_1 - x_2|\xi_0 = 8\pi \left(\frac{\sigma}{c}\right)^2 \frac{D_s}{D_{ds}} \tag{48}$$

There is also the case of $y = 1$. A solution will be $x = 0$ (and $x = 2$,), which is a singularity of the lens equation. This marks the boundary between the configurations with one, and two images. But, as will be shown, the image at $x = 0$ will have zero magnification, and so never be visible.

The sign of μ simply represents the parity of the image (mirror inversion); $|\mu|$ gives the magnification/change in brightness. So, assume we have a source with an intrinsic, or unlensed luminosity of l_0. Hence there will be two images, at $x_1 = y + 1$, and $x_2 = y - 1$, with luminosities

$$l_1 = \left|\frac{y+1}{y}\right| l_0 = \left(\frac{1}{y} + 1\right) l_0, \tag{49}$$

and

$$l_2 = \left|\frac{y-1}{y}\right| l_0 = \left(\frac{1}{y} - 1\right) l_0, \tag{50}$$

respectively, the first image always being the brighter of the two.

It is impossible to observe the actual magnification of an image, as we will not have an opportunity to observe the source unlensed. We can observe l_1 and l_2, and solving Eq. 49 and Eq. 50 for l_0 gives

$$l_0 = 2(l_1 - l_2). \tag{51}$$

Hence the intrinsic luminosity of the source and the two image magnifications can be readily calculated if the SIS model is assumed.

9 Distances

A complete summary of the use of different distances in gravitational lensing problems is given in Turner *et al.*[20]. Defining D_{os}, D_{od}, and D_{ds} as in Section 2, where d is the angular diameter distance. These distances may be calculated from the Dyer-Roeder equation,

$$(z+1)(\Omega z+1)\frac{d^2D}{dz^2}+\left(\frac{7}{2}\Omega z+\frac{1}{2}\Omega+3\right)\frac{dD}{dz}+\frac{3}{2}\tilde{\alpha}\Omega D=0, \tag{52}$$

where $\tilde{\alpha}$ is the so-called "clumpiness parameter". It is unity if the lens is a minor contribution to the total amount of smoothly distributed matter along the line of sight to the source, in which case we get the usual formulae for angular diameter distances for a FRW metric. It is zero if the lens is the only matter along the line of sight to the source, in which case D becomes the affine parameter distance (more or less) and we recover Press & Gunn's[18] treatment. Generally, the angular diameter distance D_{ij} is given by integrating Eq. 52 from z_i to z_j, with the initial conditions

$$D_{ii}=0, \tag{53}$$

and

$$\left[\frac{dD_{ij}}{dz}\right]_{z_j=z_i}=\frac{\mathrm{sgn}(z_j-z_i)}{(z_i+1)^2(\Omega z_i+1)^{1/2}}. \tag{54}$$

10 References

1. P. Schneider, J. Ehlers and E.E. Falco, *Gravitational Lensing* (Springer-Verlag, Berlin, 1992)..
2. R.D. Blandford and R. Narayan, *Ann.Rev.Astron.Astrophys*, **30** (1992) 311.
3. A. Einstein, *Sitzber. Kgl. Preuss. Acad. Wiss.*, (1915) 831.
4. A.S. Eddington, *Space, Time and Gravitation*, (Cambridge University Press, London, 1920), p 118.
5. F. Zwicky, *Phys. Rev. Lett.*, **51**, (1937) 679.
6. S.M. Carroll, W.H. Press and E.L. Turner, *Ann.Rev.Astron.Astrophys*, **30** (1992) 499.
7. C.C Dyer and R.C. Roeder, *Ap.J.*, **249**, (1981) 290.
8. P. Young, J.E. Gunn, J. Kristain, J.B. Oke and J. Westphal, *Ap.J*, **241**, (1980) 507.
9. R.D. Blandford and R. Narayan, *Ap.J.*, **310**, (1986) 568.
10. K. Subramanian and S.A. Cowling, *M.N.R.A.S.*, **219**, (1986) 333.
11. M. Birkenshaw, in *Lectures in Physics, No. 330, Gravitational Lenses*, eds J.N. Moran *et al.* (Springer-Verlag, Berlin, 1989), p59.
12. S. Pinault, *M.N.R.A.S.*, **179**, (1977) 691.

13. R.J. Bontz and M.P. Haugan, *Astro.Sp.Sci.*, **78**, (1981) 199.
14. W.L. Burke, *Ap.J.Lett*, **244**, (1981) 1.
15. P. Schneider, *Astro. Ap.*, **140**, (1984) 119.
16. S. Weinberg, *Ap.J.Lett.*, **208**, (1976) L1.
17. C.C. Dyer and R.C. Roeder, *G.R.G*, **13**, (1981) 1157.
18. W.H. Press and J.E. Gunn, *Ap.J.*,**185**, (1973) 397.
19. O.M. Blaes and R.L. Webster, *Ap.J.Lett*, **391**, (1992) L63.
20. E.L. Turner, J.P. Ostriker and J.R. Gott, *Ap.J.*, **284**, (1984) 1.

AN INTRODUCTION TO QUANTUM COSMOLOGY

DAVID L. WILTSHIRE
Department of Physics and Mathematical Physics,
University of Adelaide, S.A. 5005, Australia.

CONTENTS

1. Introduction

These lectures present an introduction to quantum cosmology for an audience consisting for a large part of astronomers, and also a number of particle physicists. As such, the material covered will invariably overlap with that of similar introductory reviews[1,2], although I hope to emphasise, where possible, those aspects of quantum cosmology which are of most interest to astronomers.

This is not an easy task – quantum cosmology is not often discussed at Summer Schools such as this where there is a large emphasis on astrophysical phenomenology, for the very good reason that the ideas involved are at present rather tentative, and quantitative predictions are thus much more difficult to arrive at. Nevertheless we should remember that not too long ago the whole enterprise of cosmology was viewed as the realm of wild speculation. Vladimir Lukash remarked that when he started out in research the advice he was given was that "cosmology was all right for someone like Zeldovich to potter around with after he had already established himself by producing a body of serious work, but it was inappropriate for a young physicist embarking on a career". Happily the situation is quite different today – I am sure the earlier lectures in this School will have convinced you that modern cosmology is a hardcore quantitative science, and that with the new technology and techniques now being developed we can expect to accurately measure all the important cosmological parameters within the next decade and thus enter into a "Golden Era" for cosmological research.

Quantum cosmology, however, still enjoys the sort of status that all of cosmology had until not so very long ago: essentially it is a dangerous field to work in if you hope to get a permanent job. I hope to convince you nevertheless that quantum cosmology represents a vitally important frontier of research, and that although it is by nature somewhat speculative, such speculations are vital if we are to understand the entire history of the universe.

On the face of it the very words "quantum" and "cosmology" do appear to some physicists to be inherently incompatible. We usually think of cosmology in terms of the very large scale structure of the universe, and of quantum phenomena in terms of the very small. However, if the hot big bang is the correct description of the universe – which we can safely assume given the overwhelming evidence described in the earlier lectures – then the universe did start out incredibly small, and there must have been an epoch when quantum mechanics applied to the universe as a whole.

There are people who would take issue with this. In the standard Copenhagen interpretation of quantum mechanics one always has a classical world in which the quantum one is embedded. We have observers who make measurements – the observers themselves are well described by classical physics. If the whole universe is to be treated as a quantum system one does not have such a luxury, and some would argue that our conventional ideas about quantum physics cease to make sense. Yet if quantum mechanics is a universal theory then it is inevitable that some form of "quantum cosmology" was important at the earliest of conceivable times, namely the Planck time, $t_{\rm Planck} = (G\hbar)^{1/2}/c^{5/2} = 5.4 \times 10^{-44}$ sec, (equivalent to 10^{19} GeV as an energy, or 1.6×10^{-35} m as a length). At such scales, where the Compton wavelength of a particle is roughly equal to its gravitational (Schwarzschild) radius, irreducible quantum fluctuations render the classical concept of spacetime meaningless. Whether or not our current efforts at constructing a theory of quantum cosmology are physically valid is therefore really a question of whether our current understanding of quantum physics is adequate for considering the description of processes at the very beginning of the universe, or whether quantum mechanics itself has to be revised at some level. Such a question can really only be answered by extensive work on the problem.

Setting aside the question of the fundamentals of quantum mechanics, let us briefly review the problems which are left unanswered in the standard hot big bang scenario. These are:

1. The *value of* $\eta = N_{\text{baryons}}/N_{\text{photons}}$: the exact value of this parameter is unexplained in the hot big bang but is crucial in determining abundances of light elements through primordial nucleosynthesis.
2. The *horizon problem*: the isotropy of the cosmic microwave background radiation indicates that all regions of the sky must have been in thermal contact at some time in the past. However, in the standard Friedmann-Robertson-Walker (FRW) models regions separated by more than a couple of degrees have non-intersecting particle horizons – i.e., they cannot have been in causal (and hence thermal) contact.
3. The *flatness problem*: given the range of possible values of the ratio of the density of the universe to the critical density at the present epoch, Ω_0, then the FRW models predict that $\Omega(t)$ must have been incredibly close to the spatially flat case $\Omega(t) \simeq 1$ at early times; e.g., assuming $\Omega_0 \gtrsim 0.1$ we find $|\Omega - 1| \lesssim 10^{-26}$ at the lepton era, and $|\Omega - 1| \lesssim 10^{-53}$ at the GUT era.
4. The *unwanted relic problem*: models of the early universe which involve phase transitions often produce copious amounts of topological defects, such as monopoles produced at the GUT scale. If one puts the numbers in one finds that the density of such relics is so great that they would exceed the critical density by such a margin in the standard FRW models that the universe should have ended long ago!
5. The *origin of density perturbations* is unexplained.
6. The *arrow of time* is a physical mystery. On the one hand the laws of physics are CPT-invariant, and on the other there is a *thermodynamic arrow of time*, as prescribed by the Second Law of Thermodynamics, and it appears to match the *cosmological arrow of time*, as prescribed by the expansion of the universe.
7. The *initial conditions* of the universe must be put in by hand, rather than being physically prescribed.

The first four problems on the list are ones that can be explained without appealing to quantum cosmology. The value of η (problem 1) is predicted by models of baryogenesis, which typically take place at the GUT scale. Problems 2–4 are solved by the inflationary universe scenario: an early phase of exponential expansion of the universe drastically changes the past light cone, thereby removing the horizon problem, while also driving Ω close to unity, and diluting unwanted relics to such very low densities that they are close to unobservable.

Problems 5–7, on the other hand, are of a nature which is beyond the scope of the inflationary universe scenario to satisfactorily explain. Inflation provides a mechanism whereby initial small "quantum" perturbations are inflated to all length scales to form a scale-free Harrison-Zeldovich spectrum, but it does not address the question as to exactly how these perturbations arise. Furthermore, a typical model of the very early universe might possess both inflationary and non-inflationary solutions, so that the precise initial conditions of the universe can be crucial for determining whether the universe undergoes a period of inflation *sufficiently long* to be consistent with observation. The length of the period of inflation is precisely the sort of quantitative result that we might hope quantum cosmology should provide. Questions such as the origin of the arrow of time might appear to be of a more philosophical nature – however, quantum cosmology should provide a calculational framework in which such questions can begin to be addressed.

1.1. Quantum cosmology and quantum gravity

Quantum cosmology is perhaps most properly viewed as one attempt among many to grapple with the question of finding a quantum theory of gravity. As a field theory general relativity is not perturbatively renormalisable, and attempts to reconcile general relativity with quantum physics have not yet succeeded despite the attentions of at least one generation of physicists. It is perhaps not surprising that the problem is such a difficult one since general relativity is a theory about the large scale structure of spacetime, and to quantise it we have to quantise spacetime itself rather than simply quantising fields that live in spacetime.

Many ideas have been considered in the quest for a fundamental quantum theory of gravity – whether or not these ideas have brought us closer to that goal is difficult to say without the benefit of hindsight. However, such ideas have certainly profoundly increased our knowledge about the nature of possible physical theories. Some important areas of research have included:

1. *Supergravity*[3–5]. Using supersymmetry, a symmetry between fermions and bosons, one can enlarge the gravitational degrees of freedom to include one or more spin–$\frac{3}{2}$ gravitinos, ψ_μ, in addition to the spin–2 graviton, $g_{\mu\nu}$. Such a symmetry can cure some but not all of the divergences of perturbative quantum general relativity*. In particular, while pure Einstein gravity is perturbatively non-renormalisable at two loops[7], or at one loop if interacting with matter[8–10], in the case of supergravity renormalisability fails only at the 3-loop level[11]▶.

2. *Superstring theory*[13]. Progress can be made if in addition to using supersymmetry, one constructs a theory in which the fundamental objects have an extension rather than being point-like: a theory of strings rather than particles. Much interest in string theory was generated in the mid 1980s with the discovery that certain string theories appear to be finite *at each order* of perturbation theory. In some

* For details of the application of perturbative techniques to quantum cosmology, both with and without supersymmetry, see [6].

▶ The status of the result concerning the 3-loop divergence of supergravity is not quite as rigorous as the other examples mentioned, as the complete 3-loop calculation has not been done. However, it is known that a 3-loop counterterm exists for all extended supergravities and there is no reason to expect the coefficient of the counterterm to be zero, making 3-loop finiteness extremely unlikely. For a review of the ultraviolet properties of supersymmetric field theories see [12].

sense stringy physics "smears out" the problems associated with pointlike interactions. The entire sum of all terms in the perturbation expansion diverges in the case of the bosonic string[14], however, and it is believed that similar results should apply to the superstring[⊛]. Furthermore, despite what many see as the mathematical beauty of string theory, there has unfortunately as yet been no definitive success in deriving concrete phenomenological predictions.

3. *Non-perturbative canonical quantum gravity*[16,17]. The fact that general relativity is not perturbatively renormalisable might simply be a failure of flat spacetime quantum field theory techniques to deal with such an inherently non-linear theory, rather than reflecting an inherent incompatibility between general relativity and quantum physics. Given the divergence of the string perturbation series mentioned above, a non-perturbative formulation of string theory would also be desirable. As a starting point a systematic investigation of a non-perturbative canonical formalism based on general relativity could provide deep insights into quantum gravity. Such a programme has been investigated by Ashtekar and coworkers, mainly from the mid 1980s onwards. One principal difference from the canonical formalism which I will describe in section 2 is that instead of taking the metric to be the fundamental object to describe the quantum dynamics, one bases such a dynamics on a *connection*, in this case an $SL(2,\mathbb{C})$ spin connection[18,19]. If one integrates this connection around a closed loop one arrives at "loop variables", which might be considered to be analogous to the magnetic flux, Φ, obtained by integrating the electromagnetic gauge potential, A_μ, around a closed loop. In the "loop representation" quantum states are represented by functionals of such loops on a 3-manifold[20,21], rather than by functionals of classical fields. Although a number of technical difficulties remain, considerable progress has been made with the Ashtekar formulation, and the reformulation of quantum cosmology in the Ashtekar framework[22–25] could be an area for some interesting future work.

4. *Alternative models of spacetime.* All the above approaches assume that the basic quantum variables, be they a metric or a connection, are defined on differentiable manifolds. Given that it is highly possible that "something strange" happens at the Planck length it is plausible that one might have to abandon this assumption in order to effectively describe the quantum dynamics of gravity. A number of ideas have been considered on these lines. One framework which has been widely used both for numerical relativity and studies of quantum gravity is that of *Regge calculus*[26 △] in which one replaces smooth manifolds by spaces consisting of piecewise linear simplicial blocks. Naturally, other possibilities for discretised spacetime structure also exist – *causal sets*[28] being another example which has not yet been so widely explored. A further possibility is *topological quantisation* whereby one replaces a manifold by a set and quantises all topologies on that set[29,30].

⊛ The question of finiteness of superstring perturbation theory is a difficult technical question, which has still to be resolved – see, e.g., [15].

△ For a brief review and extensive bibliography see [27].

Many of the above alternatives fall into the category of being attempts to construct a fundamental theory of quantum gravity. The current quantum cosmology programme is not quite as ambitious. One begins by making the assumption that whatever the exact nature of the fundamental theory of quantum gravity is, in its semiclassical limit it should agree with the semiclassical limit of a canonical quantum formalism based on general relativity alone. Thus we study the semiclassical properties of quantum gravity based solely on Einstein's theory, or some suitable modification of it.

Clearly, any predictions made from such a foundation must be treated with caution. In particular, a new fundamental theory of quantum gravity might introduce radically new physical processes at an energy scale relevant for cosmology. String theory, for example, introduces its own fundamental scale which expressed as a temperature (the *Hagedorn temperature*) is given by:

$$T_{\text{Hagedorn}} = \frac{\sqrt{\hbar c}}{4\pi k_B \sqrt{\alpha'}}, \tag{1.1}$$

the constant α' being the Regge slope parameter, which is inversely proportional to the string tension. It is not known exactly what the value of α' is; however, if T_{Hagedorn} is comparable to or significantly lower than the Planck scale, then it is clear that fundamentally "stringy" processes will be very important in the very early universe, if string theory is indeed the "ultimate" theory.

Although new fundamental physics could drastically change the predictions of quantum cosmology, I believe nevertheless that studying quantum cosmology based even just on Einstein's theory is an important activity. General relativity is a remarkably successful theory; in seeking to replace it by something better it is important that we study processes at the limit of its applicability, thereby challenging our understanding. General relativity is limited by the Planck scale – the physical arenas in which this scale is approached include: (i) very small black holes; (ii) the very early universe. Since it seems unlikely that we will ever be able to create energies of order 10^{19} GeV in the laboratory, the consequences of quantum gravity for the physics of the very early universe will remain the one way of indirectly "testing" it, at least for the foreseeable future.

It is thus important that we consider quantum gravity in a cosmological context. Even if our current attempts do not fully reach the mark, in that we do not yet have a fully-fledged quantum theory of gravity, they nonetheless constitute a vital part of the process of trying to find such a theory.

1.2. A brief history of quantum cosmology

The quantum cosmology programme which I will describe in these lectures has gone through three main identifiable phases to date:

1. *Defining the problem.* The canonical formalism, including the definition of the wavefunction of the universe, Ψ, its configuration space – superspace – and its evolution according to the Wheeler-DeWitt equation, was set up in the late 1960s[31–35].

2. *Boundary conditions.* Quantum cosmology research went into something of a lull during the 1970s but was revived in the mid 1980s when the question of putting appropriate boundary conditions on the wavefunction of the universe was treated seriously. The idea is that such boundary conditions should describe the "creation of the universe from nothing"[36,37], where *nothing* means the absence of space and time. A number of proposals for such boundary conditions emerged – two major contenders being the *"no-boundary"* proposal of Hartle and Hawking[38,39] and the *"tunneling"* proposal advocated by Vilenkin[40–42].

3. *Quantum decoherence.* The mechanism of the transition from quantum physics to classical physics ("quantum decoherence") becomes vitally important when quantum physics is applied to the universe as a whole. The issues involved have begun to occupy many researchers in the early 1990s▼.

Two other important areas of quantum cosmology (or related) research have been: (i) quantum wormholes and "baby universes"; (ii) supersymmetric quantum cosmology.

Quantum wormholes* were extremely fashionable in the particle physics community in the years 1988–1990. Such states arise when one considers topology change in the path integral formulation of quantum gravity: *quantum wormholes* are instanton solutions which play an important role in the Euclidean path integral. One deals directly with a "third quantised" formalism, (i.e., quantum field theory over superspace), which includes operators that create and destroy universes – so-called *baby universes*. Much of the excitement in the late 1980s was associated with the idea that such processes could fix the fundamental constants of nature – in particular, driving the cosmological constant to zero.

Supersymmetric quantum cosmology has emerged recently as one of the most active areas of current research♠. In considering the quantum creation of the universe we are of course dealing with the very earliest epochs of the universe's existence, at which time it is believed that supersymmetry would not yet be broken. The inclusion of supersymmetry could therefore be vital from the point of view of physical consistency.

Since the focus of this School is on cosmology, my intention in these lectures is to cover topics 1 and 2 above, and then to proceed to discuss the predictions of quantum cosmology. The third topic of quantum decoherence raises questions which have not been resolved even in ordinary quantum mechanics, since the question of decoherence really amounts to understanding what happens during the "collapse of the wavefunction". Although this is a fascinating issue it has more to do with

▼ See [43] for a review and [44] for a collection of recent papers on the subject.

* See [45] for a review.

♠ See [46] and references therein.

the fundamentals of quantum mechanics than directly with cosmology. Likewise, I will only briefly touch upon quantum wormholes and supersymmetric quantum cosmology, as these areas are still in their infancy, and one is still at the point of trying to resolve basic questions concerning quantum gravity. I hope the reader will not be disappointed by this – however, given the vast scope of quantum gravity and quantum cosmology one must necessarily be rather selective.

2. Hamiltonian Formulation of General Relativity

2.1. The 3 + 1 decomposition

In order to discuss quantum cosmology a fair amount of technical machinery is required. In the canonical formulation we begin by making a 3 + 1-split of the 4-dimensional spacetime manifold, $\mathcal{M}$, which will describe the universe, foliating it into spatial hypersurfaces, Σ_t, labeled by a global time function, t. Thus we take the 4-dimensional metric to be given by♦

$$ds^2 = g_{\mu\nu}\mathrm{d}x^\mu \mathrm{d}x^\nu = -\omega^0 \otimes \omega^0 + h_{ij}\omega^i \otimes \omega^j, \tag{2.1}$$

where

$$\begin{aligned} \omega^0 &= \mathcal{N}\mathrm{d}t \\ \omega^i &= \mathrm{d}x^i + \mathcal{N}^i \mathrm{d}t. \end{aligned} \tag{2.2}$$

Such a decomposition is possible in general if the manifold $\mathcal{M}$ is *globally hyperbolic* [47]. The quantity $\mathcal{N}(t, x^k)$ is called the *lapse function* – it measures the difference between the coordinate time, t, and proper time, τ, on curves normal to the hypersurfaces Σ_t, the normal being $n_\alpha = (-\mathcal{N}, 0, 0, 0)$ in the above coordinates. The quantity $\mathcal{N}^i(t, x^k)$ is called the *shift vector* – it measures the difference between a spatial point, p, and the point one would reach if instead of following p from one hypersurface to the next one followed a curve tangent to the normal $\boldsymbol{n}$. That is to say, the spatial coordinates are "comoving" if $\mathcal{N}^i = 0$. Finally, $h_{ij}(t, x^k)$ is the *intrinsic metric* (or *first fundamental form*) induced on the spatial hypersurfaces by the full 4-dimensional metric, $g_{\mu\nu}$. In components we have

$$(g_{\mu\nu}) = \begin{bmatrix} -\mathcal{N}^2 + \mathcal{N}^k\mathcal{N}_k & \mathcal{N}_j \\ \mathcal{N}_i & h_{ij} \end{bmatrix}, \tag{2.3}$$

with inverse

$$(g^{\mu\nu}) = \begin{bmatrix} \frac{-1}{\mathcal{N}^2} & \frac{\mathcal{N}^j}{\mathcal{N}^2} \\ \frac{\mathcal{N}^i}{\mathcal{N}^2} & h^{ij} - \frac{\mathcal{N}^i\mathcal{N}^j}{\mathcal{N}^2} \end{bmatrix}, \tag{2.4}$$

where h^{ij} is the inverse to h_{ij}, and the intrinsic metric is used to lower and raise spatial indices: $\mathcal{N}^k\mathcal{N}_k \equiv h^{jk}\mathcal{N}_j\mathcal{N}_k = h_{jk}\mathcal{N}^j\mathcal{N}^k$ etc.

♦ We use a $(- + ++)$ Lorentzian metric signature, and natural units in which $c = \hbar = 1$.

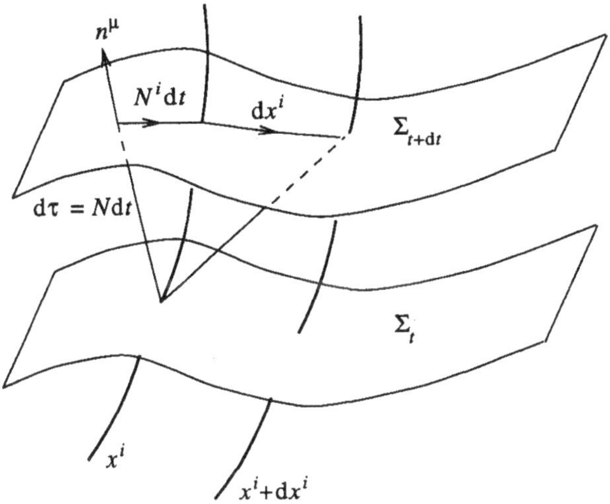

Fig. 1: The $3+1$ decomposition of the manifold, with lapse function, N, and shift vector, N^i.

One can construct an *intrinsic curvature* tensor ${}^3R^i{}_{jkl}(h)$ from the intrinsic metric alone – this of course describes the curvature intrinsic to the hypersurfaces Σ_t. One can also define an *extrinsic curvature*, (or *second fundamental form*), which describes how the spatial hypersurfaces Σ_t curve with respect to the 4-dimensional spacetime manifold within which they are embedded. This is given by

$$\begin{aligned} K_{ij} &\equiv - n_{i;j} = -\Gamma^0{}_{ij} n_0 \\ &= \frac{1}{2\mathcal{N}} \left(\mathcal{N}_{i|j} + \mathcal{N}_{j|i} - \frac{\partial h_{ij}}{\partial t} \right), \end{aligned} \tag{2.5}$$

where a semicolon denotes covariant differentiation with respect to the 4-metric, $g_{\mu\nu}$, and a vertical bar denotes covariant differentiation with respect to the 3-metric, h_{ij}: $\mathcal{N}_{i|j} \equiv \mathcal{N}_{i,j} - \Gamma^k{}_{ij}\mathcal{N}_k$ etc.

For a given foliation of $\mathcal{M}$ by spatial hypersurfaces, Σ_t, it is always possible to choose *Gaussian normal coordinates*, in which

$$ds^2 = -dt^2 + h_{ij} dx^i dx^j. \tag{2.6}$$

These are comoving coordinates ($\mathcal{N}^i = 0$) with the additional property that t is the proper time parameter ($\mathcal{N} = 1$). This is the standard "gauge choice" that is made in classical cosmology, and in such coordinates $K_{ij} = -\dot{h}_{ij}$, where dot denotes differentiation with respect to t. In making the $3+1$ decomposition, however, we are only free to make a specific choice of coordinates such as (2.6) *after* variation of the action if we want to be sure to obtain Einstein's equations, and thus we must retain the lapse and shift function for the time being.

2.2. The action

A relevant action for use in quantum cosmology is that of Einstein gravity plus a possible cosmological term, Λ, and matter, given by

$$S = \frac{1}{4\kappa^2}\left[\int_{\mathcal{M}} d^4x\sqrt{-g}\,({}^4R - 2\Lambda) + 2\int_{\partial\mathcal{M}} d^3x\sqrt{h}\,K\right] + S_{\text{matter}}, \tag{2.7}$$

where $\kappa^2 = 4\pi G = 4\pi m_{\text{Planck}}^{-2}$, $K \equiv K^i{}_i$ is the trace of the extrinsic curvature, and for many simple models the matter is specified by a single scalar field, Φ, with potential, $\mathcal{V}(\Phi)$,

$$S_{\text{matter}} = \int_{\mathcal{M}} d^4x\sqrt{-g}\left(-\frac{1}{2}g^{\mu\nu}\partial_\mu\Phi\partial_\nu\Phi - \mathcal{V}(\Phi)\right). \tag{2.8}$$

The boundary term[48] in (2.7) does not of course contribute to the classical field equations, and this term is usually omitted in a first course on general relativity. However, in quantum physics we are often interested in phenomena which occur when the classical field equations do not apply, that is "off-shell", and thus it is vitally important to retain the surface term. The simple matter action (2.8) given here should simply be seen as being representative of the type of matter action one might consider. Although the example given by (2.8) is sufficient for studying many inflationary universe models, many other alternatives might also be of interest, such as extra matter from a supergravity multiplet, or matter corresponding to the low-energy limit of string theory. In the latter case, if one works in the "string conformal frame" it is also necessary to alter the gravitational part of the action, as one characteristic of string theory is the presence of the scalar dilaton, Φ, which couples universally to matter (at least perturbatively). In that case (2.7), (2.8) would be replaced by

$$\begin{aligned} S = \frac{1}{4\kappa^2}\Bigg[\int_{\mathcal{M}} d^4x\sqrt{-g}e^{-2\Phi}\left({}^4R + 4g^{\mu\nu}\partial_\mu\Phi\partial_\nu\Phi - 8\mathcal{V}(\Phi) + \ldots\right) \\ + 2\int_{\partial\mathcal{M}} d^3x\sqrt{h}e^{-2\Phi}K\Bigg], \end{aligned} \tag{2.9}$$

where the ellipsis denotes any additional matter degrees of freedom, and we have allowed for the possibility of the generation of a dilaton potential, $\mathcal{V}(\Phi)$, via some non-perturbative symmetry breaking mechanism. However, for simplicity the specific examples we will deal with here will be confined to models of the type (2.7), (2.8).

We now wish to express (2.7), (2.8) in terms of the variables of the 3 + 1 split. One can show that

$$ {}^4R = {}^3R - 2R_{nn} + K^2 - K^{ij}K_{ij}, \tag{2.10} $$

where ${}^3R$ is the Ricci scalar of the intrinsic 3-geometry, and

$$ R_{nn} \equiv R_{\alpha\beta}n^\alpha n^\beta = -K^{ij}K_{ij} + K^2 + (n^\alpha K + a^\alpha)_{;\alpha}, \tag{2.11} $$

with $a^\alpha \equiv n^\beta n^\alpha{}_{;\beta}$. Combining (2.7), (2.10) and (2.11), and noting that the boundary integral involving a^α vanishes identically since $n^\alpha a_\alpha = 0$, we obtain

$$ S \equiv \int \mathrm{d}t\, L = \frac{1}{4\kappa^2}\int \mathrm{d}t\, \mathrm{d}^3x \mathcal{N}\sqrt{h}\left(K_{ij}K^{ij} - K^2 + {}^3R - 2\Lambda\right) + S_{\text{matter}}. \tag{2.12} $$

As in the Hamiltonian formulation of field theory we define canonical momenta in the standard fashion

$$ \pi^{ij} \equiv \frac{\delta L}{\delta \dot{h}_{ij}} = -\frac{\sqrt{h}}{4\kappa^2}\left(K^{ij} - h^{ij}K\right), \tag{2.13} $$

$$ \pi_\Phi \equiv \frac{\delta L}{\delta \dot{\Phi}} = \frac{\sqrt{h}}{\mathcal{N}}\left(\dot{\Phi} - \mathcal{N}^i\Phi_{,i}\right), \tag{2.14} $$

$$ \pi^0 \equiv \frac{\delta L}{\delta \dot{\mathcal{N}}} = 0, \tag{2.15} $$

$$ \pi^i \equiv \frac{\delta L}{\delta \dot{\mathcal{N}}_i} = 0. \tag{2.16} $$

The fact that the momenta conjugate to $\mathcal{N}$ and $\mathcal{N}_i$ vanish means that we are dealing with *primary constraints* in Dirac's terminology[49,50].

If we use $\mathcal{N}$, $\mathcal{N}_i$, h_{ij}, Φ and their conjugate momenta as the basic variables we obtain◇ a Hamiltonian

$$ \begin{aligned} H &\equiv \int \mathrm{d}^3x \left(\pi^0\dot{\mathcal{N}} + \pi^i\dot{\mathcal{N}}_i + \pi^{ij}\dot{h}_{ij} + \pi_\Phi\dot{\Phi}\right) - L \\ &= \int \mathrm{d}^3x \left(\pi^0\dot{\mathcal{N}} + \pi^i\dot{\mathcal{N}}_i + \mathcal{N}\mathcal{H} + \mathcal{N}_i\mathcal{H}^i\right), \end{aligned} \tag{2.17} $$

where

$$ \mathcal{H} = \frac{\sqrt{h}}{2\kappa^2}\left(G^{\hat{0}}{}_{\hat{0}} - 2\kappa^2 T^{\hat{0}}{}_{\hat{0}}\right) $$

◇ Details of all missing steps in this section will be provided upon request in a plain brown envelope.

$$= 4\kappa^2 \mathcal{G}_{ijkl}\pi^{ij}\pi^{kl} - \frac{\sqrt{h}}{4\kappa^2}\left({}^3R - 2\Lambda\right) + \tfrac{1}{2}\sqrt{h}\left[\frac{\pi_\Phi^{\,2}}{h} + h^{ij}\Phi_{,i}\,\Phi_{,j} + 2\mathcal{V}\right], \quad (2.18)$$

$$\mathcal{H}^i = \frac{\sqrt{h}}{2\kappa^2}\left(G^{\hat{0}\hat{\imath}} - 2\kappa^2 T^{\hat{0}\hat{\imath}}\right)$$
$$= -2\pi^{ij}{}_{|j} + h^{ij}\Phi_{,j}\,\pi_\Phi\,, \quad (2.19)$$

and

$$\mathcal{G}_{ijkl} = \tfrac{1}{2}\sqrt{h}\left(h_{ik}h_{jl} + h_{il}h_{jk} - h_{ij}h_{kl}\right) \quad (2.20)$$

is the *DeWitt metric*[33]. The hats in (2.18) and (2.19) denote orthonormal frame components of the Einstein and energy-momentum tensors. In terms of these variables the action (2.12) then becomes

$$S = \int \mathrm{d}t\,\mathrm{d}^3x\left(\pi^0\dot{\mathcal{N}} + \pi^i\dot{\mathcal{N}}_i - \mathcal{N}\mathcal{H} - \mathcal{N}_i\mathcal{H}^i\right). \quad (2.21)$$

If we vary (2.21) with respect to π^{ij} and π_Φ we obtain their defining relations (2.13) and (2.14). The lapse and shift functions now act as Lagrange multipliers; variation of (2.21) with respect to the lapse function, $\mathcal{N}$, yields the *Hamiltonian constraint*

$$\mathcal{H} = 0, \quad (2.22)$$

while variation of (2.21) with respect to the shift vector, $\mathcal{N}_i$, yields the *momentum constraint*

$$\mathcal{H}^i = 0. \quad (2.23)$$

From (2.18) and (2.19) it is clear that these constraints are simply the (00) and ($0i$) parts of Einstein's equations. In Dirac's terminology these are *secondary* or *dynamical constraints*.

The 3+1 decomposition of our spacetime looks to be very counterintuitive to the usual ideas of general relativity. This is so by choice. Einstein described the l.h.s. of his field equations which encodes the 4-dimensional geometry as a "hall of marble", which does not encourage people to tamper with it. However, to quantise spacetime we must do just that: we must deconstruct spacetime and replace it by something else. Thus far we have not done that – we have simply split our 4-dimensional manifold into a sequence of spatial hypersurfaces, Σ_t. Time is the natural variable to base this split upon since it plays a special role in quantum mechanics – it is a parameter rather than an operator.

Classically the evolution of one spatial hypersurface to the next is completely well-defined, (provided that the manifold, $\mathcal{M}$, is globally hyperbolic), and given initial data, h_{ij}, Φ, on an initial hypersurface, Σ, we can use the Cauchy development to stitch the hypersurfaces Σ_t together to recover the 4-dimensional manifold, $\mathcal{M}$. To quantise the theory, however, we want to perform a path integral over *all* geometries, not just the classically allowed ones. Thus we must consider sequences of geometries at the quantum level which *cannot* be stitched together in a regular Cauchy development to form a 4-manifold which solves Einstein's equations. (See Fig. 2.) We must therefore abandon Einstein's "hall of marble" – spacetime is no longer a fundamental object.

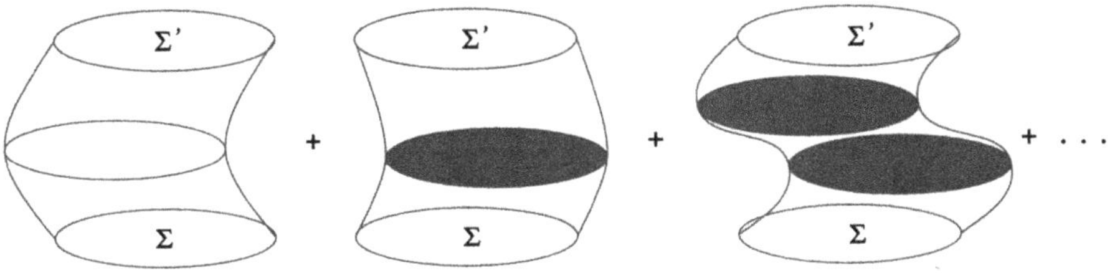

Fig. 2: Quantum geometrodynamics: in addition to the classical Cauchy development from Σ to Σ' (left), the path integral includes a sum over all 4-manifolds which interpolate between the initial and final configurations. The weighting by e^{iS} means that the greater the number of classically forbidden 3-geometries (shaded slices) is in the interpolating 4-manifold, the smaller its contribution is to the path integral.

As was mentioned in the Introduction the "deconstruction" of spacetime that we adopt here is probably the most conservative choice we could make. Even though we abandon the notion of spacetime in discussing the quantum dynamics of gravity, our fundamental objects are still defined on regular 3-manifolds, Σ. A more radical departure would be to replace these spatial hypersurfaces by some more general set by using, for example, the ideas of Regge calculus, causal sets or topological quantisation.

3. Quantisation

3.1. Superspace

As a prelude to the canonical quantisation of gravity let us first introduce the relevant configuration space on which the quantum dynamics will be defined.

Consider the space of all Riemannian 3-metrics and matter configurations on the spatial hypersurfaces, Σ,

$$\text{Riem}\,(\Sigma) \equiv \{h_{ij}(x), \Phi(x) \mid x \in \Sigma\}\,. \tag{3.1}$$

This is an infinite-dimensional space, on account of the label $x = \{x^i\}$, which specifies the point on the hypersurface, but there are a finite number of degrees of freedom at each point, $x \in \Sigma$. In fact, we are really interested in geometry here and configurations which can be related to each other by a diffeomorphism, i.e., a coordinate transformation, should be considered to be equivalent since their intrinsic geometry is the same. Thus we factor out by diffeomorphisms on the spatial hypersurfaces and identify *superspace*★ as

$$\frac{\mathrm{Riem}\,(\Sigma)}{\mathrm{Diff}_0(\Sigma)}\,,$$

where the subscript zero denotes the fact that we consider only diffeomorphisms which are connected to the identity. This infinite-dimensional space will be the basic configuration space of quantum cosmology.

The DeWitt metric (2.20) then provides a metric on superspace which we can write as

$$\mathcal{G}_{AB}(x) \equiv \mathcal{G}_{(ij)(kl)}(x), \tag{3.2}$$

where the indices A, B run over the independent components of the intrinsic metric h_{ij}:

$$A, B \in \{h_{11}, h_{12}, h_{13}, h_{22}, h_{23}, h_{33}\}.$$

The DeWitt metric has a $(-+++++)$ signature at each point $x \in \Sigma$, regardless of the signature of the spacetime metric, $g_{\mu\nu}$. To incorporate all the degrees of freedom, we also need to extend the range of the indices A, B to include the matter fields, by appropriately defining $\mathcal{G}_{\Phi\Phi}(x)$ (and other components if more than one matter field is present), thereby obtaining the full *supermetric*.

An inverse DeWitt metric, $\mathcal{G}^{AB}(x) = \mathcal{G}^{(ij)(kl)}(x)$, can be defined by the requirement

$$\mathcal{G}^{(ij)(kl)}\mathcal{G}_{(kl)(mn)} = \tfrac{1}{2}\left(\delta^i{}_m\delta^j{}_n + \delta^i{}_n\delta^j{}_m\right), \tag{3.3}$$

which gives

$$\mathcal{G}^{(ij)(kl)} = \tfrac{1}{2}\sqrt{h}\left(h^{ik}h^{jl} + h^{il}h^{jk} - 2h^{ij}h^{kl}\right). \tag{3.4}$$

3.2. Canonical quantisation

We will perform canonical quantisation by taking the wavefunction of the universe, $\Psi[h_{ij}, \Phi]$, to be a functional on superspace. Unlike the case of ordinary quantum mechanics, the wavefunction, Ψ, does not depend explicitly on time here. This is related to the fact that general relativity is an "already parametrised" theory – the original Einstein-Hilbert action is time-reparametrisation invariant. Time is contained implicitly in the dynamical variables, h_{ij} and Φ.

According to Dirac's quantisation procedure[49] we make the following replacements for the canonical momenta

$$\pi^{ij} \to -i\frac{\delta}{\delta h_{ij}}, \quad \pi_\Phi \to -i\frac{\delta}{\delta \Phi}, \quad \pi^0 \to -i\frac{\delta}{\delta \mathcal{N}}, \quad \pi^i \to -i\frac{\delta}{\delta \mathcal{N}_i}\,. \tag{3.5}$$

★ The use of the terminology "superspace" for the configuration space of quantum cosmology predates the discovery of supersymmetry, and of course is completely different from the combined manifold of commuting and anticommuting coordinates which is called "superspace" in supersymmetric theories.

and then demand that the wavefunction is annihilated by the operator version of the constraints. For the primary constraints we have

$$\begin{aligned}\hat{\pi}\Psi &= -i\frac{\delta\Psi}{\delta\mathcal{N}} = 0,\\ \hat{\pi}^i\Psi &= -i\frac{\delta\Psi}{\delta\mathcal{N}_i} = 0,\end{aligned} \tag{3.6}$$

which implies that Ψ is independent of $\mathcal{N}$ and $\mathcal{N}^i$. The momentum constraint yields

$$\hat{\mathcal{H}}^i\Psi = 0 \qquad \Rightarrow \qquad i\left[\frac{\delta\Psi}{\delta h_{ij}}\right]_{|j} = 2\kappa^2\hat{\mathcal{T}}^{\hat{0}\hat{\imath}}\Psi. \tag{3.7}$$

Using (3.7) one can show[51] that Ψ is the same for configurations $\{h_{ij}(x), \Phi(x)\}$ which are related by a coordinate transformation in the spatial hypersurface, Σ, which accords with the rationale for factoring out by diffeomorphisms in our definition of superspace. Finally, the Hamiltonian constraint yields

$$\hat{\mathcal{H}}\Psi = \left[-4\kappa^2\mathcal{G}_{ijkl}\frac{\delta^2}{\delta h_{ij}\delta h_{kl}} + \frac{\sqrt{h}}{4\kappa^2}\left(-{}^3R + 2\Lambda + 4\kappa^2\hat{\mathcal{T}}^{\hat{0}\hat{0}}\right)\right]\Psi = 0, \tag{3.8}$$

where for our scalar field example

$$\hat{\mathcal{T}}^{\hat{0}\hat{0}} = \frac{-1}{2h}\frac{\delta^2}{\delta\Phi^2} + \tfrac{1}{2}h^{ij}\Phi_{,i}\Phi_{,j} + \mathcal{V}(\Phi). \tag{3.9}$$

Eq. (3.8) is known as the *Wheeler-DeWitt equation*[32,33]. In fact, it is not a single equation but is actually one equation at each point, $x \in \Sigma$. It is a second-order hyperbolic functional differential equation on superspace. On account of factor-ordering ambiguities it is not completely well-defined, although there exist "natural" choices[33,52] of ordering for which the derivative pieces become a Laplacian in the supermetric.

3.3. Path integral quantisation

An alternative to canonical quantisation which perhaps better accommodates an intuitive understanding (c.f. Fig. 2) of the quantisation procedure is the path integral approach. Path integral techniques in quantum gravity were pioneered in the late 1970s[53,54]. The starting point for this is Feynman's idea that the amplitude to go from one state with an intrinsic metric, h_{ij}, and matter configuration, Φ, on an initial hypersurface, Σ, to another state with metric, h'_{ij} and matter configuration Φ' on an final hypersurface, Σ', is given by a functional integral of e^{iS} over all 4-geometries, $g_{\mu\nu}$, and matter configurations, ϕ, which interpolate between the initial and final configurations

$$\langle h'_{ij}, \Phi', \Sigma' | h_{ij}, \Phi, \Sigma\rangle = \sum_{\mathcal{M}} \int \mathfrak{D}g\,\mathfrak{D}\phi\, e^{iS[g_{\mu\nu},\phi]}. \tag{3.10}$$

In ordinary quantum field theory in flat spacetime the path integral suffers from the difficulty that since the action $S[g_{\mu\nu}, \phi]$ is finite the integral oscillates and therefore fails to converge. Furthermore, the solution which extremises the action is given by solving a hyperbolic equation between initial and final boundary surfaces – which is not a mathematically well-posed problem, and may have either no solutions or an infinite number of solutions. To deal with this problem one performs a "Wick rotation" to "imaginary time" $t \to -i\tau$ and considers a path integral formulated in terms of the Euclidean action, $I = -iS$. The action is then positive-definite, so that the path integral is exponentially damped and should converge. Also the problem of finding the extremum becomes that of solving an elliptic equation with boundary conditions, and this is well-posed.

One may attempt to apply a similar approach to quantum gravity, replacing S in (3.10) by the Euclidean action⊞ $I[g_{\mu\nu}, \phi] = -iS[g_{\mu\nu}, \phi]$, and taking the sum in (3.10) to be over all metrics with signature $(++++)$, which induce the appropriate 3-metrics h_{ij} and h'_{ij} on the past and future hypersurfaces. This approach to quantum gravity has had some important successes – most notably, it provides: (i) an elegant way of deriving the thermodynamic properties of black holes[55–57]; and (ii) a natural means for discussing the effects of gravitational instantons[58–60]. Gravitational instantons have been found to provide the dominant contribution to the path integral in processes such as pair creation of charged black holes in a magnetic field[61–63], and therefore provide a means of calculating the rates of such processes semiclassically.

The problems associated with the Euclidean approach to quantum gravity are considerable, however. Firstly, unlike ordinary field theories• such as Yang-Mills theory the gravitational action is not positive-definite[64], and thus the path integral does not converge if one restricts the sum to real Euclidean-signature metrics. To make the path integral converge it is necessary to include complex metrics in the sum[64,65]. However, there is no unique contour to integrate along in superspace[66–68] and the result one obtains for the path integral may depend crucially on the contour that is chosen[66]. Furthermore, the measure in (3.10) is ill-defined. It is really only in the last ten years that mathematicians have succeeded in making path integration in ordinary quantum field theory rigorously defined. Clearly, we may have to wait some time before the same can be said of path integrals in quantum gravity.

The physicists' approach is to set aside the issues involved in making the formalism rigorous and to see what can be learned nevertheless. We thus take the

---

⊞ Strictly speaking one should call this the Riemannian action, since "Euclidean" spaces are those which are *flat*, whereas *curved* manifolds with $(++++)$ signature are known as *Riemannian spaces*. However, the terminology "Euclidean" action which has been taken over from flat space quantum field theory seems to have stuck, despite the fact that we are of course no longer dealing with $\mathbb{R}^4$.

• The action for fermi fields in ordinary quantum field theory is not positive-definite, but that is not a problem since one can treat them as anticommuting quantities so that the path integral over them converges.

wavefunction Ψ of the universe on a hypersurface, Σ, with intrinsic 3-metric, h_{ij}, and matter configuration, Φ, to be defined[38,39] by the functional integral

$$\Psi\left[h_{ij},\Phi,\Sigma\right]=\sum_{\mathcal{M}}\int \mathfrak{D}g\,\mathfrak{D}\phi\, e^{-I[g_{\mu\nu},\phi]}. \tag{3.11}$$

where the sum is over a class of 4-metrics, $g_{\mu\nu}$, and matter configurations, ϕ, which take values h_{ij} and Φ on the boundary Σ. Alternative definitions of the wavefunction have been proposed. In particular, Linde[69] has argued that one should Wick rotate with the opposite sign, i.e., $t \to +i\tau$ instead of $t \to -i\tau$ as above, leading to a factor e^{+I} instead of e^{-I} in (3.11). Alternatively, one could stick with a Lorentzian path integral[70], with e^{iS} instead of e^{-I} in (3.11). In any case, in order to achieve convergence of the path integral[⊖] it is necessary to include complex manifolds in the sum, which somewhat obscures the distinctions between these alternative proposed definitions of Ψ. The real distinction between the alternative definitions arises when it comes to imposing boundary conditions, thereby restricting the 4-manifolds included in the sum in (3.11). For example, one could view the Euclidean sector as being the appropriate sector of the quantum theory in which an "initial" boundary condition on Ψ should be imposed, which would make (3.11) the natural starting point, as is the case for the no-boundary proposal[38,39]. Alternative boundary conditions would favour the Lorentzian path integral[70].

The path integral definition of the wavefunction (3.11) is consistent with the earlier definition based on canonical quantisation to the extent that wavefunctions defined according to (3.11) can be shown to satisfy the Wheeler-DeWitt equation (3.8) provided that the action, the measure and the class of paths summed over are invariant under diffeomorphisms[71].

In the canonical quantisation formalism any particular solution to the Wheeler-DeWitt equation will depend upon the specification of boundary conditions on the wavefunction. In the path integral formulation the particular solution of the Wheeler-DeWitt equation that one obtains will similarly depend on the contour of integration chosen in superspace, and the class of 4-metrics one sums over in (3.11). Unfortunately it is not known how the choice of contour and class of paths prescribes the boundary conditions on the wavefunction in the general case, although it can be answered for specific models. The question of boundary conditions is naturally of prime importance for cosmology, and we shall return to this question in §4.

3.4. Minisuperspace

In practice to work with the infinite dimensions of the full superspace is not possible, at least with the techniques that have been developed to date. One useful

⊖ Linde's suggested modification [69] to (3.11) yields a convergent path integral for the scale factor, which is all that one needs in the simplest minisuperspace models, but does not lead to convergence if one includes matter or inhomogeneous modes of the metric.

approximation therefore is to truncate the infinite degrees of freedom to a finite number, thereby obtaining some particular *minisuperspace* model. An easy way to achieve this is by considering homogeneous metrics, since as was observed earlier for each point $x \in \Sigma$ there are a finite number of degrees of freedom in superspace.

The results we shall obtain by this approach will be somewhat satisfying in that they do appear to have some predictive power. However, the truncation to minisuperspace has not as yet been made part of a rigorous approximation scheme to full superspace quantum cosmology. As they are currently formulated minisuperspace models should therefore be viewed as toy models, which we nonetheless hope will capture some of the essence of quantum cosmology. Since we are simultaneously setting most of the field modes and their conjugate momenta to zero, which violates the uncertainty principle, this approach might seem rather *ad hoc*. However, in classical cosmology homogeneity and isotropy are important simplifying assumptions which do have a sound observational basis. Therefore it is not entirely unreasonable to hope that a consistent truncation to particular minisuperspace models with particular symmetries might be found in future♣.

Let us thus consider *homogeneous* cosmologies for simplicity. Instead of having a separate Wheeler-DeWitt equation for each point of the spatial hypersurface, Σ, we then simply have a single Wheeler-DeWitt equation for all of Σ. The standard FRW metrics, with

$$h_{ij}\mathrm{d}x^i\mathrm{d}x^j = a^2(t)\left[\frac{\mathrm{d}r^2}{1-kr^2} + r^2\left(\mathrm{d}\theta^2 + \sin^2\theta\mathrm{d}\varphi^2\right)\right], \qquad k = -1, 0, 1, \qquad (3.12)$$

are of course one special example. In that case our model with a single scalar field simply has two minisuperspace coordinates, $\{a, \Phi\}$, the cosmic scale factor and the scalar field. Many more general homogeneous (but anisotropic) models can also be considered. Indeed all such models can be classified◁ and apart from the FRW models the other cases of interest are: (i) the *Kantowski-Sachs models*[77,78], which have a 3-metric

$$h_{ij}\mathrm{d}x^i\mathrm{d}x^j = a^2(t)\mathrm{d}r^2 + b^2(t)\left(\mathrm{d}\theta^2 + \sin^2\theta\mathrm{d}\varphi^2\right), \qquad (3.13)$$

and thus three minisuperspace coordinates, $\{a, b, \Phi\}$; and (ii) the Bianchi models.

The *Bianchi models* are the most general homogeneous cosmologies with a 3-dimensional group of isometries. These groups are in a one–to–one correspondence with 3-dimensional Lie algebras, which were classified long ago by Bianchi[79]. There are nine distinct 3-dimensional Lie algebras, and consequently nine types of Bianchi cosmology. The 3-metric of each of these models can be written in the form

$$h_{ij}\mathrm{d}x^i\mathrm{d}x^j = h_{ij}(t)\omega^i \otimes \omega^j, \qquad (3.14)$$

♣ For some discussions of the validity of the minisuperspace approximation see [72–75].
◁ See [76] for a review.

where ω^i are the invariant 1-forms associated associated with the isometry group. The simplest example is Bianchi I, which corresponds to the Lie Group $\mathbb{R}^3$. In that case we can choose $\omega^1 = \mathrm{d}x$, $\omega^2 = \mathrm{d}y$, and $\omega^3 = \mathrm{d}z$, so that

$$h_{ij}\mathrm{d}x^i\mathrm{d}x^j = a^2(t)\mathrm{d}x^2 + b^2(t)\mathrm{d}y^2 + c^2(t)\mathrm{d}z^2, \tag{3.15}$$

and the minisuperspace coordinates are $\{a, b, c, \Phi\}$. If we take the spatial directions to be compact such a universe will have toroidal topology. In the special case that $a(t) = b(t) = c(t)$ we retrieve the spatially flat ($k = 0$) FRW universe.

The most complicated, and possibly the most interesting, Bianchi model is Bianchi IX, associated to the Lie group $SO(3, \mathbb{R})$. In this case the invariant 1-forms may be written as

$$\begin{aligned} \omega^1 &= -\sin\psi\, \mathrm{d}\theta + \sin\theta\cos\psi\, \mathrm{d}\varphi, \\ \omega^2 &= \cos\psi\, \mathrm{d}\theta + \sin\theta\sin\psi\, \mathrm{d}\varphi, \\ \omega^3 &= \cos\theta\, \mathrm{d}\varphi + \mathrm{d}\psi, \end{aligned} \tag{3.16}$$

in terms of the Euler angles, (ψ, θ, φ), on the 3-sphere, S^3. The spatial sections of the geometry resulting from (3.14), (3.16) have the topology of S^3, but are only spherically symmetric in the special case that $h_{11}(t) = h_{12}(t) = \ldots h_{33}(t)$, which corresponds to the $k = +1$ FRW universe. Geometrically the spatial hypersurfaces can thus be thought of as squashed, twisted 3-spheres [see Fig. 3]. Bianchi IX has played an important role in classical cosmological studies of the initial singularity – it is the basis of the so-called "mixmaster universe"[80,81]. As a classical dynamical system Bianchi IX is extremely interesting because it appears to be chaotic, but only just on the verge of being so. Over the years there has been much debate as to whether Bianchi IX is or is not chaotic, and this seems to have been recently resolved by an explicit demonstration that it is not integrable■[82]. The corresponding minisuperspace model will have six independent coordinates in addition to the scalar field coordinate.

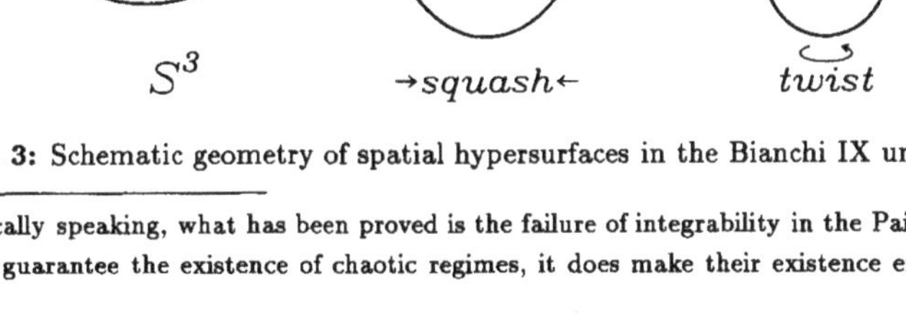

Fig. 3: Schematic geometry of spatial hypersurfaces in the Bianchi IX universe.

■ Technically speaking, what has been proved is the failure of integrability in the Painlevé sense. While this does not guarantee the existence of chaotic regimes, it does make their existence extremely probable.

Let us now consider the minisuperspace corresponding to an arbitrary homogeneous cosmology. We will assume that the minisuperspace is of dimension n, and will denote the minisuperspace coordinates by $\{q^A\}$. Since $\mathcal{N}^i = 0$ by assumption, it follows from the definitions (2.5) and (3.4) that

$$\mathcal{G}^{ijkl}\dot{h}_{ij}\dot{h}_{kl} = 4\sqrt{h}\mathcal{N}^2\left(K_{ij}K^{ij} - K^2\right), \tag{3.17}$$

and consequently the Lorentzian action (2.12) now takes the form

$$S = \int \mathrm{d}t \left[\frac{1}{2\mathcal{N}}\mathcal{G}_{AB}(q)\dot{q}^A\dot{q}^B - \mathcal{N}\mathcal{U}(q)\right], \tag{3.18}$$

where

$$\mathcal{G}_{AB}\mathrm{d}q^A\mathrm{d}q^B = \int \mathrm{d}^3x \left[\frac{1}{8\kappa^2}\mathcal{G}^{ijkl}\delta h_{ij}\delta h_{kl} + \sqrt{h}\delta\Phi\delta\Phi\right], \tag{3.19}$$

is the *minisupermetric*, which is now of finite dimension, n, and

$$\mathcal{U} = \int \mathrm{d}^3x\sqrt{h}\left[\frac{1}{4\kappa^2}\left(-{}^3R + 2\Lambda\right) + \mathcal{V}(\Phi)\right]. \tag{3.20}$$

The action (3.18) is simply equivalent to that for a "point particle" moving in a "potential" $\mathcal{U}$. Variation of (3.18) with respect to q^A thus yields a geodesic equation with force term,

$$\frac{1}{\mathcal{N}}\frac{d}{dt}\left(\frac{\dot{q}^A}{\mathcal{N}}\right) + \frac{1}{\mathcal{N}^2}\Gamma^A{}_{BC}\dot{q}^B\dot{q}^C = -\mathcal{G}^{AB}\frac{\partial\mathcal{U}}{\partial q^B}, \tag{3.21}$$

where $\Gamma^A{}_{BC}$ are the Christoffel symbols determined from the minisupermetric, while variation of (3.18) with respect to $\mathcal{N}$ yields the Hamiltonian constraint

$$\frac{1}{2\mathcal{N}^2}\mathcal{G}_{AB}(q)\dot{q}^A\dot{q}^B + \mathcal{U}(q) = 0. \tag{3.22}$$

The general solution to (3.21), (3.22) will have $2n - 1$ independent parameters, one of which is always trivial in the sense that it corresponds to a choice of origin of the time parameter. In studying any particular minisuperspace model we must take care to check that what we have done above is consistent, as it does not always follow that substituting a particular ansatz into an action before varying it will yield the same result as substituting the same ansatz into the field equations obtained from variation of the original action. Eqs. (3.21) and (3.22) should correspond respectively to the (ij) and (00) components of the original Einstein equations, while the $(0i)$ equation is trivially satisfied in the present case.

Quantisation is greatly simplified because now that our configuration space is finite-dimensional we are effectively dealing with the quantum *mechanics* of a constrained system. The canonical momenta and Hamiltonian are respectively given by

$$\pi_A = \frac{\partial L}{\partial \dot{q}^A} = \frac{\mathcal{G}_{AB}\dot{q}^B}{\mathcal{N}}, \tag{3.23}$$

$$H = \pi_A \dot{q}^A - L = \mathcal{N}\left[\tfrac{1}{2}\mathcal{G}^{AB}\pi_A\pi_B + \mathcal{U}(q)\right] \equiv \mathcal{N}\mathrm{H}. \tag{3.24}$$

The π_A are related to the canonical momenta (2.13)–(2.16) defined earlier by integration over the 3-volume of the hypersurfaces of homogeneity in (3.19). In terms of the new variables the action (3.18) and Hamiltonian constraint (3.22) are respectively

$$S = \int \mathrm{d}t \left[\pi_A \dot{q}^A - \mathcal{N}\mathrm{H}\right], \tag{3.25}$$

$$\tfrac{1}{2}\mathcal{G}^{AB}\pi_A\pi_B + \mathcal{U}(q) = 0. \tag{3.26}$$

Under canonical quantisation (3.26) yields the Wheeler-DeWitt equation

$$\hat{\mathrm{H}}\Psi = \left[-\tfrac{1}{2}\nabla^2 + \mathcal{U}(q)\right]\Psi = 0, \tag{3.27}$$

where

$$\nabla \equiv \frac{1}{\sqrt{-\mathcal{G}}}\partial_A\left[\sqrt{-\mathcal{G}}\,\mathcal{G}^{AB}\partial_B\right] \tag{3.28}$$

is the Laplacian operator of the minisupermetric. In arriving at (3.27) we have made an explicit "natural choice" of factor ordering[33,52] in order to accommodate the factor-ordering ambiguity. This choice is favoured by independent minisuperspace calculations of the prefactor, using zeta function regularisation and a scale invariant measure, which can then be related to factor ordering dependent terms through a semiclassical expansion of the Wheeler-DeWitt equation[83].

An alternative "natural choice"[35,84–86] of factor ordering would yield a *conformally invariant* Wheeler-DeWitt equation,

$$\hat{\mathrm{H}}\Psi = \left[-\tfrac{1}{2}\nabla^2 + \frac{n-2}{8(n-1)}\mathcal{R} + \mathcal{U}(q)\right]\Psi = 0, \tag{3.29}$$

where $\mathcal{R}$ is the scalar curvature obtained from the minisupermetric.

3.5. The WKB approximation

In view of the difficulties associated with solving the Wheeler-DeWitt equation in general, the best we can realistically hope for in many minisuperspace models is to look for appropriate approximate solutions in the semiclassical limit[⌘], in which

$$\Psi \simeq \sum_n \Psi_n \equiv \sum_n A_n e^{-I_n}, \tag{3.30}$$

[⌘] The semiclassical limit of the full superspace Wheeler-DeWitt equation has been treated more formally by a number of authors. See, e.g., [87], [88] and references therein.

where the sum is over saddle points of the path integral, the $\mathcal{A}_n$ being appropriate (possibly complex) prefactors. In general we might expect to find regions in which the wavefunction is exponential, $\Psi \simeq e^{-I}$, and regions in which it is oscillatory, $\Psi \simeq e^{iS}$. The latter could be viewed as the wavefunction of a universe in the classical "Lorentzian" or "oscillatory" region, while the former would correspond to a universe in a classically inaccessible "Euclidean" or "tunneling" region. As has already been mentioned, the sum in (3.30) will in general contain a number of saddle points with an action, I_n, which is neither purely real nor purely imaginary.

Our own universe is of course Lorentzian at late times, and therefore the only minisuperspace models which can be of direct physical relevance are those for which the Wheeler-DeWitt equation does possess approximate solutions of the oscillatory type. Approximate solutions of this type can be obtained by performing a WKB expansion, for which purpose it is necessary to restore $\hbar$ in the minisuperspace Wheeler-DeWitt equation (3.27). If we assume that each component Ψ_n satisfies (3.27) separately, then

$$\begin{aligned} 0 = \hat{\mathrm{H}}\Psi_n &= \left[-\tfrac{1}{2}\hbar^2\nabla^2 + \mathcal{U}\right]\mathcal{A}_n e^{-I_n/\hbar} \\ &= e^{-I_n/\hbar}\left\{\left[-\tfrac{1}{2}\left(\boldsymbol{\nabla} I_n\right)^2 + \mathcal{U}\right]\mathcal{A}_n + \hbar\left[\boldsymbol{\nabla} I_n\cdot\boldsymbol{\nabla}\mathcal{A}_n + \tfrac{1}{2}\mathcal{A}_n\nabla^2 I_n\right] + \mathrm{O}(\hbar^2)\right\}, \end{aligned} \tag{3.31}$$

where the dot implies contraction with the minisupermetric $\mathcal{G}^{AB}$. The $\mathrm{O}(\hbar^0)$ and $\mathrm{O}(\hbar)$ terms give two equations for I_n and $\mathcal{A}_n$. If we decompose I_n into real and imaginary parts according to $I_n = \mathrm{R}_n - i\mathrm{S}_n$ then the real and imaginary parts of the $\mathrm{O}(\hbar^0)$ term in (3.31) give

$$-\tfrac{1}{2}\left(\boldsymbol{\nabla}\mathrm{R}_n\right)^2 + \tfrac{1}{2}\left(\boldsymbol{\nabla}\mathrm{S}_n\right)^2 + \mathcal{U} = 0, \tag{3.32}$$

$$\boldsymbol{\nabla}\mathrm{R}_n \cdot \boldsymbol{\nabla}\mathrm{S}_n = 0. \tag{3.33}$$

Provided that the imaginary part of the action varies much more rapidly than the real part, i.e., $(\boldsymbol{\nabla}\mathrm{R}_n)^2 \ll (\boldsymbol{\nabla}\mathrm{S}_n)^2$, then (3.32) is the Lorentzian Hamilton-Jacobi equation for S_n:

$$\tfrac{1}{2}\mathcal{G}^{AB}\frac{\partial \mathrm{S}_n}{\partial q^A}\frac{\partial \mathrm{S}_n}{\partial q^B} + \mathcal{U}(q) = 0. \tag{3.34}$$

Comparison of (3.34) with (3.27) suggests a strong correlation between coordinates and momenta, and invites the identification

$$\pi_A = \frac{\partial \mathrm{S}_n}{\partial q^A}. \tag{3.35}$$

If we differentiate (3.34) w.r.t. q^C we obtain

$$\tfrac{1}{2}\mathcal{G}^{AB}{}_{,C}\frac{\partial \mathrm{S}_n}{\partial q^A}\frac{\partial \mathrm{S}_n}{\partial q^B} + \mathcal{G}^{AB}\frac{\partial \mathrm{S}_n}{\partial q^A}\frac{\partial^2 \mathrm{S}_n}{\partial q^B \partial q^C} + \frac{\partial \mathcal{U}}{\partial q^C} = 0. \tag{3.36}$$

If we define a minisuperspace vector field

$$\frac{\mathrm{d}}{\mathrm{d}s} \equiv \mathcal{G}^{AB}\frac{\partial S_n}{\partial q^A}\frac{\partial}{\partial q^B}, \tag{3.37}$$

then combining (3.35), (3.36) and (3.37) we obtain

$$\frac{\mathrm{d}\pi_C}{\mathrm{d}s} + \tfrac{1}{2}\mathcal{G}^{AB}{}_{,C}\,\pi_A\pi_B + \frac{\partial \mathcal{U}}{\partial q^C} = 0, \tag{3.38}$$

which after raising indices is the same geodesic equation (3.21) obtained earlier provided we identify the parameter, s, with the proper time on the geodesics.

We can now solve the equation given by the $\mathrm{O}(\hbar)$ term of (3.31). Since $|\nabla \mathrm{R}_n| \ll |\nabla \mathrm{S}_n|$ it follows that the terms involving R_n can be neglected, and thus

$$\nabla \mathrm{S}_n \cdot \nabla \mathcal{A}_n \equiv \mathcal{G}^{AB}\frac{\partial \mathrm{S}_n}{\partial q^A}\frac{\partial \mathcal{A}_n}{\partial q^B} \equiv \frac{\mathrm{d}\mathcal{A}_n}{\mathrm{d}s} = -\tfrac{1}{2}\mathcal{A}_n\nabla^2\mathrm{S}_n, \tag{3.39}$$

which may be readily integrated. We thus obtain a first-order WKB wavefunction

$$\Psi_n = \mathcal{C}_n \exp\left(i\mathrm{S}_n - \tfrac{1}{2}\int \mathrm{d}s\,\nabla^2\mathrm{S}_n\right), \tag{3.40}$$

where $\mathcal{C}_n$ is an arbitrary (complex) constant to be appropriately normalised, and we have reverted to natural units in which $\hbar = 1$.

The wavefunction (3.40) could be considered to be the analogue of the wavefunction for coherent states in ordinary quantum mechanics,

$$\psi_n(x,t) = c_n e^{ip_n x}\exp\left(\frac{-(x-\bar{x}_n(t))^2}{\sigma^2}\right), \tag{3.41}$$

which describes a wave packet which is "peaked" about a classical particle trajectory, $\bar{x}_n(t)$, and which thus roughly "predicts" classical behaviour. This becomes problematic, however, if we consider a superposition, $\psi = \sum_n \psi_n$, of such states since interference between different wave packets will in general destroy the classical behaviour. In order to interpret the total wavefunction as saying that the particle follows a roughly classical trajectory, $\bar{x}_n(t)$, with probability $|c_n|^2$, it is necessary that a *decoherence* mechanism should exist which renders this quantum mechanical interference negligible[89].

The issue of quantum decoherence is clearly also of great importance in quantum cosmology, since in order to interpret Ψ in (3.30) in a similar fashion a similar mechanism must exist. It has been recently argued, furthermore, that decoherence is a necessary feature of the WKB interpretation of quantum cosmology, since without decoherence the existence of chaotic cosmological solutions would lead to a breakdown of the WKB approximation[90]. This is analogous to similar problems with the commutativity of the limits $t \to \infty$ and $\hbar \to 0$ in ordinary quantum mechanics when applied to chaotic systems.

The issues involved in decoherence pose complex conceptual questions for the fundamentals of quantum mechanics itself, quite apart from the problems specific to quantum cosmology[⊠]. Here we will merely assume that such a mechanism exists,

⊠ For further details see, e.g., [91]–[94] and references therein.

and we will take the view that Ψ can be considered to "predict" a classical spacetime if there exist WKB-type solutions (3.40), which yield a strong correlation between π^A and q^A according to (3.35). The sense in which the minisuperspace positions and momenta are "strongly correlated" can be made more precise through the use of quantum distribution functions, such as the Wigner function[91,95]. By use of the Wigner function one may show[95] that wavefunctions of the oscillatory type, $\Psi \sim e^{iS}$, predict a strong correlation between coordinates and momenta, whereas wavefunctions of the type $\Psi \sim e^{-I}$, which are also typical minisuperspace solutions, do not. Such exponential wavefunctions can thus be considered as describing universes in a purely quantum "tunneling" regime, before the quantum to classical transition. We will interpret wavefunctions, $\Psi \sim e^{iS}$, as corresponding to classical spacetime, or rather a set of classical spacetimes as S is a first integral of the equations of motion.

3.6. Probability measures

Given a solution, Ψ, to the Wheeler-DeWitt equation it is necessary to construct a probability measure in order to make predictions. One central question in quantum cosmology is how one should construct such a measure.

The minisuperspace Wheeler-DeWitt equation (3.27) is a second-order equation very much like the Klein-Gordon equation in ordinary field theory, and it readily follows from (3.27) that the current[33]

$$\mathcal{J} = -\tfrac{1}{2}i\left(\overline{\Psi}\nabla\Psi - \Psi\nabla\overline{\Psi}\right) \tag{3.42}$$

is conserved: $\nabla \cdot \mathcal{J} = 0$. The similarity to the Klein-Gordon current extends to the fact that the natural inner product[33] constructed from $\mathcal{J}$ is not positive-definite and so gives rise to negative probabilities. In quantum field theory this is not a problem since one can split the wavefunction up into positive and negative frequency components which correspond to particles and anti-particles. However, as has already been mentioned there is no well-defined notion of positive frequencies in superspace on account of its lack of symmetries[96]. A further problem is that many natural wavefunctions would have zero norm with this definition. For example, the no-boundary wavefunction is real and gives $\mathcal{J} = 0$.

The similarity of Ψ to the Klein-Gordon field has suggested to many people that one should turn Ψ into an operator, $\hat{\Psi}$, thereby introducing quantum field theory on superspace, or "third quantisation". One then arrives at operators which create and annihilate universes. However, as we do not perform measurements over a statistical ensemble of universes it is not clear how we can arrive at sensible probabilities using such a formulation.

The difficulties with the Klein-Gordon current of course led Dirac to introduce the Dirac equation, and it is worth mentioning that a similar resolution of the problem is available in supersymmetric quantum cosmology. In particular, one can go to a theory which includes fermionic variables by considering quantum cosmology based on supergravity◄ rather than the purely bosonic Einstein theory. The constraints of supergravity, which may be viewed as the Dirac square root of the constraints of general relativity[98,99], are reducible to first-order equations. Furthermore, this also translates into simplifications in homogeneous minisuperspace models – the appropriate constraint equation which determines the quantum evolution of the wavefunction can be considered to be the Dirac square root of the Wheeler-DeWitt equation[100–103]. As a result it is possible to construct[100] a Dirac-type probability density which is conserved by the equation $\mathcal{Q}\Psi = 0$, where $\mathcal{Q}$ is the supercharge.

Another alternative to the question of the probability measure is to use $|\Psi|^2$ directly as a probability measure[52,104], by defining the probability of the universe being in a region, $\mathcal{A}$, of superspace by

$$\mathcal{P}(\mathcal{A}) \propto \int_{\mathcal{A}} |\Psi|^2 \, {}^*1 \tag{3.43}$$

where ^{*1} is the volume-element on superspace, $*$ being the Hodge dual in the supermetric. This definition of a necessarily positive-definite probability density works very well for homogeneous minisuperspaces, for which the volume form ^{*1} is independent of $x \in \Sigma$. This is perhaps not surprising since as was observed in §3.4 the assumption of homogeneity reduces the problem to one of quantum mechanics, and $|\Psi|^2$ is of course the probability density in conventional quantum mechanics.

Problems with the definition (3.43) do arise since even in some simple examples the wavefunction is not normalisable, but instead $\langle\Psi|\Psi\rangle = \infty$. One further problem is that whereas in ordinary quantum mechanics $|\Psi|^2$ describes a probability density in configuration space – i.e., the space of particle positions – in quantum cosmology the configuration space is (mini)superspace and time is implicitly contained in the (mini)superspace coordinates. These coordinates cannot therefore be thought of as the mere analogues of particle positions. As a result the recovery of the conservation of probability and the standard interpretation of the quantum mechanics for small subsystems is not necessarily straightforward in the approach based on (3.43), and may involve understanding some subtle questions about the role of time in quantum gravity.

Ultimately a formulation such as (3.43) which is based on absolute probabilities may not be required since it is impossible to measure statistical ensembles of universes and thus all we can really test are *conditional probabilities* rather than absolute probabilities. For example, a relevant testable probability might be the probability, $\mathcal{P}(\mathcal{A}|\mathcal{B})$, of finding Ψ in a region $\mathcal{A}$ of superspace given that Ψ started in another region $\mathcal{B}$ of superspace. Page[2,105] has explored the construction of conditional probabilities in quantum cosmology without the use of absolute probabilities.

◄ In fact, a naïve first-order Hamiltonian formulation for the minisuperspaces of homogeneous cosmologies was found early on [97], but until the development of supergravity there was no natural interpretation of the Dirac-type constraint equation obtained.

Clearly the issues surrounding the choice of probability measure involve some deep conceptual problems which may perhaps get to the heart of the broader conceptual basis of quantum gravity. Such issues have been discussed by a number of authors in the context of quantum cosmology[2,43,104–107] and I will not address them here in any detail.

For the purposes of examining how we might hope to make predictions from the proposed boundary conditions of §4.1,4.2 we shall merely consider quantum cosmology in the WKB limit. In this limit it follows from (3.40) that each of the components Ψ_n in (3.30) has a conserved Klein-Gordon-type current (3.42) given by

$$\mathcal{J}_n \simeq |\mathcal{A}_n|^2 \nabla S_n, \tag{3.44}$$

which flows very nearly along the direction of the classical trajectories. The current conservation law $\nabla_A \mathcal{J}_n{}^A = 0$ implies that

$$d\mathcal{P} = \mathcal{J}_n{}^A d\Sigma_A, \tag{3.45}$$

is a conserved probability measure on the set of trajectories with tangent ∇S_n, where $d\Sigma_A$ is the element of a hypersurface, Σ, in minisuperspace which cuts across the flow and intersects each curve in the congruence once and only once. One finds that for a pencil of trajectories near the classical trajectory the probability density (3.45) is positive-definite. Vilenkin has argued[104] that positive-definiteness of the probability measure is really only required in the semiclassical limit, as this is the only limit in which we obtain a universe accessible to observation where the conventional laws of physics apply. Therefore given that the current (3.42) works in the WKB limit, this is all that is needed if we are content that "probability" and "unitarity" are only approximate concepts in quantum gravity. One can also show[2,52] that the manifestly positive definition (3.43) yields essentially the same result as (3.45) in the WKB limit.

3.7. Minisuperspace for the Friedmann universe with massive scalar field

Let us now apply our results to the particularly simple case of a homogeneous, isotropic universe with a single scalar field, with a potential which allows for inflationary behaviour. A quadratic potential is possibly the simplest example with this property, and thus has been much studied in quantum cosmology.

For convenience we will introduce a numerical normalisation factor $\sigma^2 = \kappa^2/(3V)$ into the metric, where V is the 3-volume of the unit hypersurface – e.g., $V = 2\pi^2$ for the 3-sphere, $k = +1$. We are considering *closed universes only*, which requires making topological identifications for the $k = 0, -1$ cases, so that V remains finite. In place of (2.1), (2.2) and (3.12) we then have a metric

$$ds^2 = \sigma^2 \left\{ -\mathcal{N}^2 dt^2 + a^2(t) \left[\frac{dr^2}{1 - kr^2} + r^2 \left(d\theta^2 + \sin^2\theta d\varphi^2 \right) \right] \right\}. \tag{3.46}$$

We also take the scalar field action to be normalised by

$$S_{\text{matter}} = \frac{3}{\kappa^2} \int_{\mathcal{M}} \mathrm{d}^4x\sqrt{-g}\left(-\frac{1}{2}g^{\mu\nu}\partial_\mu\phi\partial_\nu\phi - \frac{V(\phi)}{2\sigma^2}\right). \tag{3.47}$$

Thus $\mathcal{N} \to \sigma\mathcal{N}$, $a \to \sigma a$, $\Phi = \sqrt{3}\phi/\kappa$ and $\mathcal{V} = 3V/(2\kappa^2\sigma^2)$ relative to our earlier definitions, and a, ϕ and V are now dimensionless. As for the general homogeneous minisuperspace discussed in the previous section, we use a gauge in which $\mathcal{N}_i = 0$. From (3.46) it follows that ${}^3R = \frac{6k}{\sigma^2 a^2}$ and $K_{ij} = -\frac{\dot{a}}{\sigma\mathcal{N}a}h_{ij}$, and thus the action takes the form (3.18) with a minisupermetric

$$\mathcal{G}_{AB}\mathrm{d}q^A\mathrm{d}q^B = -a\mathrm{d}a^2 + a^3\mathrm{d}\phi^2, \tag{3.48}$$

and potential

$$\mathcal{U} = \tfrac{1}{2}\left(a^3V(\phi) - ka\right). \tag{3.49}$$

Alternatively, it is sometimes useful to express (3.48) in terms of a conformal gauge

$$\begin{aligned}\mathcal{G}_{AB}\mathrm{d}q^A\mathrm{d}q^B &= \mathrm{e}^{3\alpha}\left(-\mathrm{d}\alpha^2 + \mathrm{d}\phi^2\right), &(3.50)\\ &= -(4uv)^{-1/4}\mathrm{d}u\mathrm{d}v, &(3.51)\end{aligned}$$

where $\alpha = \ln a$, or alternatively in null coordinates

$$\begin{aligned} u &= \tfrac{1}{2}\mathrm{e}^{2(\alpha-\phi)} = \tfrac{1}{2}a^2\mathrm{e}^{-2\phi},\\ v &= \tfrac{1}{2}\mathrm{e}^{2(\alpha+\phi)} = \tfrac{1}{2}a^2\mathrm{e}^{2\phi}.\end{aligned} \tag{3.52}$$

The canonical momenta are given by

$$\pi^0 = 0, \qquad \pi_a = \frac{-a\dot{a}}{\mathcal{N}}, \qquad \pi_\phi = \frac{a^3\dot{\phi}}{\mathcal{N}}, \tag{3.53}$$

and the classical equations of motion yield

$$\mathsf{H} = \tfrac{1}{2}\left(-\frac{a\dot{a}^2}{\mathcal{N}^2} + \frac{a^3\dot{\phi}^2}{\mathcal{N}^2} - ka + a^3V\right) = 0, \tag{3.54}$$

$$\frac{1}{\mathcal{N}}\frac{\mathrm{d}}{\mathrm{d}t}\left(\frac{\dot{a}}{\mathcal{N}}\right) + \frac{2a\dot{\phi}^2}{\mathcal{N}^2} - aV = 0, \tag{3.55}$$

$$\frac{1}{\mathcal{N}}\frac{\mathrm{d}}{\mathrm{d}t}\left(\frac{\dot{\phi}}{\mathcal{N}}\right) + 3\frac{\dot{a}\dot{\phi}}{a\mathcal{N}^2} + \tfrac{1}{2}\frac{\mathrm{d}V}{\mathrm{d}\phi} = 0, \tag{3.56}$$

which are equivalent to the geodesic equation (3.21) with metric (3.48) and potential (3.49). The lapse function is not of physical relevance classically since we can choose an alternate proper time parameter, $\mathrm{d}\tau = \mathcal{N}\mathrm{d}t$, or equivalently choose a gauge $\mathcal{N} = 1$ in (3.54)–(3.56) so that t is the proper time. One of these equations depends on the other two by virtue of the Bianchi identity, as is always the case in general relativity. We thus effectively have two independent differential equations in two unknowns, one first order and one second order, or equivalently an autonomous system of three first order differential equations. The solution therefore depends on three free parameters, but as mentioned above one of these amounts to a choice of the origin of time which is of no physical importance. Thus there is a two-parameter family of physically distinct solutions.

The classical solutions cannot be written in a simple closed form except for certain special values of k and the potential $V(\phi)$, which unfortunately does not even include the quadratic potential$^{\triangleright}$ $V(\phi) = m^2\phi^2$. The qualitative property of the solutions may nonetheless be determined by studying the 3-dimensional phase space. Instead of choosing a particular potential, however, let us suppose that we are in a region of the phase space for which V can be approximated by a constant. Such conditions are more or less met in the "slow-rolling approximation"[109] of inflationary cosmology, in which $|V'/V| \ll 6$ and $|V''/V| \ll 9$. In this approximation the dynamics is described by setting $\dot{\phi} \simeq 0$ in (3.54) and (3.55), and setting $\ddot{\phi} \simeq 0$ in (3.56). In approximating the nearly flat region of the potential V by a cosmological constant we ignore the slow change of the scalar field determined from (3.56). We thus obtain a simplified model which possesses classical inflationary solutions, provided the constant V is chosen to be *positive*. This will serves as a useful test model for quantum cosmology.

In the case that V is constant, eq. (3.56) integrates to give $\dot{\phi} = Ca^{-3}$, where C is an arbitrary constant, and it therefore follows that the Friedmann equation (3.54) can be written in terms of an elliptic integral in $a^2(\eta)$,

$$\eta = \tfrac{1}{2}\int_0^{a^2} \frac{dz}{\sqrt{Vz^3 - kz^2 + C^2}}, \tag{3.57}$$

where η is the conformal time parameter defined by $d\eta = a^{-1}dt$. It is thus possible to express the general solution in terms of elliptic functions. Since the properties of such functions are not very transparent perhaps, we can alternatively plot the 2-dimensional phase space – e.g., in terms of the variables $\dot{\phi}$ and $\dot{\alpha}$, as in Fig. 4 in order to understand the qualitative features of the solutions. There are four critical points, at $(\dot{\phi}, \dot{\alpha}) = \left\{\left(0, \pm 1/\sqrt{V}\right), \left(\pm 1/\sqrt{2V}, 0\right)\right\}$, the first two being nodes, $A_\pm$, which are endpoints for all values of the spatial curvature, k, and the latter two saddle points, $B_\pm$, for the $k = +1$ solutions.

The general solution for the spatially flat case ($k = 0$), which corresponds to the bold separatrices shown in Fig. 4, is given by[110]

$$\begin{aligned} a &= C_a \left|\sinh \tfrac{3}{2}\sqrt{V}t\right|^{1/3} \left|\cosh \tfrac{3}{2}\sqrt{V}t\right|^{1/3}, \\ \phi &= \tfrac{1}{3}\ln\left|\tanh \tfrac{3}{2}\sqrt{V}t\right| + C_\phi, \end{aligned} \tag{3.58}$$

$\triangleright$ The Einstein equations for the FRW universe with a massive scalar field can be solved approximately [108] in various limits, however.

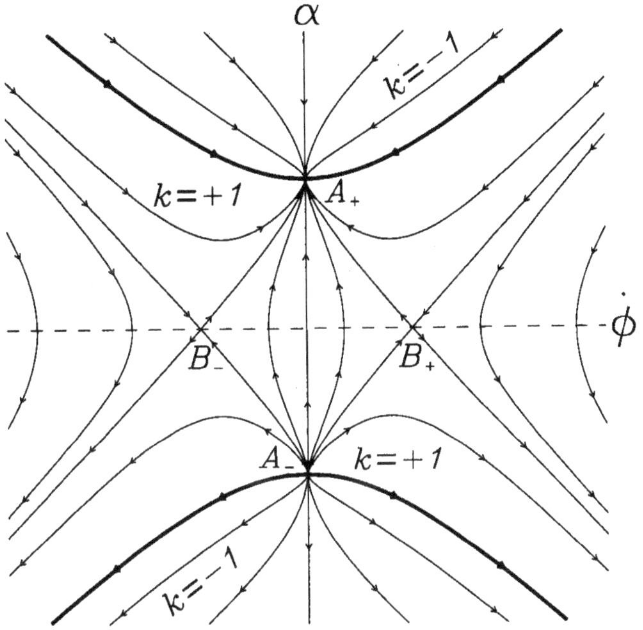

Fig. 4: The 2-dimensional phase plot of $\dot{\alpha} = \dot{a}/a$ versus $\dot{\phi}$ for the simplified model with constant potential, V.

in closed form, where C_a and C_ϕ are arbitrary constants. At late times the solutions (3.58) have an exponential scale factor, $a \to C_a \exp(\sqrt{V}\,t)$ as $t \to \infty$, and constant scalar field $\phi \to C_\phi$. Furthermore, one can see from Fig. 4 that all the $k = -1$ solutions in the upper half-plane, and a number of the $k = +1$ solutions are also attracted to the point A_+ with a similar inflationary behaviour at late times. (The corresponding point A_- on the $k = 0$ separatrix corresponds to the time-reversed solution, with an inflationary phase as $t \to -\infty$.) The simplified universe corresponding to Fig. 4 of course is far from being the complete picture, as the model does not allow for any exit from inflation. However, Fig. 4 illustrates the typical situation that a given model will possess regimes with inflationary behaviour and regimes with non-inflationary behaviour. In Fig. 4 the $k = +1$ solutions to the right and left of the separatrices that pass through $B_\pm$ fall into the latter category, for example. The situation becomes even more involved when one considers the full 3-dimensional phase space for some particular potential $V(\phi)$.

The case of the $k = +1$ solutions in Fig. 4 illustrates the general feature that classical dynamics are highly dependent on initial conditions. In order to obtain a sufficiently long inflationary epoch to overcome the problems mentioned in the Introduction, (of order 65 e-folds growth in the scale factor), the initial values of ϕ

and $\dot{\phi}$ must be restricted to a particular region of the phase space. In particular, $\dot{\phi}$ must be small initially. Classically, there is no *a priori* reason for one choice of initial conditions over any other choice, unless further ingredients are added. The degree to which inflationary initial conditions are preferred relative to other initial conditions – i.e., how probable is inflation? – is precisely the sort of question that we might therefore hope quantum cosmology could answer.

It is possible to attempt to solve this question without resorting to quantum cosmology. To do this one must construct a measure on the set of all universes[111,112], and then compare the number of inflationary solutions with a sufficiently long exponential phase to the number of other solutions. Preliminary results[112] seemed to indicate that almost all models with a massive scalar field undergo a period of inflation. However, a more careful analysis[108] revealed that the answer is ambiguous, as both the set of inflationary solutions and the set of non-inflationary solutions have infinite measure.

Alternatively, if as we expect the universe began in some sort of tunneling process or similar transition from a quantum regime, then we could expect the "initial" classical parameters to be determined, at least in a probabilistic fashion, from more fundamental quantum processes. The question of the most probable state of the universe is then pushed back a level and becomes: "what is a typical wavefunction for the universe?"

In the context of the present minisuperspace model, therefore, we can proceed by quantising the Wheeler-DeWitt equation (3.54), to obtain

$$\begin{aligned}\hat{H}\Psi = \left[-\tfrac{1}{2}\nabla^2 + \mathcal{U}\right]\Psi &= \tfrac{1}{2}\left[\frac{1}{a^3}\left(a\frac{\partial}{\partial a}a\frac{\partial}{\partial a} - \frac{\partial^2}{\partial\phi^2}\right) - ka + a^3V(\phi)\right]\Psi \\ &= \tfrac{1}{2}e^{-3\alpha}\left[\frac{\partial^2}{\partial\alpha^2} - \frac{\partial^2}{\partial\phi^2} - ke^{4\alpha} + e^{6\alpha}V(\phi)\right]\Psi \\ &= (4uv)^{1/4}\left[2\frac{\partial^2}{\partial u\partial v} - \frac{k}{2} + (uv)^{1/2}V\right]\Psi \\ &= 0\end{aligned} \tag{3.59}$$

in terms of the various sets of coordinates given earlier. In general boundary conditions will have to specified in order to solve (3.59). However, we can consider the approximate form of the WKB solutions without considering boundary conditions for the time being.

We will confine ourselves to regions in which the potential can be approximated by a cosmological constant, as in the analysis of Fig. 4, so that we can drop the term involving derivatives with respect to ϕ in (3.59), thereby obtaining a simple 1-dimensional problem which is amenable to a standard WKB analysis. The first order WKB wavefunction (3.40) which solves (3.59) in this approximation is

$$\Psi(a,\phi) \simeq \begin{cases} \dfrac{\mathcal{B}(\phi)}{a\left(a^2V(\phi)-k\right)^{1/4}} \exp\left[\dfrac{\pm i}{3V(\phi)}\left(a^2V(\phi)-k\right)^{3/2}\right], & a^2V > k, \quad (3.60) \\ \dfrac{\mathcal{C}(\phi)}{a\left(k-a^2V(\phi)\right)^{1/4}} \exp\left[\dfrac{\pm 1}{3V(\phi)}\left(k-a^2V(\phi)\right)^{3/2}\right], & a^2V < k. \quad (3.61) \end{cases}$$

If V is positive, as was assumed above, then oscillatory type solutions will thus exist for large values of the scale factor, while the exponential type solutions will exist for small values of the scale factor if $k = +1$.

The oscillatory solutions are of the form $\Psi \sim e^{iS}$ (neglecting the prefactor), where S satisfies the Hamilton-Jacobi equation (3.34). Comparing this to the Hamiltonian constraint (3.26) we find a strong correlation (3.35) between momenta and coordinates. For large scale factors, $a^2V \gg |k|$, so that $S \simeq \pm\frac{1}{3}a^3\sqrt{V}$. In this limit (3.23), (3.35) and (3.50) thus yield

$$\begin{aligned} \pi_\alpha = \frac{\partial S}{\partial \alpha} \quad &\Rightarrow \quad \dot{\alpha} \simeq \pm\sqrt{V}, \\ \pi_\phi = \frac{\partial S}{\partial \phi} \quad &\Rightarrow \quad \dot{\phi} \simeq 0, \end{aligned} \tag{3.62}$$

which correspond in fact to the inflationary points, $A_\pm$, of Fig. 4. The oscillatory wavefunction thus "picks out" classical inflationary universes.

Since the minisupermetric (3.50) is conformal to 2-dimensional Minkowski space in the coordinates (α, ϕ), it is convenient to represent it by a Carter-Penrose conformal diagram (see Fig. 5). In each case we plot $(p-q)$ horizontally and $(p+q)$ vertically, where $\tan p = \alpha + \phi$, and $\tan q = \alpha - \phi$. The boundary consists of points corresponding to past timelike infinity, $\mathfrak{i}^- = \{(a,\phi)\,|\,a = 0, \phi \text{ finite }\}$, future timelike infinity, $\mathfrak{i}^+ = \{(a,\phi)\,|\,a = \infty, \phi \text{ finite }\}$, left and right spacelike infinity, $\mathfrak{i}^0_{L,R} = \{(a,\phi)\,|\,a = \text{ finite }, \phi = \pm\infty\}$; and past and future null boundaries, $\mathfrak{I}^-_{L,R} = \{(a,\phi)\,|\,a = 0, \phi = \pm\infty\}$ and $\mathfrak{I}^+_{L,R} = \{(a,\phi)\,|\,a = 0, \phi = \pm\infty\}$. In each case the subscript L (left) is associated with $\phi \to -\infty$, and the subscript R (right) with $\phi \to +\infty$. The approximate region for which oscillatory WKB solutions exist is shown in Fig. 5(a,b) for the approximate minisuperspace with a cosmological constant, in Fig. 5(c,d) for $V(\phi) = m^2\phi^2$, and in Fig. 5(e,f) for potentials, $V(\phi)$, typically found in higher-derivative gravity theories and in string theory with supersymmetry breaking.

Naturally it is of interest to know whether the inflationary WKB wavefunctions are ***typical*** solutions to the Wheeler-DeWitt equation. To determine a typical wavefunction for the universe, we need to make a choice of boundary conditions for Ψ when solving (3.59).

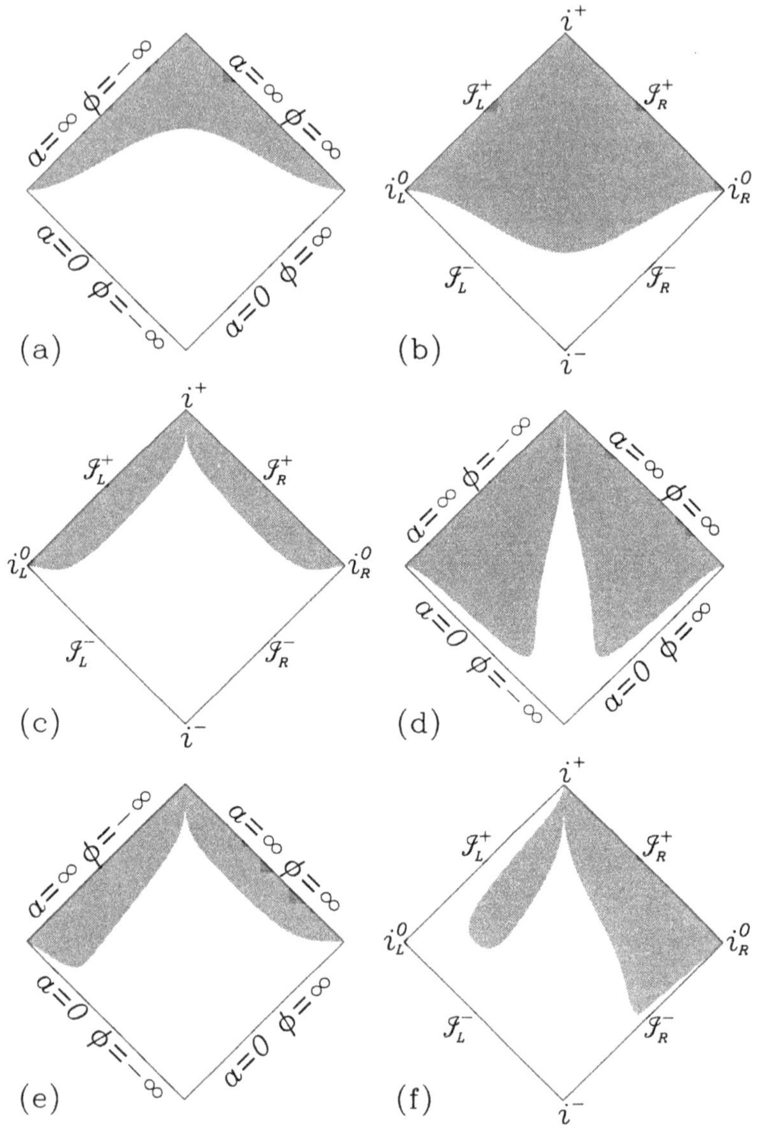

Fig. 5: Conformal diagrams of the 2-dimensional minisuperspace. The region where oscillatory WKB solutions exist, as given by the rough criterion $a^2V > 1$, is shaded for various potentials: (a) $V = 0.25$ (const); (b) $V = 4$ (const); (c) $V = 0.25\phi^2$; (d) $V = 25\phi^2$; (e) $V = \left[1 - e^{-\phi/f}\right]^2$ with $f = 1.5$; (f) $V = 4\sinh^2\phi \exp\left[-fe^{-2\phi}\right]$ with $f = 0.1$.

4. Boundary Conditions

The specification of boundary conditions for the Wheeler-DeWitt equation may seem a disappointment, as it might appear that we are just replacing an arbitrary initial choice of parameters which describe the classical evolution of the universe by an arbitrary initial choice of parameters which describe its quantum evolution. However, if quantum mechanics is a universal theory then it must have applied at the earliest epochs of the existence of the universe, in which case it is natural that the quantum dynamics precedes the classical dynamics. This justifies a quantum boundary condition for the universe as being more fundamental than a classical one. In any case, the only alternative to choosing quantum boundary conditions would be that mathematical consistency might be enough to guarantee a unique solution to the Wheeler-DeWitt equation, as DeWitt originally hoped[33]. If the experience gained from the study of minisuperspace models translates to superspace, then this would not appear to be the case, however.

The question naturally arises as to whether there should be some natural boundary condition, which once and for all determines the quantum evolution of the universe at early times, or alternatively whether the nature of the quantum dynamics might be somewhat indifferent to such choices. Deep conceptual problems are involved in trying to make headway with this question. Unlike other situations in quantum physics, where boundary conditions are readily specified by the symmetry of particular problems, such as spherical symmetry in the case of the hydrogen atom, the origin of the universe poses a situation in which all intuition must be abandoned and we can at best proceed on aesthetic grounds alone.

Having made a choice of boundary condition, we can of course solve the Wheeler-DeWitt equation and study the physical consequences for the evolution of the universe. However, without some additional principle we should by rights study many different boundary conditions before we can begin to have any confidence about the predictions made. To arrive at a principle which would circumvent this problem is an immense challenge: it would more or less amount to an additional law of physics which must be appended to the others which describe the quantum evolution of the universe. The situation might be considered to be the same as trying to describe the phase transition from gas to liquid if all the physical phenomena that we knew about related to the gaseous phase only. The universe appears to have undergone a phase transition when it was formed, but the only experience we have available involves the "after" state of the universe alone.

Progress can of course only be made by attempting to define natural boundary conditions for the wavefunction of the universe, and examining the consequences. This became an important activity in the 1980s. I will only discuss the two most studied boundary condition proposals, the "no-boundary proposal" and the "tunneling proposal". However, other proposals have been put forward, including the "all possible boundaries proposal" of Suen and Young[113] and the "symmetric initial condition" of Conradi and Zeh[114,115].

4.1. The no-boundary proposal

The proposal of Hartle and Hawking[38,39] is that one should restrict the sum in the definition of the wavefunction of the universe (3.11) to include only compact Euclidean 4-manifolds, $\mathcal{M}$, for which the spatial hypersurface Σ on which Ψ is defined forms the only boundary, and only matter configurations which are regular on these geometries. The universe then has no singular boundary to the past, as is the case for the standard FRW cosmology. The sum (3.11) thus includes manifolds such as those shown in Fig. 6, but not those shown in Fig. 2. As Hawking[39] puts it: ***the boundary conditions of the universe are that it has no boundary.***

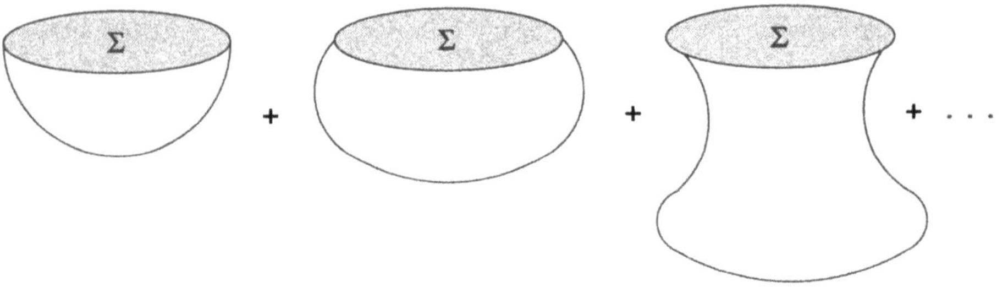

Fig. 6: Geometries allowed by the Hartle-Hawking no-boundary proposal.

Intuitively, what Hartle and Hawking had in mind in formulating this proposal was to get rid of the initial singularity by "smoothing the geometry of the universe off in imaginary time". For example, whereas a surface with $\sqrt{h} = 0$ would be singular in a Lorentzian signature metric, this is not necessarily the case if the metric is of Euclidean signature, as can be seen from the example of S^4 shown in Fig. 7. Ideally, the no-boundary proposal should tell us what initial conditions to set when we take manifolds with $\sqrt{h} \to 0$ or any similar limit consistent with the proposal. If the limit is taken at an initial time $\tau = 0$, the no-boundary proposal would lead to conditions on $h_{ij}(x,0)$, $\phi(x,0)$ and their derivatives. In practice, quantum cosmology is rarely studied beyond the semiclassical approximation, in which $\Psi \simeq \mathcal{A}e^{-I_{\rm cl}}$, where $I_{\rm cl}$ is the classical (possibly complex) action evaluated along the solution to the Euclidean field equations. In the semiclassical approximation one therefore works only with boundary conditions on the metric and matter fields which correspond to the no-boundary proposal at the classical level. In particular, we demand: (i) that the 4-geometry is closed; and (ii) that the saddle points of the functional integral correspond to regular solutions of the classical field equations which match the prescribed initial data on Σ.

One question which is not explained by the no-boundary proposal is the choice of a contour of integration for the path integral. As was mentioned earlier, the path integral over real Euclidean metrics does not converge, and thus it is necessary to include complex metrics to make the path integral converge. Such metrics will generally include ones which are neither truly Euclidean nor truly Lorentzian, and thus

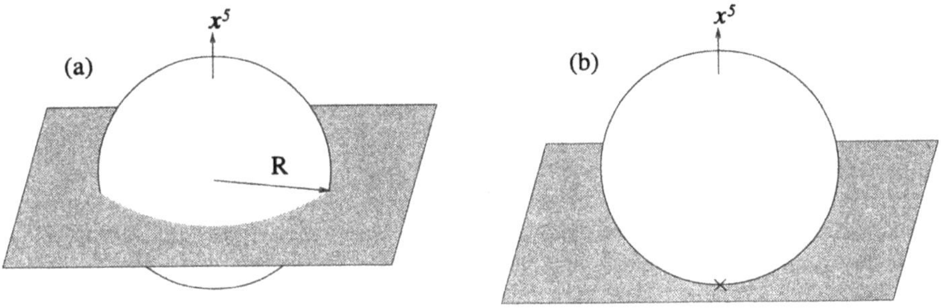

Fig. 7: Slicing a 4-sphere of radius R embedded in flat 5-dimensional space: (a) a surface $|x^5| = a < R$ intersects the 4-sphere in a 3-sphere of non-zero radius; (b) when $|x^5| = R$ the 3-sphere shrinks to zero radius but there is no singularity of the 4-geometry.

the naïve picture of a compact Euclidean geometry sewn onto a Lorentzian one, which is suggested by "smoothing the geometry of the universe off in imaginary time", is not completely accurate. In general, one might expect a truly complex metric to interpolate the Euclidean and Lorentzian ones, and in general the initial geometry might be only *approximately Euclidean* and the final geometry only *approximately Lorentzian*[116]. Unfortunately the criteria for achieving convergence of the path integral do not single out a unique contour of integration, and the no-boundary proposal does not appear to offer any further clues as to how the contour should be chosen[117].

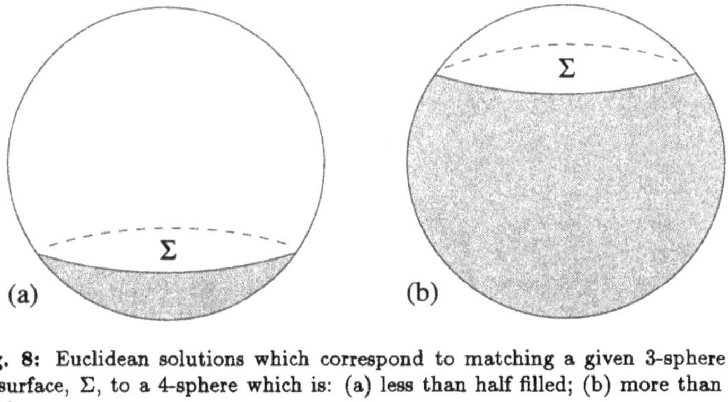

Fig. 8: Euclidean solutions which correspond to matching a given 3-sphere hypersurface, Σ, to a 4-sphere which is: (a) less than half filled; (b) more than half filled.

Non-uniqueness of the contour of integration is already a problem in the simplest conceivable non-trivial minisuperspace model, namely the $k = +1$ FRW universe with a cosmological term and no other matter – the "de Sitter minisuperspace" model. At the semiclassical level one can calculate $e^{-I_{cl}}$ by the steepest descents method. Hartle and Hawking discussed this in their original "no boundary" paper[38], and argued heuristically that one particular saddle point would yield the dominant contribution to the path integral, namely the saddle point corresponding to the classical Euclidean solution which matches the S^3 hypersurface Σ to a less than half filled 4-sphere, rather than the solution which matches S^3 to a more than half filled 4-sphere (see Fig. 8). However, their argument did not stand up to a more rigorous analysis. Halliwell and Louko[66] found a means of evaluating the minisuperspace path integral exactly, and thereby explicitly determined convergent contours. They showed that this simple model possesses inequivalent contours for which the path integral converges. These pass through different saddle points and lead to different semiclassical wavefunctions, Ψ. There are thus many different no-boundary wavefunctions, each corresponding to a different choice of contour. The problem persists in more complicated models[67,68]. Since different no-boundary wavefunctions could lead to different physical predictions, the ambiguity associated with the choice of contour would appear to be the most significant problem with the no-boundary proposal which still needs to be resolved.

Let us return to the minisuperspace example of §3.7 and examine the implications of imposing the Hartle-Hawking boundary condition. For simplicity we will specialise to $k = +1$ models. The other values of k have also been discussed in the literature.[2,105,118]

The minisuperspace no-boundary wavefunction for the $k = +1$ models is given by

$$\Psi_{\mathrm{HH}}[a, \phi] = \int^{(a,\phi)} \mathfrak{D}\mathrm{a}\, \mathfrak{D}\Phi\, \mathfrak{D}\mathcal{N}\, e^{-I[\mathrm{a},\Phi,\mathcal{N}]}, \tag{4.1}$$

where▲

$$I = \tfrac{1}{2}\int_0^{\tau_f} d\tau \mathcal{N}\left[\frac{-\mathrm{a}}{\mathcal{N}^2}\left(\frac{d\mathrm{a}}{d\tau}\right)^2 + \frac{\mathrm{a}^3}{\mathcal{N}^2}\left(\frac{d\Phi}{d\tau}\right)^2 - \mathrm{a} + \mathrm{a}^3 V\right]. \tag{4.2}$$

The integral is taken over a class of paths which match the values

$$\mathrm{a}(\tau_f) = a, \qquad \Phi(\tau_f) = \phi, \tag{4.3}$$

on the final surface, and the origin of the Euclidean time coordinate, τ, has been chosen to be zero.

There are two approaches we can take to solving the Wheeler-DeWitt equation (3.59): either (i) attempt to interpret the Hartle-Hawking boundary condition directly in terms of boundary conditions of Ψ on minisuperspace; or (ii) take a saddle-point approximation to the path integral (4.1). For consistency these two approaches should agree.

Let us first consider (3.59) directly. Hawking and Page[52,119] have argued that one can approximate the Hartle-Hawking boundary condition in minisuperspace by saying that in an appropriate measure one should have $\Psi = 1$ when $a \to 0$ with ϕ

▲ I shall use a different font to distinguish the minisuperspace coordinates in the functional integral, (a, Φ), from their boundary values, (a, ϕ), in the 3-geometry on the hypersurface Σ.

regular, and $\Psi = 1$ also along the past null boundaries, in order to provide sufficient Cauchy data to solve (3.59) everywhere in the (α, ϕ) plane. It is possible to solve (3.59) exactly for a massless scalar field[86,120], (i.e., $V = 0$), but exact solutions are not known for the massive scalar field. Approximate solutions can be found in various regimes[52,119,121], however.

Firstly, since Ψ must be regular as $a \to 0$, we see that Ψ must be independent of ϕ in this limit, $\frac{\partial \Psi}{\partial \phi} \simeq 0$, in order to overcome the divergence of the a^{-3} factor (using coordinates (a, ϕ)) in (3.59). Furthermore, for $a^2 V \ll 1$ the approximation $V \simeq 0$ is a good one[⋈], and for $k = \pm 1$ (3.59) becomes a Bessel equation in the variable $\frac{1}{2}a^2$. For $k = +1$, which is the case of interest to us here, we therefore obtain the solution[52,119]

$$\Psi \simeq \mathrm{I}_0(\tfrac{1}{2}a^2), \tag{4.4}$$

where $\mathrm{I}_0(z) \equiv \sum_{n=0}^{\infty} \frac{1}{(n!)^2} \left(\frac{z}{2}\right)^{2n}$ is the zero order modified Bessel function, and the normalisation has been fixed to satisfy the boundary condition $\Psi \to 1$ as $a \to 0$. For large a (with $a^2 V \ll 1$),

$$\Psi \sim \frac{1}{\sqrt{\pi}\, a} \exp\left(\tfrac{1}{2}a^2\right) \left[1 + \mathrm{O}(a^{-2})\right], \tag{4.5}$$

i.e., the wavefunction is of exponential type. It therefore agrees with the WKB approximation (3.61) in the limit $a^2 V \ll 1$ for large a provided we take the $(-)$ solution of (3.61) with normalisation

$$\mathcal{C}(\phi) = \frac{1}{\sqrt{\pi}} \exp\left(\frac{+1}{3V(\phi)}\right). \tag{4.6}$$

Let us now consider the limit $V(\phi) \gg 1$ but avoid regimes in which the ϕ dependence of (3.59) is significant by assuming that V is approximately constant, as in the approximation of Fig. 4, so that the ϕ derivative can still be neglected. The term involving the spatial curvature k in (3.59) is now negligible compared to the last term and can also be neglected, yielding the solution[52,119]

$$\Psi \simeq \mathrm{c}(\phi)\, \mathrm{J}_0(\tfrac{1}{3}a^3\sqrt{V}), \tag{4.7}$$

where $\mathrm{J}_0(z) \equiv \sum_{n=0}^{\infty} \frac{(-1)^n}{(n!)^2} \left(\frac{z}{2}\right)^{2n}$ is the zero order ordinary Bessel function. Since the spatial curvature term in (3.59) is always dominant for $a \to 0$, the approximation which led to (4.7) no longer applies in that limit, so the factor c cannot be normalised by the boundary condition $\Psi(0, \phi) = 1$. For large a, however,

$$\Psi \sim \frac{\mathrm{c}}{\sqrt{2\pi \mathrm{S}}} \cos\left(\mathrm{S} - \frac{\pi}{4}\right), \tag{4.8}$$

⋈ We assume that $V(\phi)$ grows less strongly than $e^{6|\phi|}$ as $|\phi| \to \infty$, i.e., $|V'/V| < 6$, so that this approximation remains valid for arbitrarily large $|\phi|$.

where $S = \frac{1}{3}a^3\sqrt{V}$, which is a superposition of the two oscillatory WKB modes (3.60) for large S. Using the WKB connection formula to match the (−) solution of (3.61) to the oscillatory region, we find agreement with the asymptotic limit (4.8), provided

$$c = \sqrt{\frac{2\pi}{3}}C = \sqrt{\tfrac{2}{3}}\exp\left(\frac{1}{3V(\phi)}\right). \tag{4.9}$$

The ϕ-dependent corrections to these wavefunctions, which result from perturbations in directions in which $\dot{\phi} \not\simeq 0$ have been discussed by Page[2,119,122].

Let us now consider the alternative method of determining Ψ by making a saddle-point approximation to the path integral. In the semiclassical approximation the wavefunction takes the form

$$\Psi \sim \exp\left(-I_{\rm cl}(a,\phi)\right), \tag{4.10}$$

where $I_{\rm cl}$ denotes the Euclidean action (4.2) evaluated at a classical (possibly complex) solution to the Euclidean field equations

$$\frac{1}{\mathcal{N}^2}\left(\frac{da}{d\tau}\right)^2 - \frac{a^2}{\mathcal{N}^2}\left(\frac{d\phi}{d\tau}\right)^2 - 1 + a^2V(\phi) = 0, \tag{4.11}$$

$$\frac{1}{\mathcal{N}a}\frac{d}{d\tau}\left(\frac{1}{\mathcal{N}}\frac{da}{d\tau}\right) + \frac{2}{\mathcal{N}^2}\left(\frac{d\phi}{d\tau}\right)^2 + V(\phi) = 0, \tag{4.12}$$

$$\frac{1}{\mathcal{N}}\frac{d}{d\tau}\left(\frac{1}{\mathcal{N}}\frac{d\phi}{d\tau}\right) + \frac{3}{\mathcal{N}a}\frac{da}{d\tau}\frac{d\phi}{d\tau} - \frac{1}{2}\frac{dV}{d\phi} = 0, \tag{4.13}$$

which follow from (3.54)–(3.56) by replacing $t \to -i\tau$. We will henceforth restrict ourselves to the gauge in which $\frac{d\mathcal{N}}{d\tau} = 0$, i.e., $\mathcal{N} = \text{const}$, in which case the functional integral over $\mathcal{N}$ in (4.1) must be replaced by an ordinary integral.

The Hartle-Hawking boundary condition demands that

$$a(0) = 0, \qquad \left.\frac{d\phi}{d\tau}\right|_0 = 0. \tag{4.14}$$

To see this consider the Euclidean 4-metric which is given by

$$ds^2 = \mathcal{N}^2 d\tau^2 + a^2(\tau)d\Omega_3{}^2. \tag{4.15}$$

The Hartle-Hawking boundary condition requires that we close this 4-geometry in a regular fashion as $\tau \to 0$. This is achieved if $a \sim \pm\mathcal{N}\tau$ as $\tau \to 0$, since (4.15) is then the same as the metric of the 4-sphere in spherical polar coordinates in this limit. This suggests that we demand $a(0) = 0$ and $\frac{1}{\mathcal{N}}\left.\frac{da}{d\tau}\right|_0 = \pm 1$. However, the second condition is guaranteed by the constraint equation (4.11) if the first condition is imposed. The second condition of (4.14) is obtained by noting that (4.13) will give a regular solution for ϕ in the limit $\tau \to 0$ only if $\left.\frac{d\phi}{d\tau}\right|_0 = 0$, since the middle term of (4.13) diverges otherwise.

If as before we make the simplifying assumption that V can be approximated by a cosmological constant, and $\frac{d\phi}{d\tau} \simeq 0$, then it follows that there exist solutions satisfying the boundary conditions (4.3)and (4.14), which may be written

$$\mathfrak{a}(\tau) \simeq \frac{a \sin(\mathcal{N}\sqrt{V}\,\tau)}{\sin(\mathcal{N}\sqrt{V}\,\tau_f)}, \tag{4.16}$$

where

$$\sin^2(\mathcal{N}\sqrt{V}\,\tau_f) = a^2 V, \tag{4.17}$$

which follows from solving the constraint (4.11). If $a^2V < 1$ then (4.17) will give an infinite number of real solutions for the constant $\mathcal{N}$, which may be conveniently parametrised

$$\mathcal{N} = \mathcal{N}_{n\pm} \equiv \frac{1}{\sqrt{V}\,\tau_f}\left[(n+\tfrac{1}{2})\pi \pm \cos^{-1}(a\sqrt{V})\right], \quad n \in \mathbb{Z}, \tag{4.18}$$

with $\cos^{-1}(a\sqrt{V})$ taken in the principal range $(0, \frac{\pi}{2})$. Substitution of (4.18) in (4.16) then gives

$$\mathfrak{a}(\tau) \simeq (-1)^n V^{-1/2} \sin(\mathcal{N}\sqrt{V}\,\tau). \tag{4.19}$$

In addition to the real solutions (4.18) of (4.17), there will also be complex solutions if $a^2V > 1$.

Using the solution (4.19) it is straightforward to evaluate the classical action (4.2). For $n = 0$, for example, we find

$$I_{\pm} = \frac{-1}{3V(\phi)}\left[1 \pm \left(1 - a^2V(\phi)\right)^{3/2}\right]. \tag{4.20}$$

If we substitute (4.19) into (4.15) we obtain the metric of the 4-sphere, and thus the classical solutions correspond to matching a given 3-sphere to 4-sphere(s). Furthermore, for $n = 0$ we find $\left.\frac{d\mathfrak{a}}{d\tau}\right|_{\tau_f} = \mp\mathcal{N}\sqrt{1-a^2V}$, so that the $(-)$ solution of (4.18) and (4.20) has $\left.\frac{d\mathfrak{a}}{d\tau}\right|_{\tau_f} > 0$, which corresponds to matching the 3-geometry to a less than half filled 4-sphere (Fig. 8(a)), while the $(+)$ solution similarly corresponds to the more than half filled 4-sphere case (Fig. 8(b)). Values of $n > 0$ would appear to give cases in which the 4-geometry pinches off to zero a number of times and then "bounces back" resulting in linear chains of contiguous 4-spheres[123,124]. However, this interpretation of the saddle points as corresponding to universes with bounce solutions does not appear to remain valid[116] once one considers complex solutions to the field equations (4.11)–(4.13).

As was discussed earlier the no-boundary condition does not prescribe a unique contour of integration, and thus it is not completely clear which of the above saddle points should be included in the semiclassical wavefunction (3.30). Halliwell and Hartle have shown[117] that the points with $n < 0$, which have a negative lapse function, lead to difficulties with the recovery of quantum field theory in curved spacetime from quantum cosmology, and thus one might hope that these points should be avoided in the contour of integration. However, no clear grounds present themselves for omitting the other saddle points. For the purpose of making predictions from the no-boundary proposal we shall therefore make a choice by assuming that the contour is such that the solution corresponding to the less than half filled 4-sphere, with action I_-, provides the dominant contribution. (This is the choice that Hartle and Hawking originally made[38].) Neglecting the prefactor, we therefore obtain a no-boundary wavefunction

$$\Psi_{\rm HH}(a,\phi) \propto \exp\left(\frac{1}{3V(\phi)}\right) \exp\left[\frac{-1}{3V(\phi)}\left(1-a^2V(\phi)\right)^{3/2}\right], \qquad (4.21)$$

in the region $a^2V < 1$. Using the WKB matching procedure one can show that the corresponding solution in the region $a^2V > 1$ is

$$\Psi_{\rm HH}(a,\phi) \propto \exp\left(\frac{1}{3V(\phi)}\right) \cos\left[\frac{1}{3V(\phi)}\left(a^2V(\phi)-1\right)^{3/2} - \frac{\pi}{4}\right], \qquad (4.22)$$

which is the superposition of the two WKB components of (3.60):

$$\begin{aligned} \Psi_{\rm HH} &= \Psi_- + \Psi_+, \\ \Psi_\pm &\propto \exp\left(\frac{1}{3V(\phi)}\right) \exp\left\{\pm i\left[\frac{1}{3V(\phi)}\left(a^2V(\phi)-1\right)^{3/2} - \frac{\pi}{4}\right]\right\}. \end{aligned} \qquad (4.23)$$

One may observe that (4.21) and (4.22) agree with the solutions (4.4) and (4.7) found earlier by direct examination of the Wheeler-DeWitt equation in the appropriate limits (4.5) and (4.8). Thus although the saddle point corresponding to the less than half filled 4-sphere does not appear to be picked out in any special way by the path integral, it is favoured by the "approximate" boundary condition[52,119] that $\Psi \to 1$ as $a \to 0$.

One must add the caveat that the approximate boundary condition should be amended if one is to consider genuinely complex solutions of the Euclidean field equations (4.11)–(4.13). This issue has been considered by Lyons[116]. In general one must analytically continue the boundary condition (4.14) demanded by the no-boundary proposal. Although the simple picture of matching a real Euclidean solution to a real Lorentzian solution at the junction is no longer maintained, one nonetheless finds solutions which are initially approximately Euclidean and at late times are approximately Lorentzian, with classical inflationary behaviour[116].

4.2. The tunneling proposal

An alternative approach advocated by Vilenkin is that the boundary condition for the wavefunction, Ψ, should be such as to embody the notion that the universe "tunnels into existence from nothing" without making such specific restrictions on the "initial" geometry as the Hartle-Hawking proposal does. Conceivably there are many possible ways in which such a notion could be translated mathematically into a definition of the wavefunction, Ψ, and indeed many alternative formulations of the tunneling proposal have been put forward$^{\between}$. Some early versions were phrased in a similar fashion to the no-boundary proposal: in particular, Vilenkin[70] proposed defining the wavefunction by a functional integral over Lorentzian metrics which interpolate between a given matter configuration, Φ, and 3-geometry, h_{ij}, and a vanishing 3-geometry, $\varnothing$, lying to its past

$$\Psi\left[h_{ij},\Phi,\Sigma\right]=\sum_{\mathcal{M}}\int_{\varnothing}^{(h,\Phi)}\mathfrak{D}g\,\mathfrak{D}\phi\,e^{iS[g_{\mu\nu},\phi]}. \tag{4.24}$$

Vilenkin has also given an alternative formulation of the tunneling proposal in terms of a boundary condition on superspace[40,41] rather than a restriction on manifolds included in the path integral. In order to formulate boundary conditions on superspace it is necessary to consider its boundary, which can be thought of as consisting of 3-metric and matter configurations for which the 3-curvature is infinite, or $|\Phi|\to\infty$ etc. As we have already seen from the example of S^4 (Fig. 7), not all singular 3-geometries will correspond to singular 4-geometries, as it is possible to obtain a singular 3-geometry by a degenerate slicing of the 4-geometry. Therefore we should distinguish points on the boundary of superspace which correspond to genuine singularities of the 4-geometry from those that correspond to degenerate slicings*. We call the former the *singular boundary* of superspace, and the latter the *non-singular boundary.*

The tunneling proposal of Vilenkin[41] is that *the wavefunction, Ψ, should be everywhere bounded, and at singular boundaries of superspace Ψ includes only outgoing modes, i.e., those that carry a flux out of superspace.* Thus ingoing modes can only enter at the nonsingular boundary. This definition is somewhat vague as there is no obvious rigorous definition of positive and negative frequency modes in superspace due to the fact that it possesses no Killing vectors[96], and thus there is no clear notion which modes are "ingoing" and which are "outgoing". Furthermore, the structure of superspace is not completely understood, and it has not been rigorously shown that its boundary can be split into singular and non-singular parts.

The tunneling proposal has in fact been formulated with the minisuperspace WKB approximation in mind, in which case the notion of the boundary of minisuperspace and the notions of ingoing and outgoing modes are more clearly defined. Since each oscillatory WKB mode $\Psi\sim e^{iS_n}$ has a current (3.44), we can classify the modes as ingoing or outgoing according to the direction of ∇S_n on the surface in question. Heuristically, the idea underlying the Vilenkin boundary condition is that the ensemble of universes described by Ψ should not include any universes

$\between$ Linde's proposal [69] embodies a similar philosophy. However, it gives a different wavefunction to Vilenkin's "tunneling" wavefunction in simple minisuperspace models [125].

$\star$ This distinction can be made more precise using Morse functions [42].

contracting down from infinite size[40], but only those that correspond to "tunneling from nothing". As we shall see, the "outgoing flux" condition[41] accords with this notion at least in the case of simple minisuperspace models.

The outgoing-flux version of the tunneling proposal agrees with the path integral formulation (4.24) in the case of the simplest de Sitter minisuperspace[42,66], but the two versions would not appear to be equivalent in general[68]. Whereas the no-boundary wavefunction fixes the initial data but leaves the contour of path integration ambiguous, the tunneling proposal fixes the contour of integration but leaves some ambiguity in the specification of the initial data, even when the outgoing-flux condition is imposed[68,126].

Consider the minisuperspace model of §3.7. In the diagrams of Fig. 5 all surfaces $\mathcal{I}^{\pm}_{L,R}$ and the points $i^0_{L,R}$ and i^+ will be part of the singular boundary, whereas the point i^-, which corresponds to $a \to 0$ ($\alpha \to -\infty$) with ϕ finite, is the only point of the nonsingular boundary. Of course, the condition that $a \to 0$ is not enough to guarantee regularity of the 4-geometry (4.15), and in general one might not expect boundary points to cleanly fall into the category of the "singular boundary" or the "nonsingular boundary". However, as we have already observed in the last section, in the present minisuperspace model the additional requirement for regularity that $\frac{1}{\mathcal{N}}\frac{\mathrm{d}a}{\mathrm{d}\tau}\big|_0 = \pm 1$ is guaranteed by the constraint equation (4.11) if $a(0) = 0$ is imposed.

The oscillatory WKB region, as shaded in the conformal diagrams of Fig. 5 and Fig. 9, is always bounded by i^0_R, $\mathcal{I}^+_R$ and i^+, and in all cases except that of Fig. 5(f)$^{\nabla}$ the oscillatory region is bounded by i^0_L and $\mathcal{I}^-_L$ also. The wavefunction in this region is given by a superposition of terms e^{iS_n}, where S_n is a solution to the Hamilton-Jacobi equation (3.34), which in terms of the variables (α, ϕ) may be written

$$-\left(\frac{\partial S}{\partial \alpha}\right)^2 + \left(\frac{\partial S}{\partial \phi}\right)^2 + U(\alpha,\phi) = 0, \tag{4.25}$$

where $U(\alpha,\phi) \equiv 2e^{3\alpha}\mathcal{U}(\alpha,\phi) = e^{4\alpha}\left(e^{2\alpha}V(\phi) - 1\right)$, and for convenience in what follows we have suppressed the index n. The characteristics of (4.25) satisfy

$$\frac{d\alpha}{2S_{,\alpha}} = \frac{d\phi}{-2S_{,\phi}} = \frac{dS}{2U} = \frac{d(S_{,\alpha})}{U_{,\alpha}} = \frac{d(S_{,\phi})}{U_{,\phi}}. \tag{4.26}$$

Thus each $S(\alpha,\phi)$ describes a congruence of classical paths with

$$\frac{d\phi}{d\alpha} = -\frac{S_{,\phi}}{S_{,\alpha}} = \frac{\mp S_{,\phi}}{\sqrt{S_{,\phi}{}^2 + U}}. \tag{4.27}$$

$^{\nabla}$ Unlike the other cases the potential of Fig. 5(f), which corresponds to a potential in which supersymmetry is broken through gaugino condensation in string theory, vanishes as $\phi \to -\infty$. This limit is the "weak coupling limit" of string theory.

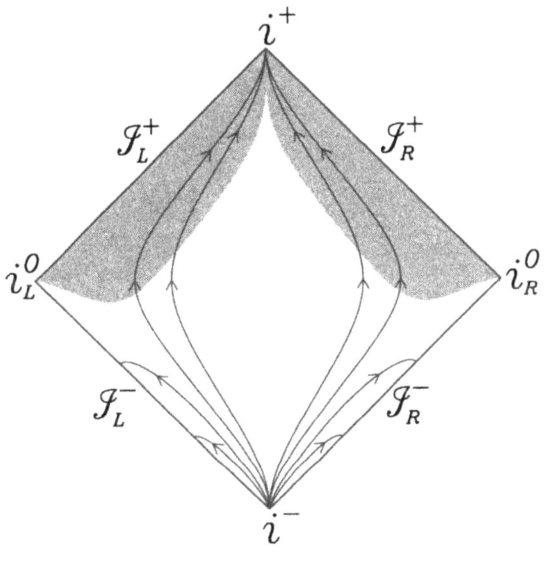

Fig. 9: Some schematic probability flows consistent with the tunneling proposal. An indicative oscillatory WKB region is shown (shaded area) for the potential $V = \phi^2$.

Since $\mathrm{U} > 0$ in the oscillatory region (assuming $V(\phi) > 0$), it follows that these integral curves satisfy $\left|\frac{d\phi}{d\alpha}\right| < 1$, i.e., they are "timelike" in the minisuperspace coordinates and have an endpoint at $\mathfrak{i}^+$. Since $\pi_\alpha \equiv -\frac{e^{3\alpha}\dot{\alpha}}{\mathcal{N}} = S_{,\alpha}$ it follows that the WKB modes correspond to expanding universes ($\dot{\alpha} > 0$) with $\pi_\alpha = S_{,\alpha} < 0$, or contracting universes with $\pi_\alpha = S_{,\alpha} > 0$, if we assume[⊘] $\mathcal{N} > 0$. Since all paths originate at finite values of α the latter solutions will extend from $\mathfrak{i}^+$ into the interior of the minisuperspace, i.e., they are "ingoing modes" and are to be excluded on account of the tunneling boundary condition. This boundary condition only allows the expanding solutions which are "outgoing" at $\mathfrak{i}^+$. In the approximation in which (4.5) applies it follows that the tunneling proposal demands that only the modes with $\Psi \sim \exp\left[\frac{-i}{3V(\phi)}\left(e^{2\alpha}V(\phi) - 1\right)\right]$ are admitted in the oscillatory region. The tunneling wavefunction is thus complex in the oscillatory region, in contrast to the

⊘ This choice is a matter of convention and the opposite choice would reverse the roles of the "ingoing" and "outgoing" modes. Equivalently, the coordinate t is an arbitrary label from the point of view of general relativity, and it is a matter of convention whether we choose t to increase or decrease towards the "future"; the "future" being defined by the expansion of the universe (the cosmological arrow) or the increase of entropy (the thermodynamic arrow). In terms of Fig. 4 it amounts to an arbitrary choice between A_+ and A_- as representing the "late-time behaviour" of the inflationary solutions.

no-boundary wavefunction (4.22) which is real.

To extend the tunneling wavefunction into the "tunneling region" (the unshaded regions of Fig. 5 and Fig. 9), we can use the WKB matching procedure. Of course, there will also be additional solutions that remain entirely in the tunneling region and never cross into the oscillatory region. In the case of the null boundaries, $\mathcal{J}^-_{L,R}$, which border this region we observe that as $\alpha \to -\infty$ (3.59) becomes the wave equation in the (α, ϕ) coordinates, and thus the solutions are asymptotically null which leads to a notion of "ingoing" and "outgoing" modes, so that the tunneling condition can be imposed. Some possible probability flows consistent with the tunneling proposal are shown in the conformal diagram of Fig. 9.

5. The Predictions of Quantum Cosmology

Given the many problems and uncertainties in the quantum cosmology programme that have been discussed above, one could easily form the opinion that it is premature to talk about the predictions that quantum cosmology makes. Nevertheless, it is important to investigate the types of predictions we might expect quantum cosmology to make about the universe, as well as realistically evaluating the limitations of the these predictions. I will concentrate on three key areas in this section. Other topics, most notably the variability of the constants of nature, have been discussed recently by Vilenkin[125,127].

5.1. The period of inflation

The construction of a suitable probability measure should allow us to answer questions such as whether inflation is a feature of a typical universe. As we saw in §3.6, the construction of a general probability measure is problematic. Thus we shall limit our discussion to the oscillatory WKB limit in which case the Klein-Gordon-type current (3.42) is adequate.

The issue of the duration of the period of inflation has been a point of some debate between proponents of the no-boundary wavefunction and proponents of the tunneling wavefunction. Consider the canonical minisuperspace model of §3.7. We have seen that in the WKB limit the Hartle-Hawking boundary condition gives rise to a wavefunction (4.22) in the oscillatory region, which is strongly peaked about the set of classical solutions (3.62) that correspond to an inflating universe: $a(t) \propto e^{\sqrt{V}\,t}$, $\phi(t) \simeq \phi_0 = \text{const.}$

Vilenkin[40–42,70] has also studied the minisuperspace model of §3.7, but with a different choice of factor ordering in the Wheeler-DeWitt equation (3.59). He concludes that similarly to (4.22) the tunneling boundary condition leads to a WKB wavefunction

$$\Psi_{\text{V}}(a,\phi) \propto \exp\left(\frac{-1}{3V(\phi)}\right) \exp\left[\frac{-i}{3V(\phi)}\left(a^2V(\phi)-1\right)^{3/2} + \frac{i\pi}{4}\right] \tag{5.1}$$

in the oscillatory region, which is also peaked about the classical trajectories of the inflationary universes (3.62).

Both the no-boundary and tunneling wavefunctions thus predict inflation, at least in the context of this simple minisuperspace model. The important question is *how much* inflation do the models predict? This will be largely decided by the value, ϕ_0, of the scalar field with which the universe "nucleates" in the semiclassical regime.

In order to study the probability flux arising from (3.42) in the WKB limit it is necessary to focus on one particular WKB component. The no-boundary wavefunction (4.22) is of course real, and the resulting current (3.42) is identically zero. However, $\Psi_{\rm HH}$ is the superposition (4.23) of two WKB components, which will correspond to contracting and expanding universes. As discussed in §3.5 it is assumed that a decoherence mechanism exists so that the interference between the two components is negligible[89], and we can therefore assume that the universe is peaked about one or other WKB component for the purposes of determining the probability measure. The difference between the no-boundary and tunneling wavefunctions may therefore not seem great once decoherence to a classical universe is assumed. In particular, if we take the "outgoing" WKB component in the no-boundary case then the only significant difference between $\Psi_{\rm HH}$ and $\Psi_{\rm T}$ is the ϕ-dependent part of the prefactor, $\exp\left(\frac{\pm 1}{3V(\phi)}\right)$, which is obtained from boundary conditions set in the tunneling region. This difference will have ramifications for the probability flux, however.

One question which we might hope to answer in our minisuperspace model would be: given that a Lorentzian universe nucleates, what is the probability that it inflates by a sufficient amount (~ 65 e-folds) to solve the problems of the standard cosmology mentioned in §1? The answer to this would involve integrating the probability flux (3.45) on the surface separating the tunneling and oscillatory regions, which is roughly given by $a^2 V(\phi) = 1$. However, our discussion here is limited by the fact that our WKB approximation applies to trajectories with $\dot{\phi} \simeq 0$. For such trajectories the probability current, $\mathcal{J}$, points chiefly in the direction of the a-coordinate in minisuperspace. We can therefore attempt to answer the question approximately by evaluating the probability current (3.44) on a surface Σ in minisuperspace with $a = \text{const}$. For sufficiently large values of a such surfaces will lie almost entirely within the oscillatory regime (see Fig. 10). On these surfaces we obtain a probability flux

$$d\mathcal{P} = \mathcal{J} \cdot d\Sigma \propto \begin{cases} \exp\left(\frac{+2}{3V(\phi)}\right) & \text{no-boundary wavefunction, } \Psi_{\rm HH}, \\ \exp\left(\frac{-2}{3V(\phi)}\right) & \text{tunneling wavefunction, } \Psi_{\rm V}. \end{cases} \tag{5.2}$$

Although the integral of (5.2) may diverge for particular potentials $V(\phi)$ of interest, this should not be viewed as a problem since, as was emphasised in §3.6, questions in quantum cosmology can only refer to conditional rather than absolute probabilities.

In the present case let us assume that inflation occurs for large values of the scalar field, as is the case for potentials of the "chaotic" type, $V = \lambda\phi^{2p}$. There will then be a minimum value of the scalar field, $\phi_{\rm suff}$, for which sufficient inflation is obtained. For $V = \lambda\phi^{2p}$ a universe with $\phi = \phi_0$ initially will undergo

$$N_e \simeq \frac{3}{2p}\left[{\phi_0}^2 - \tfrac{2}{9}p(2p-1)\right] \tag{5.3}$$

e-folds of inflation⊡, so that in the case of the quadratic potential ($p = 1$) we find $\phi_{\rm suff} \gtrsim 6.6$, for example. The relevant conditional probability for sufficient inflation on an $a =$ const surface is then given by

$$\mathcal{P}\left(\phi_0 > \phi_{\rm suff} \mid \phi_1 < \phi_0 < \phi_2\right) = \frac{\displaystyle\int_{\phi_{\rm suff}}^{\phi_2} \mathrm{d}\phi_0 \exp\left(\frac{\pm 2}{3V(\phi_0)}\right)}{\displaystyle\int_{\phi_1}^{\phi_2} \mathrm{d}\phi_0 \exp\left(\frac{\pm 2}{3V(\phi_0)}\right)}, \tag{5.4}$$

where the (+) case refers to the no-boundary wavefunction and the (−) case to the tunneling wavefunction, and the values ϕ_1 and ϕ_2 are respectively lower and upper cutoffs on the allowed values of ϕ, which must be determined by physical criteria. A minimum cut-off might be expected for a variety of physical reasons, such as avoiding classical universes which rapidly recollapse.

In their original investigations of the question of the duration of the period of inflation for the quadratic scalar potential, Hawking and Page[52] took $\phi_2 = \infty$, in which case both integrals in (5.4) are dominated by large values of ϕ, leading to a probability $\mathcal{P} \simeq 1$, and thus a "prediction" of inflation. However, this result has been criticised by Vilenkin[41] since for $k = +1$ we have $\mathcal{V} = \frac{9}{16} m_{\rm Planck}^4 V$ so that values of $m\phi \gtrsim 4/3$ are in excess of the Planck scale, and the semiclassical approximation will no longer apply. Vilenkin suggested that an upper cutoff, ϕ_2, should be introduced at the Planck scale. If this is the case then provided the lower cutoff ϕ_1 is sufficiently close to zero, we would find that the integral in the denominator of (5.4) becomes very large in the case of the no-boundary wavefunction (+ sign), leading to $\mathcal{P} \ll 1$, whereas this would not be the case for the tunneling wavefunction (− sign), and thus the latter would predict more inflation.

Introducing a cutoff at the Planck scale might be deemed a rather arbitrary procedure, since without any knowledge of Planck scale physics it is impossible to be sure whether the "real" answer is better approximated by the introduction of a cutoff or not. However, it has been argued on the basis of investigations at the 1-loop level that the wavefunction is damped at large values of ϕ by quantum corrections[128,129]. This renders the wavefunction normalisable and would justify a cutoff, ϕ_2, near the Planck scale.

⊡ In the slow-rolling approximation [109] it follows from (3.54) and (3.56) (with $\mathcal{N} = 1$) that the number of e-folds is $N_e = 6\int_{\phi_e}^{\phi_0} \mathrm{d}\phi\, V/V'$, where ϕ_0 is the initial value and ϕ_e the final value of the scalar field at the end of the inflationary epoch. For $V = \lambda\phi^{2p}$ we have $N_e = \frac{3}{2p}\left({\phi_0}^2 - {\phi_e}^2\right)$. The value of ϕ_e can be estimated from the limit set by $|V'/V| \ll 6$ and $|V''/V| \ll 9$.

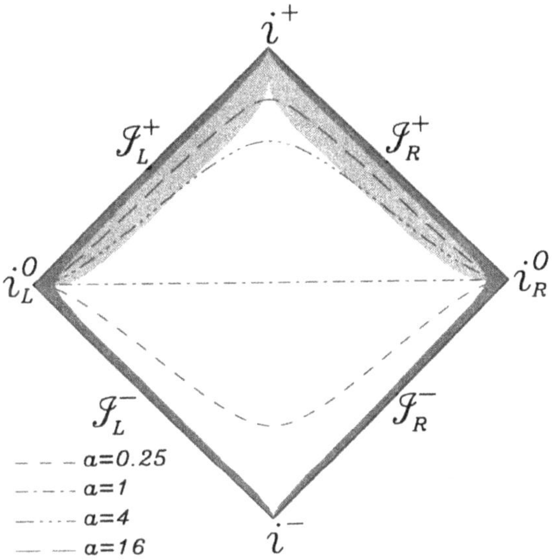

Fig. 10: Conformal diagram for $V = 0.04\phi^2$. The oscillatory region, given roughly by $a^2V > 1$, is lightly shaded. Lines a = const are superimposed. For very large values of ϕ these lie almost entirely in the oscillatory region. The region of ϕ-values excluded by a Planck scale cutoff is darkly shaded.

Given that the no-boundary wavefunction apparently yields a wavefunction peaked around the lower cutoff, ϕ_1, it is important to determine what a reasonable value of this cutoff should be. This issue has been considered by Grishchuk and Rozhansky[130], and also more recently by Lukas[131], who have conducted numerical investigations to analyse the behaviour of the caustic[132] in the (a, ϕ) plane which separates the Euclidean and Lorentzian solutions. This is illustrated in Fig. 11, where classical solutions to the Euclidean field equations (3.54)–(3.56) (with $\mathcal{N} = 1$) are shown. The solutions beginning at $a = 0$ initially follow lines with $\dot{\phi} \simeq 0$, but the approximation eventually breaks down when the trajectories curl back and recollapse with $a \to 0$ and $|\phi| \propto -\ln a \to \infty$. They thus represent a flow from $\mathfrak{i}^-$ to $\mathfrak{I}^-$ in the conformal diagrams. The solutions cross each other on the caustic, which for large values of ϕ corresponds to $a^2V(\phi) = 1$, in accordance with our expectation from the WKB approximation.

To the right of the Euclidean solutions in Fig. 11 we would find Lorentzian solutions with $\dot{\phi} \simeq 0$ sufficiently far away from the caustic[132]. These solutions are not depicted here. The "nucleation of a universe" would thus correspond to a solution of the Wheeler-DeWitt equation which was initially peaked about a classical Euclidean trajectory with $\dot{\phi} \simeq 0$, and which then crossed over the caustic to be peaked about a corresponding trajectory with $\dot{\phi} \simeq 0$ in the Lorentzian region.

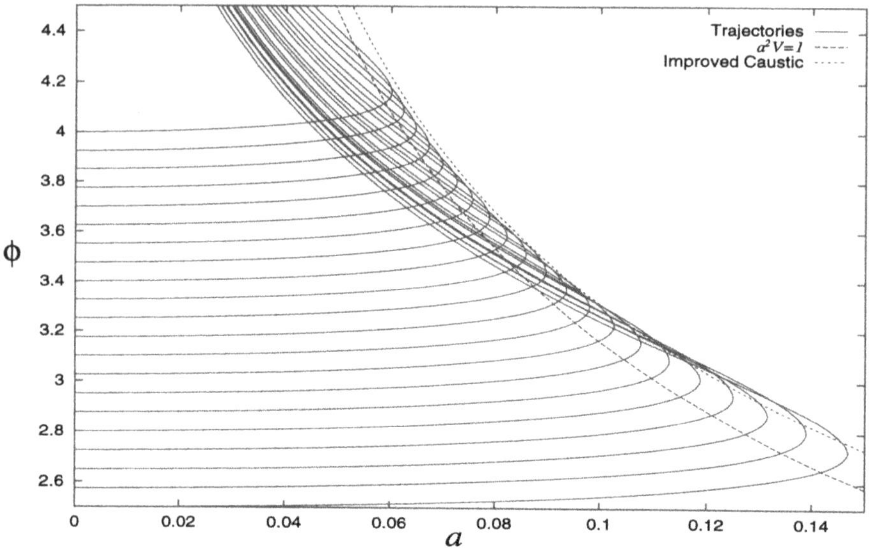

Fig. 11: Classical Euclidean trajectories for the potential $V = \phi^4$. The approximate caustic $a^2V = 1$ is indicated by a dashed line, and the improved caustic by a dotted line. (From [131].)

While crossing the caustic, the wavefunction would be peaked about a complex solution which was neither truly Euclidean nor truly Lorentzian.

It is evident from Fig. 11 that for small values of ϕ the caustic begins to deviate from the curve $a^2V = 1$. In fact, one finds[130,131] that for sufficiently small values of ϕ Lorentzian universes never nucleate at all. Accordingly, a portion of the shaded region in the conformal diagrams should be excised about the $\phi = 0$ axis in the vicinity of ι^+. The cutoff, ϕ_*, at which this occurs has been estimated by Lukas[131] as being

$$\phi_* \simeq p + \left\{2p\left[1 - (1 + e^{-1})^{-1/2}\right]\right\}^{1/2}, \tag{5.5}$$

in the case of the chaotic potentials, $V = \lambda\phi^{2p}$. Provided that the interpretation of (5.4) remains valid♠, we should therefore set $\phi_1 = \phi_*$ as the lower bound in the integrals. In the case of the quadratic potential, $V = m^2\phi^2$, we then have $\phi_1 \simeq 1.5$, and using the values of $\phi_{\rm suff}$ and ϕ_2 found earlier it is straightforward to

♠ As a cautionary note, one should observe that since the approximation of the caustic by the curve $a^2V = 1$ breaks down for small ϕ, the use of the surfaces $a =$ const which was assumed in (5.4) may not be appropriate for small ϕ, and ideally one should determine the probability flux across the caustic itself. Of course, in a more careful analysis one would consider complex solutions to the field equations in line with [116], rather than real Euclidean and real Lorentzian solutions with a junction condition. One would hope that this would not alter the conclusions of the analysis much.

check that for typical values of $m \ll 1$, (5.4) gives $\mathcal{P} \ll 1$ for Ψ_{HH} and $\mathcal{P} \sim 1$ for Ψ_V in accordance with the earlier discussion.

The above analysis would appear to indicate that the no-boundary wavefunction effectively "predicts" a value of $\phi \simeq \phi_*$, which in the case of the quadratic potential is unfortunately less than ϕ_{suff}. However, the calculation is model-dependent, and if one could find a potential for which $\phi_* \gtrsim \phi_{suff}$ then both Ψ_{HH} and Ψ_V would yield $\mathcal{P} \sim 1$. If we compare (5.3) and (5.5) (for $N_e \simeq 65$) we see that in the case of the chaotic potentials, $V = \lambda\phi^{2p}$, this requirement is equivalent to $p \gtrsim 62$. However, it requires an enormously small value of λ to keep such a V below the Planck scale and this is not promising. Lukas[131] has also estimated the value of ϕ_* for some other potentials but did not find any candidates with $\phi_* \gtrsim \phi_{suff}$. However, given the model-dependence of the calculations it cannot be ruled out that some other potential, or the coupled effects of two or more scalar fields, might give $\phi_* \gtrsim \phi_{suff}$ and thereby make inflation a "prediction" of the no-boundary wavefunction.

5.2. The origin of density perturbations

It is expected that the anisotropies in the cosmic microwave background radiation, which were first definitively observed in 1992, have their origin in quantum fluctuations in the very early universe. Such fluctuations can be described using the formalism of quantum field theory in curved spacetime. Since the era of quantum cosmology is in a sense prior to that in which quantum field theory in curved spacetime is applicable, one would hope to trace the origin of the primordial perturbations back to quantum cosmology. Indeed, this can be done[133–135] and I will very briefly outline the main results.

With homogeneous minisuperspaces as a starting point, one can add small inhomogeneous perturbations to the metric and matter fields:

$$h_{ij}(x,t) = a^2(t)\,(\Omega_{ij} + \varepsilon_{ij}), \tag{5.6}$$
$$\Phi(x,t) = \Phi_0(t) + \delta\Phi(x,t), \tag{5.7}$$
$$\mathcal{N}(x,t) = \mathcal{N}_0(t) + \delta\mathcal{N}(x,t), \tag{5.8}$$
$$\mathcal{N}_i(x,t) = 0 + \delta\mathcal{N}_i(x,t). \tag{5.9}$$

Here we have restricted attention to the $k = +1$ FRW minisuperspace model, the subscript zero denotes the unperturbed quantities, and Ω_{ij} is the unperturbed standard round metric on the 3-sphere. The perturbations can be expanded in terms of spherical harmonics on the 3-sphere[133]. If one substitutes the ansatz (5.6)–(5.9) into the classical action (2.12) and expands to quadratic order one obtains an action which can be split into the original minisuperspace action, S_0, and an additional action, S_2, quadratic in the perturbations,

$$S = S_0[q^A, \mathcal{N}_0] + S_2[q^A, \mathcal{N}_0, \varepsilon_{ij}, \delta\Phi, \delta\mathcal{N}, \mathcal{N}_i], \tag{5.10}$$

and a corresponding Hamiltonian

$$H = \mathcal{N}_0 \left(\mathsf{H}_0 + \int d^3x\, \mathcal{H}_2 \right) + \int d^3x\, \delta\mathcal{N}\, \mathcal{H}_1 + \int d^3x\, \delta\mathcal{N}_i\, \mathcal{H}^i \,, \tag{5.11}$$

where H_0 is the unperturbed Hamiltonian (3.54), and the terms $\mathcal{H}_1$ and $\mathcal{H}_2$ are linear and quadratic in the perturbations respectively. There is now a non-trivial momentum constraint at each point $x \in \Sigma$, $\mathcal{H}^i(x) = 0$, while the Hamiltonian constraint splits into a piece linear in the perturbations, $\mathcal{H}_1(x) = 0$, plus a homogeneous piece

$$\mathsf{H}_0 + \mathsf{H}_2 \equiv \mathsf{H}_0 + \int d^3x\, \mathcal{H}_2 = 0. \tag{5.12}$$

We may quantise (5.12) in the standard fashion to obtain a modified Wheeler-Dewitt equation

$$\hat{\mathsf{H}}\Psi = \left[-\tfrac{1}{2}\nabla^2 + \mathcal{U}(q) + \hat{\mathsf{H}}_2 \right] \Psi = 0, \tag{5.13}$$

in place of (3.27), where the Laplacian is still defined in terms of the original minisuperspace coordinates according to (3.28), and $\hat{\mathsf{H}}_2$ is a second order differential operator which results from quantisation of the homogeneous part of the perturbations. In the case of pure scalar field modes, for example, the perturbations may be decomposed as

$$\delta\Phi(x,t) = \frac{1}{\sigma} \sum_{nml} \mathsf{f}_{nlm}(t) Q^n{}_{lm}(x), \tag{5.14}$$

where the $Q^n{}_{lm}$ satisfy the 3-dimensional Laplace equation on S^3,

$$^{(3)}\Delta Q^n{}_{lm} = -(n^2 - 1) Q^n{}_{lm}, \tag{5.15}$$

and one finds[1,89,133]

$$\hat{\mathsf{H}}_2 = \tfrac{1}{2} \sum_{nml} \left\{ \frac{-1}{a^3} \frac{\partial^2}{\partial \mathsf{f}^2_{nlm}} + \left[(n^2-1)a + m^2 a^3 \right] \mathsf{f}^2_{nlm} \right\} \tag{5.16}$$

in the case of the quadratic scalar potential.

It is possible to find solutions to the Wheeler-DeWitt equation (5.13) in which the minisuperspace coordinates, q^A, are treated semiclassically in the WKB approximation, while the perturbations are treated quantum mechanically. One can show[1,89,133,134] that the solutions take the form

$$\Psi = \mathcal{C}(q,\Phi)\, e^{i S_0(q,\Phi)}\, \tilde{\psi}, \tag{5.17}$$

where S_0 satisfies the unperturbed Hamilton-Jacobi equation (3.34), the prefactor $\mathcal{C}$ depends only on the unperturbed minisuperspace coordinates, and the functions $\tilde{\psi}$ satisfy the functional Schrödinger equation

$$i \frac{\partial \tilde{\psi}}{\partial t} = \hat{\mathsf{H}}_2 \tilde{\psi}. \tag{5.18}$$

The different modes of the scalar perturbations (5.14) do not interact, for example, and in this case

$$\tilde{\psi} = \prod_{nml} \psi_{nlm}(t, f_{nlm}), \tag{5.19}$$

where each mode ψ_{nlm} separately satisfies (5.17) with $\hat{H}_2$ given by (5.16).

The wavefunction (5.17) is thus peaked about classical trajectories, with corrections, $\tilde{\psi}$, which satisfy the functional Schrödinger equation along these trajectories. This is in fact precisely the starting point for the treatment of matter modes in a curved spacetime background using the formalism of quantum field theory in curved spacetime. Thus the above result is important in that it demonstrates that quantum cosmology is consistent with the standard approach to the quantum treatment of cosmological perturbations. As an added bonus the imposition of a boundary condition on Ψ, such as the no-boundary or tunneling condition, will result in the choice of particular solutions of the functional Schrödinger equation, and consequently a particular vacuum state for the matter modes. Both the no-boundary condition and the tunneling condition pick out[41,136–139] a de-Sitter invariant state known as the "Euclidean" or Bunch-Davies vacuum[140,141], which is the state that is often assumed in cosmological calculations of density perturbations. In fact, this state is picked out by many boundary conditions[137,138], and thus could be regarded as a natural quantum state for matter in quantum cosmology.

One may solve the functional Schrödinger equation and study the growth[133] of the modes. As discussed above, the results obtained are the same as those which are found if one begins with quantum matter fields in de-Sitter invariant vacuum states in background inflationary spacetimes[142,143]. In particular, one obtains a scale-free spectrum of density perturbations which act as seeds for the formation of galaxies and other structures in the universe. Such a spectrum accords well with the spectrum deduced from the COBE measurements of cosmic microwave anisotropies[144].

5.3. The arrow of time

The question of the origin of the arrow of time in the face of the CPT-invariance of the laws of physics is one of the deepest unresolved conceptual issues in theoretical physics, and it has been the subject of more than one major conference[44,145]. The question of the nature of time assumes prime importance in quantum gravity and quantum cosmology[♡]. One might argue that if the basic laws of physics are time symmetric, then the arrow of time could have its origin in a boundary condition for the wavefunction of the universe.

There are many different arrows of time observed in the universe. Some of these such as the psychological arrow of time (we remember only the past), and

♡ The conceptual issues surrounding the nature of time in quantum cosmology would entail a series of lectures in themselves. For further discussion see, e.g., [24], [146]–[148] and references therein.

the electromagnetic arrow of time (the choice of retarded as opposed to advanced solutions of Maxwell's equations), could well be argued to be consequences of the thermodynamic arrow of time, which arises from the second law of thermodynamics. However, the expansion of the universe provides an alternate cosmological arrow of time which is not obviously directly related to the increase of entropy. Whether such a relationship does exist is a question which might potentially be resolved by quantum cosmology.

In 1985 Hawking[149] proposed that the thermodynamic arrow and cosmological arrows of time were correlated: that is to say if the universe were spatially closed then entropy would decrease in the contracting phase[⊙]. In arriving at this proposal he was influenced by the fact that the no-boundary wavefunction is CPT-invariant, and also by some early studies of simple minisuperspace models which possessed quasi-periodic solutions[151] which "bounced" instead of recontracting to a singularity as $a \rightarrow 0$. However, Hawking soon changed his mind on the issue, which he terms his "greatest mistake"[152]. This change of mind was brought about by a number of factors. Firstly, Page pointed out that CPT-invariance of the no-boundary wavefunction does not imply CPT-invariance for an individual WKB component of the wavefunction, which would correspond to the history of a classical universe[153]. Furthermore, minisuperspace models which were subsequently studied, such as that of the Kantowski-Sachs universe, were found not to admit bounce solutions but always possessed singularities to the future[154,155]. Finally, as was mentioned in §4.1, the bounce solutions do not feature even in simple minisuperspace models once one considers the contribution of complex metrics[116]. In general, the approximate minisuperspace boundary condition that $\Psi \rightarrow 1$ as $a \rightarrow 0$ must be altered to allow for approximately Euclidean metrics which also contribute a rapidly oscillating component to the wavefunction[116,156].

Hawking, Laflamme and Lyons[156] have recently argued that a thermodynamic arrow of time results from the imposition of the Hartle-Hawking boundary condition. More precisely, they have considered the evolution of primordial fluctuations as outlined in §5.2, but accounting for the "approximately Euclidean" geometries[116] which appear to be required if one considers complex metrics. They find[156] that the gravitational wave perturbations have an amplitude that remains in the linear regime and is roughly time-symmetric about the time of maximum expansion. Such perturbations cannot be said to give rise to an arrow of time. Density perturbations behave differently, however. They start out small but grow large and become non-linear as the universe expands, and moreover this growth continues during the contracting phase of the spatially closed universe. This growth of inhomogeneity therefore provides an arrow of time which could be considered to be an essentially thermodynamic arrow. Since it does not match the cosmological arrow in the contracting phase, the only reason for the coincidence of the two arrows in the present epoch would appear to be an anthropic one. In particular, the conditions that would

⊙ This idea was first suggested by Gold [150], and has resurfaced a number of times in different contexts.

prevail in a contracting phase would appear to preclude the existence of life[152,156]. Thus the fact that we are around to make observations means we must find ourselves in a cosmological epoch in which the two arrows coincide.

It should be added that the debate about whether the cosmological and thermodynamic arrows of time coincide has not yet been closed, however, and it is still maintained by Kiefer and Zeh[157] that the boundary condition on the Wheeler-DeWitt equation must be such that the thermodynamic arrow would reverse in a recontracting universe.

6. Conclusion

I hope to have shown you that although research in quantum cosmology is still rather speculative and open-ended, its framework nonetheless has the potential not only to provide answers to questions surrounding the origin and early evolution of the universe, but also to help us unravel the mysteries of quantum gravity. Quantum cosmology is a field in which a great deal remains to be done, and the results of §5 should be regarded as a hopeful indication of the types of predictions we might hope to make. It is too early to draw definitive conclusions about the relative merits of the various boundary condition proposals. The results of §5.1 ostensibly favour the "tunneling proposal". However, this conclusion is only based on a few simple models, and therefore some caution must be exercised.

As a final note, it is worth mentioning that recent results show that supersymmetry provides a means of restricting possible boundary conditions for the Wheeler-DeWitt equation, or the corresponding Dirac square-root equation. The supergravity constraint equations for various homogeneous minisuperspace models appear to be so restrictive that they only pick out the most symmetric quantum states[102,103,158]. In particular, simple analytic solutions for Ψ in the supersymmetric Bianchi-IX minisuperspace have been found[102] in the empty and filled fermion sectors, which have a natural interpretation as[158] wormhole states[121,159], or as[160] Hartle-Hawking no-boundary states. Some doubts were initially cast on the relevance of these solutions, as the states appear to have no counterpart in 4-dimensional supergravity[161,162]. However, more recent work[46] on supersymmetric minisuperspaces corresponding to Bianchi "class A"[76] models indicates that infinitely many physical states with finite (even) fermion number can be found, and these are direct analogues of physical states in full supergravity. This result shows that such minisuperspace models are likely to be physically very important for quantum cosmology.

Acknowledgement I would like to thank A. Lukas for giving me permission to use Fig. 11, which originally appeared in [131].

References

1. J. J. Halliwell, in *Quantum Cosmology and Baby Universes*, eds. S. Coleman, J. B. Hartle T. Piran and S. Weinberg, (World Scientific, Singapore, 1991), p. 159.
2. D. N. Page, in *Proceedings of the Banff Summer Institute on Gravitation, August 1990*, eds. R. B. Mann and P. S. Wesson, (World Scientific, Singapore, 1991).
3. P. van Nieuwenhuizen, Phys. Rep. **68** (1981) 189.
4. P. West, *Introduction to Supersymmetry and Supergravity*, (World Scientific, Singapore, 1986).
5. A. Salam and E. Sezgin (eds.), *Supergravities in Diverse Dimensions*, (World Scientific, Singapore, 1989).
6. G. Esposito, *Quantum Gravity, Quantum Cosmology and Lorentzian Geometries*, (Springer *Lecture Notes in Physics* **m12**, Berlin, 1992).
7. M. H. Goroff and A. Sagnotti, Phys. Lett. **160B** (1985) 81; Nucl. Phys. **B266** (1986) 709.
8. G. 't Hooft and M. Veltman, Ann. Inst. H. Poincaré **20** (1974) 69.
9. S. Deser and P. van Nieuwenhuizen, Phys. Rev. **D10** (1974) 411.
10. S. Deser, P. van Nieuwenhuizen and H. S. Tsao, Phys. Rev. **D10** (1974) 3337.
11. S. Deser, J. H. Kay and K. S. Stelle, Phys. Rev. Lett. **38** (1977) 527.
12. P. S. Howe and K. S. Stelle, Int. J. Mod. Phys. **A4** (1989) 1871.
13. M. B. Green, J. H. Schwarz and E. Witten, *Superstring theory*, (Cambridge University Press, 1987).
14. D. J. Gross and V. Periwal, Phys. Rev. Lett. **60** (1988) 2105.
15. S. Davis, Preprint DAMTP-R/94/27-rev (1994), hep-th/9503231.
16. A. Ashtekar, *Lectures on Non-perturbative Canonical Gravity*, (World Scientific, Singapore, 1991).
17. C. Rovelli, Class. Quantum Grav. **8** (1991) 1613.
18. A. Sen, J. Math. Phys. **22** (1981) 1781.
19. A. Ashtekar, Phys. Rev. Lett. **57** (1986) 2244; Phys. Rev. **D36** (1987) 1587.
20. C. Rovelli and L. Smolin, Phys. Rev. Lett. **61** (1989) 1155; Nucl. Phys. **B331** (1990) 80.
21. M. P. Blencowe, Nucl. Phys. **B341** (1990) 213.
22. H. Kodama, Prog. Theor. Phys. **80** (1988) 1024; Phys. Rev. **D42** (1990) 2548.
23. C. Soo and L. N. Chang, Int. J. Mod. Phys. **D3** (1994) 529.

24. L. Smolin and C. Soo, Nucl. Phys. **B449** (1995) 289.
25. J. Louko, Phys. Rev. **D51** (1995) 586.
26. T. Regge, Nuovo Cimento **19** (1961) 558.
27. R. M. Williams and P. A. Tuckey, Class. Quantum Grav. **9** (1992) 1409.
28. L. Bombelli, J. Lee, D. Meyer and R. D. Sorkin, Phys. Rev. Lett. **59** (1987) 521.
29. C. J. Isham, Class. Quantum Grav. **6** (1989) 1509.
30. C. J. Isham, Y. Kubyshin and P. Renteln, Class. Quantum Grav. **7** (1990) 1053.
31. J. A. Wheeler, in *Relativity, Groups and Topology*, eds. C. DeWitt and B. S. DeWitt, (Gordon and Breach, New York, 1963), p. 315.
32. J. A. Wheeler, in *Batelles Rencontres*, eds. C. DeWitt and J. A. Wheeler, (Benjamin, New York, 1968), p. 242.
33. B. S. DeWitt, Phys. Rev. **160** (1967) 1113.
34. C. W. Misner, Phys. Rev. **186** (1969) 1319; in *Relativity*, eds. M. Carmeli, Finkler and L. Witten, (Plenum, New York, 1970), p. 55.
35. C. W. Misner, in *Magic Without Magic: John Archibald Wheeler*, ed. J. Klauder, (W.H. Freeman, San Francisco, 1972), p. 441.
36. S. W. Hawking, in *Astrophysical Cosmology*, eds. H. A. Brück, G. V. Coyne and M. S. Longair, (Pontifica Academia Scientarium *Scripta Varia* **48**, Vatican City, 1982) p. 563.
37. A. Vilenkin, Phys. Lett. **117B** (1982) 25; Phys. Rev. **D27** (1983) 2848.
38. J. B. Hartle and S. W. Hawking, Phys. Rev. **D28** (1983) 2960.
39. S. W. Hawking, Nucl. Phys. **B239** (1984) 257.
40. A. Vilenkin, Phys. Rev. **D33** (1986) 3560.
41. A. Vilenkin, Phys. Rev. **D37** (1988) 888.
42. A. Vilenkin, Phys. Rev. **D50** (1994) 2581.
43. J. B. Hartle, in *Quantum Cosmology and Baby Universes*, eds. S. Coleman, J. B. Hartle T. Piran and S. Weinberg, (World Scientific, Singapore, 1991), p. 65.
44. J. J. Halliwell, J. Perez-Mercader, and W. H. Zurek (eds.), *The Physical Origins of Time Asymmetry*, (Cambridge University Press, 1994).
45. A. Strominger, in *Quantum Cosmology and Baby Universes*, eds. S. Coleman, J. B. Hartle, T. Piran and S. Weinberg, (World Scientific, Singapore, 1991), p. 269.
46. A. Csordás and R. Graham, Phys. Rev. Lett. **74** (1995) 4129.
47. R. M. Wald, *General Relativity*, (Chicago University Press, 1984).

48. J. W. York, Phys. Rev. Lett. **28** (1972) 1082.

49. P. A. M. Dirac, Proc. R. Soc. Lond. **A246** (1958) 326; 333.

50. A. Hanson, T. Regge and C. Teitelboim, *Constrained Hamiltonian Systems*, (Accademia Nationale dei Lincei, Rome, 1976).

51. P. W. Higgs, Phys. Rev. Lett. **1** (1958) 373.

52. S. W. Hawking and D. N. Page, Nucl. Phys. **B264** (1986) 185.

53. G. W. Gibbons and S. W. Hawking, Phys. Rev. **D15** (1977) 2752.

54. S. W. Hawking, in *General Relativity: An Einstein Centenary Survey* (Cambridge University Press, 1979), p. 746.

55. J. B. Hartle and S. W. Hawking, Phys. Rev. **D13** (1976) 2188.

56. G. W. Gibbons and S. W. Hawking, Phys. Rev. **D15** (1977) 2738.

57. G. W. Gibbons and M. J. Perry, Proc. R. Soc. Lond. **A358** (1978) 467.

58. T. Eguchi and A. J. Hanson, Phys. Lett. **74B** (1978) 249.

59. G. W. Gibbons and S. W. Hawking, Phys. Lett. **78B** (1978) 430; Commun. Math. Phys. **66** (1979) 291.

60. G. W. Gibbons and C. N. Pope, Commun. Math. Phys. **66** (1979) 267.

61. G. W. Gibbons, in *Fields and Geometry*, ed. A. Jadczyk, (World Scientific, Singapore, 1986).

62. D. Garfinkle and A. Strominger, Phys. Lett. **256B** (1991) 146.

63. H. F. Dowker, J. P. Gauntlett, S. B. Giddings and G. T. Horowitz, Phys. Rev. **D50** (1994) 2662.

64. G. W. Gibbons, S. W. Hawking and M. J. Perry, Nucl. Phys. **B138** (1978) 141.

65. K. Schleich, Phys. Rev. **D36** (1987) 2342.

66. J. J. Halliwell and J. Louko, Phys. Rev. **D39** (1989) 2206.

67. J. J. Halliwell and J. Louko, Phys. Rev. **D40** (1989) 1868.

68. J. J. Halliwell and J. Louko, Phys. Rev. **D42** (1990) 3997.

69. A. D. Linde, Zh. Eksp. Teor. Fiz. **87** (1984) 369 [Sov. Phys. JETP **60** (1984) 211]; Lett. Nuovo Cimento **39** (1984) 401; Rep. Prog. Phys. **47** (1984) 925.

70. A. Vilenkin, Phys. Rev. **D30** (1984) 509; Nucl. Phys. **B252** (1985) 141.

71. J. J. Halliwell and J. B. Hartle, Phys. Rev. **D43** (1991) 1170.

72. K. V. Kuchař and M. P. Ryan, Phys. Rev. **D40** (1989) 3982.

73. S. Sinha and B. L. Hu, Phys. Rev. **D44** (1991) 1028.

74. F. D. Mazzitelli, Phys. Rev. **D46** (1992) 4758.

75. A. Ishikawa and T. Isse, Mod. Phys. Lett. **A8** (1993) 3413.

76. M. A. H. MacCallum, in *General Relativity: An Einstein Centenary Survey* eds. S. W. Hawking and W. Israel, (Cambridge University Press, 1979), p. 533.

77. A. S. Kompaneets and A. S. Chernov, Zh. Eksp. Teor. Fiz. **47** (1964) 1939 [Sov. Phys. JETP **20** (1965) 1303].

78. R. Kantowski and R. K. Sachs, J. Math. Phys. **7** (1966) 443.

79. L. Bianchi, Mem. di Mat. Soc. Ital. Sci. **11** (1897) 267.

80. C. W. Misner, Phys. Rev. Lett. **22** (1969) 1071.

81. V. A. Belinskii, E. M. Lifschitz and I. M. Khalatnikov, Usp. Fiz. Nauk. **102** (1970) 463 [Sov. Phys. Usp. **13** (1970) 745].

82. A. Latifi, M. Musette and R. Conte, Phys. Lett. **194A** (1994) 83.

83. J. Louko, Ann. Phys. (N.Y.) **181** (1988) 318.

84. J. J. Halliwell, Phys. Rev. **D38** (1988) 2468.

85. I. G. Moss, Ann. Inst. H. Poincaré **49** (1988) 341.

86. D. N. Page, J. Math. Phys. **32** (1991) 3427.

87. T. Banks, Nucl. Phys. **B249** (1985) 332.

88. S. P. Kim, Phys. Rev. **D52** (1995) 3382.

89. J. J. Halliwell, Phys. Rev. **D39** (1989) 2912.

90. E. Calzetta and J. J. Gonzalez, Phys. Rev. **D51** (1995) 6821.

91. S. Habib and R. Laflamme, Phys. Rev. **D42** (1990) 4056.

92. E. Calzetta, Phys. Rev. **D43** (1991) 2498.

93. B. L. Hu, J. P. Paz and S. Sinha, in *Directions in General Relativity, Vol. I*, eds. B. L. Hu, M. P. Ryan and C. V. Vishveshwara, (Cambridge University Press, 1993), p. 145.

94. B. L. Hu, in [44], p. 475.

95. J. J. Halliwell, Phys. Rev. **D36** (1987) 3626.

96. K. V. Kuchař, J. Math. Phys. **22** (1981) 2640.

97. M. P. Ryan, *Hamiltonian Cosmology* (Springer *Lecture Notes in Physics* **13**, Berlin, 1972).

98. C. Teitelboim, Phys. Rev. Lett. **38** (1977) 1106.

99. R. Tabensky and C. Teitelboim, Phys. Lett. **69B** (1977) 453.

100. A. Macías, O. Obregón and M. P. Ryan, Class. Quantum Grav. **4** (1977) 1477.

101. P. D. D'Eath and D. I. Hughes, Phys. Lett. **214B** (1988) 498; Nucl. Phys. **B378** (1992) 381.

102. R. Graham, Phys. Rev. Lett. **67** (1991) 1381; Phys. Lett. **277B** (1992) 393.

103. P. D. D'Eath, S. W. Hawking and O. Obregón, Phys. Lett. **300B** (1993) 44.
104. A. Vilenkin, Phys. Rev. **D39** (1989) 1116.
105. D. N. Page, Phys. Rev. **D34** (1986) 2267.
106. J. B. Hartle, Phys. Rev. **D37** (1988) 2818, **D43** (1991) 1434.
107. D. N. Page, Phys. Rev. Lett. **70** (1993) 4034.
108. S. W. Hawking and D. N. Page, Nucl. Phys. **B298** (1988) 789.
109. E. W. Kolb and M. S. Turner, *The Early Universe*, (Addison-Wesley, Reading, Mass., 1990).
110. D. L. Wiltshire, Phys. Rev. **D36** (1987) 1634.
111. M. Henneaux, Lett. Nuovo Cimento **38** (1983) 609.
112. G. W. Gibbons, S. W. Hawking and J. M. Stewart, Nucl. Phys. **B281** (1987) 736.
113. W. M. Suen and K. Young, Phys. Rev. **D39** (1989) 2201.
114. H. D. Conradi and H. D. Zeh, Phys. Lett. **154A** (1991) 321.
115. H. D. Conradi, Phys. Rev. **D46** (1992) 612; Class. Quantum Grav. **12** (1995) 2423.
116. G. W. Lyons, Phys. Rev. **D46** (1992) 1546.
117. J. J. Halliwell and J. B. Hartle, Phys. Rev. **D41** (1990) 1815.
118. G. W. Gibbons and L. P. Grishchuk, Nucl. Phys. **B313** (1989) 736.
119. D. N. Page, in *Quantum Concepts in Space and Time*, eds. R. Penrose and C. J. Isham, (Oxford University Press, 1986) p. 274.
120. D. J. Kaup and A. P. Vitello, Phys. Rev. **D9** (1974) 1648.
121. S. W. Hawking and D. N. Page, Phys. Rev. **D42** (1990) 2655.
122. D. N. Page, Class. Quantum Grav. **7** (1990) 1841.
123. J. J. Halliwell and R. C. Myers, Phys. Rev. **D40** (1989) 4011.
124. I. Klebanov, L. Susskind, and T. Banks, Nucl. Phys. **B317** (1989) 665.
125. A. Vilenkin, in *Proceedings of the International School of Astrophysics 'D. Chalonge', Eriche, 1995*, ed. N. Sanchez, (Kluwer, Dordrecht), to appear.
126. J. Louko and T. Vachaspati, Phys. Lett. **223B** (1989) 21.
127. A. Vilenkin, Phys. Rev. Lett. **74** (1995) 846.
128. A. O. Barvinsky and A. Yu. Kamenshchik, Class. Quantum Grav. **7** (1990) L181.
129. A. O. Barvinsky, Phys. Rep. **230** (1993) 237.
130. L. P. Grishchuk and L. V. Rozhansky, Phys. Lett. **234B** (1990) 9.
131. A. Lukas, Phys. Lett. **347B** (1995) 13.
132. L. P. Grishchuk and L. V. Rozhansky, Phys. Lett. **208B** (1988) 369.

133. J. J. Halliwell and S. W. Hawking, Phys. Rev. **D31** (1985) 1777.
134. S. Wada, Nucl. Phys. **B276** (1986) 729.
135. J. J. Halliwell and P. D'Eath, Phys. Rev. **D35** (1985) 1100.
136. R. Laflamme, Phys. Lett. **198B** (1987) 156.
137. S. Wada, Phys. Rev. Lett. **59** (1987) 2375.
138. T. Vachaspati, Phys. Lett. **217B** (1989) 228.
139. T. Vachaspati and A. Vilenkin, Phys. Rev. **D37** (1988) 898.
140. T. S. Bunch and P. C. W. Davies, Proc. R. Soc. Lond. **A360** (1978) 117.
141. B. Allen, Phys. Rev. **D32** (1985) 3136.
142. S. W. Hawking, Phys. Lett. **115B** (1982) 295.
143. A. H. Guth and S. Y. Pi, Phys. Rev. Lett. **49** (1982) 1110; Phys. Rev. **D32** (1985) 1899.
144. K. M. Gorski, G. Hinshaw, A. J. Banday, C. L. Bennett, E. L. Wright, A. Kogut, G. F. Smoot, and P. Lubin, Astrophys. J. **430** (1994) L89.
145. T. Gold (ed.), *The Nature of Time*, (Cornell University Press, Ithaca, 1967).
146. A. Higuchi and R. M. Wald, Phys. Rev. **D51** (1995) 544.
147. S. C. Beluardi and R. Ferraro, Phys. Rev. **D52** (1995) 1963.
148. D. Marolf, Class. Quantum Grav. **12** (1995) 2469.
149. S. W. Hawking, Phys. Rev. **D32** (1985) 2489.
150. T. Gold, in *La Structure at l'Évolution de l'Univers, 11th Solvay Conference*, (Edition Stoops, Brussels, 1958), p. 81; Am. J. Phys. **30** (1962) 403.
151. S. W. Hawking and Z. C. Wu, Phys. Lett. **151B** (1985) 15.
152. S. W. Hawking, Vist. Astr. **37** (1993) 559; and in [44], p. 346.
153. D. N. Page, Phys. Rev. **D32** (1985) 2496.
154. R. Laflamme and E. P. S. Shellard, Phys. Rev. **D35** (1987) 2315.
155. R. Laflamme, *Time and Quantum Cosmology*, (Ph.D. thesis, University of Cambridge, 1988); in [44], p. 358.
156. S. W. Hawking, R. Laflamme and G. W. Lyons, Phys. Rev. **D47** (1993) 5342.
157. C. Kiefer and H. D. Zeh, Phys. Rev. **D51** (1995) 4145.
158. P. D'Eath, Phys. Rev. **D48** (1993) 713.
159. S. W. Hawking, Phys. Rev. **D37** (1988) 904.
160. R. Graham and H. Luckock, Phys. Rev. **D49** (1994) R4981.
161. B. de Wit, H. Nicolai and H. J. Matschull, Phys. Lett. **318B** (1994) 115.
162. S. M. Carroll, D. Z. Freedman, M. E. Ortiz and D. N. Page, Nucl. Phys. **B423** (1994) 661.